*Arid Lands in Perspective*

# ARID
# LANDS
# IN PERSPECTIVE

Including AAAS Papers on
Water Importation Into Arid Lands

**Editors:**
**WILLIAM G. McGINNIES**
**BRAM J. GOLDMAN**

**THE AMERICAN ASSOCIATION FOR
THE ADVANCEMENT OF SCIENCE**
Washington, D. C.

**THE UNIVERSITY OF ARIZONA
PRESS**
Tucson, Arizona

## EDITORIAL BOARD

THE AMERICAN ASSOCIATION FOR THE ADVANCEMENT OF SCIENCE
THE UNIVERSITY OF ARIZONA PRESS

# FOREWORD

ARID LANDS IN PERSPECTIVE is the first of a continuing number of publications to be compiled under the leadership of the Advisory Committee for Arid Lands Research of the University of Arizona and the Committee on Arid Lands of the American Association for the Advancement of Science. Articles in this book, as in future books, represent the combined efforts of many people with varied orientations to summarize aspects of current research and knowledge relevant for the multitudes attempting to inhabit Earth's warm arid areas, noted for their imbalance of natural resources.

The habitation of semiarid or arid areas, even on the barest subsistence level, involves special adaptations in almost every aspect of life, including social, economic, political, and esthetic pursuits. It demands of the inhabitants a degree of cooperation much higher than that needed for a similar subsistence level in less rigorous environments. Living and working on the edge of the desert in a semiarid area that lacks abundance of many natural resources (not the least important of which is water), the faculty members of the University of Arizona have since the latter part of the nineteenth century been studying many of the problems of such habitation.

The Southwestern and Rocky Mountain Division was organized as part of the American Association for the Advancement of Science in 1920. In 1951 members of this organization, recognizing the peculiar nature of planning for long-range use of arid areas, organized their first Arid Lands Symposium. The Committee on Desert and Arid Zones Research was soon organized as a permanent part of the Division, and it has held annual symposia on regional problems at the Division's meetings since 1955. Members of this Committee and other individuals in this Division also recognized the need for national and international cooperation in understanding and solving problems involved in arid-land living, and through their efforts the national organization of the American Association for the Advancement of Science created the Committee on Arid Lands in 1965 to work on a national and international level.

The American Association for the Advancement of Science, because of its multi-disciplinary character, is by nature suited for association with the types of investigations covered by the term "arid-lands studies." When faculty of the University of Arizona began seeking a way to publish papers from people of all nations involved in such studies, it therefore seemed a natural step for the Committee on Arid Lands of the national organization of the American Association for the Advancement of Science and the Advisory Committee on Arid Lands Research of the University of Arizona to meet and resolve to accomplish as much as possible in a joint effort.

An editorial Board of Directors, three from each parent organization, was appointed to organize and oversee publications. The University of Arizona Press became the active publishing collaborator for bringing the project into being, as joint publisher with the American Association for Advancement of Science. The University has also contributed publications management services from its Office of Arid Lands Studies. In addition to the work done by its three national board members, the Committee on Arid Lands has seen to it that papers presented at its symposium held during the annual meeting of the American Association for the Advancement of Science were prepared for publication in the present volume. This book is the outgrowth of a determined and cooperative spirit in these two organizations, and it is part of our plan to follow this book with many similar ones, as long as they continue to shed new light upon arid-lands problems and solutions.

TERAH L. SMILEY
*Chairman, Editorial Board*

## A Note From the Editors:

In building this book we have sought to serve those with broad interests in the arid lands. Scientific specialists have found that they cannot achieve mastery in a narrow field without understanding better the neighboring disciplines, the whole system that constitutes a desert environment; authorities such as government officials who have broad responsibilities for desertic regions have found that they need more specific information than the generalizations so often served up about the nature and the future of the arid lands. We have tried to provide for the enthusiasms, professional concerns, and planning requirements of several classes of potential readers — the highly technical as well as the nonspecialized readers.

WILLIAM G. McGINNIES
BRAM J. GOLDMAN

Office of Arid Lands Studies
The University of Arizona
Tucson, Arizona, U. S. A.

# CONTENTS

# SYMPOSIUM
## Water Importation Into Arid Lands
*Jay M. Bagley and Terah L. Smiley, editors*

*Arid Lands in Perspective*

# PROBLEMS FACING ARID-LAND NATIONS

MICHEL BATISSE

Natural Resources Research Division
UNESCO, Paris, France

# ABSTRACT

The diversity of situations in arid-land countries comes not only from climatic conditions, ranging from extreme aridity to semiaridity, from the different origins of limited water supplies, or from other environmental factors such as soils, mineral deposits, and presence of the sea, but also from socioeconomic conditions, particularly from the present level of development and the types of social structures such as those associated with pastoralism or irrigation.

In spite of this great diversity, a number of basic problems, some very old, are characteristic of arid lands. The history of land use in these regions shows the precariousness of life and the fragility of the balance of arid environments. Therefore various principal mistakes should be avoided in modern development. With regard to nomadism, romantic approaches should be replaced by acceptance of the trend toward permanent settlement and by appropriate measures to facilitate this transformation. Another problem concerns the danger of relative over-population and of excessive grazing leading to erosion in semiarid lands.

Difficulties inherent in the practice of irrigation include regulation of water supply, silting, water losses, salinity, waterlogging, drainage, soil deterioration, and waterborne diseases. Improvement of existing irrigation schemes might be preferable to the creation of new ones. The cost of investment for modern arid-land development is high.

The action generally required in arid lands includes, first, compilation of a thorough inventory of environmental resources. Integrated surveys are of particular value. A well-defined development policy involving sound regional planning and appropriate investment is necessary. While much knowledge exists about possible arid-land development, research is still required. Finally, a special and quite important need exists for education and training for any modern form of arid-land development.

A tentative analysis of future prospects in arid lands leads to the conclusion that the present trends in arid parts of industrialized countries might show the direction in which developing countries will also move. During a long transitional period toward modern development, the latter countries, to obtain intensive production, will have to improve their irrigation, grazing, and dry-farming practices, while expanding their industrialization and various forms of tourism from developed countries. The conclusion is one of moderate optimism for those arid countries that will be able to follow this pattern.

# PROBLEMS FACING ARID-LAND NATIONS
## Michel Batisse

A mere look at the world map of distribution of warm arid climates prepared for Unesco by Peveril Meigs in 1951 (*1*), which still constitutes a satisfactory basis for small-scale climatic classification, shows that some sixty countries are affected—to a greater or lesser extent—by aridity. That is approximately half the countries in the world. The diversity of physical, biological, social, economic, and political situations in these sixty countries is a fair cross section of the diversity in the entire world. Attempting, therefore, to analyze and summarize the problems facing arid-land nations appears immediately as an almost impossible task. The choice of problems, of examples, of suggestions, of generalizations for inclusion herein is of necessity somewhat arbitrary. The reader may well find in his own experience situations that do not fit exactly with the analysis proposed here. Our objective cannot be to comprehend all possible cases but to try to find out whether there are some general trends in the development of arid-land countries.

## DIVERSITY OF SITUATIONS

Since there is not one single type of arid-land country with one single problem called *aridity,* our first effort should be to analyze and classify the types of situations that may occur (*2*).

### Environmental Diversity

A first and most important subclassification concerns the climate itself, which is well illustrated on the previously mentioned map. On the one hand there are the extremely arid areas where there may be no rainfall at all for a whole year, and which correspond to the *true deserts* (for example; Central Sahara, Rub al Khali of Arabia, coastal desert of Peru and Northern Chile, and Namib desert). On the other hand there are the semiarid areas, which receive a substantial amount of rainfall (although not enough), which allow grazing and dry-farming, and which are well exemplified by the Fertile Crescent of the Middle East, the *sahelian zone* of Africa from Dakar to the Sudan, the North East of Brazil, the Argentinian Chaco, or the Great Plains of the United States. In between these two types are other arid areas, ranging from those that still receive some rain, like the Hauts Plateaux of Algeria or the steppes of Iran and Central Asia, and where ex-

tensive grazing is practiced, to the vast regions surrounding the true deserts, like Mauritania, most of Arabia, Central Australia, South West Africa or southwestern parts of the United States, where only pure nomadism or oasis irrigation is possible.

Detailed study of climatic conditions leads in fact to much more refined classification and cartography than mentioned above. For instance, not only are mean temperatures, precipitation, and solar radiation important, but also such factors as minimal and maximal temperatures, actual and potential evaporation, air moisture, and occurrence and duration of a dry season. The subarid mediterranean climate, with rainfall concentrated in winter and spring, is fundamentally different from the subarid tropical climate, such as that of the sahelian zone of Africa with rainfall concentrated in summer. Similarly, the occurrence of winter frost considerably differentiates the Anatolian and Iranian steppes from mediterranean regions with the same amount of rainfall (*3*). In fact, because of the existence of thresholds, comparatively small temperature or rainfall variations in arid environments lead to important differences in plant behavior, resulting in major changes in agricultural practices and potentialities.

For the layman, aridity means shortage of water, and there is obvious relationship between water and climate. The availability of water resources in an arid country is not, however, always directly related to local climatic conditions, and two areas with similar climates may show great differences in water resources. An extreme example of this is when a major river, flowing from a more humid area, crosses an arid country. Without the Nile, Egypt would be a true desert. The Colorado, the Syr-Daria, and the Indus rivers play similar roles. A somewhat parallel example of exogenous water supply is the case of groundwater aquifers, which may contain important amounts of water resulting from infiltration from distant humid areas, or even from ancient infiltrations during more humid geological periods. The sizable groundwater aquifers of the Sahara probably belong to the latter kind.

But the water situation in an arid area may vary considerably, even when this water is related to present and local precipitation, according to the geomorphology, the soils, and the natural storage capacity. The existence of many oases is based on what might be considered small topographic accidents, where water comes from underground as in the Sahara, or from a small river as in Damascus.

The latter examples show that aside from climate and water, other elements of the environment play an important role in differentiating arid areas. Among these, relief and soils deserve special mention. A flat arid plain will present very different problems from those of a mountainous region, and the historical developments in plains have been very different from those in mountains. Similarly, arid-land soils, although often characterized by accumulation of soluble salts and calcium carbonate in the upper part of the profile, can present a great diversity in composition, texture, and water-retaining capacity, so that their value for plant production can greatly differ within a distance of a few miles. Interesting soils resulting from clay may be found close to sterile sandy soils. Moreover, as a result of climatic variations, paleosols of great potential value formed under more humid previous conditions may occur in arid areas such as the Hoggar in the Sahara.

The combination of climate, water, soil, and geomorphology, together with the history of land use, gives to each arid area a particular *facies* with a particular vegetation and animal life. Vegetation, indeed, because it reflects most elements of the natural environment, can itself provide a fairly good classification of arid lands. There are, however, certain other environmental factors that might play a decisive role in arid-land development and that can be considered as accidental. These include, for instance, the presence—or absence—of mineral deposits, of energy sources such as oil, coal, or hydropower, of tourism amenities such as sandy beaches or archeological remains. They include also the presence of the sea or of brackish waters that might be used for desalination.

## Variety of Socioeconomic Conditions

So far we have considered only the variety of situations resulting from the variety of physical or biological conditions. From this point of view, certain parts of Arizona or New Mexico correspond fairly well to the environment in a country like Jordan. Yet two local situations could hardly be farther apart because of the extreme contrast in the socioeconomic conditions.

An easy way to describe this contrast is to say that in one case we have a developed country and in the other case a developing country. This, however, is merely an approximate over-simplification, not only because many *developed* countries are *developing* more rapidly themselves than are the developing ones, but because we are confronted in fact with all possible shades and transitional cases between advanced development and backward underdevelopment; and these shades and transitions occur not only between countries but equally within a single country.

Besides this somewhat misleading but fundamental concept of level of development, many related socioeconomic factors do play a major role in the diversification of situations in arid lands. For a given type of semiaridity, for instance that occurring in the Rajasthan area in India, which is climatically comparable to the northern zone of Central Australia, the basic contrast is between an *old* densely populated and overgrazed country and a *new* sparsely populated country with modern grazing and livestock management. The difference, of course, is also there between *underdeveloped* India and *developed* Australia. But it is much more in fact a difference between the historical heritage left by an ancient human occupation and the virgin potentialities of a new land.

Another well-known source of contrast is the age-old opposition between the nomad and the settler. So much has been said about this that there is no need to dwell upon it, except perhaps to suggest that the cultural traditions of nomadic tribes may not disappear readily when the tribes settle down, be it by their own will or under constraint. The universal contrast between the city and the rural dweller tends to be replaced in arid lands by the contrast between the mobility of the pastoralist and the extreme sedentariness of the remaining population, which consists not only of city dwellers but also of sedentary farmers.

Intensive irrigated agriculture in oases or along rivers is probably the origin of the remarkable urban concentrations that have characterized arid lands throughout history. What is important here is that this duality between extensive pastoralism on the one hand, and intensive agriculture with urban and industrial development on the other, persists in most arid countries and creates an important difference between them, according to whether the accent is on the former, as for instance in Mali, or on the latter, as in Egypt.

A somewhat related distinction has also to be made between those countries where the entire territory is arid, for example Libya, and those that are only partly affected by aridity. It is clear that the United States or the Soviet Union would not be what they are if their territories were entirely arid; at the same time it is clear that if this were the case their entire development policies would have to proceed from this fact. The remark may seem trivial, but a number of predominantly arid countries concentrate their efforts entirely on their nonarid areas, neglecting the remaining areas to a point that might well be detrimental to the overall development of the nations.

# BASIC PROBLEMS OF ARID LANDS

The preceding analysis points forth a large number of important parameters that diversify the situations existing in arid-land countries. Indeed, if one were to take all possible cases resulting from the combinations of these parameters, one would find that this leads to a larger number of such hypothetical cases than there are arid-land countries. The only merit of our analysis therefore may be to point out how dangerous it can be to oversimplify classifications and to conclude too rapidly that what is valid in one place should also be valid in another if both are labeled *arid*.

Having insisted on the diversity of situations, which necessarily results in a diversity of problems, we must now see what the arid lands have in common and more precisely what assets and what difficulties they offer to the ingenuity of man.

The two major and basic assets are simple to enunciate: plenty of sunshine and plenty of space. The one major and basic difficulty is usually referred to as lack of water. It should in fact be qualified more accurately as the sporadicity and/or variability of the endogenous water available to soils, plants, animals, and men. The problems of arid lands could in a way be reduced to the question of how to combine the abundance of space and sunshine with the scarcity of water to the advantage of man.

This is not a new problem. Ever since man has left indications of his way of life, we find that he has faced the problem and that he has progressively evolved a variety of answers, which, however, have not basically changed for thousands of years. The history of these efforts in the use and development of arid lands runs almost parallel with the history of civilizations (4). The lessons of this long struggle are here for us to take into account. It is, of course, debatable whether history can provide any lessons at all, and it is equally debatable whether anybody is ever prepared to take such lessons into serious consideration. However, one cannot but be struck by a recurrent phenomenon: the particular precariousness of arid-land civilizations. This precariousness obviously is due to the inherent fragility of the balance of arid environments and to the corollary fragility of the relations between man and these environments.

## Vanishing Nomadism

All traditional forms of arid-land use show fragility. Fragility is evident in nomadic life, which demands painful search over long distances for problematic grazing patches and dubious water-points, accompanied by the almost unavoidable raiding practices against sedentary populations. Fragility also describes the basic uncertainty of dry farming with the recurrent calamity of droughts and the dangers of erosion. But fragility characterizes equally well the delicate balance of irrigated lands with the constant threat of salinity or waterlogging, the plague of waterborne diseases, and the need for clockwork precision control of water sluices, which itself presupposes stability of political and administrative structures. And fragility permeates finally to urban life in cities, which also depends on permanency of water supply, maintenance of waterworks, adequacy of external food supply, protection for wind erosion, and many other factors that can easily be disrupted in the course of history.

Such types of difficulties in arid-land use still exist today and illustrate the main mistakes which have to be avoided in any modern development. A brief review of these problems can be made here.

Let us first consider the case of nomadism, which is so characteristic of arid lands. The major difficulty here is perhaps romanticism. Whether this attitude has its roots in the Bible, or in the more recent stories of Charles de Foucauld and Lawrence of Arabia, it appears that Western civilization has a *soft spot* for the nomads. At the same time, nomadism in true deserts and in arid lands constitutes a remarkable adaptation to ecological conditions and is probably the only method of utilization of these *desert* areas that does not require capital investment.

Traditional nomadic life is, however, not compatible with modern civilization and with the search for security that—perhaps unfortunately—characterizes our tamed societies. Raiding is no longer acceptable, and political insecurity is not tolerated by central governments that now have the responsibility—and the possibility—of extending authority to cover the entire territory that is internationally recognized as coming under their sovereignty. Conversely, in the few cases where no drastic pressure is imposed upon them, the nomads themselves, when they have the choice, seem to prefer to live a less romantic but more secure sedentary life. The conclusion of this is that efforts to maintain pure nomadism are bound to fail.

The problem is to face this social transformation and to take the required measures. Unfortunately, in many cases, the nomads of the desert merely become the sub-proletariat of the slums of the great cities of African, Middle-Eastern, or Latin-American arid-land nations. Solutions to this problem are difficult and could vary greatly from place to place. But in all cases there is a need for integrating socially, economically, and politically the former nomad; this requires energetic action from the very government that may have forced him to abandon his traditions. Where development takes place in the very areas of nomadic movement, integration of the nomads can be facilitated, as for instance where mining and oil-field exploitation exist.

## Over-population in Semiarid Lands

In semiarid areas, seminomadic pastoralism—associated possibly with dry farming or irrigation—remains a most important form of land use. In these areas, because *good* years with a fair amount of rainfall occur, there is a tendency to act as if no drought would come. Yet drought does come, because drought years in semiarid areas are natural phenomena, just as are floods in river basins. The consequences of severe droughts—as of floods—are all the more important where they occur less frequently, and where therefore people do not consider them an inescapable fact in their lives. But the consequences become tragic when large populations are involved. The images of famine in South Asia are known to all. The unreliability of production could, of course, be alleviated by storage and distribution, if the population were not too great and if the administration were organized on that basis. But man does not appear to be statistically minded, and the main danger in semiarid areas is probably overpopulation. This statement requires some qualification.

Overpopulation of land is a relative concept. On a rich alluvial or volcanic soil in temperate or tropical zones, land with 200 persons per square kilometer may not be overpopulated from a land-use point of view. But in semiarid areas, land with 20 persons per square kilometer, in association with a considerable number of cattle, might well be grossly overpopulated. In these areas, where plant and water resources are scarce, and where expensive grazing is the major (if not the sole) form of land use, over-pressure by human population and by animals can easily occur. The Nord Este sertao of Brazil or the Rajasthan region of India are typical examples of areas where the present human population and number of cattle, although comparatively not large, is clearly too great for the primary resource capacity of the land under present land-use conditions. When such a situation occurs, the dangers are severe.

Let us remember that deforestation, cultivation of wheat on fragile soils, ensuing erosion, and finally grazing of goats, are the successive stages through which man, by long and intensive occupation, has transformed the shores of the Mediterranean into what has been called a *man-made desert*. As a matter of fact, over-pressure by human population and even more by animals can be very localized around water-points or villages and yet lead to serious long-term consequences. It is striking in this respect to see how quickly the surroundings of natural or man-made water-points in semiarid regions of Western or Eastern Africa are denuded of vegetation, and compacted by stamping, thus leading to rapidly expanding eroded areas.

Erosion in semiarid areas can also be the result of erroneous agricultural practices. The effects of wheat cultivation in Roman times around the Mediterranean have just been mentioned. Two thousand years later the same mistakes in the American Great Plains resulted in the *dust bowl,* providing a spectatular illustration of the fragility of semiarid environments and of the need for careful, extended field experimentation before any large-scale development is undertaken.

## The Difficult Art of Irrigation

Remarks similar to those given immediately above apply also to the practice of irrigation. Since transformations imposed on the environment are even more drastic in the case of irrigation, the dangers are greater and the problems multifarious. Irrigation is always a difficult art, and nothing would be more useful—yet more impossible—than an objective study of the many failures of irrigation schemes. But financial, social, and political implications of large-scale irrigation projects are such that only official successes can be recorded.

Even now, in spite of the experience acquired, irrigation projects are still essentially visualized from engineering and financial angles only. An area is selected and properly leveled, a large dam is built, channels are constructed, and water is distributed. The regularity of water supply to the reservoir, however, is often a neglected issue, and many reservoirs remain half empty. The problem is difficult, since arid zones are characterized by a strong variability of water resources from one year to the other, while irrigation requires a well-regulated supply. The dams should normally therefore have a storage capacity three or four times higher than the annual mean, this probably leading to excessive costs and considerable evaporation losses. If the reservoir is not filled with water, it may well soon be filled with silt. This may occur all the more rapidly since sediment transport is naturally high in arid-land rivers, and since grazing and deforestation as well as urban and industrial developments in the upper basin may lead to accelerated erosion.

Nevertheless, once these difficulties have been more or less solved, the engineer usually considers that he has completed his task. Yet irrigation problems only begin here. Transfer of water to the crop is still achieved through somewhat empirical means, and fully satisfactory solutions have still to be solved. The most striking feature in this process is the heavy waste of water usually involved in those very areas where water is scarce. It is often said that in most irrigation schemes more than 50 per cent of the available water is lost by evaporation or percolation *before* it reaches the plants. This loss of water in the channels is not only regrettable in itself, but it may have extremely serious effects on the whole irrigation process when the water table, fed by the infiltrations, comes up close to the plant roots and leads to their rotting through waterlogging.

The quality of the water applied over fields is as important as its quantity. Any natural water contains a certain amount of dissolved salts, and in the arid zones, because of high evaporation, salt concentration in surface or ground-water is usually higher than in other climatic zones. The irrigation process itself leads to salt accumulation, since much of the irrigation water is evaporated from the soil, leaving its salt content behind. Soil salinization is the plague of irrigation. In semiarid areas, natural precipitation may be sufficient to leach this salt downwards. However, when rainfall is insufficient, the only solution for leaching is to spread additional quantities of water over the fields. But the excess of water, with a high salt content, may progressively raise the water table, leading to waterlogging. The combination of salinity and waterlogging have had dramatic consequences, as for instance in West Pakistan where until recently some 100,000 acres were lost from cultivation each year through this double process. The answer is, of course, drainage, using more or less refined techniques. But irrigation schemes still tend persistently to underestimate the drainage requirements for safe and permanent operation, because drainage is expensive, and probably also because the need for drainage will become apparent only after some years.

It would be extremely interesting to find out the total acreage of existing inefficient irrigation schemes with poor drainage or excessive water losses. Improving these old schemes would in many cases lead to far-better economic and social results than creating new ones from scratch. Unfortunately, it seems that often a hidden convergence of ill-conceived interests leads governments, bankers, and engineers to prefer large expensive new dams and projects—which can be inaugurated with great publicity—to less spectacular small-scale structures or improvements that, although they do not attract publicity, can produce lasting results at little cost. The advantages of such a nonspectacular approach may be significant from a technical point of view; a series of small dams, or the pumping of ground-water, may for instance lead to smaller evaporation losses and easier silt control than would result from one large dam. But the main advantages of this approach are of socioeconomic nature. It is too often forgotten that an irrigation scheme should be designed primarily for people.

The Egyptian fellah or the Punjabi peasant learns irrigation from the moment he is born. But the nomad of Central Asia or the cattle-breeder of the Niger valley has no knowledge of, and no interest in, this sedentary and muddy technique. Adaptation by the population to irrigated agriculture is a basic prerequisite for the success of new irrigation projects. Spectacular failures in all continents have resulted from ignorance of this fact. For this reason, local improvement of existing irrigation facilities and their progressive extension is more likely to be successful than launching of large-scale new ones. Such improvements or extensions will be within the immediate understanding and control of the people concerned, and might involve only modest and progressive investment.

When new land is brought under irrigation, modern techniques of leveling, of channel lining, of drainage, of water pumping, and of distribution are now available so that irrigation has become a very sophisticated practice. The control of salinity and the availability of water to individual plants (at the right times and in the right quantities) are still subjects of controversy and raise a number of problems. The extremely rapid transformation of organic matter in newly irrigated soils in hot climates may lead to serious deterioration of the best arid soils while, paradoxically, poor soils with low organic matter content may present a relatively more favorable response to irrigation when cultivated crops improve their physical properties. On the other hand, the prevention of waterborne diseases such as bilharzia and malaria from spreading into newly irrigated areas is a problem that has not yet received a fully satisfactory solution. Irrigation aims in fact at creating completely artificial ecosystems; these can be much more productive than natural ecosystems, but they have much less plasticity, and their inherent fragility makes them difficult to manage properly.

The result of all this is that a modern new irrigation project necessarily requires great care and heavy investment, and most of this investment has to be made in a lump sum. This brings us to a final and most important difficulty with which development is confronted in arid lands: high cost of investment for doubtful economic returns. Irrigation is an example of this, since, in good economic practice, its merits should be weighed against the cost of importing the same products grown in more humid areas. In many cases the balance sheet works out against irrigation, and some economists would go so far as to say that with the low price of food products irrigation is not an economically sound proposition. Others, of course, argue in favor of irrigation, at least for industrial crops. All of them, however, neglect the fact that in many countries irrigation cannot be judged by economic criteria only; social and political considerations must be taken into account. High cost of investment in arid lands is also exemplified by the problem of transport and communication facilities in non-irrigated areas. In these areas, distances are great, maintenance is expensive, and density of population is low. This obviously results in a high investment per capita (as compared with more densely populated humid areas), which heavily affects all production costs.

## ACTION REQUIRED IN ARID LANDS

This *bird's eye view* of the diversity of situations and of the multiplicity of problems in arid lands, while showing that each local activity has to be specifically suited to local conditions, points nevertheless to certain major lines of action that appear to be of general value to arid-land development. These major requirements relate to inventory of land resources, to policy, to research, and to education.

### Inventory of Resources

For practical reasons, arid areas have been comparatively less explored than more densely populated areas. The first requirement for arid-land development is that the nation know thoroughly its territory and that it have a general assessment of its potential resources and amenities. All countries have already done a great deal in this direction, but much remains to be done, especially since no inventory of resources can ever be considered as final, in view of constant progress in science, in surveying methods, and in the technology of resource use.

Much effort has been devoted to topography and general cartography, but much remains to be done in the field of geomorphology and cartography of land forms, which should play a proper role in planning arid-land development. Geological studies are comparatively easy in barren lands, and localization of mineral indices and interesting structures for possible deposits should be made systematically. Geological studies should also be oriented toward understanding the structure of aquifers. Systematic inventory of all water resources is a most essential element in the knowledge of an arid area.

Because the weather is usually clear, making aviation traffic easy, meteorology and climatology do not often receive the attention they deserve in arid lands. Hence, proper rainfall recording, and temperature, radiation, and wind measurements are insufficient. Stream-gauging and study of non-permanent streams as well as general appreciation of groundwater resources are obviously required, but there is a general lack of reliable data concerning surface water and even more as regards groundwater. In this respect it is essential to record water quality as well as quantity, since it is water quality that often constitutes the main limiting factor in irrigation.

A small-scale soil map will indicate the major soil types—including saline soils—while more detailed surveys will identify soil capability or point out erosion susceptibility or water-holding capacity. Small-scale vegetation maps, with indication of forests that still remain, will also constitute a useful background document for the understanding of ecological relationships. Floristic and faunistic studies will indicate the existing species, samples of which should be carefully collected and preserved for future reference. Finally, the inventory of the environment should analyze, and possibly present on maps, existing demographic and land-use features and socioeconomic structures. Special mention should be made of the necessary assessment of amenities such as natural sites, archeological remains and interesting wild life, and of possibilities for tourism, recreation, and health centers.

While a great deal of the above *inventory* actually exists in most countries, there is everywhere a need to extend it further, to review it, and to undertake more detailed evaluations in selected areas. For these purposes, matters of methodology become important. In view of the very character of arid lands, with clear skies and sporadic plant cover, aerial photography gives remarkable results in most of the conventional surveys listed. These methods are equally essential in another type of survey particularly suited to arid lands: the *integrated* surveys of natural environments and land resources (5). Such surveys have been developed in different ways in different countries, but they have received early and systematic attention in the arid zones of Australia (6). The major advantage of integrated surveys—in which teamed scientists from different disciplines define jointly, on the basis of aerial photographs and field studies, natural landscape patterns and land types and evaluate their possible use—is that main inter-related elements of the environment are considered together, and not separately, thus providing a sound classification of land for assessment of development potentialities and problems. The method can concentrate on physiographic features in deserts (7) but can incorporate in other areas ecological features, and even socioeconomic features, as long as they have a sufficient degree of stability.

### A Policy for Development

If the first requirement is to study the territory and assess its potential resources, the next step in arid-land development is to define a policy. To date, much development in arid areas has taken place without policy and without planning, but in such areas this extremely "liberal" approach does not often lead to sound results and can hardly be reconciled with modern techniques and large-scale investments. The first policy question for government or local authorities in an arid area may be whether or not to develop. The answer is simple in a country that is totally arid and has no other alternative than develop as well as it can. But the question is relevant to countries that have only a small percentage of arid areas.

Once development is decided, the choice of a particular policy will have to be based on the features of the environment as highlighted by the resources inventory, on human factors, and on the overall economic development policy of the country. A policy will involve, first, a choice between various possibilities in regional planning. It will also involve investment and action. There can be no hard-and-fast rule in this matter which calls upon so many environmental, social, economic, legal, and political constraints, and the process of decision at all levels, from the government to the farmer, will always be a complex one (8).

It can, however, be stressed—as emphasized earlier—that mistakes in arid lands are always costly and often fatal. Any development plan should therefore pay the greatest attention to safeguarding soils and waters, to maintaining their permanent productivity, and to preserving the quality of the total environment. Such principles, whether they relate to range management, to intensive irrigation, or to urban and industrial development, can be applied only if strict rules are followed in the implementation of the development policy, which in turn normally requires the creation of suitable institutions for guidance and control. It may be considered as most unfortunate, but individual uncontrolled initiative in the use of environmental resources—be it pumping of ground water or grazing of sheep—is nowadays even less permissible in arid lands than elsewhere.

### Arid-Land Research Problems

Much knowledge exists about possible development of arid land and is simply waiting to be applied. Developing countries in particular could make better use of this existing knowledge. These arid-land countries are at an advantage over humid tropical countries, as developed countries have had much more experience with problems of aridity than on humid tropical problems. Research, nevertheless, is everywhere the mainspring of progress, and much research is and will continue to be required on the potentialities of arid-zone development.

Of the numerous research problems which deserve attention, a few examples might be quoted. Some relate to fundamental questions such as improving the efficiency of photosynthesis, which might give a decisive advantage to areas of abundant sunshine, or discovering the evolution of present climates and the existence of climate thresholds in plant germination and growth. Many of these problems are of an interdisciplinary character, such as agroclimatological studies where correlations are sought between climatic factors and crop yield and behavior (9), or studies on the tolerance of plants to heat, to drought, or to salinity. The results already obtained in a Unesco-sponsored project in Tunisia on the utilization for irrigation of water with relatively high salt-content open the door to another interesting line of applied research (10).

Many studies will have to be devoted to the proper use or reuse of water (11). These may include the correct understanding of large-scale hydrological systems such as those of the Chad Basin, or the Northern Sahara aquifers now being studied. More and more attention should be paid to the systems analysis of water-resources use and management, thus attempting to provide satisfactory answers to the delicate problem of choosing between alternative uses of limited water supplies (12). Similarly, a considerable amount of scientific and technological research is required to reduce water waste, on such topics as evaporation control, analysis of plant requirements in irrigation, watershed management, industrial or agricultural reuse of water, and hydroponics.

Many other areas of research and experimentation are of interest to arid-land nations. All of them, for instance, are concerned with technological progress in desalination of sea or brackish water, and they should at least keep track of developments in this most important field. All of them are interested in experimental work designed to improve land-use practices in dry areas, whether it relates to selection of varieties, soil fertility, windbreaks, reforestation, or animal husbandry. All of them are interested in architectural, technological, or medical studies aimed at improving living conditions in hot, dry climates. All of them, therefore, have a definite interest in cooperating with each other and in keeping in touch with all forms of arid-zone scientific development.

Inventory of land resources, policy and planning studies, as well as research, require specific institutions. This applies equally to arid and nonarid countries, but there appears to be a strong case in arid areas, particularly when the entire country is not arid, for interdisciplinary centers where integrated surveys and oriented research studies can be carried out, and where teams of specialists can be trained at the same time. Such *desert institutes* exist in a number of countries; the Central Arid Zone Research Institute of Jodhpur (India), the

Negev Research Institute of Beersheba (Israel), and the Natural Resources Research Institute of Abu-Ghraib (Iraq) can be quoted as examples of such an approach in developing countries.

### Education: The Key to Development

In all arid-land nations, the final broad area for action is education. Education may well in fact be the real *key to arid-zone development* (13). The reasons for this appear clearly in the previously given analysis of problems. Whether we are concerned with semiarid or with arid areas, whether the intention is to transform the way of life of existing populations or to establish new settlements with populations coming from other areas, we are dealing with a fragile environment that requires careful management. The permanent settlement of nomads, which means for them a complete change in tradition, cannot be successful if it is not accompanied by a serious education program that will facilitate their adaptation and integration to a sedentary society and give them the necessary minimal skills. Populations practicing seminomadic pastoralism and dry farming in semiarid areas can, only through appropriate education, develop more productive methods of animal breeding, introduce new crop varieties, avoid erosion, establish storage facilities for drought periods, and organize markets and transportation for commercialization of their products. In the case of modern irrigated agriculture, which requires a high level of organization and technical skill, the role of education is absolutely fundamental; and, as pointed out, considerable risk exists when an irrigation scheme involves people who have not received adequate training for this purpose.

Nonagricultural forms of arid-land development, such as mining, industry, tourism, and all activities of urban life require no less education and skill than similar activities in other climatic areas. In fact they require perhaps a little more, since it is particularly important to impress upon the entire population the basic fragility of the environment, so that it behaves at all times in a way compatible with the conservation of its resources. One hopes that water waste of all kinds and useless deterioration of soils and vegetation can be reduced in this manner.

General education will not, of course, differ essentially in substance from that given elsewhere, and the content of technical education will depend on particular environmental conditions and local needs. Two seemingly contradictory considerations must, however, be kept in mind. The first one, already stressed, is that because of the sophistication of techniques, modern development in any arid land requires a higher level of education and skill. The second is that to date education in many arid lands, because of scarcity of population and low living-standards, has generally been of a lower level than in the nonarid lands, even within the same country. This situation places a great responsibility on all arid-land nations, whether developed or developing.

## FUTURE PROSPECTS FOR ARID LANDS

Any attempt to predict what might happen in the future is always risky and presumptuous. This is particularly true in relation to arid lands, at a time when man has shown great ability to control environmental conditions when he so desires. The problems of living on the moon, or even in an ordinary jet plane, are much more difficult than those of living in the hottest and driest arid lands. Everything is feasible in the latter regions if we are prepared to bear the cost. Technological progress—and sometimes scientific breakthrough—may well radically change tomorrow the price of basic commodities, so that present reasoning on the comparative advantages of living here or there may be completely modified. It is obvious, for instance, that a really cheap way of desalting sea water—which may not be out of technological reach—would entirely alter predictions on the future of arid lands. Similarly, a method for efficient use of solar energy would have far-reaching consequences.

Meanwhile, however, some prudent assessment of the foreseeable future is required, simply because no policy and no planning would be possible without some prospective views on future developments. The lessons of history, which enable certain mistakes to be avoided and which draw attention to certain recurrent problems, are not of great help in predicting the future because the nature of our present technological civilization and our power of intervention in natural processes are clearly without historical precedent. In this respect, one lesson can be really useful: the experience of the *developed* industrialized arid-land nations for the benefit of those which are *developing*, because it shows certain trends—good or bad—which may not be universal but which at least correspond to modern technological civilization.

### Prosperous Arid Lands

It can hardly be denied that the people of developed countries have recently been greatly attracted to desert areas. This is the case in Western Europe, with millions of tourists moving southward every year to the beaches of the Mediterranean and the historic arid lands of the Middle East and North Africa. But the case of the United States—and to some extent Australia—is even more interesting, since the movement takes place within the same country and leads to settlement.

So much has been written about this recent phenomenon that it will suffice to recall the trends it reveals; this indeed can be summarized in a nutshell, namely, taking advantage of the two assets of arid lands, space and sunshine. This is, of course, possible now only through a combination of factors, including individual mobility (automobile), air conditioning, high investments (including long-distance transport of water), light industries, intensive irrigated agriculture, an important sector of *tertiary services,* good intrastructure of roads, communications, hospitals, and lastly, but essential, high general level of education.

The result might be described as an urban industrialized type of society, using, however, a large amount of open space, where high efficiency in water use is possible (although not achieved so far). Such a society will thrive on sophisticated industrial production (such as electronics or aircraft), on high-efficiency agricultural production (*industrialized* irrigation or cattle breeding, possibly hydroponics), on a complex system of services (for example: banking, insurance, tourism, health and retirement centers, research centers). It will require stable and complex management for such basic services as water supply (including reuse or desalting), air conditioning, and nature conservation (including pollution control). Such a society as this could be equally well envisaged in arid or in semiarid lands, the only difference between the two being probably in the different emphasis placed on irrigation versus animal breeding.

The preceding outline, which corresponds already to existing situations in Southern California and Arizona, boils down to a system of *megalopolis* combined with concentrated and highly productive agriculture. This is not different from the pattern that is emerging in more humid areas, so that arid and semiarid lands, given the attraction they have, do not present insurmountable handicaps in developed countries as long as high investment and sophisticated management are possible.

## The Long Road to Development

Unfortunately, the future does not look so bright for the arid lands in developing countries. It could, of course, be argued that soon or later, as long as their development proceeds, they will reach a similar pattern. This may be so, but in the most optimistic hypothesis, their problem for the immediate future is that of a long transitional phase toward modern development. In spite of the vast number of books and conferences devoted to it, economic and social development still remains an obscure process. It appears to start in isolated spots where efforts have been concentrated, and it requires careful nursing— essentially capital formation, investment and technical skill—before some elements of lasting success

take root. For this reason, the limited human and capital resources in developing countries should be invested where they are likely to be most productive and profitable. Investment in a new irrigation project or in improving grazing lands is not often likely to compete with investment in industry or education, and the more arid areas of a developing country are therefore not likely to receive priority attention, unless, of course, they contain valuable mineral deposits.

If one accepts this outline for developed arid lands, not as an objective but as a probable trend, a first and simple rule for developing countries would be to think twice before embarking on operations that appear in straight conflict with this trend. The constraints of production—particularly food production—may, however, lead to certain temporary actions that may have to be abandoned in the long run. Irrigation of poor soils with water of poor quality is a case in point; it may be meaningless in fifty years time, but it may immediately double the standard of living in certain underdeveloped areas.

Very schematically, one can first visualize, in both arid and semiarid lands in developing countries, an important development of intensive irrigation. A better appraisal of water resources, both in quantity and in quality, a careful study of the type of wanted products and the time they are wanted, and a proper organization of marketing and transportation should facilitate this trend. Because of the rainfall, however, which reduces the amount of water required and improves leaching, this irrigation will be easier, cheaper, and therefore more likely to develop successfully in semiarid areas. Irrigation in arid areas in developing countries should normally remain localized, to serve, for instance, the needs of a mining and industrial community, unless of course an important source of exogenous water exists.

In a number of semiarid areas, extensive grazing as well as livestock farming can be developed. This may require comparatively little investment and particularly applies to the vast semiarid grasslands of western, central, and eastern Africa. The problems there are selection, adaptation of numbers to grazing capacity, installation of water-points, definition of grazing rights, and, above all, organization of marketing and of refrigeration and transport facilities. In other semiarid areas where population and animal stock are already much beyond the capacity of the environment, major improvements will be slow and difficult. The destruction of rodents and pests, the reduction in number of cattle combined with the amelioration of cattle quality, conservative dry-farming and tree planting, especially if some development of craftsmanship and industry can be induced, could assist progress in these areas and accelerate the transformation from a subsistence economy toward a market or planned economy.

Unfortunately, no magic formula can be advanced to unveil the path along which arid-land developing countries will actually go in the near future. One thing is certain: they are on the move. It is striking to note the amount and degree of transformation in many of these countries in the past twenty years; it seems largely due to their increased contacts with the rest of the world. They are indeed playing their part—in spite of the tremendous obstacles raised by nature, ignorance, poverty, and tradition—in the rapid and unprecedented changes that are taking place everywhere.

In this race for progress, the arid-land developing countries may well be ahead of other developing countries in the humid tropics, which once appeared to be better endowed by nature. From the agricultural point of view, their potentialities are real as regards livestock production and intensive irrigation. The difficulties of these two techniques can be surmounted. While humid subtropical areas appear to be at an advantage over arid areas for agricultural production, the latter seem definitely in a better position than the equatorial areas and the cold areas.

With regard to the nonagricultural sectors, arid-land developing countries have certain advantages over other developing countries that might prove decisive. Many of them, for instance, are comparatively close to the industrialized temperate zones. With air transport safe and easy (with relatively cheap airport construction), with sandy beaches, scenic amenities, and healthful climate, they can offer to the populations of smoggy and crowded industrial cities a yearly ration of sun and open space and pleasant opportunities for rest and retirement. With administrative leadership, appropriate investment and foreign aid, they can, of course, initiate or strengthen industrialization based on local energy sources and raw materials—including minerals and fibers. They are, however, here at no particular advantage over other developing countries. With energetic expansion of education and training, they can progressively diversify their economy and raise their standard of living. While probably maintaining for a long time a larger agricultural sector, they could indeed follow the broad pattern that is taking shape in arid regions of industrialized countries. But in this direction they still have a long and difficult way to go.

# REFERENCES

1. MEIGS, P.
   1953    World distribution of arid and semi-arid homoclimates. *In* Reviews of research on arid zone hydrology. Unesco, Paris. Arid Zone Programme 1:203-210.

2. HILLS, E. S. (ed.)
   1966    Arid lands: a geographical appraisal. Methuen and Co., Ltd., London; Unesco, Paris. 461 pp.

3. UNESCO
   1963    Bioclimatic map of the Mediterranean zone, explanatory notes. Unesco, Paris. Arid Zone Research 21. 60 pp.

4. STAMP, L. D. (ed.)
   1961    A history of land use in arid regions. Unesco, Paris. Arid Zone Research 17. 388 pp.

5. BATISSE, M.
   1966    Etudes integreés du milieu naturel. Publications of the ITC-Unesco Centre for Integrated Surveys, Delft, Netherlands—S.1–3-32.

6. CHRISTIAN, C. S., and G. A. STEWART
   1964    Methodology of integrated surveys. Unesco, Paris. Natural Resources Research. Proceedings Toulouse Conference.

7. MITCHELL, C. W., and R. M. S. PERRIN
   1966    The subdivision of hot deserts of the world into physiographic units. International Symposium on Photo Interpretation, 2nd, Paris, 1966, Transactions 4(1):89-106.

8. WHITE, G. F.
   1966    Deserts as producing regions today. pp. 421-437. *In* Arid lands: a geographical appraisal. Methuen and Co., Ltd., London and Unesco, Paris.

9. PERRIN DE BRICHAMBAUT, G., and C. C. WALLÉN
   1963    A study of agroclimatology in semi-arid and arid zones of the Near East. World Meteorological Organization, Geneva. Technical Note 56.

10. VAN HOORN, J. W.
   1966    Research on the utilization of saline water for irrigation in Tunisia. Unesco, Paris. Nature and Resources 2(2):1-5.

11. BATISSE, M.
   1967    Research and training for water resources development. International Conference on Water for Peace, Washington, 1967, Paper 692.

12. WHITE, G. F.
   1960    Alternative uses of limited water supplies. pp. 411-421. *In* The problems of the arid zone, proceedings of the Paris Symposium. Unesco, Paris. Arid Zone Research 18.

13. AMIRAN, D. H. K.
   1965    Arid zone development: a reappraisal under modern technological conditions. Economic Geography 41(3):189-210.

# PLAYA VARIATION

JAMES T. NEAL

Department of Geography
United States Air Force Academy, Colorado, U. S. A.

# ABSTRACT

*Playa* is the general term used to describe the generally barren depressions, or the lowest portions of desert basins, that periodically collect runoff and sediments; they occur in all of the major arid zones on the earth. There is a wide range in the type, number, and distribution of playas—from vast floors of formerly permanent lakes situated in large structural basins (U. S. and Chile) to small, wind- and animal-eroded depressions (pans of South Africa).

Playas have an important place in man's use of arid lands. Natural evaporite products are mined in many places for potash, salt, and other minerals. Playas occur in flatlands that are often potentially suitable for agriculture and grazing, and they are an important element in the basin hydrologic system.

Variation in climate from that existing at present, especially during Pleistocene and post-Pleistocene time, has had profound effects in the development of basin features and in the character of the hydrologic systems found in individual basins. Water entering the basins as ground-water may flow into, through, or around a playa; and some, all, or none of it may discharge through the playa surface. The amount and composition of groundwater discharge usually determines the thickness and mineralogy of saline constituents at the playa surface.

There is varying nomenclature from one arid region to the next, and few of the terms are correlative, largely because of regional variations in the playas. The similarities of surface properties suggest that standardized *playa surface descriptors* should be used that relate to physical and environmental properties in the basins. When these descriptors are combined with topographic, hydrologic, geologic, and climatologic information, an adequate understanding and developmental history of an individual playa can be gained.

Similarities and variations in the physical properties of playas are produced by various factors, and the general characteristics of playas in the United States, Iran, Australia, Chile, North Africa, and elsewhere range widely.

# PLAYA VARIATION
## James T. Neal

## INTRODUCTION

*Playa* is the geological term used in the United States to describe a dry lake bed or an ephemeral lake situated in an arid environment. It is the Spanish word for *shore* or *beach,* but it has lost its original meaning in the United States. The terms *clay pan, salt pan, salt flat, alkali flat, salina,* and *salt marsh* are often used colloquially and reveal compositional variations. In North America alone playas number in the thousands; several hundred are ten square kilometers or larger in area. Playas or similar features occupying depressions or topographic lows are common landforms in other desert areas of the world, but they are known by local terms such as *sebkha* or *shott* (North Africa), *pan* (South Africa), *kavir* (Iran), *salar* (Chile), and *salina* (Mexico). Even though playas in the United States are somewhat atypical, the term *playa* has received wide usage, and will be used in this paper in its widest context.

Playas are the flattest of all landforms, often sloping only one-half meter per kilometer, or even less. They generally occupy the lowest portion of a basin and may periodically collect rainfall runoff. The term *playa lake* is frequently used to describe a flooded playa. Microrelief on a playa seldom exceeds one meter; it is commonly less than ten centimeters.

Playas have been used over the centuries for a variety of applications. Where trafficable they provide excellent overland transportation routes. The extraction of salt for domestic consumption dates back hundreds of years and continues to the present in North Africa and the Middle East. Being located in the lower and more moist portions of desert basins, the playas often border on areas used for agriculture. Numerous playas overlie valuable accumulations of potash and other natural evaporite products. In recent years the significance of playas has been recognized in hydrologic studies related to water balance, and they have been studied as indicators of climatic change. More recently military organizations have used them as emergency landing fields, and land automotive speed trials have been conducted on the vast salt flats at Lake Eyre, Australia, and Bonneville Salt Flats, Utah. Playas have also been considered for use as land-recovery areas for manned space vehicles, and for extremely large antenna arrays. In spite of this wide usage, playas have been one of the least understood of terrestrial landforms.

The factors responsible for playa development vary from region to region, and within a particular region, depending on the relations of the influencing elements of surface and ground-water, composition, climate, topography, and geologic history. Pronounced surface changes have been noted on playas over relatively short intervals of time and demonstrate that they cannot be regarded as stable features of the desert landscape. These changes are often caused in part by subtle variations in the amount and flow of both surface and ground-water, or by changes in evaporation, wind, and temperature. In some cases the effects of man's recent activity are evident; elsewhere they are cumulative and probably represent longer-term changes caused by climatic variation. In order to understand fully the physical nature of arid environments, and to exploit the natural advantages that playas provide, it is necessary to develop basic comparative data, so that both short- and long-term predictions can be made regarding their future stability.

Although relatively small in total area, the ubiquitous occurrence of playas in the world's arid regions, combined with their unique natural properties, make them an important element in understanding and utilizing arid lands. The intent of this paper is to identify and compare regional aspects of playa geology. Emphasis must of necessity be placed on those areas best known to this writer; accordingly, North America, Australia, Iran, and Chile receive the greatest attention herein.

## THE PHYSICAL ENVIRONMENT

### Location and Present Climate

The major arid zones of the world are located about latitudes 30°N and S, where large areas are dominated by high-pressure systems and subsiding air all the year round. It is in these low-latitude deserts that the vast majority of playas are concentrated. As a general rule the annual rainfall averages less than 300 millimeters, and the potential lake evaporation is in excess of ten times this amount. The annual amount and distribution of rainfall in these areas is characteristically erratic. The evaporation rate is dependent on temperature, wind, and humidity; it decreases greatly during winter months, mainly because of lower average temperature. Mean daily summer temperatures frequently may exceed 30°C. The physical environment and boundaries of the thirteen major desert areas in which playas are located (Table 1) have been summarized elsewhere (*1*). According to this climatic summary, playas are located in *extremely arid* and *arid* desert regions, but also in the *semiarid* as well, and in borderline areas, some of which were more arid in times past.

TABLE 1

**WORLD DESERT AREAS CONTAINING PLAYAS**

| | |
|---|---|
| North American | Arabian |
| Atacama-Peruvian | Iranian |
| Monte-Patagonian | Turkestan |
| Sahara | Thar |
| Kalahari-Namib | Takla-Makan |
| Somali-Chalbi | Gobi |
| Australian | |

### Basin Geomorphology

The essential natural requirements for a playa are a topographically enclosed basin, having interior drainage, in an environment where evaporation exceeds precipitation. The basins in which playas occur originate in a variety of ways; however, each arid region usually has its own distinctive characteristics. Whatever the basin origin, the individual character of the playas will vary depending on hydrologic, climatic, sedimentologic, and other factors that will be discussed in subsequent sections. A summary of principal basin types is shown in Table 2; it is based on the processes that produce them.

its shape (*5*); that is, wind- or water-eroded basins will display directionally oriented dunes, shore features, and the basin shape itself. Structural basins are frequently elongate and usually parallel with the dominant structural axis. Similarly, shoreline and interdune basins are usually elongate and parallel to the coast or dune axis. Animal, meteorite, volcano, and solution basins are more nearly round in comparison with the other varieties.

The size of a playa is a direct function of the processes that produce it. The smallest animal wallows measure in tens of meters across, whereas some of the larger pluvial lake playas exceed 50,000 square kilometers. An average size of all basins that contain playas would probably approach 5 to 10 square kilometers.

### Surface Runoff and Evaporation

The water which infrequently floods playas to shallow depths arrives there by several means; the amount is controlled by a number of factors. In the United States, torrential mountain rainstorms may drain down the valley slopes and flood a playa, without a drop of rain falling on the playa; however, if rains are gradual, much of the water will be absorbed by the alluvial slopes. It then enters the basin as ground-water. The size of the basin, the size of the playa relative to the basin, and the position of the playa within the basin must also be considered. Generally speaking, playas are rather impervious to rainwater infiltration, especially when they are clay-rich, whereas the sands, gravels, and coarser materials of the surrounding

TABLE 2

**PRINCIPAL TYPES OF PLAYA BASINS**

| Basin Type | Representative Area |
|---|---|
| Structural (faulted or downwarped) | Great Basin, Nevada (Fig. 1); Iran |
| Deflation (wind eroded) | North Africa (Figs. 2 & 4) |
| Glacial (pluvial) | Great Salt Desert, Utah (Fig. 3) |
| Solution (subsidence) | Southern High Plains, West Texas |
| Fluviatile (river systems; floodplains) | Salinaland, Western Australia (*3*) |
| Shoreline (raised coastlines) | Coastal Salt Marshes (world wide) |
| Interdune (depressions between dunes) | Arabia; North Africa (Fig. 4) |
| Mass movement (alluvial fan damming) | Broadwell Playa, California (*4*) |
| Animal (wallows) | High Plains, Texas; Kalahari Desert, South Africa |
| Meteorite (craters) | Meteor Crater, Arizona |
| Volcanic (craters and calderas) | Zuni Salt Lake, New Mexico |

An individual basin need not fall into a single category; in fact several processes may act to shape a basin and then in varying degree continue to act during basin developmental stages. For example, several of the Great Basin playas in the United States are situated in structural valleys that contained Pleistocene lakes and which also experienced previous periods of intensive wind erosion.

The genesis of a playa is frequently reflected in

alluvium have much greater capacity for transporting moisture.

Because many playa clays are relatively impervious to downward percolation, flood waters must largely be dissipated by evaporation or by wind, which can easily blow shallow sheets of water over long distances. The average annual evaporation from open lakes in the principal playa areas of the United States ranges from 100 to 225 centi-

Fig. 1. Delamar Valley, Nevada, U. S. A. A large playa is situated in this structural basin, formerly the site of a large Pleistocene lake.

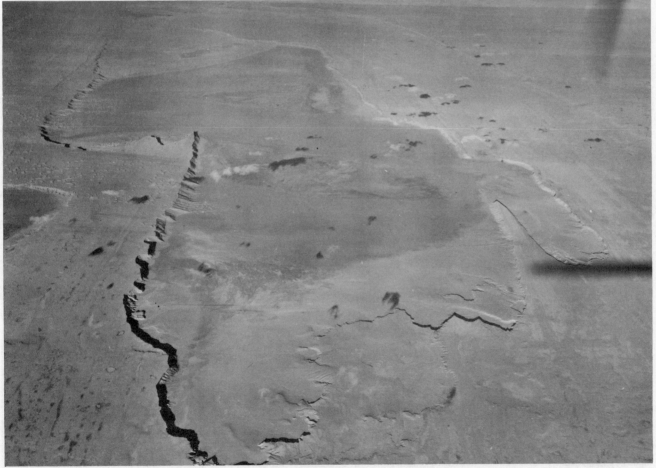

Fig. 2. Large, shallow deflation basin on the Morocco-Spanish Sahara border, near Cape Juby, in essentially horizontal Tertiary sediments. The basin fill is derived from erosion of its margins and may also be related to dune movement from the left side. The basin fill is most likely sand and silt, with a moist zone in the central portions, as indicated by the darker tones (2).

Fig. 3. Shoreline features of Pleistocene Lake Bonneville, Wah Wah Valley, Utah, U. S. A. A large clay playa (*left*) is located in the lower portions of this structural basin. Sevier Lake (*upper right*) is a saline playa that discharges ground-water whereas the adjacent Wah Wah Playa (*left*) does not.

Fig. 4. Small interdune playas, western Algeria. The surficial materials on the playas are distinctly different from those in the dunes and may be derived from the pre-dune floor (2). The amount of saline residue on the surface of these playas is dependent on the amount of ground-water discharge that is occurring.

meters, which is many times greater than the rainfall. The standard deviation of lake evaporation data in playa areas averages about 12 centimeters annually. It is important to note that some three-quarters of the evaporation occurs between May and October (Fig. 5).

interval the maximum pan evaporation for a single day was 1.01 inches (2.5 centimeters) when maximum air temperature was 109°F (43°C), wind movement was 220 miles (355 kilometers) in 24 hours, and mean humidity 54 per cent. Using a lake-pan evaporation coefficient of 0.60, maximum lake

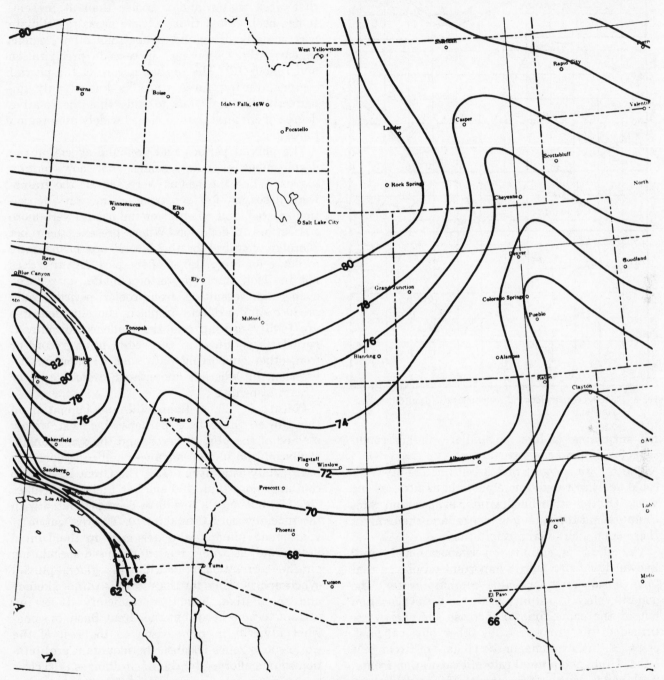

Fig. 5. Average May-October evaporation as per cent of annual, in western United States (6).

Only a very few playas have afforded an opportunity for detailed measurement of evaporation of flood water. Silver Lake, California, contained water for a period of 18 months following a Mojave River flood in March 1938. Records of evaporation, temperature, humidity, and wind were taken by Blaney (7) during this period (Fig. 6). During this

evaporation was 0.61 inches (15.5 millimeters) for a single day.

The playa at Indian Springs, Nevada, was flooded to a depth of approximately 45 centimeters on September 21, 1963; it had completely evaporated by January 20, 1964. This observation closely parallels Blaney's evaporation rates at Silver Lake and is

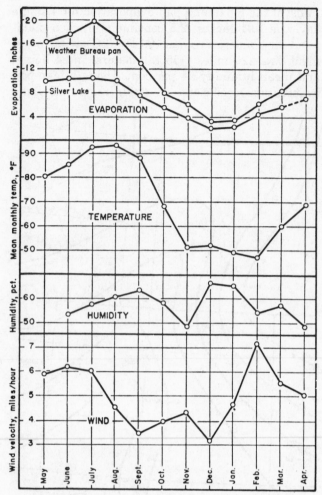

Fig. 6. Monthly evaporation and meteorological data, Silver Lake, California (7).

not surprising in light of similar weather conditions for that time of year.

These data suggest that 8 centimeters of water could be evaporated from a playa in as little as one week. On the other hand, winter evaporation rates could extend the time to 6 weeks, and then only if there is no additional precipitation.

The effect of salinity on evaporation is well known; it is particularly important on playas that have a salt crust or saline capillary water. The general effect of salinity is to reduce evaporation but at the same time to increase the energy returned to the atmosphere by other physical processes (8). For example, in the climate of Great Salt Lake, Utah, the annual rate of evaporation from a fresh-water surface is about 150 centimeters, whereas it is less than 100 centimeters from a lake at salt saturation (9). Evaporation from a brine surface with a salt crust will be reduced to less than 50 centimeters.

## Past Climate

The arid zones in which playas are situated experienced climatic conditions different from present ones during the Pleistocene, and even more recently;

this fact is well documented and attested to by a convergence of evidence from many disciplines (10, 11). Beyond the limits of Pleistocene continental glaciers (which in North America extended to latitude 38° and in Europe to 47°) there existed zones that experienced pluvial climates, or areas that were wetter and/or cooler than at present. It has been stated that in some areas the climate was actually drier and cooler than other regions concurrently receiving increased precipitation and runoff (12). The precise nature of the pluvial regimen during these times is known only imperfectly, but it is fair to state that the relative degree of climatic change varied widely from region to region.

The pluvial periods had profound effects on the development of many basins that now contain playas. When the basins were wetter, the framework necessary for lacustrine sedimentary deposition existed, and when drier the basins were more susceptible to deflation. Where present, the relict shoreline features of the basins are prima facie evidence for larger lakes of the past. But this does not in itself imply a wetter climate, as reduced evaporation resulting from cooler periods could produce similar effects. In places shoreline features are totally lacking and the evidence for pluvial periods is obtained in the sedimentary record or from other supporting data such as snowline depression in adjacent mountains, relict soils, or stream trenching.

Pleistocene lakes, Bonneville and Lahontan, in the United States are perhaps the most widely studied of the pluvial lakes, and the synchroneity of glacial and pluvial advances and recessions has been firmly established (13). The Great Basin also contains more than 100 smaller Pleistocene lakes (14, 15), but only a few have been correlated with the glacial record thus far (10). The realization of widespread pluvial lake existence in the United States and elsewhere has often caused speculation on the correlation of world wide glacial-pluvial synchroneity. Thus far the record is rather sketchy and speculative, if not contradictory. It is important to note here that the Great Basin is somewhat atypical in comparison to the rest of the world's arid zones because the mountain precipitation source afforded by the Basin Ranges is so close to the basins.

Estimates of the increase in rainfall that would be necessary to sustain several permanent lakes in the United States have varied from about 25 to 50 centimeters (16, 17). These estimates must consider the relative importance of precipitation, evaporation, runoff, and the seasonal distribution of each of these factors. Because they are estimates, caution must be exercised in their use.

The pluvial climates may have been drier in

some areas (12). There is also widespread evidence that glacial and postglacial periods of increased wind activity existed in the Mojave Desert (18), Australia (19), North Africa (20), and elsewhere.

Many glaciologists agree that since about 1850 most of the world's glaciers have receded and that the rate has increased since about 1920. This interval follows the *little ice age* (21), which has contained numerous cyclic advances and retreats of glaciers since about 1600, when data were first recorded (22, 23). Since about 1850 the climate is believed to have been characterized by warmer temperatures, greater evaporation, and perhaps less precipitation.

By definition, playas are dry most of the time, but even in fairly recent years, and particularly since about A.D. 1600, some playas became wet lakes because of increased precipitation and lower temperature. The desiccation of many perennial lakes and the formation of playas in the past 50 to 100 years appear to be associated with the arid conditions following the *little ice ages* (24, 25).

During the last several decades, glaciers in western United States have had renewed growth after several decades of accelerated recession. Concurrently, a general rise in water levels was noted in some enclosed lakes in Oregon and Washington. Thus, like glaciers, the enclosed lakes rise or diminish with changes in precipitation and evaporation. Consistent trends have been noted: high levels in the 1870's and 1880's and again in the early 1900's, then extreme desiccation until the middle and late 1930's when some basins were completely dry. To the southwest in Nevada, Walker and Pyramid lakes have fallen 20 and 15 meters, respectively, between 1927 and 1959, although some of this is probably due to increased usage by man (25).

It is apparent that playas are sensitive to meteorological fluctuations; there appears to be a correlation with the *little ice age* and the years that followed it.

If the present period of increasing warmth and glacier recession continues, the effects on basin hydrology and associated playa morphology will be profound, and the conditions that prevailed during the mid-post-Pleistocene might return (26).

There have been attempts to predict contemporary climatic trends, based on expected sunspot developments (27) or on apparent meteorological trends. Butzer and Twidale (28) have stated that reliable explanations for the comparatively minor but ecologically important fluctuations within the general atmospheric circulation are lacking, and that useful suggestions on the future arid-zone climatic trends cannot be offered until this lack is remedied.

Past climates have been important in the emplacement and development of pluvial lakes, but the distribution, number, history, and geological factors range widely from desert to desert; no general scheme of past climatic fluctuation exists that is acceptable in more than one region (28).

## CLASSIFICATION AND NOMENCLATURE OF PLAYAS

### Previous Classification

Numerous attempts have been made to classify playas (Table 3); each of them intimates that surface conditions and groundwater discharge are related. Problems arise where playa surfaces are transitional, or where more than one category is present in a single playa. Such limitations are often found in systems for classifying natural phenomena. Classifying playas becomes difficult when groundwater conditions are transitional, when there is a range of hydrologic conditions sufficient to produce diverse surface effects, and when hydrology is not the only controlling factor. The terms used in Table 3 have been largely applied to the United States.

### Variation in Playa Nomenclature

Ideally, one interested in comparing the major classes of playas from one arid region to the next should have at his disposal a table of comparative

TABLE 3

### SUMMARY OF PLAYA CLASSIFICATION (29)

| Author | Playa Type | | | | | |
|---|---|---|---|---|---|---|
| Foshag (30) (1926) | Moist playa (*) | | Dry playa | | | |
| Thompson (31) (1929) | Moist playa | | Dry playa | | Crystal-body playa | |
| Jaeger (32) (1942) | Salt pan | | Salt-free loam or Dry playa | | Lime pan or fflalkpfannen | |
| Stone (4) (1956) | Moist type | | Dry type | | Compound | Artificial | Crystal-body playa |
| | Salt encrusted | Clay encrusted | Clay pan | Lime pan | — | — | — |

(*) After D. G. Thompson, oral communication before the Geological Society of Washington, April 28, 1920.

terms, especially one that would indicate the fundamental character. Unfortunately, this is not feasible for a variety of reasons. First, much of the world's arid zones have not been mapped geomorphologically in detail, and the playas have often been delineated only by their areal boundaries. Secondly, classification of playa types has suffered because of the lack of a commonly accepted basis for classifying them (such as basin type, hydrologic character, or surface characteristics). Thirdly, the local nomenclature is frequently misused (by the local inhabitants and on regional maps) and will also vary sufficiently within one region so as to render the terms almost useless. Finally, regional variations in the physical characteristics are often distinctive enough to make a generalized nomenclature scheme of dubious value.

One solution to the pitfalls described above is essentially to ignore regional terminology and to concentrate on delineating the physical characteristics in a more descriptive manner that would include basin characteristics and hydrology, composition, microrelief, and geologic history. In this approach the individual gains the proper understanding without being burdened with cumbersome terminology. Nonetheless, this paper would be incomplete if some of the more commonly used regional nomenclature was not included (Table 4).

In the following sections of this paper then, emphasis will be placed on explaining how playa surfaces vary. Accordingly, the next section will explain the variation in physical properties and mineralogy; subsequent sections will discuss the hydrologic framework that controls the circulation, depth, and composition of ground-water in and between basins, and the general aspects of regional playa variation. It is these aspects that enable the geologist to distinguish one type of playa from another.

## SURFACE FEATURES OF PLAYAS

### General Aspects

Playas frequently exhibit the relief features of preexisting lakes, and thus have many of the features of wet lakes in temperature climates. Shorelines (Fig. 3), islands (Figs. 7 and 8), spits and bars, and peripheral dunes (Fig. 10) may be observed. The important difference lies in the fact that these playas are now subject to the arid cycle of erosion, whereas lakes in humid-climate areas are not. Thus, many original lacustrine features have been obliterated.

Fig. 7. Bedrock island in playa. Ibrahimabad Kavir, Iran.

The edges of most barren playas are characterized by an increase in vegetation, sediment size, and slope as alluvial slopes are encountered. The change from playa to alluvial slope is often abrupt (Fig. 1), but it can be gradual and without definitive demarcation lines. A veneer of gravel is frequently found around the playa periphery. Vegetated mounds, around which blowing sand accumulates, are frequently noted.

TABLE 4

## REGIONAL PLAYA NOMENCLATURE

| Locale | General Terms | Clay-silt Playas | Saline Playas |
|---|---|---|---|
| United States | Playa, dry lake, alkali flat | Dry playa, clay playa | Salt flat, salt marsh, salina |
| Mexico | Laguna, salina | Laguna | Salina |
| Chile | — | — | Salina (moderate salt), salar (much salt) |
| Australia | Playa, lake | Clay pan | Salt pan, salina |
| Russia | Pliazh | Takir | Tsidam |
| Mongolia | Gobi, nor | Takyr | Tsaka, nor |
| Iran | Daryacheh | Daqq | Kavir |
| South Africa | Pan, vloer, mbuga | Clay pan, kalpfannen (lime) | Salt pan, kalahari |
| North Africa | Sebkha | Garaet, qarat | Sebkha, chott |
| Arabia | — | Khabra | Mamlahah (salt flat), sabkhah (coastal salt flat) |
| Jordan | Ghor | Qa | — |
| Iraq | Hawr | Faydat | Sabkhat |
| India | Rei | — | — |
| Pakistan | Hamun | — | — |

Fig. 8. Ridge of Miocene (?) bedrock, 10 meters about floor of Sabzevar Kavir, Iran. Photo by D. B. Krinsley, U. S. Geological Survey.

Fig. 10. Sand dunes along western shore of Lake Frome, South Australia. Photo by D. B. Krinsley, U. S. Geological Survey.

Fig. 9. Salt river in central part of Sabzevar Kavir, a groundwater discharging playa in northeastern Iran. Photo by D. B. Krinsley, U. S. Geological Survey.

Vegetation seldom grows on the central areas of playas, but is normally seen on the periphery. The type of vegetative growth may be dependent on the salt content of the playa surface (33) or related to soil and groundwater conditions.

## Playa Surfaces

All playa surfaces are flat; periodic flooding serves as an effective leveling agent. Slope changes are usually so gradual that they often go unnoticed by the human observer. Actual profiles, although few, show slopes of 1 per cent and less to be common. Variations in sediment size and composition and the availability and chemistry of surface and ground water produce a variety of surface forms (Table 5). A brief description and photographic documentation of these surfaces follows.

The *hard, dry crust* is the "cleanest" of all playa surfaces. It has virtually no microrelief except that created by mud shrinkage, which is usually negligible. Most surfaces of this type are so hard that automobile traffic will not noticeably rut them (Figures 11-14). The typical surface consists of small sunbaked mud polygons, about 10 centimeters across, which are often light gray-buff. Minor amounts of calcium carbonate or sodium chloride are common surface constituents. A shine or glaze may be observed on some of these surfaces; it is apparently related to a high degree of fine particle orientation. Frequently a flaky or shredded condition develops when drying mud polygons curve upward. Hard, dry crusts most always have some fine material—usually clay and/or fine silt—which serves to bind the particles together into a compact mass. Some playas that once were lakes contain as

TABLE 5

**PLAYA SURFACE DESCRIPTORS**

| Surface Type | Example |
| --- | --- |
| Hard, dry crusts (usually clay-silt) | Ibrahimabad Kavir, Iran (Fig. 11) |
| (a) Smooth | |
| (b) Flaky | |
| Hard evaporite crusts (usually salt) | |
| (a) Smooth | Bonneville Salt Flats, Utah (Fig. 19) |
| (b) Rugged | Salar de Atacama, Chile (Fig. 20) |
| Soft, friable surfaces (often containing salt) | |
| (a) Smooth | Little Salt Lake, Utah (Fig. 17) |
| (b) Puffy | Harper Lake, California (Fig. 15, 16) |
| Soft, sticky-wet (usually with salt surfaces) | |
| (a) Smooth | |
| (b) Salt encrusted | Humboldt Salt Marsh, Nevada (Fig. 22) |

much as 90 per cent clay and silt; most have much less, however. Salt may be present in these crusts in small amounts, but where it exceeds about 5 per cent there will be strength degradation. Some soft, silt-sand crusts (Fig. 18) are more closely related genetically to hard clay crusts, but in a purely descriptive light they are soft and friable, much in the same manner as those that are soft because they contain salt. Hard, dry crusts that are free of salt and rich in granular or sandy components favor the growth of vegetation (Fig. 14), although it is uncommon on these surfaces.

*Soft, friable surfaces* are one variety of "self-rising ground" *(4, 31)*, a puffy condition of the soil (Figs. 15-17). On these playas the capillary fringe is often at the surface, but the upper 15 centimeters of soil are generally dry and loosely consolidated.

A "lumpy" surface with a 5 to 8 centimeter micro-relief is common. An automobile can be driven over some of these surfaces, but only with difficulty on others. These surfaces are frequently salt-stained (Fig. 17) and in places appear to grade into soft, sticky-wet surfaces. They are usually light brown in color but vary with changing salt and moisture content. These surfaces normally reflect capillary discharge of ground-water, as the salt in the near-surface sediments accumulates through evaporative concentration with time. They may contain 25 per cent and more of soluble salt in the upper 30 centimeters, but drop to less than 5 per cent a meter below. Where groundwater levels are declining, the source of this evaporative concentration is removed, and with time the surface salts can be flushed out during infrequent flooding, thereby

Fig. 11. Hard, dry crust of clay and silt, Ibrahimabad Kavir, Iran.

Fig. 13. Hard, dry crust with cobbles of scoriaceous basalt along periphery. Broadwell Playa, California.

Fig. 12. Hard, dry crust of sandy-silt. Soil shrinkage is reduced in coarser materials with the result that mud cracks are less prominent here, as compared with figure 11.

Fig. 14. Hard, dry crust; Playas Lake, New Mexico. Surface is sufficiently friable and free of salt to permit bunch grass to grow, whereas grass is totally lacking in figures 11-13.

Fig. 15. Soft, friable surface of clay-silt, Harper Playa, California.

Fig. 17. Smooth surface of soft, friable silt with white salt on top; Little Salt Lake, Utah.

Fig. 16. Close-up view of Figure 15, showing irregular, puffy nature of playa surface.

Fig. 18. Soft silt playa near Lacalle in Aguas Blancas District, Chile. Photo by G. E. Stoertz, U. S. Geological Survey.

causing a conversion to hard, dry crust (*34*). These surfaces correspond to the zardeh in Iran, and elsewhere are referred to as puffy *solonchak* soils.

*Soft, sticky-wet surfaces* are usually found where the water table is at or near the surface. They may be very smooth, water-saturated sediments, or they may contain appreciable quantities of salt and develop pressure-ridge polygons (Fig. 22). All gradations between these varieties may be observed. Variations are frequently encountered as one traverses some playas from edge to center; the latter usually contains higher concentrations of salt. Here, vehicular traffic of any sort is virtually impossible. As these surfaces are moist, the color is usually dark brown, but where salt is present the surface whitens (Figs. 22, 27). Playas in interdune depressions often have surfaces of this type, but because they normally contain a predominance of sand, they are less sticky than playas rich in clay.

*Hard evaporite crusts* are thick accumulations of salt, gypsum, or other material on the surface, generally having been deposited through long intervals of evaporative concentration of near-surface

brines (Figs. 19-22). They range from several centimeters to many meters in thickness (Figs. 23-25). Varying amounts of fine sediment may be intermixed, and there is a correlation between the strength of the crust and the amount of salt it contains. The position of the saturated zone beneath the crust is also a determining factor. Where the salt is relatively pure and the saturated zone is a meter or more deep, the crusts are extremely hard. Smooth salt crusts (Fig. 19) are uncommon, but where they occur a pavement rivaling the hard, dry crust (Fig. 11) in strength results. Salt-ridged and rugged crusts (Figs. 20, 21, 23) are the most common; variations in composition and microrelief produce many diverse forms. Microrelief may range from a few centimeters to one meter. Salt pressure polygons, or thrust polygons, frequently are seen on these surfaces and may be as much as 15 meters between ridges. These surfaces are usually associated with water tables very near the surface and thus often grade into soft, sticky-wet surfaces when the salt content is low. Vehicular traffic is usually extremely difficult, if at all possible, but

Fig. 19. Smooth salt crust at Bonneville Salt Flats, Utah. Salt pressure polygons are seen forming; they project upward about 3 centimeters. Standing water is visible in distance. Shoreline terraces of former Lake Bonneville are well outlined on hill.

Fig. 22. Salt-encrusted sticky-wet surface at Humboldt Salt Marsh, Nevada. Salt pressure ridges are seen forming here; between ridges the salt crust is about 1 centimeter thick. Footprints are 5 centimeters deep.

Fig. 20. Salar de Atacama, Chile. Microrelief in this rugged perennial salt crust averages about 30 centimeters here. Photo by G. E. Stoertz, U. S. Geological Survey.

Fig. 21. Close-up view of "salt flowers" forming in contraction cracks in salt crust.

## TABLE 6
## MINERALS OF EVAPORITE CRUSTS

| *Chlorides* | |
| --- | --- |
| Halite | NaCl |
| Sylvite | KCl |
| *Sulphates* | |
| Gypsum | $CaSO_4 \cdot 2H_2O$ |
| Anhydrite | $CaSO_4$ |
| Epsomite | $MgSO_4 \cdot 7H_2O$ |
| Thenardite | $Na_2SO_4$ |
| Glauberite | $Na_2Ca(SO_4)_2$ |
| Mirabilite | $Na_2SO_4 \cdot 10H_2O$ |
| Bloedite | $Na_2Mg(SO_4)_2 \cdot 4H_2O$ |
| *Carbonates* | |
| Calcite | $CaCO_3$ |
| Dolomite | $CaMg(CO_3)_2$ |
| Natron | $Na_2CO_3 \cdot 10H_2O$ |
| Gaylussite | $Na_2Ca(CO_3)_2 \cdot 5H_2O$ |
| Trona | $Na_3H(CO_3)_2 \cdot 2H_2O$ |
| Pirssonite | $Na_2Ca(CO_3)_2 \cdot 2H_2O$ |
| *Nitrates, Borates* | |
| Soda-niter | $NaNO_3$ |
| Niter | $KNO_3$ |
| Colemanite | $Ca_2B_6O_{11} \cdot 5H_2O$ |
| Borax | $Na_2B_4O_7 \cdot 10H_2O$ |
| Ulexite | $NaCaB_5O_9 \cdot 8H_2O$ |
| Searlesite | $Na_2O \cdot B_2O_3 \cdot 4SiO_2 \cdot 2H_2O$ |

in unusual cases, such as the Bonneville Salt Flats, there is no problem.

The previous three categories of playa surfaces have a wide range of characteristics, and they are each related to the amount and chemical quality of groundwater discharge. Because many playas having these surfaces are commercially exploited for their natural saline products, it is appropriate to list the more common minerals that they contain (Table 6). There has been more scientific study of saline mineralogy and solution geochemistry in playas than any other aspect because of this economic interest (*35-40*). New minerals and new precepts are continually being identified.

Fig. 23. Northwest part of Salar de Maricunga, Chile, showing salt polygons on salt crust subject to seasonal flooding. Photo by G. E. Stoertz, U. S. Geological Survey.

Fig. 24. Salt crust on the large salt lake near Shiraz, Iran. Water is found 3 centimeters beneath the surface here; salt crust is 12 centimeters thick. Salt is harvested here for domestic consumption in Shiraz.

Fig. 25. Salt quarry at Salar Grande, Chile, showing prominent vertical fractures and silty layer near surface. Photo by G. E. Stoertz, U. S. Geological Survey.

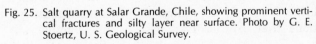

Fig. 26. Humboldt Salt Marsh, a groundwater-discharging playa in northern Nevada, U. S. A. Aerial view taken in spring of year shows an inner zone of standing water (*1*), springs (*2*), phreatophyte zone (*3*), and zonation of playa surfaces. Compare with Figure 27. Approximate scale: 1/55,000.

*Mixed surfaces* are the rule rather than the exception. The preceding discussion and that which follows demonstrate how some surfaces grade into each other and how others may change abruptly, or with time.

### Migration of Playa Surfaces

Table 5 is not intended as a classification of playas or of playa surfaces; it is only a summary of the principal variations that are observed on most of the world's playas.

It is important to realize that the playa surfaces that have been described may be highly ephemeral. A single desert cloudburst that floods a playa can completely rearrange and significantly alter the surficial nature of playa sediments (*24*). This is particularly true of those surfaces that have relatively small amounts of salt. Comparative views of the same area at Humboldt Salt Marsh, Nevada, are shown in Figures 26 and 27 to demonstrate this principle.

Pronounced surface changes occurred at Harper

Playa, California, following extensive flooding during the winter of 1965-66. The predominant surface prior to the flooding (Fig. 28) was soft, friable silty clay, but after flooding the reconstitution of this material, combined with the dissolution of salt, resulted in a hard, dry crust (Fig. 29). The reconstituted surface shows a substantial increase in strength, although some areas that had changed were showing strength degradation in late 1967, concurrent with an increase in salt accumulation. The observations show that playa surfaces with relatively low salt contents are highly subject to change (34). The amount of capillary discharge taking place appears to influence the permanency of change that occurs. Variation in detrital mineral composition appears relatively less important in controlling the playa surface development than soluble salt content; however, a higher granular content of detrital components tends to favor salt accumulation. Similar changes to those at Harper Playa were noted at Sabzevar Kavir, Iran, after two successive years (41).

Fig. 27. Same area as Figure 26; photo taken in dry season of another year. Note changes in distribution of surface types. White area is salt crust.

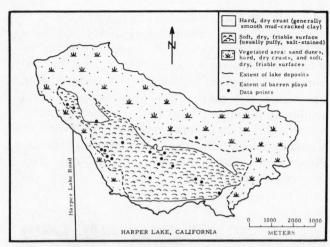

Fig. 28. Map of playa surface conditions, October 1962, Harper Lake, California.

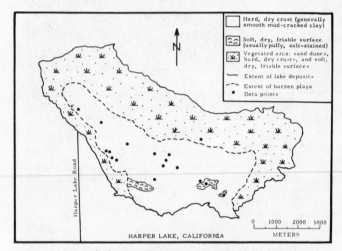

Fig. 29. Map of playa surface conditions, November 1966, Harper Lake, California.

## HYDROLOGIC CHARACTERISTICS OF PLAYA BASINS

### General Aspects

I have shown in the section "Surface Features of Playas" that capillary (and also artesian) discharge of ground-water plays a major role in the development of playa surface characteristics. The manner in which the basins receive, transport, and ultimately discharge ground-water is therefore significant in understanding playa variation.

The topographic divides of the groundwater basins containing playas are usually situated within the boundaries of the groundwater divide. Individual playas are frequently located within a single groundwater basin, as in the United States, Iran, and Chile. On the other hand, in the arid zones of northern and southern Africa (42), Central Asia, interior Arabia, and Australia, many of the regional basins are thousands of square kilometers in extent and contain numerous playas.

Most playas discharge ground-water in varying amounts, but some are totally dry and completely devoid of any interaction with ground-water. There is no precise depth known above which ground-water rises by capillarity and discharges at the surface, but it is likely in excess of 8 meters. The sediment type controlling capillarity and the depth to which evaporation extends vary widely in each geographic region.

For most hard, dry crusts of clay and silt, there is either negligible groundwater discharge or else the ground-water circulates out of or past the topographic basin, prohibiting the accumulation of saline constituents at the surface. There is evidence also that many compact clay crusts inhibit capillary rise and discharge of ground-water (43).

There is a great diversity in the chemical quality of ground water in basins that contain playas (44). As a general rule, potable water can be found in most of the basins; however, in the basins of the Kara Kum Desert not a drop of potable water is to be found over an area of hundreds of thousands of square kilometers (45). The potability and palatability of water are largely determined by its anion concentration. A concentration of more than 500 parts per million (ppm) of dissolved solids is normally not recommended for human consumption, but irrigation can be effective at many times this amount, and it is not uncommon for livestock to consume water having a concentration of 3,000 ppm.

Some attempts (30) have been made to classify groundwater basins on the basis of the dominant anion (for example, chloride, sulfate, carbonate). This is a useful way of characterizing basins in a particular region where there is a range in quality of water. There are at least three sources of the dissolved solid constituents: (a) rock weathering, (b) volcanic emanations and thermal springs, (c) and pre-existing lakebeds.

A common attribute in basins containing groundwater-discharging playas is the increase of dissolved solids in the ground-water below and adjacent to the playa, whereas elsewhere in the basin system potable water can be found. Although the surface increase in saline constituents can be explained by evaporative concentration, the processes controlling the overall system are poorly understood (46).

There is concern in the Great Basin of the United States and elsewhere that water that originated as recharge during pluvial stages of the Pleistocene is being pumped from aquifers in arid basins. When this water is exhausted, unpotable water may have to be desalinized.

In areas of saline ground-water, the conduct of agriculture and grazing is obviously restricted. One way of overcoming the problem is to channel rainfall runoff and use a playa as the storage area. This practice has met with limited success in the United States and has received serious consideration in the Turkestan Desert (*45*), where the takirs (clay-floored playas) limit infiltration sufficiently to form good storage areas of fresh water.

## Groundwater Discharge

A common mode of groundwater discharge in many playa basins occurs in the form of springs and/or transpiration from phreatophytes around the playa periphery (Fig. 26). This situation exists often because a transmissibility and slope contrast is encountered between the coarser gravels of the alluvial slopes and the finer-grained playa sediments. In these cases the piezometric surface is higher than the playa surface.

More important on the whole is discharge resulting from capillary rise of ground-water, and direct evaporation from the piezometric surface, which in many salt-crusted playas is coincident with the surface. The evaporative concentration of saline minerals is evidence that groundwater discharge is occurring (*29*).

Figure 30 (*47*) shows that playas and closed valleys that discharge ground-water can be divided into Class I, total discharging playas; Class II, partial discharging playas; Class III, bypass discharging playas; and Class IV, a combination of Classes I, II, and III.

Class I playas (total discharging) are relatively uncommon in the Basin and Range Physiographic Province of the United States, but more common in Iran and Chile. Estancia Playa, New Mexico, and Willcox Playa, Arizona, appear to be of this type. Playas Valley and Lake Lucero, New Mexico,

may either be types IA or one of the bypass types (Class II). Doty (*48*) indicates the possibility that Playas Valley may be an internal-groundwater-discharging type, Class IA.

Class II is characterized by a groundwater divide enclosing a large area that may be many thousands of square kilometers in area. Groundwater discharge does not occur in one playa, as in Class I, but occurs in two or more separated areas that need not be playas. Many combinations of discharge are possible for Class II; for example, discharge can occur in two or more playas or in a playa and adjoining valley with outgoing drainage. Some of the discharge of water can also occur in bypass playas through capillary discharge. The Class II type of playa appears to be controlled by a basement of relatively high transmissibility. The best-known occurrence of Class II playas is in southwest Nevada, where a large area enclosing Mercury, Nevada, and the surrounding area is underlain by Paleozoic carbonates of high relative transmissibility. In this area there is interbasin circulation of ground-water (*49*).

Class III playas, particularly Class IIIA, are relatively common; they are called *bypass playas* because part of the ground-water is discharged in the playa and part continues in a down-gradient direction of the piezometric surface and is discharged outside of the playa area. Most discharge from these playas occurs through the capillary fringe and through phreatophytes. A minor amount of discharge occurs through springs and directly through the surface. The playas in Hachita and Animas

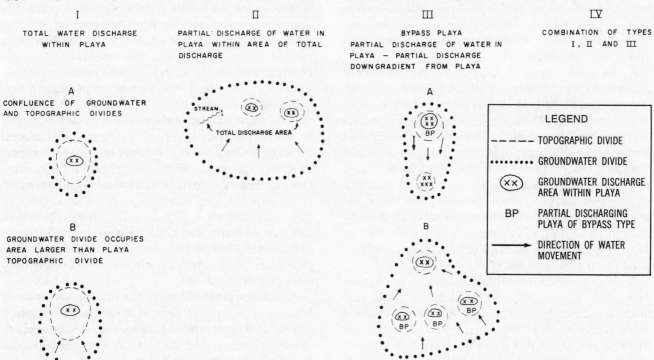

Fig. 30. Types of groundwater-discharging playa based on relation of groundwater divide to topographic divide and on position of area of groundwater discharge within groundwater divides.

valleys, New Mexico, and many Nevada playas belong to Class IIIA. In the case of Hachita Valley, the topographic divide and groundwater divide nearly coincide. Part of the Hachita discharge occurs from the playa surface; the remainder of the water moves southward where it is discharged at Ojo de los Mosquitos in Mexico.

Class IV playas are a combination of playa attributes of Classes I, II, and III. Many combinations of types are possible for Class IV. A common type could be a combination of Class I and Class III. In this case, a deeper zone of confined water is overlain by a poorly permeable confining bed through which only small amounts of water percolate upward.

## Playa Surface Features Controlled by Groundwater Levels

The position of the saturated water level beneath the playa surface is important because it controls the base level of erosion and to some extent that of aggradation; thus, fluctuations of the saturated zone can accelerate or diminish these processes. The interdune basins of North Africa (Fig. 4) and Central Asia persist because they are perennially moist at the surface, thereby precluding further deflation below this level.

In evaporite crusts with near-surface water levels, a lowering of the piezometric surface may result in erosion of the playa materials to the new level of the water-pressure surface. During this erosion cycle, solution processes assume greater importance because the evaporite products formed in groundwater-discharging playas are commonly soluble salts and carbonates (Table 6). This cycle of erosion is accompanied by the formation of solution pits and sinkholes, some of which are several meters deep and 25 meters across (24).

A common type of deposition occurs where phreatophytes capture windblown sand and silt and form small hills known as *phreatophyte mounds* or *bush mounds*. They are found around numerous playas in the United States, Iran, Central Asia, and North Africa. Phreatophyte mounds range in height from less than 1 meter to more than 5 meters, and in places are capped by a living or dead phreatophyte. Two factors appear necessary for their formation: (a) a playa with the water table at shallow enough depth and of low enough mineralization to support phreatophytes, and (b) a steady supply of windblown sand and silt (24).

A change in the density or nature of vegetation cover can influence the morphology of playa surfaces; for example, if phreatophytes are killed in a playa where most of the discharge occurs through these plants, the water may then discharge directly at the playa surface through capillary discharge or direct discharge from the water table. Under these new discharge conditions, the original hard, smooth surface that is characteristic of dry-surfaced playas may change to the puffy character of groundwater-discharging playas.

Giant desiccation fissures are unusually large mud cracks, often 1 meter wide and 5 meters deep; they intersect to form giant polygons that are commonly 75 meters across (Figs. 31-33). These features are especially prominent in the United States and Mexico, where thick lacustrine sequences underlie many playas, but they have also been reported in Jordan, Libya, and Central Asia.

Newly formed giant polygons are found mainly on playas where the principal surface is the hard, dry crust (50). A few cases have been noted on soft, friable surfaces where there has been a steady decline in the water table. These observations suggest that giant polygons represent large-scale deep contraction caused by desiccation of fine-grained sediment. Such desiccation is occurring in many areas where water levels are falling. For example, at Rogers playa the water levels have steadily declined since the turn of this century, largely because of heavy pumping for agriculture and domestic use. Artesian conditions, which once existed in three-fourths of all wells in the area, are no longer active, and drops of 10 meters and more in water levels are common. In each case where giant fissures have formed, the piezometric head has fallen considerably lower than the playa surface and frequently beneath the clay body into sand that is encountered at depths ranging from 3 meters to more than 20 meters (Fig. 34).

Intensified periods of aridity have increased surface evaporation. Ordinarily surface evaporation is known to extend at least 6 meters deep in soil, but it can have a significant effect at depths of 20 meters in arid environments (42). Intense evaporation extending well beneath the surface appears to play an important role in the formation of giant desiccation fissures.

A delicate balance exists in the moisture profile in the near-surface layers of dry-surface playas such as Rogers. Two principal *forces* oppose each other, as shown diagrammatically on Figure 34. Evaporation ($E$) from the surface downward is countered by rising capillary and artesian pressure ($C + A$) above the saturated zone. Should either force change in relative intensity, the moisture profile will fluctuate vertically, although fine-grained clays at depth will change only gradually, taking a period of years to desiccate (or hydrate).

At Rogers playa ($C + A$) has diminished, mainly because of mining of ground water, so that when a critical moisture (and shrinkage) point was reached, cracking began. The rate of desiccation may also be significant, and shrinkage stresses could be relieved by slow creep without rupture occurring at all.

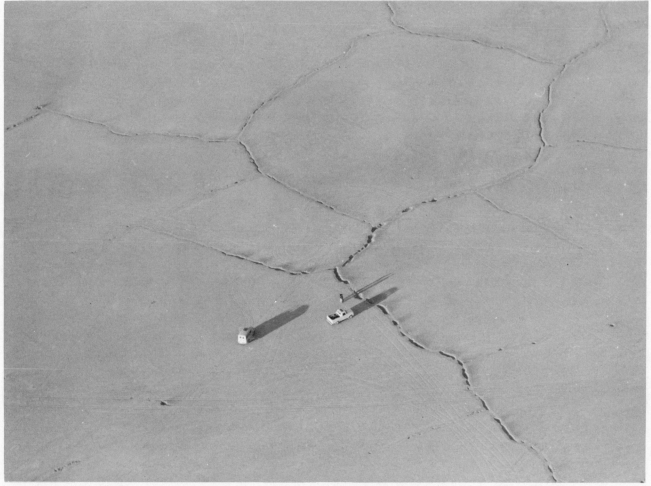

Fig. 31. Giant desiccation polygons, Rogers playa, California. Polygons average 75 meters across.

## REGIONAL ASPECTS OF PLAYA OCCURRENCE

### Australian Playas (51)

Recent strandlines immediately peripheral to the present shoreline of Lake Eyre are related to historical flooding. Otherwise, there is no evidence in Western or South Australia of beaches or of wave-cut features which might indicate that the playas had previously been occupied by more extensive lakes. This suggests that Pleistocene periods of glacial advance, which are represented by pluvial lakes in the United States and elsewhere, were not times of significantly increased precipitation in the interior of Australia. The stratigraphy at Lake Frome suggests that old dunes are buried beneath the most recent playa sediments.

Galloway (19) points out that the "immense, homogenous system of longitudinal dunes, now stabilized, extends across the heart of the continent through more than twenty degrees of latitude." Therefore, he argues, "There is good reason to postulate a considerable degree of climatic uniformity at times during the Pleistocene." Since the water table is the ultimate base for eolian

Fig. 32. Newly formed giant desiccation fissure, North Panamint Playa, California.

activity, the Lake Frome section suggests a lower water table during a period of previous eolian activity. If the older dunes are in fact Pleistocene dunes, they were formed during a period which was windier, colder, but drier. The dunes were subse-

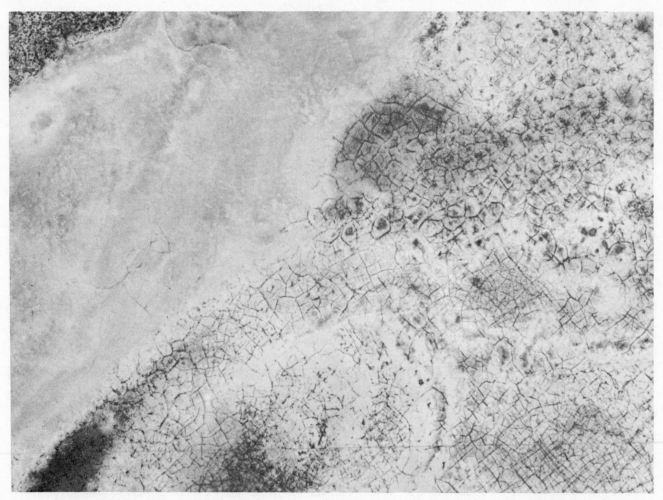

Fig. 33. Giant desiccation polygons at San Augustin Plains, New Mexico, now outlined by vegetation. Age of fissures is unknown.

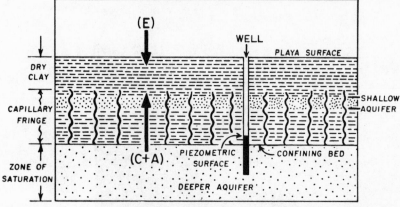

Fig. 34. Idealized hydrologic section, Rogers Playa, California. The relationship of opposing evaporative and capillary forces are shown. Giant desiccation fissures form primarily in the dry clay and capillary fringe, but may penetrate the zone of saturation.

quently covered by playa sediments during a period characterized by a higher water table. Australia may thus have experienced no true pluvial period comparable to that which has been well documented in the United States.

*Clay- and Silt-Floored Playas*

Silt- or clay-floored playas are widely distributed, being most common in the southern part of Western Australia and in South Australia. These surfaces occupy approximately 50,000 square kilometers. Virtually all of the areas are in part silt or clay floored. On maps, the larger playas are generally identified as lakes, dry lakes, or salt lakes, and the smaller ones as claypans. Frequently these different terms are used arbitrarily in identifying playas that are identical in practically all respects. Similarly, the symbols by which playas are shown on maps include solid-blue water tints, swamp and marsh symbols, cross-hatched symbols representing intermittent lakes, brown dots representing sand, and open areas without a symbol. Gregory (52) points out that the *lakes* of Western Australia are not lakes, but a series of salt claypans, and although their usually waterless condition is well understood, their title of lakes is too firmly established to be displaced. According to David (53), the name *salt lake* is also really a misnomer, since they are normally dry and they never contain any but very thin deposits of salt.

David (53) reported that there are almost 200 known salt lakes, which range in area from less than 1 to 3,000 square kilometers, and which he describes as follows: "The floor of a lake may be of solid rock, extraordinarily smooth and level;

sometimes this is covered with a thin layer of sand and silt, and exceptionally the detritus extends down to very considerable depths, as in Lakes Cowan and Dudas. In the former a bore passed through 277 feet (85m) of silt."

Waite (54) has stated that "claypans, . . . though extremely boggy in wet weather, dry with such a hard surface that foot of man or hoof of horse leave little or no impression." Madigan (55) describes the small claypans east of Lake Eyre as being "of all sizes and shapes from round ones a mile across to narrow ones 20 miles (32 km) long, occur [ring] sporadically in this undulating sea of sand. . . . The floors are absolutely flat sand-free clay surfaces, of a buff color. They were at times rather boggy, but during droughts, when the surface is crusty and gypseous, they can be crossed by motor vehicles. It appears that the eddies caused by the depression, together with the smoothness of the floor, keep the floors swept clean of sand."

## Salt-Encrusted Playas

Salt crust, characteristic of some playas and formed by the evaporation of flood-water or of ground-water, may range in thickness from a fraction of an inch to at least 18 inches. The salt crust is commonly thickest in the lowest part of the playa, but a patchy appearance is common over the entire surface. The playa surfaces are so nearly flat and so large that a particular flood will not necessarily cover the entire lake floor. The distribution of salt may then depend on where the wind happened to blow the last remaining sheet of salt water before it evaporated (56).

In many places, such as on Lake Dora, the salt crusts are a smooth, safe surface for a vehicle. Bonython (57) said that a jeep, having reached the thick salt crust of Lake Eyre (North), by running over planks laid on the soft lakebed where the gap between the land and the salt was narrowest (near Prescott Point), was able to travel at 40 miles per hour on a smooth surface, most of which was as firm as concrete.

The water table is generally very close to the surface, and beneath the salt there is commonly a thick layer of soft clay containing gypsum crystals and saturated with brine. This would afford little support to the salt crust, which would probably act somewhat like a layer of ice floating on water. A few lakes, for example Lake Macdonnell as measured in 1921, are known to have had salt crusts up to 40 centimeters thick. There is no definite information about lakes having thicker crusts.

Lake Eyre, one of the largest salt playas in the world, was investigated by Johns and Ludbrook (58) in connection with a study of evaporites and brines. They concluded that the salts were derived from "oceanic sources transported to the present situation through the agency of wind and an internal drainage system."

From a study of Lake Torrens, Johns (59) concluded that the depositional basin originated in early Tertiary and no subsequent marine incursions had occurred—the basin had gradually subsided, accumulating almost 1,000 feet (300 meters) of fresh-water lacustrine sediments, and evaporite sequences suggested a closed lake during arid intervals.

Lake Gairdner was studied by Johns (60), who reached the following conclusions:

a) The lake extended gradually westward during the Quaternary and accumulated an alternating sequence of salt and windblown sands.

b) The upper sediments contain up to 80 per cent gypsum and are underlain by sands. Halite, common throughout the sequence, forms a crust over most of the lake surface. Detrital quartz is a major component of the lake sediments, and calcite, kaolinite, pyrite, and limonite are constant minor constituents throughout.

c) Composition of the lake brines reflects origins from seawater, but the potash content of the brines is lower than that of seawater.

## Rock-Floored Playas

The area in which rock is actually exposed at the surface probably constitutes at most a small percentage of the surface of the rock-floored playas and is most commonly along their western margins. Commonly there are extensive areas of rock floor covered by a thin layer of silt or clay, the thickness of the surficial cover increasing toward the east. The water table generally lies at shallow depths beneath the surface, so the unconsolidated cover is commonly moist. The distribution of rock floors on Lake Goongarrie has been mapped in considerable detail by J. T. Jutson (61). The description of Lake Goongarrie playa that follows may exemplify the conditions that exist on parts of other playas.

Lake Goongarrie, in Western Australia, is located at latitude 30°02′S., longitude 121°13′E. The playa, which is 18 kilometers long and covers an area of about 125 square kilometers, is situated about 2 kilometers east of the railway between the towns of Comet Vale and Goongarrie, in the North Coolgardie Goldfield.

The rock floors of the playa occupy a considerable area of its western part and can be traced for a mile from west to east. They are either devoid of detritus, possess a mere film, or are covered by 5 to 8 centimeters of fine silt. Jutson (61) has described the microrelief in this way:

In places [the rock floors] are slightly furrowed, along the strike of the rocks, but this does not destroy the "billiard table" character of the floors. The actual plane of the surface may be slightly inclined or undulating, but precise levelling is needed to determine the directions of slope. So horizontal, however, are some of the rock surfaces that when rain falls (unless it be long continued) it simply rests on such surfaces without flow. . . . The occurrence of the bare rock floors [is] most striking, . . . possessing the appearance of having been recently swept by a gigantic broom.

## North American Playas

There are some 300 playas that exceed 5 square kilometers in area; almost all of them are located in the Basin and Range Physiographic Province of California, Oregon, Nevada, Arizona, Utah, and New Mexico (extending southward into Mexico). Most of these larger playas are situated in large structural basins that contained pluvial lakes during Pleistocene time (15). Lakes Bonneville and Lahontan of Utah and Nevada were of sufficient volume to cause isostatic compression of the crust in those areas; the rebound that accompanied the desiccation of these lakes is still going on (62). Because of the widespread development of pluvial lakes, there are thick lacustrine sequences of sand, silt, and clay beneath many of the present-day playas.

All of the playa surfaces listed in Table 5 are found in North America, and all of the basin types (Table 2) exist. The most important regional aspect of playa development is related to the occurrence of Pleistocene lakes in the structural basins. This may explain why deflation has not been as active as in Africa and elsewhere; that is, the conditions for deflation never existed for any great length of

time. There has been considerable discussion on the measure of wind activity in times past (18). Aside from the studies at Danby (63) and Estancia (64) playas (Fig. 35), there is a general lack of evidence that deflation has been of major importance in the development of Basin and Range playas. Sedimentation is occurring today, but it is slow. Several C$^{14}$ dates have shown net accumulation to be less than one meter per 1000 years.

About one-half of the larger playas contain hard, dry crusts, a ratio that is not noted elsewhere. Several factors help explain this. In addition to deep and through-flowing ground-water in many of the basins, there has been lowering of water levels by man (pumping) in some areas (24), both of which serve to minimize groundwater discharge and prevent evaporite accumulation. Another factor is the close proximity of the mountain recharge source, which causes more frequent floodings of the playas than in other deserts. These floodings effectively flush out and remove the surficial evaporite accumulations (34). On several playas soft, friable surfaces are converting to hard, dry crusts; the change is related to lowering of groundwater levels and to man's activity (24).

Evaporite crusts are formed where groundwater discharge occurs. Generally the evaporites are restricted to the near-surface layers, but in a few places they are more than a meter thick. Searles Lake, California, has a central core of evaporite minerals that is as much as 10 meters thick, and also an alternating sequence of lacustrine and

Fig. 35. Estancia Playa, a groundwater-discharging playa in central New Mexico. This aerial photo clearly shows the effects of past deflation, as indicated by linear erosional and depositional features in the basins. Approximate scale: 1/55,000.

evaporite deposits at depth which is related to the glacial-pluvial cycles of the Pleistocene (65). These thicker evaporite sequences are apparently related to evaporative concentration of the original Pleistocene lake. The Searles Lake "trona reef" is a linear concentration of hydrated sodium acid carbonate that is believed to have formed along a fault zone where trona-bearing spring water rose and precipitated (66). Evaporite minerals in playas are often zoned, in a manner related to their solubility and precipitation order (36, 67).

Many smaller playas are located outside of the Basin and Range Province. They are especially prominent in the southern High Plains of eastern New Mexico and West Texas, numbering in the hundreds. Numerous small, shallow depressions are located in the semiarid areas of Wyoming, Idaho, Colorado, and Kansas; these are periodically flooded and collect fine sediment. The plains of Alberta, Saskatchewan, and Manitoba in extreme southern Canada contain several dozen playas; most are situated in the clay-silt floors of former glacial lakes. Owing to their location on the fringes of the North American arid zone, the Canadian playas are moist or contain water more often than those further south. For this reason they do not contain thick evaporite crusts, but many are sufficiently saline to hinder agricultural pursuits in the basins.

The playa basins of the southern High Plains owe their origin to several diverse processes that have breached the *cap rock* in the area. A number of the larger playas of western Texas were wet lakes during the Pleistocene, but the smaller ones were formed by solution and subsidence, erosion by animals, deflation, or a combination of these (68, 69). These smaller playas are similar in size and topographic setting to the *pans* of South Africa.

The playas of Mexico are comparable in almost all respects to those of the United States. There may be a slight increase in the ratio of saline playas, but this has not been confirmed.

## Iranian Playas

The intermontane basins of the Iranian Desert contain playas similar in outward appearance to many of those in the Great Basin of the United States. The sedimentary record of the Darya-i-Namak, a large salt playa or kavir south of Tehran, shows an alternating sequence of salt and clastic facies, suggesting alternating lacustrine and desiccation stages which might be correlative with the major glacial advances and recessions of Europe (70). The central plateau of Iran, containing the vast Dasht-I-Kavir and other kavir basins, shows a lack of evidence that a pluvial climate comparable to the United States ever existed (41). Studies of snowline recession and pollen sequence (12) at

Lake Zeribar in the Zagros Mountains, southwestern Iran, have established that a treeless landscape existed from 23,000 to 11,000 years ago, indicating a cool and dry climate concomitant with worldwide Pleistocene climatic change. The effect on interior basins in the form of increased runoff has not been recognized. The postglacial sequence in the Zagros Mountains does show that the cool, dry climate of Pleistocene time gave way to a warm, dry climate, and that the level of precipitation increased slowly until 5,500 years ago (12).

The inferred pluvial sequence of the Darya-i-Namak then may be a local phenomenon, explained by increased runoff (and combined with reduced evaporation) from the nearby Elburz and Zagros Mountains. There is some evidence that the northernmost playa basins of Iran, adjacent to the Elburz Mountains, may have maintained permanent lakes during the latest Pleistocene period. Krinsley (41) cites this evidence at Sabzevar and Damghan kavirs.

The Dasht-I-Kavir, or Great Kavir, is not a playa per se but an eroded bedrock plain in which many playas occur in synclinal folds. Because of the abundance of evaporite minerals in the Miocene sediments and the near-surface ground-water, an extensive silt-evaporite crust is found. Giant compressional salt ridges a meter high and forming polygons 50 meters across are common. Otherwise, the overall Dasht-I-Kavir is generally so flat that it resembles a vast playa.

The Iranian kavirs are characteristically saline; they are more common by far than silt- and clay-floored playas (daqs). A common feature is zoning or progressively increasing salinity as one goes toward the kavir center (71).

All of the playa surfaces of Table 5 are found in Iran, and, in fact, all of them occur on a single playa, Sabzevar Kavir (41). Although Sabzevar Kavir is a perennial groundwater-discharging playa, repeated washing has created a seasonally migrating, hard, dry crust in the northernmost part. Without this washing from runoff, the saline minerals would accumulate at the surface and degrade the strength.

On the playa areas where they concentrate, the purer salts are harvested for domestic consumption (Figs. 9, 24) in several locations. The chief saline mineral is halite (NaCl), but other evaporite products are also found. As is the case in other arid regions, the groundwater-discharging playas create excessive salinity in the lower reaches of the basins. The Persian people have been aware of this problem for several thousand years and thus use the qanat or underground tunnel to obtain fresh water from the higher alluvial slopes (72).

Many of the Iranian kavirs are so extensive that repeated observation is impracticable. The use of

Fig. 36. Daryacheh-I-Tasht (*left*) and Daryacheh-I-Bakhtigan (*right*), Southern Iran, Photographed on the Gemini V Mission with 70 mm MS Ekta-chrome in Late August, 1965. This black and white reproduction of the original color transparency clearly shows the variation in water depth in these two salt lakes, ranging from water-free salt crust to deeper water (dark gray).

satellite photography (73), which will in future years be obtained routinely, is a valuable supplement to conventional aerial mapping photography and can help overcome observational limitations (Fig. 36).

## Chilean Playas (74)

Playas occur in the region of interior surface drainage north of latitude 27°30′S. Approximately 100 playas, covering a total area of about 7,300 square kilometers (about 1 per cent of Chile) occupy the lowest parts of closed basins, of which there are about 110 in northern Chile. Within the region of interior drainage, most playas are clustered in three distinct subregions which have contrasting environments: (*a*) the lower parts of the Central Valley and adjoining parts of the Coast Range, (*b*) the Atacama Basin and vicinity, and (*c*) the Andean Highlands. Closed basins containing playas in northern Chile are related to regional tectonic

movements such as uplift of the Coast Range, down-sinking or downwarping of the Central Valley and the Atacama Basin, and uplift and associated volcanism in the Andean Highlands. In the Coast Range most of the individual closed basins probably result from displacements along high-angle faults, many of which are still active.

There is great diversity in the elevation, size, and basin characteristics of the playas in the three regions. Playas in the Central Valley and Coast Range have relatively low elevations and small playas in large basins, whereas in the Andean Highlands the elevations are high and the playas are large relative to the basin size. In the Atacama Basin and vicinity, the characteristics are somewhat intermediate. The variation in elevation and playa size affects temperatures, precipitation, evaporation, and flooding characteristics (74).

Nearly all Chilean playas are at least partly covered by salt crusts or efflorescences (Figs. 20, 23, 25), but the composition of the crusts varies greatly, and often inexplicably, being closely related to the geomorphology of the individual playas and often varying considerably in different parts of a single playa. The composition of the salt crusts of specific playas is generally not well known, particularly in the Andean Highlands. Salar crusts range in thickness from less than 1 meter to more than 5 meters and are most commonly halite or gypsum (75).

The boundaries of salars as shown on most maps are commonly thought to be more closely related to groundwater conditions than to the extent of flooding by surface water. The margins of many salars coincide with the upper edge of a zone of springs, saline efflorescences, and moist ground. The term *salar* generally denotes an area of salt-encrusted or saline soil that commonly contains a playa, but which may include a wide range of other landforms or geomorphic types.

Most of the playas of northern Chile have a water table at shallow depths beneath their lowest parts, where groundwater discharge takes place by evaporation from moist salt, moist soil, or open water, and locally by transpiration from phreatophytes such as tamarugos, or from salt grass. Of the 61 most conspicuous playas, about three-fourths are classified as predominantly moist, of which 14 are predominantly salt-encrusted and 28 partly salt and partly soil; of the remaining 19 basins, 12 contain mixed playa surfaces, 6 contain perennial lakes rather than playas, and one (Salar Grande) contains a solid crystalline body of nearly pure salt (Fig. 25).

A large number of Chilean playas, particularly in the Andean Highlands, are characterized by wet margins, where soil is kept in a saturated condition by spring flow or subsurface seepage from adjoining slopes. In general, a considerable percentage of drainage reaching many of the playas is thought to be in the form of subsurface flow.

Chilean playas are generally characterized by a predominance of moist surfaces and by abundant salt. This is in marked contrast to the playas of the Great Basin (United States and Mexico), which commonly have dry silt and clay surfaces. This may result from the fact that most closed basins in Chile are either completely rimmed by rock or nearly so, resulting in a high degree of hydrologic closure; consequently groundwater discharge is almost entirely by evaporation, water tables are generally near the surface in the lowest parts of the basins, and salts generally remain within the basins. The more fundamental reason for this contrast between Chilean and Great Basin playas may be that tectonic movements and volcanism have been recently active in northern Chile, resulting in rock-rimmed basins that are relatively youthful. The Great Basin, on the other hand, is characterized by elongate ranges nearly surrounded by permeable alluvial

deposits, which underlie a part of the drainage divides around many closed basins, allowing subsurface flow into lower basins.

From a study of playas in 30 closed basins, Stoertz (74) concluded that wind has been a significant erosional or depositional agent in at least 80 per cent of the basins. Out of the 30 basins, 21 appear to contain surfaces blanketed by wind-blown soil, commonly characterized by mounds or low dunes. Twelve of the 30 basins appear to contain wind-scoured or deflated areas, and 8 show evidences of severe wind erosion. Several playas within the basins show probable evidence that wind-blown flood water has influenced the distribution of playa sediments or has promoted the corrosion of salt layers or of soil cemented by salt. The distribution of former lakeshore deposits or wave-cut benches reflects the influence of prevailing westerly or northwesterly winds in the Andean Highlands since at least the last pluvial age. Because of the near-surface water levels in most of the present playas, deflation is not as active now as in times past.

Nearly every closed basin containing a playa or a lake in northern Chile shows some evidence of more extensive flooding in the past, generally in the form of beach ridges or wave-cut benches. Former deltas along the margins of several playas are commonly graded to a level near that of the highest clearly recognizable former lake shoreline and show evidence of former much larger rivers. In general, a trend of shrinking shorelines is thought to have continued for a long period up to the present, and it appears that playas are commonly being encroached upon by alluvium. This is interpreted as a result of increasing aridity and particularly decreasing amounts of surface drainage reaching the playas.

The considerable quantity of salt within many of the Chilean basins suggests that the basins have been closed for a long time, and that the climate, though more humid at times, has at least been sufficiently arid that former lakes have seldom overflowed their divides and the salt that accumulated during a long period has not been washed out of the region of interior drainage. Extensive continental evaporite deposits in several ages of rocks dating back at least as far as mid-Tertiary age are found in northern Chile. This and other evidence suggests that the region has had a predominantly arid to semiarid climate since mid-Tertiary or earlier and that tectonic movements resulting in the formation of closed basins have been active for an equally long time (74).

## North African Playas

There are more than a thousand playas in North Africa, spanning the entire continent from Spanish Sahara to Sudan. They range in size from small depressions 1 kilometer square to extremely large playas, some exceeding 3,000 square kilometers.

The North African (Sahara) Desert as a whole is one of low relief and low elevation, and this has a bearing on groundwater occurrence and circulation. In contrast with Chile, Iran, and the United States, where there are many intermontane groundwater basins, there are considerably larger regional basins in North Africa (42). For this reason, there are many playas or topographic depressions within a single groundwater region. Many playas are areas of groundwater discharge from gradually sloping, sedimentary artesian aquifers. Diversity is the rule governing the characteristics of these secondary basins that contain playas; all combinations of surface drainage, or lack of it, are found (2). Most of the basins occur in areas of relatively flat to gently warped sediments, and few if any are associated with faulting.

In comparing the basins of North Africa with those of the United States, more differences than similarities are found, even though the playas display similar surface features (Fig. 5). Climate, topography, and geology are distinctly different than in North American basins. The climate is far more arid today, and past climates were probably never moist enough to maintain pluvial lakes comparable to those of the Basin and Range Province. The Lake Chad basin is an exception, as it still contains a permanent lake, but its extent is greatly diminished from previous dimensions. Lack of relief in the Sahara has an important effect in not providing increased runoff and sediments—a chief factor of both pluvial lake and playa development in the Basin and Range Province of North America. Thick lacustrine sequences are thus generally lacking in North Africa.

In the central and eastern portions of the Sahara, groundwater-discharging playas are most common, and saline surfaces resemble those in other regions. Hard, dry crusts appear to be relatively uncommon in the Sahara, in contrast to the United States. There is a tendency toward an increasing number of hard, dry crusts in the western Sahara (2).

The chotts or large groundwater-discharging playas of Tunisia have been studied by Coque (76), who believes the original basins were of tectonic origin. Recurring depositional and erosional cycles accompanied broad Pleistocene climatic change. Desiccation followed lowering of the piezometric surface of artesian aquifers, and erosion by deflation proceeded to the new base levels. Periodic flooding that occurs is a chief agent in controlling surficial patterns.

Garaets are playas that have little if any groundwater discharge; the ground-water lies at depths of 7 to 19 meters. Photos of these playas reveal that the surfaces are comparable to hard, dry crusts of the United States.

Many large depressions such as the Qattara are located in the Libyan Desert, a number of which are below sea level. The floors of these depressions contain saline crusts, locally called *sabakha*, indicating that groundwater discharge is occuring. It is thought that these depressions formed as the result of aeolian erosion and deflation. Counterparts of these depressions are unknown elsewhere.

Erosion and deposition by wind are far more important in North Africa than elsewhere (Figs. 2 and 4). This is due to a combination of past and present aridity, topography, and geologic setting, each contributing favorably to deflationary processes.

## South African Playas

The playas of the subcontinent of South Africa are known as *pans;* where they are concentrated the area is referred to as *panneveld*. In size the pans range from less than 1 to about 250 square kilometers. They occur in flat ground and frequently have a rim 6 to 9 meters above the floor. The playas of South Africa are located in three areas: (a) the Gazaland plain in Mozambique, (b) the belt formed by outcropping Karoo System rocks, and (c) the Kalahari Desert. In general the playas are smaller, but there is a greater density than in most other regions.

Much of the rainfall of South Africa averages about 50 centimeters, so that vegetation is more abundant than in other arid regions; however, frequent flooding or surficial saline accumulations serve to keep most of the playas relatively barren.

Several varieties of playas occur in South Africa; they have been called *calc-pans* (32), *dune-pans,* and *salt-pans.* There are also clay-floored pans and rock-floored pans. These varieties do not vary appreciably from other playas already described in other regions. The calc-pans, or lime-crusted playas, owe their existence to calcareous rocks of the Kalahari, to which the calc-pans are apparently peculiar. The lime enrichment is apparently related to both infilling of the lime-rich surrounding material and to groundwater enrichment. Lowering groundwater levels are also thought to be necessary.

The largest playas of South Africa are the Etosha and Makarikari pans, both of which are salt-crusted, groundwater-discharging playas. The Etosha Pan is without doubt formed in a tectonic basin, but the Makarikari origin is less certain (77).

Wind is important as an erosive agent in the formation of many of the pans; other agents important to a lesser extent are animals that wallow in the (temporarily wet) depressions and gradually carry away fine sediment. Where deflation has been most active, it seems clear that it is related to past climatic change.

Hundreds of small playas, locally called *pans,* occur in troughs between ridges of sand in the

Wankie National Park of extreme western Rhodesia. The linear sand ridges are persistent over an area of about 8,500 square kilometers and occur in a remarkably repetitive pattern; it is believed they are the sites of Pleistocene dunes, now eroded and having a relief of only several meters (78). The pans range in diameter from only a few meters to more than 100 meters and in depth from a few centimeters to more than 3 meters. Analysis of sediment in one of the pans showed 7 per cent coarse and medium sand, 28 per cent fine sand, 13 per cent silt, and 52 per cent clay. Thus the pans collect and hold runoff and are comparable in surface composition, but not in size, to other playas with hard, dry crusts. The origin of these pans is thought to lie in deflation of the silt-clay fraction of the trough sediments, and excavation by animals (78).

### Coastal Salt Flats

Coastal salt marshes and salt-encrusted tidal flats are common in the low latitudes, and they occupy relatively large areas in some regions, such as the Rann of Kutch in India and the Trucial Coast of Arabia. The surfaces are often identical in general appearance and relief to inland salt-encrusted playas. Many of these coastal salt flats have been cut off from the ocean by barrier ridges (79); they are shoreline basins that should be considered in a general way along with playas. Some of the larger shoreline basins contain depressions within their boundaries (80).

The important difference in these shoreline basins is that the evaporites originate from evaporative concentration of sea water, whereas inland playas accumulate evaporites through the process of ground-water discharge. There can also be additional capillary discharge of saline ground water in shoreline basins, however, where salt water has encroached inland. The widespread occurrence of these shoreline basins has caused speculation that some playas far inland have been inundated by the sea in times past (58).

## CONCLUSIONS

*Playa* is the general term that is appropriate for describing the topographic low of a desert basin, or of other depressions that periodically are flooded and accumulate sediment. Occurring in all of the world arid zones, playas have great diversity in size and character, which results from a combination of factors including climate, geology, and history. Many regional terms are used to describe playas, but it is difficult to correlate them from one arid region to another.

A number of geologic processes have formed the basins in which playas occur. Structural basins are characteristic locations for playas in North and South America and in Iran. Broad downwarps in gently dipping strata are the chief cause of basins in parts of North and South Africa. Wind-carved playa basins are best developed in North Africa, but wind erosion has also aided in basin formation in South Africa, North and South America, and elsewhere. Solution and subsidence are not recognized as being major processes, but the effects are seen locally. Small depressions in South Africa and West Texas have been at least initiated by wallowing animals. Shoreline basins along coasts contain playa-like surfaces and are found the world over.

Past climatic variation has had widespread and profound influence on playa development. The principal North and South American playas are the sites of Pleistocene pluvial lakes, and thick lacustrine sequences underlie them. Postglacial climates in North America and North Africa were at times more arid than at present, and wind erosional and depositional effects are seen in and around the playas.

Hydrologically, it is significant whether or not a playa discharges ground-water. All combinations and amounts of discharge are noted in the various regions that contain playas. Heavily discharging playas usually contain evaporite accumulations in the near-surface layers. Playas discharging little or no ground-water favor development of hard, dry crusts. Groundwater-discharging playas far outweigh the number of nondischarging playas.

Playa surfaces have similar characteristics the world over, but the basins and the processes that produced them do not. A relatively small number of "playa surface descriptors" (Table 5) can be used to characterize the surfaces. However, because different processes can produce similar surfaces, the descriptors should not be used to classify playas. To gain a complete understanding of a particular playa, the descriptor must be considered along with hydrologic, geologic, and climatic information.

The surfaces of playas are highly subject to change, especially where frequent flooding occurs. A migration of the surficial zones is frequently noted in which the pattern and character of the surface is completely changed. For this reason, mapping playa surface zones has little lasting value, except for comparative purposes. The periphery of a playa is less apt to change. For these reasons, playas are among the most "dynamic" of terrestrial landforms.

\* \* \*

## ACKNOWLEDGMENT

I thank W. S. Motts, D. B. Krinsley, and G. E. Stoertz for their assistance over the past five years and for material used in this paper.

# REFERENCES AND NOTES

1. OFFICE OF ARID LANDS STUDIES
   1968  Introduction. *In* W. G. McGinnies, B. J. Goldman, and P. Paylore, eds., Deserts of the world: an appraisal of research into their physical and biological environments. University of Arizona Press, Tucson.

2. SMITH, H. T. U.
   *In press* Photogeologic studies of desert basins in northern Africa. Report to be printed for Air Force Cambridge Research Laboratories under contract AF 19(628)-2486 with the University of Massachusetts.

3. HILLS, E. S., C. D. OLLIER, and C. R. TWIDALE
   1966  Geomorphology. pp. 53-76. *In* E. S. Hills, ed., Arid lands: a geographical appraisal. Methuen and Company, Ltd., London.

4. STONE, R. O.
   1956  A geologic investigation of playa lakes. University of Southern California, Los Angeles. (Thesis) 302 pp.

5. REEVES, C. C., JR.
   1968  Introduction to paleolimnology. Elsevier Publishing Company, New York.

6. KOHLER, M. A., T. J. NORDENSON, and D. R. BAKER
   1959  Evaporation maps for the United States. U. S. Weather Bureau, Technical Paper 37. 13 pp.

7. BLANEY, H. F.
   1957  Evaporation study at Silver Lake in the Mojave Desert, California. American Geophysical Union, Transactions 38(2):209-215.

8. HARDBECK, E. G., JR.
   1955  The effect of salinity on evaporation. U. S. Geological Survey, Professional Paper 272-A. 6 pp.

9. ADAMS, T. C.
   1964  Salt migration to the northwest body of Great Salt Lake, Utah. Science 43(3610): 1027-1027.

10. FLINT, R. F.
    1963  Pleistocene climates in low latitudes. Geographical Review 20(1):123-129.

11. CHARLESWORTH, J. K.
    1957  The Quaternary era. Edward Arnold, Ltd., London. 1700 pp.

12. WRIGHT, H. E., JR.
    1968  Climatic change in the eastern Mediterranean region. University of Minnesota, Final Report. [Contract Nonr-710(33) / Task Number 389-129]

13. MORRISON, R. B., and J. C. FRYE
    1965  Correlation of the Middle and Late Quaternary successions of the Lake Lahontan, Lake Bonneville, Rocky Mountain (Wasatch Range), Southern Great Plains, and Eastern Midwest areas. Nevada Bureau of Mines, Report 9:1-45.

14. FETH, J. H.
    1961  A new map of western coterminous United States showing the maximum known or inferred extent of Pleistocene lakes. U. S. Geological Survey, Professional Paper 424-B:110-112.

15. SNYDER, C. T., G. HARDMAN, and F. F. ZDENEK
    1964  Pleistocene lakes in the Great Basin. U. S. Geological Survey, Miscellaneous Geologic Investigations Map I-416.

16. SNYDER, C. T., and W. B. LANGBEIN
    1962  The Pleistocene lake in Spring Valley, Nevada, and its climatic implications. Journal of Geophysical Research 67(6): 2385-2394.

17. LANGBEIN, W. G.
    1962  The water supply of arid valleys in intermountain regions in relation to climate. International Association of Scientific Hydrology, Bulletin 7(1):34-39.

18. SMITH, H. T. U.
    1967  Past versus present wind action in the Mojave Desert region, California. U. S. Air Force Cambridge Research Laboratories Report AFCRL-67-0683. 25 pp.

19. GALLOWAY, R. W.
    1965  Late Quaternary climates in Australia. Journal of Geology 73(4):603-618.

20. SMITH, H. T. U.
    1963  Eolian geomorphology, wind direction, and climatic change in North Africa. Final report, contract AF 19(628)-298, Air Force Cambridge Research Laboratories (AFCRL-63-443). 49 pp.

21. PORTER, S. C., and G. H. DENTON
    1967  Chronology of neoglaciation in the North American cordillera. American Journal of Science 265:177-210.

22. DORF, E.
    1960  Climatic changes of the past and present. American Scientist 48(3):341-364.

23. SCHOVE, D. J.
    1961  Tree rings and climatic chronology. *In* Solar variations, climatic change, and related geophysical problems. New York Academy of Sciences, Annals 95:605-622.

24. NEAL, J. T., and W. S. MOTTS
    1967  Recent geomorphic changes in playas of western United States. Journal of Geology 75(5):511-525.

25. LAWRENCE, D. B., and E. G. LAWRENCE
    1961  Response of enclosed lakes to current glaciopluvial climatic conditions in middle latitude western North America. *In* Solar variations, climatic change, and related geophysical problems. New York Academy of Sciences, Annals 95:341-350.

26. DUTCHER, L. C., and H. E. THOMAS
    1967  Surface water and related climate features of the Sāḥil Sūsah area, Tunisia. U. S. Geological Survey, Water-Supply Paper 1757-F. 40 pp.

27. WILLETT, H. C.
    1951  Extrapolation of sunspot-climate relationships. Journal of Meteorology 8:1-17.

28. BUTZER, K. W., and C. R. TWIDALE
    1966  Deserts in the past. pp. 127-144. *In* E. S. Hills, ed., Arid lands: a geographical appraisal. Methuen and Company, Ltd., London.

29. SNYDER, C. T.
    1962  A hydrologic classification of valleys in the Great Basin, western United States. International Association of Scientific Hydrology, Bulletin 7(3):53-59.

30. FOSHAG, W. F.
    1926  Saline lakes of the Mojave Desert. Economic Geology 21:56-64.

31. THOMPSON, D. G.
    1929  The Mojave Desert region, California. U. S.

Geological Survey, Water-Supply Paper 578. 759 pp.

32. JAEGER, F.
   1942  Ein besonderer seentypus; die Trockenseen oder Pfannen. Geologie der Meere und Binnengewässer 6:65-103.

33. HUNT, C. B.
   1966  Plant ecology of Death Valley, California. U. S. Geological Survey, Professional Paper 509. 68 pp.

34. NEAL, J. T.
   1968  Playa surface changes at Harper Lake, California: 1962-1967. In Playa surface morphology: miscellaneous investigations. U. S. Air Force Cambridge Research Laboratories, Environmental Research Papers 283 (AFCRL-68-0133):5-30.

35. EUGSTER, H. P., and G. I. SMITH
   1965  Mineral equilibria in the Searles Lake evaporites, California. Journal of Petrology 6:473-522.

36. JONES, B. F.
   1965  The hydrology and mineralogy of Deep Springs Lake, Inyo County, California. U. S. Geological Survey, Professional Paper 502-A. 56 pp.

37. JONES, B. F.
   1966  Geochemical evolution of closed basin water in the western Great Basin. In Northern Ohio Geological Society, Symposium on Salt, 2nd, Proceedings 1:181-199.

38. FRIEDMAN, G. M.
   1966  Occurrence and origin of Quaternary dolomite of Salt Flat, West Texas. Journal of Sedimentary Petrology 36:263-267.

39. SMITH, G. I., and D. V. HAINES
   1964  Character and distribution of non-clastic minerals in the Searles Lake evaporite deposits. U. S. Geological Survey, Bulletin 1181-P. 58 pp.

40. LOMBARDI, O. W.
   1963  Observations on the distribution of chemical elements in the terrestrial saline deposits of Saline Valley, California. U. S. Naval Ordinance Test Station Report NOTS-TP-2916. 41 pp.

41. KRINSLEY, D. B.
   1968  Geomorphology of three kavirs in northern Iran. pp. 105-130. In Playa surface morphology: miscellaneous investigations. U. S. Air Force Cambridge Research Laboratories Environmental Research Paper 283 (AFCRL 68-0133).

42. AMBROGGI, R. P.
   1966  Water under the Sahara. Scientific American 214(5):21-29.

43. LANGER, A. M., and P. F. KERR
   1966  Mojave playa crusts: physical properties and mineral content. Journal of Sedimentary Petrology 36(2):377-396.

44. LANGBEIN, W. G.
   1961  Salinity and hydrology of closed lakes. U. S. Geological Survey, Professional Paper 412.

45. KUNIN, V. N.
   1967  The study of the local waters in the deserts of the U. S. S. R. (translated title). Problemy osvoyeniya pustyn' 5:40-56. Soviet Geographical Review 9(6).

46. FETH, J. H.
   1967  Reconnaissance survey of ground-water quality in the Great Basin. U. S. Geological Survey, Professional Paper 550-D:237-241.

47. Figure 30 and accompanying discussion adapted from:
   MOTTS, W. S.
   1965  Hydrologic types of playas and closed valleys and some relations of hydrology to playa geology. In Geology, mineralogy and hydrology of U. S. playas. U. S. Air Force Cambridge Research Laboratories, Environmental Research Paper 96 (AFCRL-65-266): 73-104.

48. DOTY, G. C.
   1960  Reconnaissance of ground water in Playas Valley, Hidalgo County, New Mexico. New Mexico State Engineer, Technical Report 15:14-16.

49. WINOGRAD, I.
   1962  Interbasin movement of ground water at the Nevada Test Site, Nevada. U. S. Geological Survey, Professional Paper 450-C:108-111.

50. NEAL, J. T., A. M. LANGER, and P. F. KERR
   1968  Giant desiccation polygons of Great Basin playas. Geological Society of America, Bulletin 79:69-90.

51. Material on Australian playas adapted from:
   KRINSLEY, D. B., C. C. WOO, and G. E. STOERTZ
   1968  Geological characteristics of seven Australian playas. In Playa surface morphology: miscellaneous investigations. Air Force Cambridge Research Laboratories Environmental Research Paper 283(AFCRL-68-0133):59-103.

52. GREGORY, J. W.
   1914  The lake system of Westralia. Geographical Journal 43:656-664.

53. DAVID, T. W. E.
   1950  The geology of the Commonwealth of Australia. Edward Arnold, London. 3 vols.

54. WAITE, E. R.
   1917  Results of the South Australian Museum Expedition to Strzlecke and Cooper Creeks. Royal Society of South Australia, Transactions and Proceedings 41:405-658.

55. MADIGAN, C. T.
   1945  The Simpson Desert Expedition, 1939. Society of South Australia. Transactions 69(1)118-139.

56. MADIGAN, C. T.
   1944  Central Australia. Oxford University Press. 316 pp.

57. BONYTHON, C. W.
   1956  The salt of Lake Eyre—its occurrence in Madigan Gulf and its possible origin. Royal Society of South Australia, Transactions 79:66-92.

58. JOHNS, R. K., and N. H. LUDBROOK
   1963  Investigation of Lake Eyre. South Australia Department of Mines, Report of Investigations 24. 104 pp.

59. JOHNS, R. K.
   1964  Investigation of Lake Torrens. South Australia Department of Mines. 29 pp.

60. JOHNS, R. K.
   1966  Investigation of Lake Gairdner Grids G6 and H6. South Australia Department of Mines, Report Book 62. 13 pp.

61. JUTSON, J. T.
   1918  The sand ridges, rock floors, and other as-

sociated features at Goongarrie in sub-arid Western Australia, and their relation to the growth of Lake Goongarrie, a "dry" lake or playa. Royal Society of Victoria, Proceedings, n.s., 31:113-128.

62. CRITTENDEN, M. D., JR.
    1963    Effective viscosity of the earth derived from isostatic loading of Pleistocene Lake Bonneville. Journal of Geophysical Research 68(19):5517-5530.

63. BLACKWELDER, E.
    1931    The lowering of playas by deflation. American Journal of Science, 5th Series, 21: 140-144.

64. MEINZER, O. E.
    1911    Geology and water resources of Estancia Valley, New Mexico. U. S. Geological Survey, Water-Supply Paper 275.

65. STUIVER, M.
    1965    Carbon isotopic distribution and correlated chronology of Searles Lake sediments. American Journal of Science 262:377-392.

66. MABEY, D. R.
    1956    Geophysical studies in the intermontane basins in southern California. Geophysics 21:839-853.

67. HUNT, C. B.
    1960    The Death Valley salt pan, a study of evaporites. U. S. Geological Survey, Professional Paper 400-B:456-458.

68. REEVES, C. C., JR.
    1966    Pluvial lake basins of West Texas. Journal of Geology 74(3):269-291.

69. JUDSON, S.
    1950    Depressions of the northern portion of the southern High Plains of eastern New Mexico. Geological Society of America, Bulletin 61: 253-274.

70. HUBER, H.
    1965    Personal communication, National Iranian Oil Company. Dr. Huber has also written an unpublished manuscript, "The Quaternary deposits of the Darya-i-Namak," based on oil company drilling data.

71. BOBEK, H.
    1959    Features and formation of the Great Kavir and Masileh. University of Tehran, Arid Zone Research Centre, Publication 2. 63 pp.

72. CRESSEY, G. B.
    1958    Qanats, karez, and foggara. Geographical Review 47:27-44.

73. NEAL, J. T.
    1968    Satellite monitoring of lakebed surfaces. In Playa surface morphology: miscellaneous investigations. Air Force Cambridge Research Laboratories Environmental Research Paper 283 (AFCRL-68-0133):131-149.

74. *Material on Chilean playas adapted from:*
    STOERTZ, G. E.
    1966    Playas and salars of Chile. Unpublished manuscript prepared for Air Force Cambridge Research Laboratories by Military Geology Branch, U. S. Geological Survey. 266 pp.

75. ERICKSEN, G. E.
    1963    Geology of the salt deposits and the salt industry of northern Chile. Report prepared for the United Nations Special Fund Mineral Survey Project by U. S. Geological Survey. 164 pp.

76. COQUE, R.
    1962    La Tunisie Présaharienne. National Center for Scientific Research, France. 476 pp.

77. WELLINGTON, J. H.
    1955    Southern Africa: a geographical study; Vol. I, Physical geography. Cambridge University Press. 528 pp.

78. FLINT, R. F., and G. BOND
    1968    Pleistocene sand ridges and pans in western Rhodesia. Geological Society of America, Bulletin 79:299-314.

79. KIRKLAND, D. W., J. P. BRADBURY, and W. E. DEAN, JR.
    1966    Origin of the Carmen Island salt deposit, Baja California, Mexico. Journal of Geology 74(6):932-938.

80. EVANS, G., C. G. STC. KENDALL, and P. SKIPWITH
    1964    Origin of the coastal flats, the sabkha, of the Trucial Coast, Persian Gulf. Nature 202 (4934):759-761.

# QUANTITATIVE ANALYSIS OF DESERT TOPOGRAPHY

LAWRENCE K. LUSTIG
Senior Editor, Earth Sciences
Encyclopaedia Britannica

# ABSTRACT

The quantitative analysis of desert topography may be approached from two aspects, namely, the analysis of specific landform types and regional topographic analysis. The study of specific landform types generally consists of attempts to develop quantitative form parameters that define the geometry of a given feature and to relate these parameters to the natural processes that are operative on that feature. The methods employed and the degrees of success achieved are illustrated by outlines of the treatment to date of alluvial fans, headland-bay beaches, barchan dunes, and river meanders. The form parameters involved include polynomial equations, the logarithmic spiral, the ellipse and parabola, and the sine-generated curve.

Regional analysis, in contrast, involves attempts to define terrain parameters that may be used to map regions and to determine degrees of similarity or distinctiveness between or within large areas. This approach to treating desert topography is discussed with respect to several schemes of terrain classification and is illustrated by discussion of an analysis of the Basin and Range province by trend-surface methods.

# QUANTITATIVE ANALYSIS OF DESERT TOPOGRAPHY
## Lawrence K. Lustig

## INTRODUCTION

Johannes Kepler (1) is best known for his early investigations of the nature of planetary motions in the solar system. In 1611, however, he posed a non-astronomical problem that concerned him throughout much of his life. The question was, "Why are snowflakes always six-cornered?" Lack of knowledge of crystal-symmetry relations prevented Kepler from answering this question satisfactorily. Nevertheless, the matter is instructive because it suggests that our basic understanding of a wide range of natural phenomena stems from some initial consideration of external form, or of systemic regularity.

This is also true when landforms constitute the subject of investigation. Although we know that landforms are a reflection of a balance, or tendency toward balance, between the processes that mold landforms and the materials and processes that maintain them, our knowledge of these processes and their precise effects is incomplete in many respects. In order to investigate the relationship between form and processes, it is therefore desirable to define the form geometry in a quantitative manner at the outset. The thought patterns that govern such a study parallel the procedural sequence, and both are illustrative of the basic goals of quantitative analysis of specific landforms. In general, the following steps pertain:

1) Let us consider a given type of landform.
2) Can we approximate the configuration of this form by some mathematical function or simple geometric solid?
3) If so, can we make some quantitative statement concerning the universal mean values and ranges of values of selected form parameters of this landform?
4) Can we relate any selected form parameters to parameters of the natural processes that are operative on this landform?

If these questions can be answered affirmatively, then the analysis will provide an objective classification scheme for the landform considered, an objective method for the comparison of similar features in widely separated regions, and new insight to the complex interaction of form and process. The last goal has seldom been achieved, but the principles of analysis of specific landforms will be illustrated by considering briefly a few examples of such studies that are presented in this paper.

A second type of quantitative study is the analysis of landforms on a regional basis. In such studies, the primary goal is to determine terrain parameters that are mappable and that will permit the quantitative comparison of different regions. Typically, the distribution of slope values in a given region serves as the principal parameter, and this procedure has led to the discrimination of different land *units, facets, systems,* and other designations. More elaborate schemes for determining regional similitude, or lack of same, have involved soil, climate, lithology, and vegetation factors, in addition to that of slope. In recent years, the availability of high-speed computers has permitted regional topographic distinctions to be made by trend-surface analysis and other mathematical methods, distinctions that are laborious to attempt to obtain manually. Such methods enable one to obtain magnitudes of regional variation in an objective manner. Although space limitations preclude a lengthy exposition of these techniques, I shall provide an example of the work that has been done on regional analysis later in this paper.

It may be noted that thus far no mention has been made of desert topography as such. In fact, the interested reader may think it odd that a paper purporting to treat arid lands begins with mention of snowflakes and continues with a discussion of landforms in general. There are two reasons for this. First, the analysis of form—and the relationship of form to processes—rests upon principles that are independent of climate. Second, and perhaps more important, it should be recognized that desert topography includes many physical features that simply are representative of general classes of landforms. Mountains, plains, coastal features, drainage systems, and others obviously can and do occur under any climatic regime. Even such features as alluvial fans and sand dunes, which are commonly thought of as specifically representative of desert areas, exist in a wide variety of environmental settings. This should not be construed as an argument that desert topography lacks distinctiveness but, rather, as the reason for the choice of examples that are presented in this paper. Each example is illustrative of some aspect of quantitative analysis of desert topography, despite the fact that the environmental settings are not deserts of the world in every case.

## Analysis of Individual Landforms

I have chosen four studies of individual landforms to illustrate the possibilities of quantitative analysis of desert topography in this category of endeavor. These are alluvial fans, headland-bay beaches, barchan dunes, and river-channel meanders.

Alluvial fans are sedimentary deposits that border highland areas at the mouths of the drainage basins that serve as their sediment source. They are commonly fan shaped in plan view, radiating outward from an apex at the highland front, and occurring in every desert of the world. These landforms have been studied from various points of view, and much information on fan structure, morphology, and sediment-transport processes, among other matters, is available. Recent papers by Bull (2), Anstey (3), Denny (4), Lustig (5), Melton (6), and Hooke (7) provide much of the salient data and some useful quantitative relations. In keeping with the purposes of the present paper, however, the work of Troeh is most germane (8).

Considering the gross aspects of alluvial fans, it is clear that ideally they may be best approximated by a simple geometric solid, namely a right circular cone. The equation of a circular cone may be expressed as

$$Z = P + SR \qquad (1)$$

where $Z$ is the elevation at any point on the cone, $P$ is the elevation at the central point, $S$ is the slope of the sides of the cone, and $R$ is the distance from $P$ to $Z$. Because alluvial fans possess surface curvature, a quadratic equation will provide a better fit; hence, equation (1) may be expanded to

$$Z = P + SR + LR^2. \qquad (2)$$

The rate of change of slope with radial distance from the fan apex is simply

$$\frac{dZ}{dR} = S + 2LR, \qquad (3)$$

and equations (2) and (3) can be converted to Cartesian coordinates by substitution of $(x, y)$ coordinate values for $R$. Troeh fits equation (2) to fans by topographic-map construction and subsequent computation. Essentially, the choice of three radial lines and intersecting circles that approximate contour lines on the map provide three values of $Z$. The method of simultaneous equations then permits determination of the unknowns, and the equation of best fit for the fan, or pediment, is obtained.

Troeh did not pursue the matter beyond this point, and in view of my introductory remarks it is clear that this example of study of a specific landform indicates only a means of mathematical approximation of form. If a sufficient number of fans in a given region were studied, then a general equation of best fit might well emerge. This equation would then provide a basis for comparison of fans in different regions or, alternatively, it would provide a basis for the determination of correlations between form and process parameters.

Coastal studies provide more fertile illustrative ground, perhaps because knowledge of wave and surf effects is far more advanced than is knowledge of processes on fans and pediments. Galvin's review (9) of theoretical studies of longshore currents indicates that our understanding of near-shore processes is still incomplete, but the stated comparison of marine and arid-region processes is nonetheless fair. In any event, Yasso's work on headland-bay beaches clearly is worthy of mention here (10). A headland-bay beach is a beach that lies in the lee of a headland, which is subjected to a predominant direction of wave attack. Yasso studied such beaches along the New Jersey and California coasts, neither of which is noted for aridity, but it should be mentioned that his results accord well with Silvester's observations in South Africa (11) and are probably of general applicability.

The beaches in question are characteristically concave seaward, and their radii of curvature increase with distance from the headland. Accordingly, the mathematical function chosen for form fitting in this instance was the logarithmic spiral, which may be expressed as

$$r = e^{\theta \cot \alpha}, \qquad (4)$$

where $r$ is the length of a radius vector from the center of the spiral, $\theta$ is the angle swept by $r$, and $\alpha$ is the angle between the radius vector and a tangent to the curve at that point. These relations are shown in Figure 1. The rate of change of the radius of curvature is determined from the first derivative of equation (4), namely,

$$\frac{dr}{d\theta} = \cot \alpha \, e^{\theta \cot \alpha}. \qquad (5)$$

If equation (4) is written in linear form then

$$\ln r = \theta \cot \alpha, \qquad (6)$$

and this relationship between $\ln r$ and $\theta$ can be used to determine goodness of fit. Essentially, one plots values of $\ln r_i$ and $\theta_i$, obtains the constants in the least-squares regression equation

$$\ln r = a\theta + b \qquad (7)$$

by the usual methods and then calculates a theoretical radius vector, or $r$ value, for each $\theta_i$ by

$$\rho_i = e^{a\theta_i + b}, \qquad (8)$$

where $\rho_i$ is a theoretical $r$ value. The mean-squared error for fit of a given logarithmic spiral was sub-

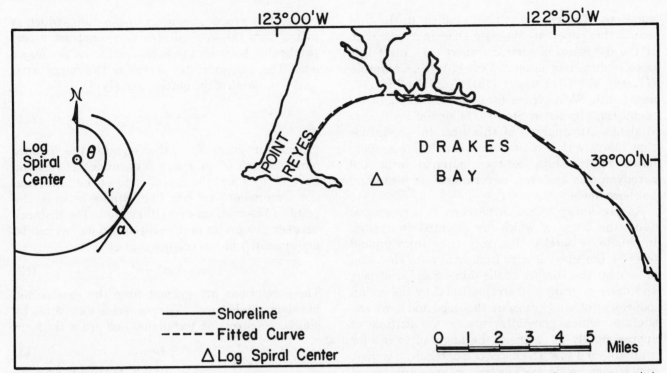

Fig. 1. The logarithmic spiral, $r = e^{\theta\cot\alpha}$, and the fit of this curve to the head-land-bay beach at Drakes Bay, California. After Yasso (*10*); symbols are explained in text.

sequently minimized by overlay adjustments on maps of the beaches investigated.

The results obtained by Yasso must be rated as excellent. Figure 1, taken from his report, shows the close accord between the form of a California beach and the fitted logarithmic spiral. Such results suggest that headland-bay beaches may well be equilibrium landforms that reflect the effects of wave refraction about a given headland. If this is so, then one utilitarian end is obvious. That is, any significant departure of a given beach from this logarithmic spiral form provides information on the incoming wave trains; either the effects of the wave train are non-uniformly distributed along the shore, or wave trains from several directions of origin must be operative. Such knowledge has obvious value in instances where direct observations of currents are unavailable. Subsequent investigation of one of the beaches involved in Yasso's study, incidentally, revealed that it was in fact an equilibrium landform (*12*).

Before leaving the subject of beaches, the observations of Cloud on beach cusps might be cited in a similar vein (*13*). The cusps in question are commonly seen, crescent-shaped indentations along beaches and are separated by *horns* or ridges of sediment. The dimensions of such cusps render them *invisible* on plan views of even moderate scale, but, as is true of the gross configuration of a given beach, these smaller landforms likewise reflect the marine processes that mold them. It is well known that the spacing of cusps along a particular beach is some-

what variable. The mean spacing of beach cusps is nevertheless a characteristic of a given beach at a given point in time. Thus, mean spacing of cusps is, again, a typical form parameter, and Cloud has made the interesting suggestion that cusp spacing may be correlative with Plateau's Rule.

This rule refers to the fact that a liquid cylinder under gravity-free conditions becomes unstable when its length exceeds $2\pi r$. Such a cylinder will then separate into segments, the lengths of which are proportional to the diameter of the cylinder. Cloud argues that this situation is not far removed from that of a breaking wave, granting the fact that the hydrodynamics of breaking waves is far from simple. Because beach cusps and their spacing are some function of breaking waves, a correlation between mean spacing, or other form parameters, and the heights of incoming waves may well exist. Thus, like the geometry of headland-bay beaches, some aspects of the distribution of beach cusps could serve to provide information on coastal waves and currents. The significance of this suggestion within the context of this paper will not be lost on those who are aware of the extensive coastlines that bound the arid regions in many parts of the world.

As a third example of the quantitative analysis of specific landforms, sand dunes of the barchan type may be considered. Investigations of these crescentic sand accumulations were undertaken first in the 1880's, but until recent years the data consisted only of published photographs, plan views, and sketches, and a few numerical values for dune dimensions. The most pertinent studies, in the context of this paper, are Bagnold's basic consideration of the dynamics of sand transport (14), investigations of dune migration in Peru (15, 16), California (17, 18), the Gobi desert (19), Saudi Arabia (20), and South West Africa (21), and an unpublished manuscript by Roger Smith (22), one of my former students. An analysis of this literature indicates that much work remains to be done before soundly based relationships between barchan form and aerodynamics emerges; nevertheless, a beginning has been made.

As previously stated, a barchan is a crescentic dune, the horns of which are oriented downwind. In profile, a barchan typically rises from ground level at the upwind edge to a maximum elevation at or near the slipface of the dune. Sand transport and dune migration is accomplished by the eolian movement of sand grains up the slope and down the slipface, which generally retains an attitude of approximately 32 degrees, the angle of repose for grains of sand size. The horns of a barchan, of course, may be symmetric, asymmetric, or largely destroyed from time to time, depending on changes in wind frequency, magnitude, and direction. Given this basic barchan configuration, however, a number of quantitative form parameters are obvious, and these have been introduced or used in one or more of the papers previously cited.

Dune migration rates are commonly expressed by the empirical relation

$$\frac{1}{D} = a + b(H) \qquad (9)$$

or occasionally in the form

$$HD = k, \qquad (10)$$

where, in either case, $H$ is barchan height, $D$ is the linear distance of migration per year, and $a, b, k$ are constants for a given area of study. With respect to barchan geometry, a variety of parameters has been employed. Among these are length of the windward slope, maximum height of slipface, maximum height of dune, distance between the horns, inclination of the windward slope, and others. These have been used or correlated in various combinations in attempts to better define dune geometry. Length-to-width ratios have been used for classification schemes, sand volumes have been calculated, and empirical relations between horn width and crest height have been established, for example. The only effort to treat overall barchan shape, however, is Smith's work (22) on the fitting of the ellipse and parabola to plan views of barchans that were obtained from published literature or available aerial photographs.

The fitting of an ellipse to a given barchan is accomplished by determining three points on the curve. The axis of symmetry of the dune, which is a line from the upwind apex to a constructed perpendicular between the horns, serves as the major axis. The perpendicular serves as the minor axis. In the equation of the ellipse, namely

$$\frac{x^2}{a^2} + \frac{y^2}{b^2} = 1, \qquad (11)$$

the semi-major axis $a$ is then equal to the distance along the axis of symmetry of the dune, between the upwind apex and the intersection of the lines, and the semi-major axis $b$ is the distance between the point of intersection and either horn. The distance between the origin of the ellipse and its foci can be determined from the relationship

$$c = \sqrt{a^2 - b^2}. \qquad (12)$$

These relations are evident from the constructed fits shown in Figure 2. The parabolae were fitted by similar construction techniques and are of the form

$$x^2 = 4py. \qquad (13)$$

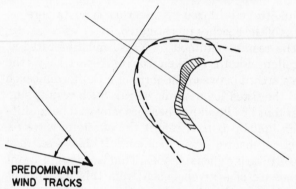

- - - FITTED PARABOLA X²=4PY

||||||| BARCHAN SLIPFACE

PREDOMINANT
WIND TRACKS

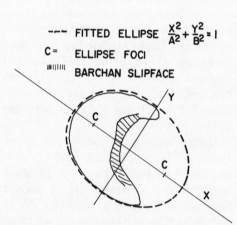

- - - FITTED ELLIPSE $\frac{X^2}{A^2} + \frac{Y^2}{B^2} = 1$

C = ELLIPSE FOCI

||||||| BARCHAN SLIPFACE

Fig. 2. Fit of the parabola $x^2 = 4py$, and the ellipse $x^2/a^2 + y^2/b^2 = 1$, to plan view of a barchan dune in the Gobi Desert (19), after Smith (22).

No statistical tests of goodness of fit were made; overlays of constructed curves were adjusted to plan views of the barchans to minimize areas of dunes that occurred inside or outside a given fitted curve. The examples shown in Figure 2, however, indicate that the fits obtained by this method are quite good. Smith concluded that the ellipse generally provided a better approximation of barchan shape than did the parabola, particularly in those instances of horn convergence downwind. The dunes in coastal deserts, however, appeared to be slightly more diverse in form; within a given area, certain barchans were fitted best by ellipses, whereas others were fitted best by parabolae. Dunes that could not be approximated by either curve may be atypical or may reflect a recent destructive wind regime; the data are insufficient for definitive statements on this point.

In all the studies of dunes to date, considerations of the form geometry, occasionally in combination with data on sand size and bulk density, are followed by discussions of the inferred wind regimes in the areas of study. It is fair to say that an enormous gap still exists in relating the form geometry to dynamic factors in the quantitative sense. Some authors have tended to stress Bagnold's (14) identity as explanative in this sense. The indentity is

$$C = \frac{q}{\gamma H}, \qquad (14)$$

where $q$ is the discharge of sand across a dune, $\gamma$ is the specific weight of sand, $H$ is dune height, and $C$ is the rate of dune migration. But one must not lose sight of the fact that this identity still fails to relate wind regimes to dune shape, which is, or should be, one of the primary goals of most studies. I would not tend to conclude from the results to date that the quantitative analysis of dunes is hopeless, but, rather, that the analysis of dune morphology and the factors that govern that morphology represent a most promising area for future endeavors. In addition, aside from the existing gap of knowledge, namely relating form and process, it should be noted that much work on the geometry of dune types other than barchans still remains to be done. Sand mountains, star-shaped dunes, and sigmoidal types would certainly be amenable to such an approach. As evidence of this statement, dunes of the last-named type, sigmoidal dunes, are instructive because they may best be described as S-shaped or meandering in plan view. The geometry of meanders has been treated in a variety of ways and some of the work accomplished may well prove relevant in studies of appropriate dunes.

When meanders are mentioned in connection with desert topography, the reference, of course, is to meandering dry washes or portions of drainage systems and not to sand dunes. This is indeed the intention here, but before pursuing this topic it is important to note that meandering is extremely common in the realm of physical phenomena. The atmospheric jet streams, their oceanic counterparts, lightning, major features of the earth's crust, certain rill systems on the moon (23), and the flow of water droplets down a pane of glass all seem to reflect the adage that nature abhors a straight line. Large-scale fracture systems, in fact, provide the only notable exception that readily comes to mind.

With respect to river systems, the work done on meanders falls into three categories. The earlier, unsophisticated view was that meandering streams reflected the anthropomorphic attributes of age; that is, rivers that meandered were thought to be old, tired, sluggish, and the like. These invalid assumptions were ultimately replaced by determinations of empirical relations by many authors. The work of Leopold and Wolman (24) and Hack and Young (25), for example, provided information on meander wavelength and amplitude, sinuosity, and other attributes of naturally meandering streams, and related such factors to hydrologic and lithologic controls. The utility of such an approach is illustrated by Dury's work (26-28) on misfit streams, which was based on observed differences between the meandering patterns of streams and of their enclosing valleys.

Bagnold investigated the energy dissipation in flow around curves from the theoretical viewpoint, using knowledge gained from flow conditions in closed pipes (29). He found that the resistance to flow was a minimum when the ratio of the mean radius of curvature to pipe diameter had a value between two and three. This range of values also prevails in natural channels and tends to lead to the conclusion that natural streams are following paths of least resistance, or paths requiring the least energy expenditure when their course is meandering. Bagnold's study was basic to the problem and, in a sense, provided one-half of the solution. The remainder largely stems from Von Schelling's exposition of random walk problems (30,31).

A random-walk path between two points is based on the assumption that at each unit increment of distance, or step along the path, there is an equal probability of a step forward or a step to either side of the straight-line path. This means, of course, that any random-walk path between two points will be sinuous; a straight-line path is the least likely occurrence because this would result only if all the steps were identical in direction. The lengths of the sinuous random-walk paths treated by Von Schelling were defined by

$$s = \frac{1}{\sigma} \int \frac{d\varphi}{\sqrt{2(\alpha - \cos \varphi)}}, \qquad (15)$$

where $s$ is arch length of path, $\sigma$ is standard deviation, $\varphi$ is the angle of deviation from a straight-line path, and $\alpha$ is a constant of integration. This constant can be set equal to cos $\omega$, and $\omega$ is then defined as the maximum angle that the path makes with the mean direction of path. For the most frequently occurring random-walk path

$$\sum \frac{(\Delta\varphi)^2}{\Delta s} = \text{a minimum}, \qquad (16)$$

where $\Delta\varphi$ and $\Delta s$ are unit increments of $\varphi$ and $s$ as previously defined. This means that the most probable meander path is one for which the variance, or mean squared deviation in direction, is a minimum.

Knowledge of these results enabled Langbein and Leopold to develop a sine-generated curve and a method of fitting to the geometry of natural meanders (32). This curve is generated by the relation

$$\varphi = \omega \sin \frac{s}{M} 2\pi, \qquad (17)$$

where $\varphi$ is the angle of deviation at location $s$, $\omega$ is again the maximum angle between the meandering path and the mean direction of path, and $M$ is the total length of path. An example of the excellent fits of this curve to natural meanders is shown in Figure 3. The hydraulic variables in meandering reaches of several natural rivers were also investigated, and it was found that the mutual adjustment of such factors as depth and velocity of flow, slope, bed shear, and the friction factor was such that the overall variability was less than in straight reaches of rivers.

Langbein and Leopold, hence, conclude that the meandering pattern is the most stable pattern in natural river channels; it is also the most probable. One might say that of the examples presented thus far, quantitative work on river meanders is the most satisfactory. We can define the geometry of meanders, we can relate meanders to the parameters of flow, to some extent, and our basic understanding of meandering as such is certainly advanced. Moreover, there is little question that the principles and approaches to the problem, as briefly outlined here, can be translated to study of such major desert rivers as the Nile, Tigris-Euphrates, Indus, Helmand, and others. Nevertheless, certain problems still exist. Central to these problems is lack of sufficient studies of meander-fitting to state with precision the degree of departure from the most probable meander path that can be said to fall within the statistical range of this path. That is, departure from the ideal form may indicate either the presence of local constraints or the fact that the potential for future adjustment, toward the most probable path, is great. Because a most serious problem in arid regions or elsewhere is the migration of channels, it would be most useful to obtain data on the overall range of meandering paths that occur in nature and the degrees of stability that may be associated with these paths. This need for additional data will be partially met through time, but unfortunately, as I have indicated previously (33), our knowledge of desert drainage systems is exceedingly limited, save for data on the flow of major perennial streams.

The four examples of studies of specific features cited above should suffice to give the reader an outline of the nature of such studies, their shortcomings, and some intimation of their potential practical value. Other examples could be provided or, alternatively, each of the four discussed might have been treated in greater detail. It is preferable, however, to provide some information on the nature of regional studies of desert topography, and the remainder of this paper will be devoted to this topic.

## REGIONAL TOPOGRAPHIC ANALYSIS

As stated in the introductory section of this paper, regional topographic analysis is basically an attempt to obtain meaningful terrain parameters that are mappable and that will provide a means for determining the degree of similarity that exists between or within large regions. A number of schemes have been used to accomplish this end, a few of which will be discussed here. First, and in some senses the simplest, is the classification of terrain based solely on the criterion of slope. Savigear, who is a leading exponent of this approach, has devised a mapping

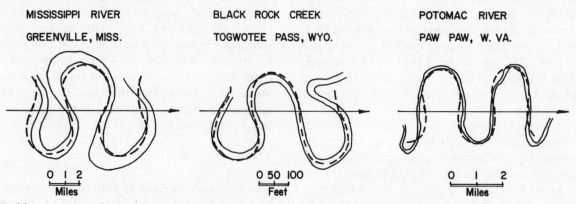

Fig. 3. Fit of the sine-generated curve, $\varphi = \omega \sin s/M2\pi$, to natural river meanders. After Langbein and Leopold (32); symbols are explained in text.

system that rests upon the fact that the ground surface in any location consists of plane and curved surfaces, termed facets and segments, that are discontinuous (34). The discontinuities are breaks in slope that are abrupt or gradual, but it is argued that these are recognizable in any case. The ground surface in this system, then, is subdivided into units that are termed *flats, slopes,* and *cliffs,* depending upon their inclination, *facets* and *segments,* as previously defined, *discontinuities,* and *micro*-units, which are features that are too small to be represented at the scale employed for mapping. A set of symbols is used to represent these units on the map, and one thus obtains, in general, a convenient pictorial view of the terrain in any given area.

Pitty (35) has discussed this method with particular reference to the subjectivity involved in the recognition of slope discontinuities. He argues that a useful modification would be the introduction of unit lengths, or distances, of measurement on the ground. In any event, one could, conceivably, make comparisons of the ground surface between or within regions on the basis of slope values alone, if such a procedure is deemed suitable for the purposes that exist.

Most regional analyses, however, are not based upon the sole criterion of slope but upon a variety of factors. One of the systems that should be mentioned in this regard is termed *geomorphological mapping.* Such mapping is international in scope, primarily because the International Geophysical Union and Unesco support the program. Maps of various areas have become increasingly abundant (36, 37) and, in fact, a map of Europe at a scale of 1:500,000 currently is planned (38). Geomorphological maps are also directed toward a depiction of relief features on a regional basis. Unfortunately, however, proponents of the method hold that in addition to depicting relief, or landforms, the maps should include information on the age and origin of all the features and deposits that occur in a given region. In order to accomplish this end, a vast array of symbols and colors has been proposed to account for the nearly infinite combinations of type of feature, age, and supposed origin that occur in nature. Ultimately, of course, some international agreement on symbols will be attained (38, 39) but the essential difficulty that cannot be overcome, as I have indicated elsewhere (33), is that the genesis of landforms cannot be mapped. It is a subjective quantity that is a function of the beliefs of a given investigator.

The practice of the C.S.I.R.O. of Australia must be considered preferable to geomorphological mapping, as such. Although directed toward a classification of terrain in terms of potential land use, the scheme of this agency consists of assembling information on the topography, geology, geomorphology, soils, and vegetation of a given area, and then categorizing these data to define subregions on a semi-quantitative basis. A recent example of such work is a report on the Isaac-Comet area in Queensland (40).

The military, of course, have long been interested in the development of quantitative parameters to define desert terrain and to determine analogous regions within this environment. In the United States, terrain-analog folios of the major deserts of the world have been prepared by combining geometry, ground (lithology and soils), and vegetation factors (41), taking the Yuma, Arizona, area as the standard for comparison. There are a number of difficulties involved (33), but the folios represent an admirable effort.

Similarly, the goal of the Military Engineering Experimental Establishment in the United Kingdom (42, 43) is to determine "recurrent landscape patterns," or analogous areas, in all parts of the world. Although not identical with this official effort, the work of Mitchell and Perrin (44) is illustrative of the results. These authors define basic physiographic units, termed *facets,* on the basis of rock type, morphology, and slope, and larger units, termed *land systems,* on the basis of the most commonly included facets. They have classified most of the desert areas of the world by means of identification of these facets and systems on aerial photographs and in the field.

Each of the efforts described above has advantages and disadvantages. The merit of any given scheme of terrain classification is to some extent dependent upon the goals involved and the resources that are at hand to achieve those goals. As a general comment, however, it is obvious that some undesirable subjectivity is inherent in several of these approaches. Although mathematical methods cannot enhance initially poor data, in certain respects they do insure that objective conclusions will be derived from the initial data. There are in existence a number of interesting applications of Fourier analysis, for example, to the problem of classifying micro-relief features (45, 46). I should like to conclude this paper, however, with a summary of one of my own investigations that involves another analytical method (47).

This work consisted of a trend-surface analysis of topographic data from the Basin and Range province of the western United States. It was not undertaken solely to quantify topography or to classify terrain but was begun as a means of investigating the geological history of the region and the question of the degree of topographical diversity that can be tolerated within a *physiographic province,* as this term and been defined. Eleven topographical parameters were defined, namely, (1) area of ranges to total area, (2) range length, (3) range width, (4) range height, (5) range relief, (6) range volume, (7) cumulative length of trends (straight-line escarpments), (8) cumulative deviation of trends, (9) range width to length, (10) range width to height, and (11) range length to height. Measurements of these parameters were obtained from the 46 1:250,000 topographic sheets that cover the Basin

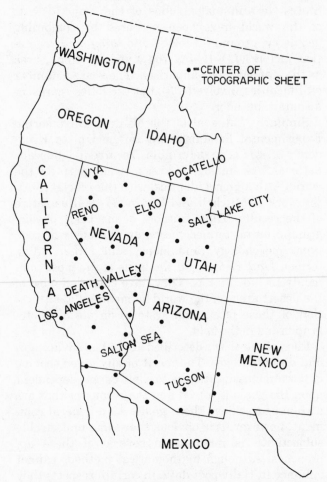

Fig. 4. Map of part of the western United States showing the locations of the 1:250,000 topographic sheets that cover the Basin and Range province.

and Range region (Figure 4), and the centers of these sheets served as a grid system for plotting mean values of each parameter and subsequent computer analysis.

Trend-surface analysis, as the term in commonly used, consists of the fitting of polynomial surfaces to a data set in an $x$, $y$, $z$, coordinate system by least-squares methods. It is an extremely useful quantitative technique that may best be understood by consideration of a two-dimensional analogy. If an investigator is confronted with data that consist of two variables, then three choices of data presentation are available. He can simply plot the values as a scatter diagram on arithmetic graph

paper, he can construct a straight line through such values by visual means, or he can obtain the unique line of best fit to the data by the statistical procedure of the least-squares method. The essential point is that because the latter is the only means of obtaining the unique line of best fit, it is an objective method. The same is true of the presentation of topographic or other data in a three-dimensional system. One is attempting to map the distribution of $z$-values that are a function of $x$, $y$, geographic coordinates, where, in the case at hand, the $z$-values are the mean values of the topographic parameters and the $x$, $y$ coordinates are the centers of the sheets. The $z$-values can, of course, be contoured manually, but this is a subjective method that is akin to the visual construction of a straight line in the two-dimensional case.

Trend surfaces, then, are surfaces of best fit that correspond to polynomial equations in an $x$, $y$, $z$, coordinate system. A first-degree surface is a plane whose equation is linear and is of the form

$$Z = A + Bx + Cy. \tag{18}$$

A second-degree surface is a paraboloid, and its equation contains the zeroth and first-degree terms and appropriate quadratic terms, namely,

$$Z = A + Bx + Cy + Dx^2 + Exy + Fy^2. \tag{19}$$

A third-degree equation is oscillatory or doubly sinusoidal in space, and its equation consists of the linear and quadratic terms above (19) and four additional cubic terms, namely,

$$Z = A + Bx + Cy + Dx^2 + Exy + $$
$$Fy^2 + Gx^3 + Hx^2y + Ixy^2 + Jy^3. \tag{20}$$

One can continue to write equations of higher degree by the addition of terms, of course, but in the study under consideration the topographic data were fitted by trend surfaces of first, second, and third degree only. The geometric relations of first-, second-, and third-degree equations for two and three variables are shown in Figure 5.

The most rapid method of generating the surfaces of best fit is to use computers, and the program I employed produced the desired trend surfaces and several useful statistical error measures. Among these measures, the variance that is not explained by the fitted surface is important. It is expressed as

$$U = \sum_{i=1}^{N} (z_{i(o)} - z_{i(c)})^2, \tag{21}$$

where $z_{i(o)}$ is an observed $z$-value, or the original data, $z_{i(c)}$ is the corresponding computed $z$-value on the surface of best fit of a given degree, and $U$ is the unexplained variance. The variance that is explained by a given surface is a measure of goodness

2 VARIABLES          3 VARIABLES

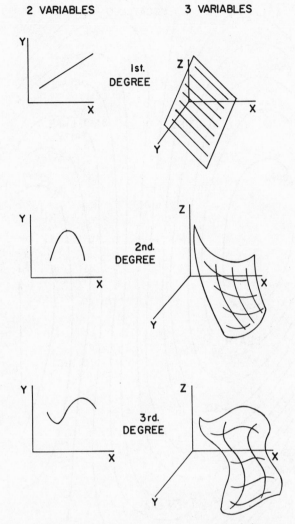

Fig. 5. Geometric relations of first, second, and third degree equations for two and three variables.

of fit and is simply the difference between the total variance and the unexplained variance. That is,

$$E = V - U, \qquad (22)$$

where $E$ is the variance explained, $U$ is the unexplained variance, and $V$ is total variance, which is defined by

$$V = \sum_{i=1}^{N} (z_i - \bar{z})^2, \qquad (23)$$

where $z_i$ represents the original data values and $\bar{z}$ is the mean of these values.

Ratios of $E$ to $V$ were calculated for each surface of given degree that was fitted to each of the eleven topographic parameters. These ratios express the explained variance as a percentage and, when combined with correlation-coefficient values, they permit one to interpret the analytical results with some confidence. In the Basin and Range study, the goodness of fit improved as trend surfaces of increasing degree were used. In other words, the best fits were achieved by the third-degree surfaces for all parame-

ters. This need not be invariably true, because fitting always depends on the nature of the initial data. Investigation of dipping beds in the subsurface, for example, might well show that first-degree equations provide the best fit simply because the structure approximates a plane surface. In any event, the correlation coefficients associated with the fits of third-degree trend surfaces in my study ranged from 0.602 for the range volume parameter to 0.858 for the cumulative deviation of trends within the province. These parameters, indeed each of the topographic parameters mentioned, cannot be too informative to the reader because the methods of computation and the definitions have not been explained. Space limitation precludes such explanation here but a complete exposition is available (47) for those who wish additional details.

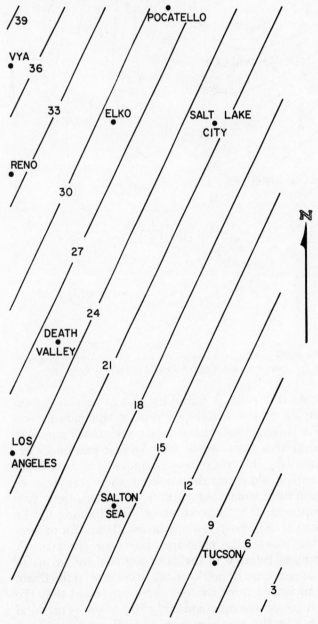

Fig. 6. First-degree trend-surface map of the area of ranges to total area ratio. Contour interval is three per cent.

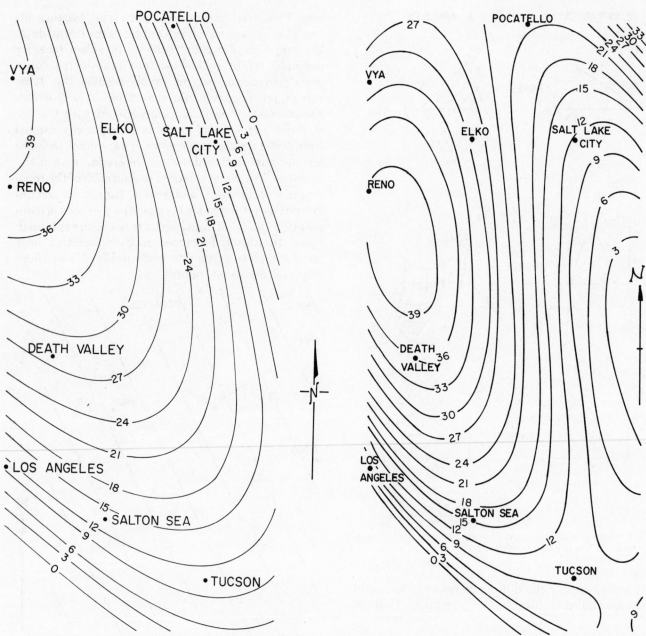

Fig. 7. Second-degree trend-surface map of the area of ranges to total area ratio. Contour interval is three per cent.

Fig. 8. Third-degree trend-surface map of the area of ranges to total area ratio. Contour interval is three per cent.

At this point it would be best to provide an example of the results for one of the parameters. For this purpose, I have chosen the area of ranges to total area ratio, which was the first parameter examined. This ratio was computed for each topographic sheet by determining the total area of mountain ranges, or portions of ranges, that were represented by a given sheet, and dividing by the area of that sheet. It might be noted that this parameter essentially measures the area occupied by ranges, relative to the area occupied by basins in the province. Remarks in the literature on the Basin and Range province long have suggested that the ranges occupy approximately 50 per cent of the total area in the northern part and 20 per cent in the southern part.

Although the absolute values of such percentages

will vary with the measurement criteria that are used, the initial data in this study showed that values of 50 per cent or more occurred for only two sheets and that values of less than 10 per cent persist along the southern margin of the province. The trend surface maps, of course, reflect the absolute values of the initial data that were used in their computation. Figure 6 shows the first-degree trend map of this parameter. It consists of a series of parallel lines because it is the equation of a plane of best fit (equation 18 and Figure 5). It should be noted that the regional values range from three per cent in the south to 39 per cent in the northwest. The magnitude of the regional variation is therefore 1300 per cent.

The second-degree trend surface is a paraboloid of best fit (equation 19 and Figure 5), and the third-degree surface is oscillatory (equation 20 and Figure 5). These maps are shown in Figures 7 and 8, respectively, and clearly indicate the regional distinctions that may be made. The northwestern part of the Basin and Range area, the locus of which is the 39-per cent contour closure on Figure 8, also was distinguishable on trend maps of the remaining 10 topographical parameters, by reason of the occurrence of similar closures. Some degree of similitude between the northeastern and southern parts of the province was also discernible when the trend results for all parameters were considered.

The technique of trend-surface analysis proved highly successful in this study. The present paper is not the proper vehicle for expressing the conclusions, but, in general, the analysis of topography permitted inferences to be drawn on the age of ranges in various parts of the province, the distribution of fans and pediments, and the nature and magnitude of the regional variations that apparently can exist within an area considered to be a physiographic province and, hence, one in which the topography should be relatively homogeneous.

# CONCLUSION

In summary, the quantitative analysis of desert topography has been illustrated by examples of studies of individual landforms and by reference to regional schemes of investigation. The examples presented here in either category are not exhaustive, and with respect to regional analysis I do not wish to foster the impression that the methods of trend-surface analysis or Fourier analysis are the sole means of treatment. A wide array of mathematical methods is available and, in certain instances, these may be better suited to a given problem. It may be concluded, however, that work in all categories of topographic analysis can be improved and advanced. The individual landform studies to date have largely failed to relate form parameters to process parameters. Moreover, analysis of many types of features, all of which have obvious geometric analogs, has not yet been attempted. As a last point, it may be said that even in the most successful studies of individual features there has been a failure to treat a sufficient number of cases to insure that the population of a given form has been adequately sampled.

The regional studies that have been cited require little additional elaboration. Objectivity is a clear prerequisite of such work in all its phases, namely defining units, obtaining initial data, and treating those data. On the basis of my own experience I would favor trend surface or other suitable analytical techniques for such data treatment. If the goal of regional analysis primarily is to determine topographic similitude or degrees of diversity between or within regions, then mathematical methods provide the most objective means of attaining these ends.

# REFERENCES

1. KEPLER, J.
   1966 The six-cornered snowflake. Clarendon Press. (Translated from the 1611 Latin edition by Colin Hardie)
2. BULL, W. B.
   1964 Geomorphology of segmented alluvial fans in western Fresno County, California. U. S. Geological Survey Professional Paper 352-E.
3. ANSTEY, R. L.
   1965 Physical characteristics of alluvial fans. U. S. Army Natick Laboratories, Technical Report ES-20.
4. DENNY, C. S.
   1965 Alluvial fans of the Death Valley region, California and Nevada. U. S. Geological Survey Professional Paper 466.
5. LUSTIG, L. K.
   1965 Clastic sedimentation in Deep Springs Valley, California. U. S. Geological Survey Professional Paper 352-F.
6. MELTON, M. A.
   1965 The geomorphic and paleoclimatic significance of alluvial deposits in southern Arizona. Journal of Geology 73:1-38.
7. HOOKE, R. L.
   1967 Processes on arid-region alluvial fans. Journal of Geology 75(1):438-460.

8. TROEH, F. R.
   1965 Landform equations fitted to contour maps. American Journal of Science 263:616-627.
9. GALVIN, C. J., JR.
   1967 Longshore current velocity: a review of theory and data. Reviews of Geophysics 5:287-304.
10. YASSO, W. E.
    1965 Plan geometry of headland-bay beaches. Journal of Geology 73:702-714.
11. SILVESTER, R.
    1960 Stabilization of sedimentary coastlines. Nature 188(1):467-469.
12. STRAHLER, A. N.
    1966 Tidal cycle of changes in an equilibrium beach, Sandy Hook, New Jersey. Journal of Geology 74:247-268.
13. CLOUD, P. E., JR.
    1966 Beach cusps: response to Plateau's Rule? Science 154:890-891.
14. BAGNOLD, R. A.
    1941 The physics of blown sand and desert dunes. Methuen, London.
15. FINKEL, H. J.
    1959 The barchans of southern Peru. Journal of Geology 67:614-647.
16. HASTENRATH, S. L.
    1967 The barchans of the Arequipa region, south-

Hastenrath (cont)
    ern Peru. Zeitschrift für Geomorphologie 11:300-331.

17. LONG, J. T., and R. P. SHARP
    1964  Barchan-dune movement in Imperial Valley, California. Geological Society of America, Bulletin 75:149-156.

18. NORRIS, R. M.
    1966  Barchan dunes of Imperial Valley, California. Journal of Geology 74:292-306.

19. HÖRNER, N. G.
    1957  Some notes and data concerning dunes and sand drift in the Gobi desert: the Sino-Swedish Expedition, Publication 40. Statens Etnografiska Museum, Stockholm. 40 p.

20. KERR, R. C., and J. O. NIGRA
    1952  Analysis of eolian sand control. American Association of Petroleum Geologists, Bulletin 36:1541-1573.

21. STENGEL, H. W.
    1963  Wasserwirtschaft in Südwestafrika. Afrika-Verlag der Kreis, Windhoek.

22. SMITH, R. S. U.
    1967  Elliptical approximation of barchan-dune plans. Tucson, Arizona. (Unpublished ms)

23. LINGENFELTER, R. E., S. J. PEALE, and G. SCHUBERT
    1968  Lunar rivers. Science 161:266-269.

24. LEOPOLD, L. B., and M. G. WOLMAN
    1957  River channel patterns; braided, meandering, and straight. U. S. Geological Survey Professional Paper 282-B.

25. HACK, J. T., and R. S. YOUNG
    1959  Intrenched meanders of the North Fork of the Shenandoah River, Virginia. U. S. Geological Survey Professional Paper 354-A.

26. DURY, G. H.
    1964  Principles of underfit streams. U. S. Geological Survey Professional Paper 452-A.

27. DURY, G. H.
    1964  Subsurface exploration and chronology of underfit streams. U. S. Geological Survey Professional Paper 452-B.

28. DURY, G. H.
    1965  Theoretical implications of underfit streams. U. S. Geological Survey Professional Paper 452-C.

29. BAGNOLD, R. A.
    1960  Some aspects of the shape of river meanders. U. S. Geological Survey Professional Paper 282-E.

30. VON SCHELLING, H.
    1951  Most frequent particle paths in a plane. American Geophysical Union, Transactions 32:222-226.

31. VON SCHELLING, H.
    1964  Most frequent random walks. General Electric Company, Schenectady, N. Y., Report 64GL92.

32. LANGBEIN, W. B., and L. B. LEOPOLD
    1966  River meanders—theory of minimum variance. U. S. Geological Survey Professional Paper 422-H.

33. LUSTIG, L. K.
    1968  Inventory of research on geomorphology and surface hydrology of desert environments. pp. 95-283. In W. G. McGinnies, B. J. Goldman, and P. Paylore, eds., Deserts of the

world: an appraisal of research into their physical and biological environments. University of Arizona Press, Tucson.

34. SAVIGEAR, R. A. G.
    1965  A technique of morphological mapping. Association of American Geographers, Annals 55:514-538.

35. PITTY, A. F.
    1967  Some problems in selecting a ground-surface length for slope-angle measurement. Revue de Géomorphologie dynamique 17:66-71.

36. POLISH ACADEMY OF SCIENCES
    1963  Problems of geomorphological mapping. Géographical Studies 46.

37. TRICART, J., and M. MICHEL
    1965  Monographie et carte géomorphologique de Langunillas (Andes venezuéliennes). Revue de Géomorphologie dynamique 15:1-33.

38. FRÄNZLE, O.
    1966  Geomorphological mapping. Nature and Resources 2(4):14-16.

39. GELLERT, J. F.
    1967  Poursuite des travaux en vue de la légende de la carte geomorphologique détaillée. Revue de Géomorphologie dynamique 17:129-131.

40. STORY, R., et al.
    1967  Lands of the Isaac-Comet Area, Queensland. C.S.I.R.O. Land Research Series 19. Melbourne.

41. VAN LOPIK, J. R., and C. R. KOLB
    1959  A technique for preparing desert terrain analogs. U. S. Army Corps of Engineers, Technical Report 3-506.

42. BECKETT, P. H. T., and R. WEBSTER
    1962  The storage and collation of information on terrain. Military Engineering Experimental Establishment, Interim Report. Christchurch, England.

43. BRINK, A. B., et al.
    1966  Report on the working group on land classification and data storage. Military Engineering Experimental Establishment, Report 940. Christchurch, England.

44. MITCHELL, C. W., and R. M. S. PERRIN
    1966  The subdivision of hot deserts of the world into physiographic units. Symposium International de Photo-Interpretation, IIᵉ, 4: 1/89-106.

45. FMC CORPORATION
    1964  A research study concerning the application of a Fourier series description to terrain geometries associated with ground mobility and ride dynamics. U. S. Army Engineering Waterways Experiment Station, Contract Report 3-33.

46. STONE, R. O., and J. DUGUNDJI
    1966  A study of microrelief—its mapping, classification, and quantification by means of a Fourier analysis. Engineering Geology 1(2): 89-187.

47. LUSTIG, L. K.
    1969  Trend surface analysis of the Basin and Range province and some geomorphic implications. U. S. Geological Survey Professional Paper 500-D.

# THE NEW SYSTEM OF SOIL CLASSIFICATION AS APPLIED TO ARID-LAND SOILS

KLAUS W. FLACH
Soil Survey Laboratory
Soil Conservation Service
U. S. Department of Agriculture
Riverside, California, U. S. A.

and
GUY D. SMITH
Soil Conservation Service
U. S. Department of Agriculture
Washington, D. C., U. S. A.

# ABSTRACT

Classes in the new comprehensive system of soil classification (*7th Approximation*) consist of soils of similar genesis that occur as natural units in the landscape. Although based on genetic and geographic relationships, the definitions of classes are in terms of morphological, chemical, and physical soil properties. Major soil-forming processes, such as the accumulation of organic matter or of clay, are reflected in *diagnostic surface* or *subsoil horizons* that are used repeatedly in the definitions of taxa. The system has seven categories with 10 *orders* at the highest and some 8,000 *soil series* at the lowest level. Names of taxa were coined primarily from classical roots. Developed soils in arid lands are mostly in the order **Aridisols. Aridisols** with horizons of clay accumulation are in the suborder **Argids;** those without, in the suborder **Orthids.** The presence or absence of pan horizons, sodium saturation, salinity, organic matter content, soil texture, mineralogy, and soil temperature are among the criteria for classes at lower categories.

# THE NEW SYSTEM OF SOIL CLASSIFICATION AS APPLIED TO ARID-LAND SOILS

## Klaus W. Flach and Guy D. Smith

In 1951 the Soil Survey Staff of the U. S. Department of Agriculture, with the help of many soil scientists in the United States and abroad, began work on a new system of soil classification. This system developed through various stages—or approximations—each stage being tested and discussed and revised. The seventh of these approximations was presented in book form to the 1960 International Congress of Soil Science at Madison, Wisconsin. This approximation, hence, became widely known, and some refer to the new system as the 7th Approximation (1). Two supplements to the 7th Approximation were issued, one in 1964 and the other in 1967. A revised version of the classification is being written now for publication in 1969.

This paper is a discussion of the principles of the system and its application to soils of arid lands.

## WHY A NEW SYSTEM?

The beginning of a systematic soil survey in the United States around the turn of the century led to the recognition of local kinds of soil, called soil types and series, as independent bodies. With time, the soil series were recognized as natural bodies that had unique properties and that occupied distinct positions in the landscape. While in the beginning soil series were defined very broadly, they became more and more narrowly defined with advances in soil science and growing needs for more precise information for modern, intensive land use planning. Today more than 8,000 series are recognized in the United States. Russian scientists had previously recognized that soils are natural bodies and that broad differences among soils are the result of different climatic environments. This concept led to the recognition of broadly defined great soil groups such as Chernozems or Podzols, and to the concept of the zonal soils, "soils having well-developed soil characteristics that reflect the influence of the active factors of soil genesis—climate and living organisms, chiefly vegetation"(2).

There was an obvious need to organize the multitude of soil series into broader groups of manageable size. Several attempts were made between 1927 and 1949 to relate the soil series to great soil groups in systems of soil classification. The last one of these systems was published by the Soil Survey Staff in 1949 (3).

As more soil surveys were made, more series discovered, and knowledge about soils increased, it became apparent that this system was inadequate to organize what was known about soils. The genetic definitions were ambiguous. Many soils had no place in the classification. The names of great soil groups could not be adapted to form names for classes with some properties of two great soil groups. The shortcomings of the system were particularly apparent for soils outside the experience of the originators of the system, notably the soils of the deserts and the tropics where vast areas of soils could not be classified to bring out their relationships.

## PRINCIPLES, ORGANIZATION, AND NOMENCLATURE OF THE SYSTEM OF CLASSIFICATION

To escape the difficulties encountered with the old system of soil classification, the new system presents a deliberate departure in organization, logic, and nomenclature. Those who become familiar with this new system, however, will easily recognize the heritage of the old ideas of soil classification.

Smith (4) listed the following (slightly modified) desirable attributes of a system of soil taxonomy:

a) The definitions of taxa should be operational so that they carry the same meaning to each user.

b) The classification should be a multicategory system and should have a large number of taxa in the lower categories. The lower categories must be as specific as possible about many soil properties; but the large number of taxa are more than the mind can comprehend—hence, higher categories, with taxa based on fewer properties, are necessary.

c) The taxa should be concepts of real bodies of soil that have geographic area.

d) Differentiae should be properties that can be observed in the field or that may be inferred from properties observable in the field or from data from soil science and other disciplines.

e) The classification must be capable of providing taxa for all soil in a landscape.

f) The classification should be capable of modification to accommodate new knowledge.

To meet these criteria, six categories—order, suborder, great group, subgroup, family, and series—are recognized in the new system. Table 1 shows the functions and the nomenclature system used for the categories. Groupings at each of these levels are made of series that, in the judgment of soil scientists, have important properties in common. A listing of the ten orders of the system is

TABLE 1
## CATEGORIES AND NOMENCLATURE OF THE CLASSIFICATION SYSTEM

| Category | Number | Function | Name |
|---|---|---|---|
| Orders | 10 | Large groups of soils differing mainly in the kind and degree of dominant pedogenic processes as reflected by the combinations of diagnostic horizons. | Contains a formative element and the ending *sol* (e.g., Ar-*id*-i-sol; Latin: *aridus,* to be dry). |
| Suborders | 47 | Similar to that of order but more, and more homogeneous, taxa. | Consists of the formative element of the order and a modifying element; always a two-syllable word (e.g., Argid, an Aridisol with an argillic horizon). |
| Great groups | 200 | Groups of soils with the same kinds of horizons in the same sequence, with similar soil moisture and soil temperature regimes, and with similar base status. | Suborder name and prefix; e.g., Hapl-argid, a simple (Greek: *haplos,* simple) Argid. |
| Subgroups | Numerous | The typical taxon of a great group, and intergrades either to other great groups or with aberrant properties—e.g. presence of hard rock at shallow depth. | The adjective "typic" followed by the great group name (e.g. Typic Haplargid), or the order, suborder, or great group toward which the taxon intergrades (e.g., Duric Haplargid, a Haplargid with some characteristics of a Durargid). |
| Families | Numerous | Subdivision of a subgroup based primarily on soil texture, mineralogy, and soil temperature. | The family designation follows the subgroup name; e.g., Typic Haplargid, fine loamy (texture), mixed (mineralogy), thermic (temperature) family. *Also:* name of representative series. |
| Series | — | Subdivision of family based primarily on characteristics used at higher categories, but with finer subdivisions. | Series names are local place names, usually from the locality where the soil was first described; e.g., Flagstaff, Tucumcari. |

given in Table 2. The definitions of these taxa, however, were made in terms of observable soil properties (criterion *d*). As far as possible, differentiae were chosen that are significant in themselves. Some of these are not obvious, and the definitions of most of the taxa appear arbitrary and, unless applied to real bodies of soils, incomprehensible. The criteria used are not arbitrary, however; they are based on desirable groupings of soil series and reflect the experiences and judgments that have accumulated over 60 years of soil survey work. They reflect the soil-forming factors and emphasize groupings that have been found useful.

TABLE 2
## THE ORDERS AND THEIR RELATIONS TO ARID AND SEMIARID LANDS

| | |
|---|---|
| **Entisols:** | Soils without pedogenic horizons; common in arid lands. |
| **Vertisols:** | Fine textured soils that have deep wide cracks at some time during the year; present in arid lands. |
| **Oxisols:** | Highly weathered soils with an exchange complex of extremely low activity; may occur in arid lands. |
| **Aridisols:** | Soils with pedogenic horizons extensive in arid lands and some semiarid lands. |
| **Mollisols:** | Soils with mollic horizons of semiarid or subhumid lands. |
| **Alfisols:** | Soils with a horizon of clay accumulation of high base saturation outside arid lands and without a mollic horizon; occur in semiarid lands. |
| **Ultisols:** | Soils with a horizon of clay accumulation (argillic horizon) of low base saturation; rare in semiarid lands. |
| **Inceptisols:** | Soils with a structural B (cambic) horizon but no mollic or argillic horizon; occur in semiarid lands. |
| **Spodosols:** | Soils with a subsoil horizon of aluminum and organic matter or iron accumulation; not in arid or semiarid lands. |
| **Histosols:** | Organic soils; rare in arid lands. |

Hence, the system is genetic in concept, but the definitions are strictly in terms of morphological criteria. The system is independent of genetic theory; it can be applied uniformly by soil scientists with various backgrounds, and it has a built-in mechanism that will force revisions if present criteria cause unreasonable groupings or if newly discovered facts make present groupings unsatisfactory. In this way the system does not prejudice future research, as do systems based on genetic definitions (5).

In the former system of soil classification the central concept of each class was described, but the placement of class boundaries was left to the judgment of the individual soil classifier. Such a system is workable within small areas and with a limited number of classifiers who are in personal contact with each other. If soils are to be compared over large areas, the changes are so gradual that different soil scientists will not make identical decisions; hence, one of the most significant innovations of the new system is well-defined boundary criteria.

Tables 2, 4, 5, and 6 show general definitions of taxa in the new system. Clearly, an investigator who wishes to use the system must have a thorough grounding in soil science. Since the definitions are commonly in terms of properties a soil may not have, the investigator can visualize the characteristics of a given group only if he applies the system to actual soils.

The names of the taxa are distinctive at each categorial level (Table 1). They were coined, with the help of classical philologists, by generally accepted rules of scientific nomenclature from Greek and Latin roots (and some coined roots) but without using names from a living language (6). All names of orders contain a formative element and the ending *sol;* for example, **Aridi-sol,** where *id* is the formative element. All names of suborders are two syllable words, the second syllable being the formative element of the order name, such as in **Argid.** Similar systematic conventions are used in the nomenclature of great groups, subgroups, and families.

## WHAT IS A SOIL?

Since soil, unless interrupted by nonsoil, covers the land surface as a continuum, the question arises as to what is *a soil* that we wish to classify. The authors of the classification system have created the concept of the *pedon,* which is the smallest volume of soil in which the range of thickness and degree of expression of horizons can be studied for descriptions or sampling. The concept of the pedon is akin to that of the unit cell of the mineralogist.

It was necessary, however, to assign an arbitrary minimum size of 1 square meter and a maximum size of 10 square meters to the pedon. Contiguous pedons within the range of the definition of a soil series are a *polypedon* and constitute the real thing we classify. The polypedon, however, should not be confused with the mapping unit, or phase, which is the basic cartographic unit of a detailed soil map. The mapping unit consists not only of one or more polypedons of the kind of soil for which it is named but it may also contain pockets of nonsoil and polypedons of other kinds of soil (7).

## MAJOR SOIL-FORMING PROCESSES IN ARID LANDS AND THE KINDS OF SOIL HORIZONS FORMED BY THEM

Weathering and soil formation are controlled by the action of the environment, mostly climate and organisms, on soil parent materials. In arid environments chemical weathering is slow because of the lack of water to sustain weathering reactions and

TABLE 3

### DIAGNOSTIC HORIZONS AND FEATURES

| Diagnostic Horizons | Pedologic Significance | Elements of Definition* |
|---|---|---|
| **Surface horizons** | | |
| Mollic epipedon | Surface horizon of accumulation of dark colored organic matter and of plant nutrients, primarily in a grassland environment | Dark color, moist, value and chroma of 3 or less† Thickness more than 25 centimeters More than 1 per cent organic matter More than 50 per cent base saturation Not hard and massive when dry |
| Ochric epipedon | Surface horizon with little organic matter accumulation | Higher value or chroma of 4 or more, or lower organic matter content, or lower base saturation |
| **Subsurface Horizons** | | |
| Argillic | Horizon of clay accumulation | One-fifth more clay than the overlying horizon Coatings of illuviated clay on surfaces of structural units, sand grains, or in pores |
| Natric | Horizon of clay and sodium accumulation | Same as argillic horizon and more than 15 per cent exchangeable sodium |
| Cambic | Horizon of pedogenic alteration | Evidence of soil structure (as contrasted to rock structure) or evidence of removal of calcium carbonate, or mottling |
| Calcic | Horizon of pronounced carbonate accumulation | Evidence of pedogenic carbonates, more than 15 per cent carbonate equivalent, and more than 15 centimeters thick |
| Gypsic | Horizon of pronounced gypsum accumulation | Product of per cent gypsum × thickness (cm) is 150 or more |
| Salic | Horizon of pronounced salt accumulation | More than 2 per cent salt; thickness (cm) × per cent salt is more than 60 |
| Duripan | Cementation by silica and carbonates | Opal or chalcedony coatings; does not slake completely in acid or water |
| Petrocalcic | Horizon indurated by pedogenic carbonates | Containing as much calcium carbonate as the calcic horizon and indurated; does not slake in water but slakes completely in acid |

| Other Features | Pedologic Significance | Elements of Definition |
|---|---|---|
| Usually dry | Dry for more than half of growing season | Water at more than 15 bar tension for more than half of the time the soil is warmer than 8°C |
| Moisture control section | Root zone of annual crops below zone that dries rapidly | Zone not moistened by a 2.5 centimeter rain, but moistened by a 7.5 centimeter rain |
| Durinodes | Silica cemented nodules | Weakly cemented nodules with concentric stringers of opal |
| Lithic contact | Lower limit of the soil over hard bedrock | As hard or harder than 3 on Mohs scale |
| Paralithic contact | Lower limit of soil over soft rock | Less hard than 3 on Mohs scale but can be dug with a spade only with difficulty |
| Slickensides | Evidence of churning | Shiny, intersecting, and striated surfaces |

*Greatly simplified
†Munsell notation

TABLE 4

## SOILS OF ARID LANDS

*A. Key to Orders, Suborders, and Great Groups of Soils of Arid Lands*

| | ORDER<br>Suborder<br>Great Group |
|---|---|
| Soils having more than 30 per cent clay, cracks when dry, and slickensides | VERTISOLS |
|   Vertisols that are usually dry | Torrerts |
|   (three other suborders of Vertisols are recognized) | |
| Soils with no diagnostic horizons except an ochric horizon | ENTISOLS |
|   Entisols with a texture of loamy fine sand or coarser between 24 cm and 100 cm depth | |
|     Psamments that are usually dry | Psamments |
|     (five other great groups of Psamments are recognized) | Torripsamments |
|   Entisols in which organic matter decreases irregularly with depth or is less than 0.35 per cent at 125 cm depth | Fluvents |
|     Fluvents that are usually dry | Torrifluvents |
|     (five other great groups of Fluvents are recognized) | |
|   Other Entisols | Orthents |
|     Orthents that are usually dry | Torriorthents |
|     (five other great groups of Orthents are recognized) | |
| Soils with oxic horizon | OXISOLS |
|   Oxisols that are usually dry and have color values of more than 4 in the epipedon | Torrox |
|   (three other suborders are recognized in Oxisols) | |
| Soils having an ochric epipedon that is not hard and massive, and with one or more of the following: | ARIDISOLS |
|   a. Usually dry and with an argillic, or natric, cambic, calcic, petrocalcic, duripan horizon, or | |
|   b. Electrical conductivity >2 millimhos to 125 cm if sandy, to 90 cm if loamy, to 75 cm if clayey, and a cambic, calcic, petrocalcic, gypsic, or duripan horizon, or | |
|   c. Saturated with water for 1 month or more and a salic horizon | |
|   Aridisols with no argillic or natric horizon | Orthids |
|     Orthids having a duripan | Durorthids |
|     Other Orthids saturated with water for 1 month or more that have a salic horizon above any calcic or gypsic horizon | Salorthids |
|     Other Orthids with a petrocalcic horizon | Paleorthids |
|     Other Orthids that are calcareous throughout and have either a calcic or a gypsic horizon | Calciorthids |
|     Other Orthids | Camborthids |
|   Other Aridisols with an argillic or a natric horizon | Argids |
|     Argids having a duripan and a natric horizon | Nadurargids |
|     Other Argids having a duripan | Durargids |
|     Other Argids having a natric horizon | Natrargids |
|     Other Argids having a petrocalcic horizon or an argillic horizon with >35 per cent clay and an abrupt upper boundary | Paleargids |
|     Other Argids | Haplargids |

*B. Key to Orders, Suborders, and Great Groups of Soils of Semiarid Lands\**

| Order | Suborders | Great Group | Criteria in Addition to Those Listed in footnote |
|---|---|---|---|
| Vertisols | Usterts | | |
| | Xererts | | |
| Ultisols | Ustults | | Must have less than 0.9 per cent organic carbon in the B2t horizon and less than 20 kg organic matter in the surface cubic meter. |
| | Xerults | | |
| Alfisols | Ustalfs | | Or has secondary lime in the subsoil. |
| | Xeralfs | | |
| Mollisols | Ustolls | | Or has secondary lime in the subsoil. |
| | Xerolls | | |
| Entisols | | | |
| | Psamments | | See table IV part A for definition |
| | | Ustipsamments | |
| | | Xeropsamments | |
| | Orthents | | See table IV part A for definition |
| | | Ustorthents | |
| | | Xerorthents | |
| | Fluvents | | See table IV part A for definition |
| | | Ustifluvents | |
| | | Xerofluvents | |
| Inceptisols | Ochrepts | | With ochric surface horizon, no evidence of poor drainage, and the clay fraction is not allophanic |
| | | Ustochrepts | |
| | | Xerochrepts | |

*Soils that are seasonally dry (summer) are separated from soils that are intermittently dry in all orders except Oxisols. Xer- and Ust- classes are defined as follows:

  Xer: dry for 60 consecutive days or more in more than 7 out of 10 years in all parts of the soil between 8 cm and 50 cm
  Ust: dry for 90 cumulative days or more in 7 out of 10 years in all parts of the soil between 18 cm and 50 cm

to remove the weathering products. Erosion is usually concentrated in certain parts of the landscape, leaving extensive remnants of very old land surface with highly developed soils in close association with young, unweathered, and undeveloped soils. Such old soils, notably those from which all primary minerals have been lost and converted to hydrous oxides and 1:1 silicate clay minerals, were formed largely during periods of considerably higher rainfall than today. Chemical weathering in arid environments is nevertheless probably much more important, and physical weathering less so, than was presumed previously (8).

While some properties, such as mineralogical composition, may reflect past environments, others, such as organic matter content and distribution of bases (exchangeable metal cations) in the profile, are usually rather sensitive indications of the present climate.

## DIAGNOSTIC HORIZONS

Soil properties of genetic significance are spelled out in detail throughout the system. A few, the diagnostic horizons and features, are complex and basic to the system (Table 3) and need to be discussed separately.

### Diagnostic Surface Horizons (Epipedons)

Six diagnostic surface horizons (epipedons) have been defined, but only two, the *ochric epipedon* and the *mollic epipedon*, are important for the classification of arid-land soils.

The kind and amount of organic matter in the surface soil determines the kind of epipedon formed. Most surface soils of arid lands are almost devoid of organic matter and are therefore ochric epipedons. A few soils of arid lands and many soils of semiarid lands have surface horizons containing enough organic matter and of dark enough color to be mollic epipedons. Mollic and ochric epipedons are basic to the differentiation of **Aridisols** from **Mollisols**.

### Diagnostic Subsurface Horizons

Some diagnostic subsurface horizons are not found in arid-land soils; others, such as the **cambic, argillic,** and **oxic** horizons, occur in soils of both arid and humid environments. Some, such as **salic horizons** (horizons of salt accumulation), **calic** and **gypsic horizons** (carbonate and gypsum accumulation), and **petrocalcic horizons** and **duripans**\* (lime and silica cemented pans), occur only in soils of semiarid and arid lands and are part of the differentiating criteria for these soils (Table 4, part A).

Most soils of stable surfaces in arid lands have a B horizon from which some carbonates have been removed, the structure of the parent material has been altered, some *soil structure* has formed, or some reddening has taken place. If little or no clay has moved into this horizon, it is a **cambic** horizon. If a significant amount of clay has moved into a sub-

*A few duripans may also be found in humid environments.

soil horizon, it is an **argillic** horizon, or if the horizon contains more than 15 per cent sodium, it is a **natric** horizon. Evidence of clay translocation is a higher clay content in the B as compared to the A horizon and, ideally, morphological evidence such as coatings of illuviated clay (clay skins) on structural surfaces.

At this time, these criteria are not completely satisfactory. In many soils of the desert the eluvial horizon is so thin that clay removal may have been by the action of wind or water; and many fine textured B horizons of desert soils shrink and swell so much on wetting and drying that clay skins may be destroyed or ped surfaces are so unstable that they never form. There are, however, many B horizons of intermediate clay content that have clearly identifiable clay skins and an abrupt and irregular upper boundary. Both have been taken as evidence of clay translocation in soils of humid areas. The alternative hypothesis, that fine-textured B horizons underlying coarse-textured A horizons in desert soils are the result of greater weathering intensity in the B as compared to the A, has not been proven convincingly.

The **natric** horizon is a special form of the **argillic** horizon in which more than 15 per cent of the exchange complex is saturated with sodium ions. Many natric horizons have a characteristic columnar structure, but some do not. Natric horizons tend to *seal up* when irrigated, and soils with natric horizons are difficult to reclaim.

**Oxic** horizons are highly weathered subsoil horizons of tropical areas that lack weatherable minerals and have an exchange complex of extremely low activity.

**Calcic** horizons are horizons in which significant amounts of calcium carbonate accumulated by pedologic processes. The pedogenic origin of the calcium carbonate is indicated by the distribution of carbonates as powdery fillings, concretions, or as pendants on gravel. Some calcic horizons have fine laminae of calcium carbonate near the upper boundary of the horizon. Calcic horizons must meet certain criteria of lime content and thickness. **Gypsic** horizons are similar to calcic horizons except that gypsum is the dominant mineral.

**Petrocalcic** horizons are indurated calcic horizons. They are continuous, and dry fragments do not slake in water. They have a hardness (Mohs scale) of 3 or more and cannot be penetrated by a spade or auger when dry. Some petrocalcic horizons have a laminar capping. They are also called *hard caliche* or *calcrete* by geologists. In Africa they are called *croute calcaire* and are very extensive.

**Duripans** are indurated horizons in which amorphous or cryptocrystalline silica is the primary cement. Calcium carbonate is commonly a secondary cementing agent. Some part, at least, of the duripan does not slake in water or in acid, but it may slake after repeated, alternating treatment with concentrated alkali and hydrochloric acid. Duripans are found most commonly in soils containing easily weatherable pyroclastics such as volcanic ash or tuff. Some very old soils derived from igneous parent materials in mediterranean climates also have duripans.

**Salic** horizons are horizons of accumulation of salts that are more soluble than gypsum. Sodium chloride and sulfate are the common salts in these horizons.

## CLASSIFICATION OF SOILS OF ARID LANDS

### Definition of Aridisols, Torri Suborders, and Torri Great Groups

As may be seen from Figure 1 and Table 4, part A, soils of arid lands may be in the order **Aridisols,** or in **Torri** suborders of **Oxisols** and **Vertisols,** or in **Torri** great groups of **Entisols.** While the definition of **Torri** suborders and great groups is based almost exclusively on the soil climate, the definition of **Aridisols** is based in part on the soil climate and in part on the presence or absence of certain diagnostic horizons, on the presence of salt in the soil profile, and on the structure and consistency of the surface horizon. Soils are categorized as **Aridisols** regardless of the soil

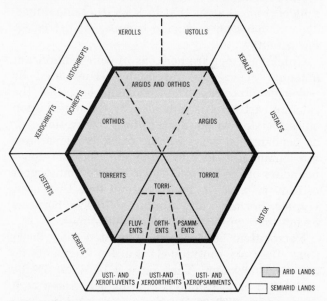

Fig. 1. Relationships among suborders and great groups of arid-land and semiarid-land soils.

climate if they have an extremely saline horizon overlying ground-water at shallow depths. On the other hand, soils are excluded from Aridisols regardless of their soil climate if they have a mollic, umbric, or histic epipedon, an oxic subsurface hori-

zon, are strongly shrinking and swelling on drying and wetting, or lack pedogenic horizons. Some of these latter soils are separated as *Torri* suborders or great groups of other orders.

These (at first sight) rather complex criteria are used to keep soils with horizons that strongly reflect an arid environment together and to use as far as possible criteria that can be identified without the need for long-term data on soil climate. Of these criteria, the requirement that **Aridisols** may not have a **mollic** epipedon is particularly important in the United States and elsewhere, since **Mollisols** and **Aridisols** have a long common boundary of great practical importance. Since the boundary is determined by the distribution of mollic surface horizons, it can be identified on the basis of soil color in the field or on the basis of organic matter content which is easily determined. As far as we know now the geographic boundaries of **Aridisols** in the United States coincide quite closely with Thornthwaite's −40 moisture index (9). **Torri** great groups of **Entisols** and **Aridisols** are defined as being dry in all parts of their moisture-control section for more than half of the active growing season (when the soil temperature is higher than 5°C) in most years.

The moisture control section, in turn, is defined as that part of the soil profile that is not moistened by a rainfall (within 24 hours) of 25 millimeters but that is moistened by a rainfall of 75 millimeters. Approximate boundaries of this section for fine loamy, coarse silty, fine silty, and clayey (see family texture classes, Table 5) soils are 10 and 30 centimeters and for sandy soils 30 and 90 centimeters. The boundaries for coarse loamy soils are intermediate. The boundaries of the moisture-control section are such that water supplied by a single summer storm that stays in the upper 10 centimeters of the profile and is usually lost by evaporation is not considered in classification.

In the absence of extensive soil moisture measurements, the moisture regime is calculated using Thornthwaite's formula (9), assuming an available moisture-holding capacity of the soil of about 7.5 centimeters. These criteria, at this time, are quite unsatisfactory. Long-range data based on the analysis of day-to-day weather records, measurements of soil moisture, and better estimates of water-holding capacity of soils (the correlation of water-holding capacity and soil texture is not as close as has been assumed in this calculation) are being accumulated and will, in time, allow better definitions and better placement of soils in the classification system. Soils in areas with monsoon and mediterranean climates cause further difficulties. To keep soils from areas such as Bombay (India) with a 4- to 5-month rainy season and 1,700-millimeter rainfall out of **Aridisols,** the definition of **Torri** subgroups and of **Argids**

TABLE 5

**FAMILY CRITERIA**

| Classes | Number of Classes | Criteria |
|---|---|---|
| 1. *Soil texture** | | |
| Uniform gravelly and stony soils | 4 | a. Packing of coarse fragments |
| | | b. Particle size distribution of <2 mm fraction |
| Uniform other soils | 6 | Particle size distribution |
| Soils with highly contrasting textures | 25 | Particle size contrast within 7.5 cm |
| 2. *Soil mineralogy** | | |
| Coarse and medium textured soils | 13 | Composition of the whole soil or of the sand fraction |
| Fine textured soils | 6 | Composition of the sand fraction |
| 3. *Soil temperature* | 8 | Mean soil temperature and contrast in soil temperature between summer and winter |
| 4. *Other classes†* | 12 | Depth, permeability, consistence, moisture equivalent, slope |

*Applies to a "control section" which consists either of the whole or part of the argillic or natric horizon if present or of an arbitrary section from 25 cm to 1 m or to a pan or lithic contact.
†Used only in certain taxa.

was amended to eliminate soils that are continuously moist for more than 90 days in most years.

The decision to exclude soils without horizon development, the **Entisols**, from **Aridisols** was based on the desire to keep all undeveloped soils in one order (otherwise undeveloped soils would have been in **Entisols** and **Aridisols**), and the decision to keep **Oxisols** and **Vertisols** out of **Aridisols** was based on the consideration that the low fertility of **Oxisols** and the shrink-swell characteristics of **Vertisols** impose drastic restrictions on the use of these soils.

### Classification of Aridisols at the Suborder and Great Group Level

**Aridisols** are subdivided into soils having an argillic horizon, **Argids**, and soils having a cambic horizon, **Orthids** (Table 4, part A).

**Argids** are generally older than **Orthids** and occur generally in older parts of a desert landscape, often in remnant surfaces of dissected fans mostly of Pleistocene age. Commonly, they have other adjuncts of maturity, such as a desert pavement, or a desert varnish, or both. Since they are commonly on sloping land, they are not used much for intensive agriculture in the western United States.

Weakly and moderately developed **Argids** without pans and without excess sodium are **Haplargids**. They are the most intensively used of the **Argids**.

They are commonly on gently sloping areas, and the clay content of the argillic horizon is low enough that the soils may drain freely; they do not present major management problems.

Two kinds of **Argids**, both representing extreme profile development, are combined in **Paleargids**. **Paleargids** have either a lime-cemented pan horizon, a petrocalcic horizon, or they have an argillic horizon containing more than 35 per cent clay and an abrupt upper boundary. Both conditions hinder root growth and water movement and present similar management problems. The two kinds of **Paleargids** are separated at the subgroup level. **Paleargids** occur in the oldest parts of a desert landscape. If there are carbonates in the parent material, or if dust containing carbonates falls on the soil, **Petrocalcic Paleargids** predominate. In other areas **Typic** (or some other subgroup) **Paleargids** are more common. **Paleargids** present serious management problems if irrigated, but commonly they are in parts of the landscape that do not lend themselves easily to large-scale intensive irrigation.

**Argids** with a natric horizon (an argillic horizon with more than 15 per cent of the exchange sites saturated with sodium) are **Natrargids** or, if a duripan underlies the natric horizon, **Nadurargids**. Many **Natrargids** (and **Nadurargids**) have a characteristic columnar structure in the subsoil and the bleached *albic* horizon of the classical Solonetz or Solodized Solonetz soils. **Natrargids** usually are on nearly level land, but their agricultural use is severely limited by the low permeability of the natric horizon. **Natrargids** that are even slightly permeable may be reclaimed by large applications of gypsum, but some seal up so completely that their reclamation is not economically feasible.

**Durargids**, as the name implies, have a duripan (within one meter of the surface). They form on parent material containing volcanic ash or other pyroclastic materials. They present management problems similar to those of **Petrocalcic Paleargids**. **Nadurargids** combine the characteristics and disadvantages of **Natrargids** and **Durargids**.

**Aridisols** that do not have an **argillic** horizon are in the suborder **Orthids**. **Orthids** have more horizon development than **Torriorthents** (or **Torrifluvents** and **Torripsamments**) and have either a cambic horizon, a duripan, a calcic horizon, a petrocalcic horizon, or a salic horizon. Most **Orthids** occur on younger parts of the desert landscape than do **Argids**, such as the lower end of fans and in recent parts of valley floors. They also occur as large bodies in areas of late Pleistocene deposits such as the Lake Lahontan basin or glaciated and periglacial areas of eastern Oregon and Washington.

**Orthids** that have only a cambic horizon and do not have a duripan, calcic, petrocalcic, or a salic horizon are **Camborthids**. They are some of the most extensive soils of the desert. They may occur on steeply sloping areas where erosion is too fast to allow formation of an argillic horizon or a pan, or on younger, level, lower parts of fans. Camborthids on level topography are usually very suitable for irrigation farming since they lack excessive amounts of salt and horizons that impede water movement.

**Orthids** with calcic horizons and lime throughout the profile are **Calciorthids. Calciorthids** are usually in highly calcareous parent material or near a source of calcareous dust, or they may have calcareous ground water at shallow depths. Chlorosis is a major agricultural problem on some **Calciorthids.**

**Orthids** with duripans are **Durorthids. Durorthids** are commonly associated with **Durargids,** but usually their duripan is less strongly developed.

**Paleorthids** are **Orthids** with a petrocalcic horizon but that have no duripan or salic horizon. They occur on older surfaces, mostly in association with **Paleargids.** The **petrocalcic** horizons of some **Paleorthids** are exhumed horizons of a buried soil. They may not truly be horizons of the modern soil and may extend to depths of many meters.

**Salorthids** are some of the most striking soils of the desert. They have a **salic** horizon, that is, a horizon with more than 2 per cent salt. The soil is considered a **Salorthid** only if there is reason to assume that the salt accumulation is from the capillary fringe of a shallow, saline groundwater table and reflects the present environment. The presence of a groundwater level above one meter depth for at least one month of the year is taken as such evidence. The salt concentration in **Salorthids** is extremely high and far above the levels that are tolerated by any crop. The high level was chosen because lower salt contents of soils can be changed rapidly through management.

### Torri Suborders and Great Groups of Entisols, Vertisols, and Oxisols

Soils of arid lands that have no distinct pedogenic horizons (**Entisols**), soils in which soil movement due to shrinking and swelling and drying and wetting has obliterated subsoil horizons (**Vertisols**), or highly weathered soils with extremely low activity of the exchange complex (**Oxisols**), are grouped with similar soils of other climatic environments. They are separated as Torri classes at the suborder and great group levels (Table 4, Part A).

Sandy soils without profile development, for example, are **Psamments.** The sandy soils in dunes on the desert are **Torripsamments.** The undeveloped soils of moist alluvial plains that are flooded periodically are **Fluvents,** and those of the desert are **Torrifluvents. Torrifluvents** are not very extensive soils in the total arid lands; but since they are level, have a deep and commonly medium-

textured profile, and are commonly near a source of irrigation water, they make up a high proportion of the irrigated soils of the desert.

The undeveloped soils of the upland are **Torriorthents,** a somewhat mixed group of soils. Some lack horizon development because they are too shallow to bedrock, some because they contain an abundance of gravel and rock, and some because they are in steep and eroded parts of the landscape. Few uses can be found for such soils in farming or ranching.

**Torrerts** are **Vertisols** in arid climates. They are not extensive soils in the United States but are extensive in the Sudan in Africa and in Australia. Only six soil series are classified in **Torrerts** in the United States. The suborder **Torrert** is not subdivided into great groups.

**Oxisols** that meet the definition of **Torri** suborders are **Torrox.** They occur in close association with soils of a slightly moister environment, **Ustox,** and are similar in morphology. **Torrox** are not known to occur in the United States, and whether this suborder will be retained will have to be decided on evidence from outside the United States.

### Classification of Aridisols at the Subgroup, Family, and Series Level

The central concept of each great group is defined at the subgroup level as a typic subgroup and is separated from soils with properties that are transitional to other great groups of the same order or to any other taxon at other categorical levels. Also, provisions are made for certain *aberrant* properties such as shallowness to hard rock.

The great group **Camborthids** and its subgroups are shown as an example in Table 6. The definition of the **Typic Camborthids** is self-explanatory. Of the other subgroups of **Camborthids,** the **Lithic Camborthid** may be mentioned as an example of a subgroup with aberrant properties and the **Durixerollic Camborthid** as an example of an intergrade to the great group **Durixerolls** of the order **Mollisols.** A **Lithic Camborthid** is a **Camborthid** with hard rock at depths of less than 50 centimeters. A **Durixerollic Camborthid** contains more organic matter than a **Typic Camborthid,** has a brittle, duripan-like horizon, and has a soil-moisture regimen characterized by the precipitation occurring mostly in the winter.

Throughout this paper we have tried to give examples of interpretations of classes at the various levels of categorical generalization. These interpretations must be very general because the soils included in these classes may differ greatly in important properties. At the family level, soil texture, mineralogy, soil temperature, and selected other criteria (Table 5) are used to create more nearly homogeneous group of soils. Unless there is redun-

TABLE 6

## THE GREAT GROUP CAMBORTHIDS AND ITS SUBGROUPS*

**Camborthids. Orthids** that

1. have a cambic horizon;
2. have no duripan that has its upper boundary within 1 meter of the surface;
3. have no salic horizon within 75 centimeters of the surface if saturated with water (i.e. within the capillary fringe) within 1 meter of the surface for 1 month or more;
4. have no calcic or gypsic horizon with an upper boundary within 1 meter of the surface unless, after the upper 18 centimeters are mixed, some subhorizon above the calcic horizon is free of carbonates and textures are as fine or finer than loamy very fine sand;
5. have no petrocalcic horizon that has its upper boundary within 1 meter of the surface.

**Typic Camborthids. Camborthids** that

a. have none of the characteristics associated with wetness (details omitted);
b. have no brittle horizon more than 15 centimeters thick with opal coatings or durinodes within 1 meter of the surface;
c. have no lithic contact within 50 centimeters of the surface;
d. have less than 1 per cent mean organic matter content in the upper 38 centimeters if the sand/clay ratio is $< 1$, or less than 0.3 per cent if this ratio is 13 or more — intermediate sand/clay ratios require intermediate organic matter content;
e. do not swell and shrink strongly on wetting and drying (less than 9 per cent).

**Aquic Camborthids.** Like Typic, except for *a*, with or without *d*

**Aquic Duric Camborthids.** Like Typic, except for *a* and *b*, with or without *d*

**Duric Camborthids.** Like Typic, except for *b*

**Duric Lithic Camborthids.** Like Typic, except for *b* and *c*

**Durixerollic Camborthids.** Like Typic, except *b* and *d*, and seasonally dry

**Durustollic Camborthids.** Like Typic, except *b* and *d*, and intermittently dry

**Lithic Camborthids.** Like Typic, except for *c*

**Lithic Ustollic Camborthids.** Like Typic, except for *c* and *d*, and intermittently dry

**Lithic Xerollic Camborthids.** Like Typic, except for *c* and *d*, and seasonably dry

**Torrertic Camborthids.** Like Typic, except for *e*, or *d* and *e*, with or without *b*, and usually dry

**Ustertic Camborthids.** Like Typic, except for *e*, or *d* and *e*, and intermittently dry

**Ustollic Camborthids.** Like Typic, except for *d*, and intermittently dry

**Xerertic Camborthids.** Like Typic, except for *e*, and *d* and *e*, and seasonally dry

**Xerollic Camborthids.** Like Typic, except for *d*, and seasonably dry

*The definition of Camborthids is complete. The definitions of Typic Camborthids and of the intergrade subgroups have been generalized.

dancy (all **Vertisols**, for example, would be in clayey families by definition) the soils are classified in the appropriate texture, mineralogy, and temperature families. A **Typic Paleargid**, given as an example in Table 7 and Figure 2, is in the *fine* (texture), *mixed* (mineralogy), *thermic* (soil temperature) family. A **Xerollic Calciorthid**, given as an example in Table 8 and Figure 3, is in the *fine loamy, carbonatic, mesic* family. A few additional criteria such as presence of carbonates to distinguish among some **Torrifluvents** are used in a few taxa. Soil families are narrowly enough defined that rather detailed recommendations concerning the use and management of the comparable phases of the series in one family can be made.

Many families of individual subgroups have only one soil series, the lowest category of the system, placed in them. Whether additional series should be recognized will depend on practical considerations. So far, criteria for setting up series within families have been empirical, and few formal conventions have been adopted. A few families still have many series in them. The *fine, mixed, mesic* family of **Typic Haplargids**, for example, has about twenty series in it. These twenty series were recognized to meet the need of local soil surveys. Lacking a detailed and comprehensive system of classification, it was not possible to avoid recognizing new series that largely duplicated series that had been established elsewhere. The series in large families are now being scrutinized and unnecessary series are being eliminated.

We pointed out before that the taxonomic class, soil series, has to be distinguished clearly from the cartographic unit, the mapping unit or phase, bearing the same name. Mapping units not only have the inclusions we discussed before, they also are commonly phases with restricted ranges in slope, stoniness, and other properties that may be very important to land use. The mapping units of small-scale, generalized, or reconnaissance maps may also be phases, or associations of phases, of families, subgroups, or of any of the higher categorical classes of the system.

## CLASSIFICATION OF SOILS OF SEMIARID AREAS

Some soils of semiarid lands (moisture index between −20 and −40) are in **Aridisols** and **Torri-** suborders and great groups of other orders, but most are in **Ust-** and **Xer-** suborders and great groups. **Ust-** and **Xer-** suborders and great groups (Table 4, part B) are defined largely in terms of their soil moisture regimes.

**Xer-** suborders and great groups are intended to group soils of Mediterranean climates. They are defined as being continuously dry for more than 60 consecutive days in all parts of the moisture control section (Table 3).

**Ust-** suborders and great groups are intended to separate soils of other semiarid areas having sporadic rainfall during all seasons and no pronounced wet season. In terms of soil moisture they are defined as being dry for 90 cumulative days or more in some subhorizon of the moisture control section.

**Mollisols** and **Alfisols** may also be placed in the **Ust-** suborders if they have a calcic horizon or a horizon of soft powdery calcium carbonate accumulation. These criteria separating Ustolls and Ustalfs from Udolls and Udalfs of more humid areas can be

TABLE 7

## PROFILE DESCRIPTION AND DATA FOR A PEDON OF TUBAC GRAVELLY LOAM, A TYPIC PALEARGID, FINE, MIXED, THERMIC FAMILY

LOCATION: Cochise County, Arizona. Vegetation: Blue grama (*Boutelua gracilis*), black grama (*B. eriopoda*), Rothrocks grama (*B. rothrockii*). Parent Material: Alluvium from andesite, rhyolite, rhyolitic tuff, quartzite and granite. Physiography: Lower end of coalescent fans. Slope: 1.6 per cent. Elevation: 1360 meters.

A11   0 to 5 cm, light brown (7.5YR 6/4) sandy loam, brown (7.5YR 4/4) when moist; weak medium platy structure; slightly hard, very friable, slightly sticky, slightly plastic; common roots; pH 5.5; abrupt smooth boundary.

A12   5 to 15 cm., reddish brown (5YR 5/3) finely gravelly sandy loam, reddish brown (5YR 4/3) when moist; weak medium subangular blocky structure; slightly hard, friable, slightly sticky, slightly plastic; many roots; pH 5.5, clear smooth boundary.

A3   15 to 18 cm, reddish brown (5YR 5/3) [dry color] finely gravelly clay loam; weak subangular blocky structure; hard, friable, sticky and plastic; many roots; pH 7.5; abrupt wavy boundary.

B21t   28 to 53 cm, reddish brown (5YR 4/3) [dry color] gravelly clay; moderate medium prismatic structure; very hard, friable, sticky, very plastic; many flattened roots between peds; smooth ped faces; pH 8.0; clear wavy boundary.

B22tca   53 to 90 cm, reddish brown (5YR 5/4) [dry color] gravelly clay with common pink (5YR 7/3) weathered pebbles; moderate fine and medium subangular blocky structure; very hard, friable, sticky, plastic; few roots; many thin clay films on ped faces; slightly effervescent with HCl in matrix, with strongly effervescent lime flecks; pH 8.2; gradual wavy boundary.

B3tca   90 to 120 cm, reddish brown (5YR 5/4) [dry color] gravelly sandy clay loam; massive; very hard, friable, sticky, plastic; few roots; few thin clay films in pores; slightly effervescent in spots; pH 8.2; gradual wavy boundary.

C   120 to 180 cm, light reddish brown (5YR 6/3) very gravelly loamy sand; massive; very hard, friable, nonsticky, nonplastic; no roots; many fine interstitial pores; noneffervescent; pH 8.2.

DIAGNOSTIC CRITERIA

Order, Aridisols: Usually dry (climatic data).

Suborder, Argids: Argillic horizon; clay increase from A to B horizon, clay skins in the B3 horizon.

Great Group, Paleargids: More than 15 per cent clay increase within 7.5 cm (A3 to B21t horizon) and more than 35 per cent clay in the 28-78 cm layer.

Subgroup, Typic Paleargids: No brittle horizon, no lithic contact, the weighted organic matter content of the 0-38 cm layer is 0.36 per cent (< 0.46 for sand/clay ratio of 4.6). No petrocalcic or natric horizon.

Family: Fine, mixed, thermic: More than 35 per cent clay and no dominant clay mineral in the upper 50 cm of the argillic horizon; the mean annual soil temperature is more than 15°C and less than 22°C, and the mean summer soil temperature differs by more than 5°C from the mean winter temperature.

| Climatic data (Willcox, Arizona) | J | F | M | A | M | J | J | A | S | O | N | D | Av. |
|---|---|---|---|---|---|---|---|---|---|---|---|---|---|
| Mean temperatures, 1899-1957 (°C) | 5.1 | 7.2 | 9.9 | 13.6 | 17.6 | 22.8 | 25.4 | 24.3 | 21.6 | 15.6 | 9.4 | 5.6 | 14.8 |
| Mean precipitation, 1899-1957 (mm) | 22 | 23 | 19 | 7 | 6 | 11 | 59 | 64 | 29 | 17 | 19 | 23 | 299 |

| Depth cm | Horizon | Sand (2-0.05) | Silt (0.05-0.002) | Clay (< 0.002) | Very coarse (2-1) | Coarse (1-0.5) | Medium (0.5-0.25) | Fine (0.25-0.1) | Very fine (0.1-0.05) | 0.05-0.02 | Int. III (0.02-0.002) | Carbonate clay | > 2 mm Pct. whole soil | 1/3 bar g/cc | Oven dry g/cc | 1/3 bar Pct. | 15 bar Pct. | Extensibility COLEf in./in. |
|---|---|---|---|---|---|---|---|---|---|---|---|---|---|---|---|---|---|---|
| 0-5 | A11 | 65.9 | 28.7 | 5.4 | 6.1 | 12.6 | 11.1 | 23.0 | 13.1 | 19.1 | 9.6 | | 19 | 1.59 | 1.60 | 11.1 | 3.9 | 0.002 |
| 5-15 | A12 | 62.3 | 28.5 | 9.2 | 7.0 | 11.7 | 9.9 | 20.7 | 13.0 | 17.3 | 11.2 | | 34 | 1.48 | 1.50 | 13.2 | 4.7 | 0.003 |
| 15-28 | A3 | 56.1 | 27.2 | 16.7 | 12.7 | 12.7 | 7.6 | 13.7 | 9.4 | 16.3 | 10.9 | | 44 | 1.43 | 1.48 | 18.0 | 7.0 | 0.011 |
| 28-53 | B21t | 26.3 | 15.0 | 58.7 | 7.5 | 7.1 | 2.8 | 5.2 | 3.7 | 6.5 | 8.5 | | 31 | 1.20 | 1.53 | 40.1 | 21.6 | 0.083 |
| 53-90 | B22tca | 41.5 | 10.4 | 48.1 | 14.1 | 12.7 | 5.6 | 6.0 | 3.1 | 4.6 | 5.8 | | 65 | 1.27 | 1.39 | 31.0 | 19.0 | 0.032 |
| 90-120 | B3tca | 61.1 | 15.2 | 23.7 | 17.0 | 19.1 | 9.6 | 10.4 | 5.0 | 7.1 | 8.1 | | 50 | 1.40 | 1.50 | 21.3 | 10.8 | 0.022 |
| 120-180 | C | 83.0 | 6.1 | 10.9 | 28.1 | 32.0 | 13.2 | 8.0 | 1.7 | 2.4 | 3.7 | | 54 | 1.38 | 1.40 | 17.6 | 6.7 | 0.006 |

| Depth cm | Organic carbon Pct. | Carbonate as CaCO3 Pct. | Ca | Mg | Na | K | Sum of bases | 5A1a NH4OAc | 8A1a Electrical conductivity mmho/cm | Exchangeable Na Pct.‡ | Gypsum Pct. | 8C1a H2O 1:1 | 8C1a H2O 1:10 | Mont. | Verm. | Mica | Kaolin. |
|---|---|---|---|---|---|---|---|---|---|---|---|---|---|---|---|---|---|
| 0-5 | 0.25 | | 2.5 | 1.2 | 1.0 | 0.8 | 5.5 | 5.2 | 0.35 | | | 5.7 | 6.5 | | | | |
| 5-15 | 0.33 | | 3.9 | 1.5 | 1.0 | 0.7 | 7.1 | 6.7 | 0.55 | | | 6.3 | 6.7 | xx | — | xx | xx |
| 15-28 | 0.43 | | 7.1 | 2.6 | 1.0 | 0.9 | 11.6 | 10.2 | 1.34 | 1 | | 7.0 | 7.3 | xx | — | xx | |
| 28-53 | 0.34 | | 26.0 | 9.5 | 2.9 | 2.4 | 40.8 | 32.6 | 2.18 | 3 | | 7.5 | 8.2 | | | | |
| 53-90 | 0.20 | 3 | 28.3 | 8.3 | 4.2 | 2.6 | 43.4 | 29.6 | 2.76 | 5 | | 7.9 | 8.6 | xx | — | xxx | tr |
| 90-120 | 0.04 | tr | 15.3 | 4.5 | 4.4 | 2.0 | 26.2 | 19.1 | 3.45 | 8 | | 7.7 | 8.6 | | | | |
| 120-180 | 0.02 | tr | 7.8 | 2.1 | 4.0 | 1.8 | 15.7 | 14.0 | 2.71 | 9 | | 7.6 | 7.8 | xxxx | — | xx | tr |

*Methods code (10)
†Coefficient of Linear Extensibility (shrink-swell coefficient)
‡Calculated from SAR

§— = not found
tr = traces
x = small

xx = moderate
xxx = large
xxxx = predominant

TABLE 8

## PROFILE DESCRIPTION AND DATA FOR A PEDON OF SANPETE GRAVELLY LOAM, A XEROLLIC CALCIORTHID, FINE LOAMY, CARBONATIC, MESIC FAMILY

LOCATION: Utah, Millard County. Vegetation: Black sagebrush (*Artemisia nova*), little rabbitbrush (*Chrysothamnus stenophyllus*), winterfat (*Eurotia lanata*), and Indian ricegrass (*Oryzopsis hymenoides*).

Parent Material: Mixed alluvium from limestone, sandstone, quartzite, rhyolite. Topography: Toe of an alluvial fan, 1½ per cent slope. Elevation: 1890 meters.

A1   0 to 8 cm, light brownish gray (10YR 6/2) fine gravelly loam, dark grayish brown (10YR 4/2) moist; weak medium platy structure; soft, very friable, nonsticky, nonplastic; few roots; strong effervescence; pH 8.4; clear smooth boundary.

B2   8 to 33 cm, pale brown (10YR 6/3) fine gravelly sandy loam, brown (10YR 4/3) moist; weak subangular blocky structure; soft, very friable, nonsticky, nonplastic; many roots; strong effervescence; secondary carbonates as pendants and coatings on underside of pebbles; pH 8.4; clear wavy boundary.

C1ca   33 to 64 cm, white (10YR 8/2) [dry color] fine gravelly loam; weak medium subangular blocky structure; very hard, friable, nonsticky, slightly plastic; many roots; violent effervescence; secondary carbonate coatings on pebbles; pH 8.6; diffuse smooth boundary.

C2ca   64 to 97 cm, white (10YR 8.3) [dry color] fine gravelly loam; massive; hard, friable, nonsticky, slightly plastic; few roots; violent effervescence; secondary carbonate coatings on pebbles; pH 8.6; diffuse smooth boundary.

C3ca   97 to 122 cm, white (10YR 8/2) [dry color] fine gravelly loam; massive; hard, friable, nonsticky, slightly plastic; violent effervescence; pH 8.6; clear wavy boundary.

IIC4   122 to 168 cm, very pale brown (10YR 7/3) fine gravelly coarse sandy loam; massive; slightly hard, friable, nonsticky, nonplastic; weak effervescence; secondary carbonate coatings on pebbles; pH 8.2; few roots.

DIAGNOSTIC CRITERIA

Order, Aridisols: Usually dry (climatic data), no mollic horizon (color value too high).

Suborder, Orthids: Cambic horizon, no argillic horizon (the clay increase in the Clea horizon is due to clay-size carbonates).

Great Group, Calciorthids: Calcic horizon (> 15 per cent CaCO₃ and evidence of carbonate movement) and calcareous throughout.

Subgroup, Xerollic Calciorthids: The weighted organic matter content of the 0-38 cm layer is 0.62 per cent (> 0.44 per cent for sand/clay ratio of 5.0), dry for more than 60 consecutive days.

Family: Fine loamy, carbonatic, mesic: In the 25-100 cm layer more than 18 per cent clay, more than 15 per cent sand, and more than 40 per cent carbonate equivalent; mean annual soil temperature is between 8 and 15°C and the mean summer soil temperature differs by more than 5°C from the mean winter temperature.

Climatic data (Precipitation: Desert Range Experiment Station; Temperature: Milford, Utah)

| | J | F | M | A | M | J | J | A | S | O | N | D | Av. |
|---|---|---|---|---|---|---|---|---|---|---|---|---|---|
| Mean temperature, 58 years(°C) | 3.8 | 0.0 | 4.2 | 8.9 | 13.7 | 18.7 | 23.4 | 22.3 | 16.9 | 10.1 | 2.8 | 2.2 | 9.6 |
| Mean precipitation, 29 years (mm) | 8 | 9 | 11 | 14 | 17 | 11 | 19 | 19 | 10 | 18 | 11 | 8 | 155 |

| Depth cm | Horizon | Sand (2-0.05) | Silt (0.05-0.002) | Clay (< 0.002) | Very coarse (2-1) | Coarse (1-0.5) | Medium (0.5-0.25) | Fine (0.25-0.1) | Very fine (0.1-0.05) | 0.05-0.02 | Int. III (0.02-0.002) | Carbonate clay | > 2 mm Pct. whole soil | 1/3 bar g/cc | Oven dry g/cc | 1/3 bar Pct. | 15 bar Pct. | Extensibility COLEf in./in. |
|---|---|---|---|---|---|---|---|---|---|---|---|---|---|---|---|---|---|---|
| 0-8 | A1 | 64.5 | 25.2 | 10.3 | 9.2 | 16.5 | 11.9 | 16.7 | 10.2 | 14.7 | 10.5 | 0 | 24 | 1.43 | 1.46 | 14.6 | 7.0 | 0.007 |
| 8-33 | B2 | 63.0 | 25.0 | 12.0 | 7.6 | 14.5 | 12.0 | 17.3 | 11.6 | 12.0 | 13.0 | 1 | 15 | 1.19 | 1.21 | 13.6 | 8.6 | 0.007 |
| 33-63 | C1ca | 39.0 | 32.2 | 28.8 | 4.3 | 8.9 | 6.9 | 10.8 | 8.1 | 9.7 | 22.5 | 16 | 38 | 1.14 | 1.18 | 33.7 | 13.3 | 0.009 |
| 63-97 | C2ca | 33.8 | 39.9 | 26.3 | 7.1 | 7.2 | 4.8 | 8.4 | 6.3 | 6.9 | 33.0 | 14 | 40 | 1.19 | 1.23 | 30.3 | 13.4 | 0.011 |
| 97-122 | C3ca | 25.9 | 42.8 | 31.3 | 6.8 | 5.1 | 3.4 | 5.8 | 4.8 | 7.0 | 35.8 | 21 | 43 | 1.22 | 1.25 | 33.7 | 12.6 | 0.007 |
| 122-168 | IIC4 | 36.7 | 38.7 | 24.6 | 7.7 | 6.9 | 4.9 | 8.6 | 8.6 | 11.9 | 26.8 | 9 | 31 | 1.26 | 1.33 | 21.3 | 12.7 | 0.017 |

| Depth cm | Organic carbon Pct. | Carbonate as CaCO₃ Pct. | 6N4c Ca 5B3a | 6O4c Mg 5B3a | 6P2a Na 5B1a | 6Q2a K 5B1a | Sum of bases | 5A1a NH₄OAc | 8A1a Electrical Conductivity mmho/cm | Exchangeable Na Pct. | Gypsum Pct. | 8C1a H₂O 1:1 | 8Cla H₂O 1:10 | Mont. | Verm. | Mica | Kaolin. |
|---|---|---|---|---|---|---|---|---|---|---|---|---|---|---|---|---|---|
| 0-8 | 0.64 | 21 | 10.2 | 1.8 | 0.9 | 2.1 | 14.6 | 12.1 | 1.31 | 5 | | 8.5 | 9.1 | x | | x | tr. |
| 8-33 | 0.61 | 21 | 11.9 | 2.3 | 0.5 | 0.7 | 15.2 | 12.8 | 0.50 | 1 | | 8.4 | 8.9 | xx | | xxx | x |
| 33-63 | 0.60 | 49 | 8.5 | 2.8 | 2.3 | 0.2 | 11.6 | 8.5 | 3.55 | 10 | | 8.2 | 9.0 | | | | |
| 63-97 | 0.26 | 57 | 10.0 | 4.0 | 6.8 | 0.2 | 9.5 | 4.7 | 13.5 | 19 | 0.1 | 8.2 | 8.7 | | | | |
| 97-122 | 0.25 | 67 | 6.0 | 1.9 | 6.6 | 0.1 | 6.4 | 3.0 | 11.5 | 21 | tr. | 8.3 | 8.9 | | | | |
| 122-168 | 0.13 | 40 | 6.3 | 4.9 | 7.6 | 0.3 | 12.3 | 9.5 | 11.5 | 25 | — | 8.3 | 9.1 | xx | x | xx | x |

*Methods code (10)
†Coefficient of Linear Extensibility (shrink-swell coefficient)

‡ — = not found
tr. = traces
x = small

xx = moderate
xxx = large
xxxx = predominant

Fig. 2. A typic Paleargid, fine, mixed, thermic family. (See also Table 7. The description and data are from another site.) Scale: feet. Photo by G. D. Smith, SCS.

used effectively in the field in establishing the long common boundary of these soils in the Great Plains. The Ustoll-Udoll boundary in the Great Plains is to the east of Thornthwaite's −20 moisture index; hence, geographically, the Ustolls include some soils that geographers would consider to be in a subhumid climate. These are separated as subgroups that intergrade to the Udolls.

A limit to the organic matter content of Ust- and Xer- suborders is used in Ultisols (Table 4, part B). It separates Ustults and Xerults from Humults and makes a more homogeneous group of the latter.

## REFERENCES AND NOTES

1. The present paper is a contribution of Soil Survey Investigations, Soil Conservation Service, U. S. Department of Agriculture, Washington, D. C. 20250, U. S. A. The discussion is based upon:
   U. S. DEPARTMENT OF AGRICULTURE, SOIL CONSERVATION SERVICE, SOIL SURVEY STAFF
      1967  Soil classification, a comprehensive system, 7th Approximation. Supplement, March 1967. 207 pp.
   The main publication was:
   U. S. DEPARTMENT OF AGRICULTURE, SOIL SURVEY STAFF, SOIL CONSERVATION SERVICE
      1960  Soil classification, a comprehensive system, 7th approximation. Washington, D. C.

2. U. S. DEPARTMENT OF AGRICULTURE
      1938  Soils and men. Yearbook of Agriculture, p. 1180.

3. THORP, J., and G. D. SMITH
      1949  Higher categories of soil classification. Soil Science 67:117-126.

4. SMITH, G. D.
      1965  Lectures on soil classification. Pedologie, Bulletin of the Belgian Society of Pedology, Special Issue No. 4.

5. CLINE, M. G.
      1963  Logic of the new system of soil classification. Soil Science 96:17-22.

6. HELLER, J. L.
      1963  The nomenclature of soils, or what's in a name? Soil Science Society of America, Proceedings 27:216-220.

7. JOHNSON, W. M.
      1963  Relation of the new comprehensive soil classification system to soil mapping. Soil Science 96:31-34.

8. REICHE, P.
      1950  A survey of weathering processes and products. University of New Mexico, Publications in Geology, #3.

9. THORNTHWAITE, C. W.
      1948  An approach toward a rational classification of climate. Geographical Review 38:55-94.

10. U. S. DEPARTMENT OF AGRICULTURE, SOIL CONSERVATION SERVICE
      1967  Soil survey laboratory methods and procedures for collecting soil samples. Soil Survey Investigations Report 1.

Fig. 3. Sampete gravelly sandy loam, a Xerollic Calciorthid, fine-loamy, carbonatic, mesic family. (See also Table 8.) Scale: feet. Photo by V. K. Hugie, SCS.

# A BIBLIOGRAPHY OF DESERT DUNES AND ASSOCIATED PHENOMENA

ANDREW WARREN
University College London
London, England

# A BIBLIOGRAPHY OF DESERT DUNES AND ASSOCIATED PHENOMENA

## Andrew Warren

This bibliography is an attempt to record the study of dunes over the last ninety years or so. It is not fully comprehensive, as can be seen by its heavy numerical bias toward references on the Sahara, the United States, and Poland. In many parts of the world there must be significant local studies that have not come to the attention of the author. Only a few studies of coastal dunes in which general points are discussed have been included, so that many works in this field have been ignored.

The bibliography shows that the study of dunes has had sporadic progress. In many instances information has been acquired from general surveys, or as a result of a passing interest. There has been little progressive research of the kind that leads to useful generalizations (there being some notable exceptions); for example, some leading authorities refer hardly at all to earlier work in the field, and some work, such as the early German interest in building generalized models of dune development, seems to have been forgotten.

Until recently there has been little concrete information about dune forms in many desert areas, and sandseas have been difficult to penetrate and to work in. This has meant that many of the conclusions about dunes have been the result of work on isolated forms or on forms which, because of their semiarid or even humid environment, could be considered atypical or even *fossil* paleoforms, for instance the *ergs morts* of the southern Sahara (e.g., Grove 1958, p. 86*), the fixed dunes of Poland (e.g., Högbom 1923, p. 82*), and the dunes of the American High Plains (e.g., Hefley and Sidwell 1945, p. 86*).

The post-war expansion of interest in desert areas, which has included the development of better desert logistics, and the enormous increase in the coverage of desert areas by aerial photography (e.g., Smith 1963, p. 93*) and by good maps, means that there is now a very large body of information on desert dunes. This information is almost embarrassing, since, as a result, it appears that there is more variety and complexity among dune forms than had been thought.

One of the obvious responses to this flood of knowledge has been the development of dune classifications. In America, for instance, there are the classification of Hack (1941, p. 86*) and of Melton (1940, p. 87*), while in Mauritania there is Monod's classification (1958, p. 87*), which has been extended by Tricart and Cailleux (1962–1963, p. 94*). Most of the classifications can be applied only to limited areas, and there is still no satisfactory universal classification.

In view of the multivariate controls on dune-forming processes, it is arguable if a classification, based as it must almost inevitably be on a limited number of criteria, is of real use to research. A more profitable line of inquiry could be the study of the characteristics, dynamics, and environmental relationships of particular dunes or dune complexes. Since the classic work of Bagnold (1941, p. 84*), it is studies such as these which have yielded the most useful results.

The characteristics of dunes have been studied in three principal ways. First, there has been the study of dune sediments. Many of these have been prompted by an interest in distinguishing depositional environments by their sedimentary characteristics, but others have been attempts to link specific dune types with certain sand size and shape parameters (Alimen *et al.* 1957, p. 84*; Bellair 1952, p. 96*). Secondly, there has been an interest in the sedimentary structures of dune deposits, which has proved very profitable and will undoubtedly stimulate further study (McKee 1966, p. 92*; McKee and Tibbitts 1964, p. 92*; Sharp 1966, p. 93*). Thirdly, a few workers have concentrated interest in dune form, and although this has been a recent development, it too seems to hold promises of results (Finkel 1959, p. 85*; McCoy *et al.* 1967, p. 92*; Hastenrath 1967, p. 91*).

Study of dune dynamics and of the environmental relationships of dunes is still hampered by lack of information about the desert environment. Where information on, for instance, desert meteorology is available or has been specially collected, the work is showing useful results. In Peru and in the Mojave, barchan movement has been correlated with wind observations (Finkel 1959, p. 85*; Long and Sharp 1964, p. 92*; Norris 1966, p. 92*), and in other parts of the southwestern United States bedding and other dune characteristics have been matched with wind measurements. Although much of the work on sand movement is experimental and not directly connected with dune studies, there has been an advance in certain directions since Bagnold's work (Belly 1964, p. 94*; Williams 1964, p. 96*). It is studies of this dynamic/statistical kind which show the greatest hope of advance in dune study.

Because of the peripheral nature of dune study,

---

*Bibliographic graphic page in *Arid Lands in Perspective*.

whereby it has been an interest of a wide variety of specialists such as mathematicians, botanists, geologists, sedimentologists, engineers, travelers, agriculturalists, foresters, and geographers, the field has been characterized in the past by a certain amount of dilettantism, although there have been as well some very solid pieces of research. The more careful and persistent studies of the last few years will probably lead to more important and useful generalizations, but not until more data has been collected about dune sediments and desert meteorology in particular and not until more sophisticated ways have been devised for measuring and comparing dune forms themselves.

In addition to recognizing the work of my own associates, I would like to express my appreciation to Miss Patricia Paylore and Miss Mary L. Moore of the University of Arizona's Office of Arid Lands Studies for the extensive library verification exercises they performed in order to fill gaps in bibliographical entries supplied from my files.

The material is arranged as follows:

# BIBLIOGRAPHY OF LITERATURE ON SAND DUNES (EXCLUDING COASTAL DUNES)

## (a) Before 1920

BAILEY, S. I.
1899 The sand-dunes of the Desert of Islay, Peru. Harvard University Astronomical Observatory, Annals 39(2):287-292.

BARCLAY, W. S.
1917 Sand dunes in the Peruvian Desert. Geographical Journal 49:53-56.

BASHIN, O. V.
1899 Die Enstehung wellenähnlicher Oberflöchenformen. Gesellschaft für Erdkunde zu Berlin, Zeitschrift, III, 34:408-424.
1900 Die Entstehung der Dünen. Centralblatt der Bauverwaltung, XX, Jahrbuch 38:251-232.
1903 Dünenstudien. Gesellschaft für Erdkunde zu Berlin, Zeitschrift 6:422-430. Abstr. *in* Geographical Journal 23:588 (1903).

BEADNELL, H. J. L.
1901 Dakhla Oasis, its topography and geology. Fara'fra Oasis, its topography and geology. The topography and geology of the Faiyum Province. Cairo.
1909a Desert sand dunes. Cairo Scientific Journal 3(34):171-172.
1909b An Egyptian oasis. John Murray, London. (Review by W. F. Hume *in* Cairo Scientific Journal 3(33):148-154)
1910 The sand dunes of the Libyan Desert. Geographical Journal 35:379-395.

BERTOLOLY, E.
1900 Kräuselungsmarken und Dünen. Münchener Geographische Studien Journal 9.

BLANFORD, W. T.
1876 On the physical geography of the Great Indian Desert. Asiatic Society of Bengal, Journal 45(2).
1877 Geographical notes on the Great Desert between Sind and Rajputana. Geological Survey of India, Memoir 10:10-21.

BOWMAN, I.
1916 The Andes of southern Peru. American Geographical Society, Special Publication 2. 336 p.

BRAUN, G.
1911 Entwicklungsgeschichtliche Studien au Europäischen Flachlandsküsten und ihren Dünen. Universität Berlin, Institut für Meerskunde, Veröffentlichungen 15. 174 p.

BUCHER, W. H.
1919 On ripples and related surface sedimentary forms. American Journal of Science 47(4):149-210, 241-269.

CARNIER, J.
1891 Les dunes au Sahara. Société Géographique Français, Comptes Rendus. 114 p.

CHOLNOKY, E. v.
1902 Die Bewegungsgesetze des Flugsandes. Földtani Közlöny 32:106-143.
1907 Uber die Lagereränderungen des Tiszabeltes. Földrajzi Közlemeneyek 35.
1910 Die oberflächengestalt des Alfold. Földrajzi Közlemeneyek 38.

CHUDEAU, R.
190- L'Aïr et la region de Zinder. Géographie 15:321-336.
1907 Excursion géologique au Sahara et Soudan. Société Géologique de France, Bulletin sér. 4, 7:319-346.
1909a Le Sahara sudanaise. Colin, Paris. 326 p.
1909b Notes géologiques sur la Mauritanie. Géographie 20:9-24.
1910 Rapport de journée dans le bassin de Tombouctou. Journal Officiel de l'Afrique Occidentale Française, Supplément.
1911 Remarques sur les dunes. A propos d'une étude de M. H. J. Lwellyn-Beadnell. Géographie 24:153-160.
1915a Excursion géologique au nord et est de Tombouctou. Société Géologique de France, Bulletin 15:95-112.
1915b L'Azaouad et le Djouf. Géographie 30:417-436.
1918 La depression du Faguibine. Annales de Géographie, p. 43-60.

COCKAYNE, L.
1911 Report on the sand dune areas of New Zealand and their geology, botany and reclamation. J. MacKay, Government Printer, Wellington. 76 p.

COFFEY, G. N.
1909 Clay dunes. Journal of Geology 17:754-755.

CONWAY, M.
1901   The Bolivian Andes. Harper, N. Y., London.
CORNISH, V.
1897   On the formation of sand dunes. Geographical Journal 9:278-302.
1899   On "Kumatology." Geographical Journal 13:624-628.
1900a   Formation des dunes de sable. Bruxelles.
1900b   On desert sand dunes bordering the Nile Delta. Geographical Journal 15:1-32.
1908   On the observation of desert sand dunes. Geographical Journal 31:400-402.
1914   Waves of sand and snow. T. Fisher-Unwin, London.
COURBIS, M.
1890   Les dunes et les eaux souterraines du Sahara. Société Géog. Français, C. R.
CZERNY, F.
1876 - Die Werkung der Winde auf die Gestaltung der
1877   Erde. Petermanns Geographische Mitteilungen, Erganzungsband 11.
DOUGLASS, A. E.
1909   The crescentic dunes of Peru. Appalachia 12:34-35.
ENOCK, C. R.
1908   Surface forms in western South America. Geographical Journal 31:684.
FALCONER, J. D.
1911   The geography and geology of northern Nigeria. Macmillan, London. 295 p.
FLAMAND, G. B. M.
1898   De l'Oranie au Gourara. Paris.
1899   La traversee de l'erg occidental (grand dunes du Sahara Oranais). Annales de Géographie 9: 231-241.
1919   Recherches géologiques et géographiques sur le "Haut-Pays" de l'Oranie et sur le Sahara (Algérie et territoires su sud). Lyons.
FLOYER, E. A.
1897   Notes on Mr. V. Cornish's paper on sand dunes. Geographical Journal 11:559-563.
FREE, E. C.
1911   (see first column, p. 95)
FULLER, M. L.
1899   Season and time elements in sand plain formation. Journal of Geology 7:452-462.
GARDE, G.
1911   Entre le Niger et le Chad. Paris.
GAUTIER, E. F.
1908   Missions au Sahara. Paris.
GERHARDT, P.
1900   Handbuch des Deutschen Dünenbaues. Parey, Berlin. 656 p.
GOLDSMID, J. G.
1897   Singing and drifting sand. Geographical Journal 9:454-455.
GRABAU, A. W.
1913   Principles of stratigraphy. Dunes: p. 551-578. Saler, N. Y.
GÜNTHER, S. v.
1907   Ein naturmodel der dünenbildung. Akademie der Wissenschaften, München, Mathematisch-Physikalische Klasse, Sitzungsberichte, p. 139-153.
HAHMANN, P.
1912   Die Bildung von Sandwellen. Annalen der Physik, ser. 4, 39:637-676.
HEDIN, S.
1896   A journey through the Takla-Makan Desert,

Chinese Turkestan. Geographical Journal 8:264-278, 356-372.
1904   The scientific results of a journey in central Asia. Stockholm.
HOLTENBURGER, M.
1913   On a genetic system of sand dunes including two new types. American Geographical Society, Bulletin 45:513-515.
HORNER, N. G.
1904   In Hedin, op. cit.
HURST, H. E.
1909   Notes on sand dunes at Tema. Cairo Scientific Journal 31:82-83.
IVTCHENKO (aka IWTSCHENKO), A.
1908   La mobilité des dunes. Annuaire Géologique de la Russie 9:224-255.
1910a   La stratification dans les dépôts éoliens. Annuaire Géologique de la Russie 12(5-6):145-153, 154-170.
1910b   Sur la morphologie des mers de barkhanes. Annuaire Géologique de la Russie 12(7-8):239-249.
JENTSCH, K. A.
1900   Die geologie der Dünen. p. 1-24. In P. Gerhardt, Handbuch des Deutschen Dünenbaues. Parey, Berlin.
JORDAN, U. W.
1876   Physische Geographie und Meteorologie der libischen Wüste. Cassel. 216 p.
JUTSON, J. T.
1918   The sand ridges, rock floors and other associated features at Goonarrie in sub-arid Western Australia. Royal Society of Victoria, Proceedings n.s. 31:113-128.
KEILACK, K.
1918   Die grossen Dünengebiete Norddeutschlands. Deutsche Geologische Gesellschaft, Zeitschrift B. Monatsbericht 69:2-19.
KING, W. H. J.
1912   Travels in the Libyan Desert. Geographical Journal 39:133-137.
1916   The nature and formation of sand ripples and dunes. Geographical Journal 47:189-209.
1918   Study of a dune belt. Geographical Journal 51:16-33.
KORN, J.
1919   Uber Dünenzüge im Torfe. K. Preussische Geologische Landesanstalt, Jahrbuch (1916) 37:147-156.
LOMAS, J.
1907   Desert conditions and the origin of the British Trias. Liverpool Geological Society, Proceedings 10:172-
MACDOUGAL, D. T.
1912   The North American deserts. Geographical Journal 39:105-123.
MALAKOWSKI, S.
1917   Les dunes anciennes des environs de Varsovie. Société des Sciences et de Lettres de Varsovie, Travaux 3(23).
MEDDLICOTT, H. B. and W. T. BLANFORD
1905   A manual of the geology of India. Calcutta. 545 p.
MEINARDUS, W.
1905   In discussion: Verhandel der fünfzehnten Deutschen Geographentages zu Danzig. Berlin.
NIEHOFF, K.
1917   Geomorphologische Karte der Niger und seinter

Nacheargebiete. 1:7,500,000. Mitteilungen aus den Deutsche Schutzgebieten 30.

OLDHAM, R. D.
1903  A note on the sand hills of Clifton, near Karachi. Geological Survey of India, Memoir 34(3):133-157.

PASSARGE, S.
1904  Die Kalahari. Berlin.
1911  Die pfannenformen Holformen der Sudafrikanschen Steppen. Petermanns Geographische Mitteilungen 57:130-135.

POMPECKJ, J. F.
1906  Barchans in southern Peru. Centralblatt für Mineralogie, Geologie und Paläontologie, p. 373.

POOL, R. J.
1913  Glimpses of the Great American Desert. Popular Science Monthly 80:209-235.

PUMPELLY, R.
1908  Turkestan. Carnegie Institution of Washington, Publication 73. 303 p.

REINCKE, J.
1903  Die Entwicklungsgeschichte der Dünen an der Westkuste von Schleswig. K. Preussische Akademie der Wissensfhaften, Berlin, Sitzungsberichte 1.

RITTER, E.
1898  Les dunes. Globe 37:22-31.

ROLLAND, G.
1881  Sur les grands dunes de sable du Sahara. Société Géologique de France, Bulletin, sér. 3, 10:30-47.

ROTH, E.
1900  Die deutschen Dünen und ihr Bau. Globus 78(3): 48-52.

SOKOLOW, N. A.
1894  Die Dünen, Bildung, Entwicklung, und innerer Bau (translated from the Russian). Springer, Berlin.

SOLGER, F.
1905  Über fossile Dünenformen in NordDeutschen Flachlande. Verhandel der fünfzehnten Deutschen Geographentages zu Danzig. Berlin.
1910a  Geologie der Dünen. In Dünenbuch. F. Enke, Stuttgart.
1910b  Studien über Nordostdeutsche Inlandünen. Forschungen zur Deutschen Landes- und Volkskunde 19(1):1-89.

STEENSRIP, K. J. V.
1894  Om Klitterns Vandrung. Dansk Geologisk Forening, Copenhagen, Meddelelser 1:1-14.

TALE, G. P.
1904  Sketch of the Baluchistan desert and part of eastern Persia. Geological Survey of India, Memoir 21.

TUTELOWSKI, P.
1913  Das Postglaziale Klima in Europe und in Nordamerika, das postglazialen wüsten und die Lossenbildung. International Geographical Congress, 11th, Stockholm, Comptes Rendus.

UDDEN, J. A.
1894  Erosion, transport and sedimentation by the atmosphere. Journal of Geology 2:318-331.

WALTHER, J.
1891  Die Denudation in der Wüste und ihr geologische Bedeutung. Königlich Sächsische Gesellschaft der Naturwissenschaften, Abhandlungen 16: 347-369.
1900  Das Gesetz und Wüstenbildung. In Gegenwart

und Vorzeit. Verlag von Quelle und Heyer, Leipzig. 421 pp.

WILMER, H. C.
1894  The dunes in south west Africa. South African Philosophical Society, Transactions 5. (Abstr. in Geographical Journal, 1894, 4:233)

YATE, A. C.
1894  Sand dunes. Geographical Journal 9:672-673.

ZITTLER, K. V.
1883  Beitrage zur Geologie und Palaeontologie der Libischen Wüste. Palaeontolographica 30.


**(b) 1920-1939**

ANGELIS, M. DE
193-  Osservazioni sur alcune sabbie della Libia.
(?)   Missione Scientifica della R. Accademia d'Italia a Cufra (1931), vol. 3, p. 49-

AUBRINIÈRES, L.
1935  Notes sur le Sahel mauritanien. Comité d'Etudes Historiques et Scientifiques de l'Afrique Occidentale Française, Bulletin 28(4):381-392.

AUFRÈRE, L.
1928  L'orientations des dunes et la direction des vents. Académie des Sciences, Comptes Rendus 187:833-835.
1929  Le problème géologique des dunes dans les desérts chauds du nord de l'ancien monde (Sahara, Arabie, Inde). Association Française pour l'Avancement des Sciences, 53e session, Le Havre, Compte Rendu, p. 393-397.
1930a  L'orientation des dunes continentales. International Geographical Congress, 12th, Cambridge, 1928, Report of Proceedings, p. 220-231.
1930b  Les dunes et les vents du Sahara. Association Française pour l'Avancement des Sciences, 54e session, Algér, Compte Rendu, p. 144-148.
1931  Le cycle morphologique des dunes. Annales de Géographie 40:362-385.
1932a  L'action conservatrice des phénomènes physico-chemiques dans le modèle des déserts accentués. Association de Géographes Français, Bulletin 57:66-69. (Observations complimentaires de M. L. Aufrère à propos de la communication de Robert Perret sur l'Ahaggar.)
1932b  Morphologie dunaire et météorologie saharienne. Association de Géographes Français, Bulletin 56:34-48.
1933  Les dunes continentales, leurs rapports avec le sous-sol, le passé géologique recent et le climat actuel. International Geographical Congress, 13th, Paris, 1931, Compte Rendu 2(1):699-711.
1934  Les dunes du Sahara algérien; notes de morphologie dynamique. Association de Géographes Français, Bulletin 83:130-142.
1935  Essai sur les dunes du Sahara algerien. Geografiska Annaler 17:481-500.

BAGNOLD, R. A.
1931  Journeys in the Libyan Desert. Geographical Journal 78:13-39, 524-535.
1933  A further journey in the Libyan Desert. Geographical Journal 82:103-129, 211-235.
1938a  Grain structure of sand dunes in relation to water content. Nature 142:403-404.
1938b  The measurement of sand storms. Royal Society of London, ser. A, Proceedings 167:282-291.

BALL, J.
1927  Problems of the Libyan Desert. Geographical Journal 70:21-38, 105-128, 209-224.

1939  Contributions to the geology of Egypt. Cairo. 308 pp.

BEADNELL, H. J. L.
1934  Libyan Desert dunes. Geographical Journal 84:337-340.

BELLAIR, P.
1938  Observations sur les dunes Sahariennes. Société Géologique de France, Comptes Rendus 17:331-332.

BERKEY, C. P., and F. K. MORRIS
1927  The geology of Mongolia. American Museum of Natural History, New York.

BINGHAM, M.
1925  Le relief du Peru. Revue de Géographie Alpine 13:679-706.

BOSWORTH, T. O.
1922  Geology of the Tertiary and Quaternary periods in the northwestern part of Peru. Macmillan, London.

BOURCART, J., and V. MALYCHEFF
1926  Premiers résultats de recherches sur les sables du Sahara. Société Géologique de France, Bulletin sér. 4, 26(4):191-208.
1927  Premiers résultats d'une étude sur le Quaternaire marocain. Société Géologique de France, Bulletin, sér. 4, 27:3-33.
1928  L'action du vent à la surface de la terre. Revue de Géographie Physique et de Géologie Dynamique 1:26-54, 194-265.

BROSSET, D.
1939  Essai sur les ergs du Sahara occidental. Institut Français d'Afrique Noire, Dakar, Bulletin 1:657-690.

BRUEL, G.
1926  Les résultats scientifiques de la mission Gossard. L'Afrique Française, Renseignements Colonials et Documents 36:139-141.

BRYAN, K.
1922  Erosion and sedimentation in Papago Country, Arizona. U. S. Geological Survey, Bulletin 730:19-90.

BYKOV, G. E.
1932  Les formes du relief de la région d'Atbassar (Kazakhstan). Gosudarstvennoe Geograficheskoe Obschchestvo, Izvestiia 64:61-75.

CAILLEUX, A.
1936  Les actions éoliennes périglaciaire en Europe. Société Géologique de France, Bulletin 6(5):102-104.

DI CAPORIACCO, L.
1936  Dune australiane e dune del deserto libico. L'Universo (Istituto Geografico Militare, Firenze) 17(9):665-667.

CAYEUX, L.
1928  Origines des sables des dunes sahariennes. International Geological Congress, 14th, Madrid, 1926, Compte Rendu pp. 783-788.

CHUDEAU, R.
1920  L'étude sur les dunes sahariennes. Annales de Géographie 29:334-    .
1921  Les dunes de Gascogne. Géographie 35:263-268.

CLAPP, F. G.
1926  In the northwestern Australian Desert. Geographical Review 16:206-231.

CLAYTON, P. A.
1933  The west side of the Gilf Kebir. Geographical Journal 81:254-259.

COBB, C.
1931  Dune sands and aolian soils in relation to present and past climatic conditions on the continent of North America. International Geographical Congress, 13th, Paris, 1931, Proceedings 2:712.

CORNISH, V.
1927  Waves of granular material formed and propelled by winds and currents. Royal Aeronautical Society, London, Monthly Notices, Supplement, July.
1928  Limits of forms and magnitude of desert dunes. Nature 121:620-622.
1934  Ocean waves and kindred geophysical phenomena. Cambridge University Press, 164 pp.

CREMA, C.
1933  Risalti crestiformi e "seghife," due tipi morfologici proprii delle regioni aride osservati in Libia. International Geological Congress, 16th, Washington, Proceedings 2:737-740.

DEWERS, F.
1935  Probleme der Flugsandbildung in Nordwestdeutschland. Naturwissenschaftlicher Verein, Bremen, Abhandlungen 29(3-4):324-366.

DOBROLOWSKI, A. B.
1924  Movement de l'air et de l'eau sur les accidents du sol. Geografiska Annaler 6:300-367.

DOUBIANSKY, V. A.
1928  The sand deserts of the south-east Karakum. Bulletin of Applied Botany, Genetics and Plant Breeding 19(4):225-

DUFOUR, A.
1936  Observations sur les dunes de Sahara meridional. Annales de Géographie 45:276-285.

DURAND, J.
1953  Le vent et sa consequence: l'érosion éolienne, facteur de formation des sols au Sahara. In Desert Research, proceedings international symposium held in Jerusalem, May 7-14, 1952, sponsored by the Research Council of Israel and Unesco. Research Council of Israel, Special Publication 2:434-437.

ENQUIST, F.
1932  The relation between dune form and wind direction. Geologiska Föreningens i Stockholm Föhandlingar 54(388):19-59.

EXNER, F. M.
1920  Zur Physik der Dünen. Akademie der Wissenschaften, Wien, Mathematisch-Naturwissenschaftliche Klasse, Abteilung I, 129(9-10):929-952.
1921  Dünen und Mäander, Wellenformen der festen Erdoberflache, deren Wachstum und Bewegung. Geografiska Annaler 3:327-335.
1927  Über Dünen und Sandwellen. Geografiska Annaler 9:81-89.

FORTH DE LANCEY, N. B.
1930  More journeys in search of Zerzura. Geographical Journal 75:49-64.

FREDRICK, E.
1932  The relation between dune form and wind direction. Geologiska Foreningens i Stockholm Föhandlingar 59:19-59.

GABRIEL, A.
1938  The southern Lut and Iranian Baluchistan. Geographical Journal 92:195-210.

GAUTIER, E. F.
1935  The Sahara, the great desert. New York.

GERASIMOV, I. P.
1933  Recent geological processes in the deserts of western Turkestan. International Geological

Congress, 16th, Washington, 1933, vol. 2:782. (Abstr.)

GHERISIMOV, J. P.
1931 De quelques formes de relief de la steppe désertique. Gosudarstvennoe Geograficheskoe Obshchestvo, Izvestiia 63(4):293-300.

GRABAU, A. W.
1920 A textbook of geology. Part 1. Heath, Boston.

HACHISUKA, M., ed.
1932 Le Sahara. Société d'Editions Géog., Marit. et Col., Paris. 167 p.

HARTNACK, W.
1925 Wanderdünen Pommerns, ihre Form und Entstehung. Druck und Verlag von Julius Abel, Greifswald. 112 pp.
1931 Zur Entstehung und Entwicklung der Wanderdünen au der deutschen Ostseeküste. Zeitschrift für Geomorphologie 6:174-217.

HELLER, S. J.
1932 Sur la morphologie de quelques formations sabloneuse des Karakoum transcaspiens. Gosudarstvennoe Geograficheskoe Obshchestvo, Izvestiia 64(4-5):387-390.

HELLER, S. J., and V. KUNIN
1933 Sur l'origine des chaînes de sable. Académie des Sciences, Leningrad, USSR, Comptes Rendus (Doklady), n.s. 2:88-91.

HILLS, E. S.
1939a The lunette, a new landform of aeolian origin. Australian Geographer 3:15-21.
1939b The physiography of northwestern Victoria. Royal Society of Victoria, Proceedings, n.s. 51(2):297-320.

HÖGBOM, I.
1923 Ancient inland dunes of north and middle Europe. Geografiska Annaler 5:113-243.

HORNER, N. G.
1927 Brattforshden. Elt värmländskt randdeltekomplex och desdyner. Sveriges Geologiska Undersökning, Årsbok 20, 1926.
1936 Geomorphological processes in continental basins of Central Asia. International Geological Congress, 16th, Washington, 1933, Part 2, pp. 723-729.

HUME, W. F.
1921 The Egyptian wilderness. Geographical Journal 58:249-276.
1925 The geology of Egypt. Ministry of Finance, Cairo.

JORRE, G.
1935 Les formes de relief dans les steppes desertiques de l'Asie Centrale russe. Annales de Géographie 44:317-321.

JÜNGST, H.
1938 Palaogeographische Auswertung der Kreuzchichtung. Geologie Meere und Binnengewässer 2:229-277.

JUTSON, J. T.
1934 The physiography of Western Australia. Geological Survey of Western Australia, Bulletin 61.

KÁDÁR, L.
1934 A study of the sand sea in the Libyan Desert. Geographical Journal 83:470-478.
1938 Die periglazialen Binnendünen des norddeutschen und polnischen Flachlandes. Congrès International de Géographie, 15e, Amsterdam, 1938, Comptes Rendus 1:167-183.

KAISER, E.
1926 Höhenschichtenkarte der Deflationslandschaft in der Namib, Sudwestafricas. Bayerische Akademie der Wissenschaften, Abhandlungen, 30.
1927 Die Diamantenwüste Südwestafrikas. Reimer, Berlin. 2 vols.

KING, L. C.
1939 South African scenery. Edinburgh. 2nd ed.

LAMARE, P.
1933 Travaux géographiques et géologiques récents sur l'Arabie meridionale. Annales de Géographie 42(240):623-630.

LENCEWICZ, S.
1922 Les dunes continentales de la Pologne. Wydmy Sródladowe Polski 11:12-59.

LEWIS, A. D.
1936 Sand dunes of the Kalahari within the borders of the Union. South African Geographical Journal 19:25-57.

LOUIS, H.
1929 Die Form der nord deutschen Bogendünen. Zeitschrift für Geomorphologie 4:7-18.

MADIGAN, C. T.
1930 An aerial reconnaissance into the southeastern portion of Central Australia. Royal Geographical Society of Australasia, South Australian Branch, Proceedings 30:83-108.
1936 The Australian sand-ridge deserts. Geographical Review 26(2):205-227.
1938 The Simpson Desert and its borders. Royal Society of New South Wales, Proceedings 71:503-535.

MONOD, T.
1928 Une traversée de la Mauritanie occidentale. Revue de Geographie Physique et de Géologie Dynamique 1:3-25, 88-106.
1936 Présentation d'un croquis géologique schématique du Sahara occidental. Société Geologique de France, Compte Rendu 10:164.

MÒNTERIN, U.
1935 Sulla trasformazione delle dune trasversali in longitudinali nel Sahara libico. Reale Accademia delle Scienze di Torino, Classe di Scienze Fisiche, Matematiche, e Naturali, Atti 70:62-80.

MOSS, R.
1938 The physics of an ice cap (sastrungi and barchans). Geographical Journal 92:215-217.

MUSSET, M.
1923 Über Sandwanderung, Dünenbildung und Veränderung an der Unterpommerischen Küste. Zeitschrift für Bauwesen, Berlin, vol. 73.

PETRUSHEVSKII, B. A.
1937 De l'origine des trainées de sable du Karakum. Gosudarstvennoe Geograficheskoe Obshchestvo, Izvestiia 69(6):956-967.

PHILBY, H. S. J.
1933 The Empty Quarter. Constable, London. 433 pp.

PRESCOTT, J. A., and C. S. PIPER
1932 The soils of the South Australian mallee. Royal Society of South Australia, Transactions 56: 118-146.

PRICE, W. A.
1933 Role of diastrophism in the topography of the Corpus Christi County, Texas. American Association of Petroleum Geologists, Bulletin 17(8): 907-962.

REMPEL, P.
1936 The crescentic dunes of the Salton Sea and their relation to vegetation. Ecology 17:347-358.

RUSSELL, R. J.
1932 Landforms of the San Gorgonio Pass, Southern California. University of California Publications in Geography 6:106-114.

RUSSELL, W. L.
1929 Drainage alignment in the western Great Plains. Journal of Geology 37(3):249-255.

SAINT-JOURS, B.
1933 Les dunes de Gascogne. Congrès International de Géographie, 13ᵉ, Paris, 1931, Compte Rendu 2(1):713-716.

SANDFORD, K. S.
1933a Geology and geomorphology of the southern Libyan Desert. Geographical Journal 82:213-219.

1933b Man and Pleistocene climate of the northwest Sudan. International Geological Congress, 16th, Washington, 2:812. (Abstr.)

1933c Past climate and early man in the southern Libyan Desert. Geographical Journal 82:219-222.

1935a Geological observations on the northwest frontiers of the Anglo-Egyptian Sudan and the adjoining part of the southern Libyan Desert. Royal Geological Society of London, Quarterly Journal 91:323-281.

1935b Sources of water in the northwestern Sudan. Geographical Journal 85:412-431.

SERRET, M.
1933 L'Ouaddaï, physionomie du pays; les facteurs d'érosion. Congrès International de Géographie, 13ᵉ, Paris, 1931, Compte Rendu 2(1):685-687.

SHAW, W. B. K.
1936 An expedition to the southern Libyan Desert. Geographical Journal 87:193-221.

SHOTTEN, F. W.
1937 The lower Bunter sandstones of north Worcestershire and east Shropshire. Geological Magazine 74:534-552.

SIDWELL, R., and W. P. TANNER
1938 Quaternary dune building in Central Kansas. Geological Society of America, Bulletin 49:139. (Abstr.)

SMITH, H. T. U.
1938 Sand dune cycle in western Kansas. Geological Society of America, Bulletin 50:1934-1935. (Abstr.)

SOLGER, F.
1920 Baobachtungen über Flugsandbildungen. Deutsche Geologische Gesellschaft, Zeitschrift 72(6/7):168-180.

TALMAGE, S. B.
1932 The origin of the gypsum sands of the Tala Rossa River. Geological Society of America, Bulletin 43:185-186.

THOMPSON, W. O.
1932 Original structures of beaches, bars and dunes. Geological Society of America, Bulletin 48:723-751.

TRIKANOS, J.
1928 Windripplen. Petermanns Geographische Mitteilungen 74(9/10):266-271.

TWENHOFL, W. H. et al
1932 Treatise on sedimentation. 2nd ed. Williams and Wilkins Co., Baltimore. 926 p.

URVOY, Y.
1933a Les formes dunaires à l'Ouest du Tchad. Annales de Géographie 42:506-515.

1933b Modele dunaire entre Zinder et le Tchad. Association de Géographes Français, Bulletin 74(69): 79-82.

1935 Terrasses et changements de climats quaternaire à l'est du Niger. Annales de Géographie 44:254-263.

1936 Structure et modele du Soudan français. Annales de Géographie 45:19-49.

WEGEMAN, C. H.
1939 Great sand dunes of Colorado. Mines Magazine 29:445-448.

WHITFIELD, C. J.
1937 Sand dunes in the Great Plains. Soil Conservation 2(9):208-209.

WILDVANG, D.
1936 Uber flugsande der ostrieschen Geest. Naturwissenschaftlicher Verein, Bremen, Abhandlungen 29(3-4):292-307.

WILCKENS, O.
1926 Die oberrheinischen Flugsande. Geologische Rundschau, Sonderband 17a:555-597.

WINGATE, O.
1934 In search of Zerzura. Geographical Journal 83:281-308.

WITTSCHELL, L.
1931 Morphological effects of sand and dust storms. Geographical Journal 77:588-589.

WOLDSTEDT, P.
1929 Das Eiszeitalta. Stuttgart.

## (c) 1940-1959

Adam, L.
1959 The physical geography of the mezofold (translated title). Földrajzi. Monog. 2, 514 p. Akad. Kaido, Budapest.

ALIMEN, H.
1953a Caractères granulométriques d'un Dépôt effectué par le vent de sable à Beni-Abbès (Algérie). Société Géologique de France, Compte Rendu 11-12:234-237.

1953b Variations granulométriques et morphoscopiques due sable le long de profils dunaires au Sahara occidental. Centre National de Recherches Scientifiques, Paris, Colloques Internationaux 35:219-235.

ALIMEN, H., M. BURON, and J. CHAVAILLON
1958 Caractères granulométriques de quelques dunes d'ergs du Sahara nord-occidental. Académie des Sciences, Paris, Compte Rendu 247:1758-1761.

ALIMEN, H., J. CHAVAILLON, and G. CONRAD
1959 Formations arides et paléosols quaternaires au Sahara nord-occidental. Société Géologique de France, Compte Rendu 5:104-105.

ALIMEN, H., J. CHAVAILLON, and J. MARGAT
1959 Contribution à la chronologie préhistorique africaine. Essai de correlation entre les dépôts quaternaires du Bassin Guir-Saoura (Sahara) et du Bassin du Tafilait (Maroc). Congrès Préhist. Fr., Monaco. 20 pp.

ALIMEN, H. and D. FENET
1954 Granulométrie de sables d'erg aux environs de la Saoura (Sahara occidental). Société Géologique de France, Compte Rendu 9-10:183-185.

ALIMEN, H., and A. F. DE LAPPARENT
1946 Dune quaternaire et dune bartonienne à Frère-en-Tardenois. Société Géologique de France, Compte Rendu 5(16):173-175.

ALIMEN, H., and M. MERCIER
1948 Topographie dunaire au sommet de l'Auversien dans le Tardenois. Académie des Sciences, Compte Rendu 226(18):2083-2085.

1951    Remaniements éoliens d'âge Bartonien dans les sables de la partie orientale du Bassin de Paris. International Sedimentological Congress, 3rd, Groningen-Wageningen, 1951, Proceedings, p. 25-41.

ALIMEN, H. *et al.*

1953    Les Chaines d'Ougarta et la Saoura. International Geological Congress, 19th, Alger, Monographie Régulière, sér. 1, 15:93-106.

1957    Sables quaternaire du Sahara nord-occidental (Saoura-Ougarta). Service de la Carte Géologique de l'Algérie, Bulletin n.s. 15, 207 pp. (Critique *in* Revue de Géomorphologie Dynamique, 1958, p. 180-182.)

ALMEIDA, F. F. M. DE

1953    Botucatú, a Triassic desert of South America. International Geological Congress, 19th, Algiers, 1952, fasc. 7:9-24.

AMBROZ, V.

1947    Sprase Pahorkatin. Sbornik Statniho Geologickeho Ustavu Ceskoslovenske Republiky XIV.

ANONYMOUS

1949    Collective investigations of dunes near Torun (translated title). Studia Societatis Scient. Torunensis. Suppl. I.

ANONYMOUS

1950    La fixation des dunes en Mauritanie. Institut Colonial, Marseilles, Cahiers Coloniaux 9:384-385.

BAGNOLD, R. A.

1941    The physics of blown sand and desert dunes. Methuen, London. (Reprinted 1954, 1960).

1951    The sand formations of south Arabia. Geographical Journal 117:78-86.

1953*a*   The surface movement of blown sand in relation to meteorology. *In* Desert research, Proceedings, International Symposium held in Jerusalem, May 7-14, 1952, sponsored by the Research Council of Israel and Unesco. Research Council of Israel, Special Publication 2:89-93.

1953*b*   Forme des dunes de sable et régime des vents. *In* Actions éoliennes. Centre National de Recherches Scientifiques, Paris, Colloques Internationaux 35:23-32.

BAYARD, LIEUT.

1947    Aspects principaux et consistence des dunes (Mauritanie). Institut Français d'Afrique Noire, Bulletin 9.

BELLAIR, P.

1940    Les sables de la dorsale saharienne et du bassin de l'Oued Rhir. Service de la Carte Géologique de l'Algérie, Bulletin sér. 5, no. 5.

1941    Etude granulométrique de quelques formations arenacées du Gourara et la Saoura. Société d'Histoire Naturelle de l'Afrique du Nord, Bulletin 32:191-196.

1949    Le Quaternaire de Tejerhi (Fezzân). Société Géologique de France, Compte Rendu 9-10: 160-162.

1951    La Ramla des Daouada (Fezzân). Institut de Recherches Sahariennes, Travaux 7:69-85.

1953    Sables désertiques et morphologie éolienne. International Geological Congress, 19th, Algiers, 1952, fasc. 7:113-118.

1953    Le quaternaire de Tejerhi (Fezzân). Institut des Hautes Etudes de Tunis, Publication Scientifique 1:9-16. (Mission au Fezzân, 1949).

BELLAIR, P., and A. JANZEIN

1952    Grand Erg oriental. Société des Sciences Naturelles de Tunisie, Bulletin.

BIROT, P., R. CAPOT-REY, and J. DRESCH

1955    Recherches morphologiques dans le Sahara central. Institut de Recherches Sahariennes, Travaux 13:13-74.

BLACK, R. F.

1951    Eolian deposits of Alaska. Arctic 3:89-111.

BOSAZZA, V. L.

1953    The palaeogeography of the Kalahari desert in southern Africa. International Geological Congress, 19th, Algiers, 1952, fasc. 7:103. (Abstr.)

1957    The Kalahari system in southern Africa and its importance in relationship to the evolution of man. *In* Pan-African Congress on Prehistory, 3rd, Livingstone, 1955, Actes p. 127-132.

BOULAINE, J.

1953    L'érosion éolienne des sols salés et la morphologie superficielle des chotts et des sebkhas. Société d'Histoire Naturelle de l'Afrique du Nord, Bulletin.

1954    Le Sebkha Ben Ziane et sa "lunette" ou bourrelet; exemple de complexe morphologique formé par la dégradation éolienne des sols salés. Revue de Géomorphologie Dynamique 5(3):102-123.

1956    Les lunettes de basses plaines oranaises: formations éoliennes argileuses liées à l'extension de sols salins; la Sebkha de Ben Ziane, la dépression de Chantrit. International Quaternary Association, 4th Congress, Rome-Pisa, 1953, Actes p. 143-151.

BRYAN, K., and F. T. McCANN

1943    Sand dunes and alluvium near Grants, New Mexico. American Antiquity 8:281-295.

BUNKER, D. G.

1953    The south-west borderlands of the Rub'al Khali. Geographical Journal 119(4):420-430.

CAILLEUX, A.

1941    Action du vent et du gel au Quaternaire dans le région bordelaise. Société Géologique de France, Bulletin 15(5):259-266.

1942    Les actions eoliennes périglaciaire en Europe. Société Géologique de France, Mémoir n.s. 21:46-

1951    Interprétation climatique des éolisations pliocènes et quaternaires en France. Société Géologique de France, Bulletin sér. 6, Compte Rendu 3-4:44-46.

1952    Observations à l'article de M. Walther sur les sables éoliens. Revue de Géomorphologie Dynamique 3(2):99.

CAPOT-REY, R.

1941    Observations géologiques à la bordure de l'Erg Occidental. Société d'Histoire Naturelle de l'Afrique du Nord, Bulletin 32:47.

1943    La morphologie de l'Erg Occidental. Institut de Recherches Sahariennes, Travaux 2:69-104.

1945    The dry and humid morphology of the Western Erg. Geographical Review 35:391-407.

1947    L'Edeyen de Mourzouk. Institut de Recherches Sahariennes, Travaux 4:67-109.

1953*a*   Recherches géographiques sur les confins algéro-libyens. Institut de Recherches Sahariennes, Travaux 10:33-73.

1953*b*   Le Sahara français. Presses Universitaires de France, Paris. 564 p.

CAPOT-REY, R. and F. CAPOT-REY

1948    Le déplacement des sables éoliens et la formation des dunes desértiques, d'après R. A. Bagnold. Institut de Recherches Sahariennes, Travaux 5:47-80.

CHENOWETH, P. A.
1952 Statistical methods applied to Trentonian stratigraphy in New York. Geological Society of America, Bulletin 63:521-560.

CHMIELEWSKA, M., and K. WASYLIKOWA
1961 Witów: Late-Pleistocene dunes and peat-bogs. International Congress on Quaternary, 6th, Warsaw, 1961, Guide-Book of Excursion C, the Lódź region, p. 75-84.

CHOUBERT, G.
1941 Le Dra et l'Irigui. Revue de Géographie du Maroc, 1:33-36.
1945 Note préliminaire sur le Pontien du Maroc. Société Géologique de France, Bulletin 15(5): 677-674.

CLAYTON, W. D.
1957 The swamps and sand-dunes of Hadejia. Nigerian Geographical Journal 1:31.

COALDRAKE, J. E.
1954 The sand-dunes of the Ninety-Mile Plain, southeastern Australia. Geographical Review 44:394-407.

COOPER, W. S.
1944 Development and maintenance of the natural profile of a transverse dune ridge. American Philosophical Society, Year Book, pp. 150-153.
1958 Coastal sand dunes of Oregon and Washington. Geological Society of America, Memoir 72. 169 p. (Reviewed by Smith [1959] q.v.)

CORNET, A.
1950 Reconnaissance géologique dans L'Erg d'Oubari et la Hamada Zeher (Fezzân). Institut de Recherches Sahariennes, Travaux 6:63-72.

COULSON, A. L.
1940 Sand Dunes of the Portland district and their relation to post-Pliocene uplift. Royal Society of Victoria, Proceedings, n.s., 52:315-335.

COURSIN, A.
1956 Etudes des barkanes à l'est de Port Etienne. TPOAF Report. Dakar. 17 pp.

CROCKER, R. L.
1946 The soil and vegetation of the Simpson Desert and its borders. Royal Society of South Australia, Transactions 70:235-258.

CVIJANOVICH, B. G.
1953 Sur le rôle des dunes en relation avec le système hydrologique de la nappe souterraine du Grand Erg. Institut de Recherches Sahariennes, Travaux 9:131-136.

DEVILLERS, C.
1948 Les dépôts quaternaires de l'Erg Tihodaine (Sahara Central). Société Géologique de France, Bulletin sér. 5, t. 18, Compte Rendu Sommaire 10:189-191.

DOSKATCH, A. G.
1948 Les étapes fondamentales du développment des idées sur le relief des deserts sableux. Académie des Sciences de l'URSS, Institut Géographique, Travaux 39:233-    .
1954 Matériaux pour la carte géomorphologique de la plaine méridionale de la Volga et de la dépression précaspienne. p. 47-87. In Recherches géomorphologiques dan la dépression précaspienne. Editions de l'Académie des Sciences de l'URSS, Moscow.

DUBIEF, J.
1943 Les vents de sable dans le Sahara français. Institut de Recherches Sahariennes, Travaux 2:11-35.

1952a The evolution of arid zones in the past and today. Unesco NS/AZ112, Paris. (Mimeo) 13 p.
1952b Le vent et le déplacement du sable au Sahara. Institut de Recherches Sahariennes, Travaux 8:123-164.
1953a Les vents de sable dans le Sahara français. In Actions éoliennes. Centre National de Recherches Scientifiques, Paris, Colloques Internationaux 35:45-70.
1953b Le climat saharien. Maroc-Medial 322-332. 8 p.

DUCHEMIN, G. J.
1958 Essai sur la protection des constructions contre l'ensablement à Port-Etienne (Mauritanie). Institut Français d'Afrique Noire, Bulletin sér. A, 20:675-686.

DYLIKOWA, A.
1958 Phase du développement des dunes aus environs de Łódź. Acta Géographica Universitatis Lodziensis 8:233-268.

EDELMAN, C. H.
1947 Les limons et les sables de couverture des Pays-Bas. pp. 303-310. In La géologie des terrains recénts dans l'ouest de l'Europe. (Session Extraordinaire, Sociétés Belges de Géologie, Sept. 19-26, 1946). M. Hayez, Bruxelles.

EDMONDS, J. M.
1942 The distribution of the Kordofan sand. Geological Magazine 79:18-30.

ENJALBERT, H.
1950 Observations morphologiques sur les Lands de Gascogne; les gorges du Ciron et le karst de Casteljaloux. Revue Géographique des Pyrénées et du Sud-Ouest 21(1):5-42.

ESCANDE, L.
1949 Ondulations de sable des modelés réduits et dunes de désert. Académie des Sciences, Paris, Compte Rendu 229(13):613-615.
1953 Similitude des ondulations de sable des modelés reduits et des dunes du désert. In Actions éoliennes. Centre National de Recherches Scientifiques, Paris, Colloques Internationaux 35:71-    .

EVANS, G. L., and G. E. MEADE
1945 Quaternary of the Texas high plains. University of Texas Publication 4401:485-502.

EYMANN, J. R.
1953 A study of sand-dunes in the Colorado and Mojave deserts. University of California (unpublished M.S. Thesis)

FEDOROVICH, B. A.
1940 Some fundamental considerations concerning the origin and development of the sand relief (translated title; English summary). Academy of Sciences of the U. S. S. R., Bulletin, Geography and Geophysics series, 6:885-910.
1948a La question de l'origine et la morphogènese du relief sableux des déserts. Akademiia Nauk SSSR, Institut Géografi, Vestnik 39:160-183.
1948b Le relief des sables d'Asie en tant qu'image de la circulation atmospherique. Problemy Fizicheskoi Geografii 13:91-109.
1956 L'origine du relief des déserts de sables actuels. p. 117-129. In Essais de Géographie. Leningrad et Moscou.

FINKEL, H. J.
1959 The barchans of southern Peru. Journal of Geology 67:614-647. [Review by W. A. Price in Geographical Review, 1959, 50(4):585-586]

FLINT, R. F.
1959　Pleistocene climates in eastern and southern Africa. Geological Society of America, Bulletin 70:343-371.

GABRIEL, A.
1957　Zur Oberflächengestaltung der Pfannen in Trockenräumen Zentralpersiens. Geographische Gesellschaft, Wien, Mitteilungun 99(2-3): 146-160.

GALON, R.
1958　Sur les dunes continentales en Pologne (translated title; French summary). p. 13-30. In R. Galon, ed., Wydmy sródladowe Polski. PWN, Warszawa.
1959　New investigations of inland dunes in Poland. Przeglad Geograficzny 31(suppl):93-110.

GERASIMOV, I. P., and J. C. MA
1958　Geneticheskie tipy pochv na territorii Kitaiskoi Narodnoi Respubliki i ikh geograficheskoe rasprostranenie. (Survey of the major genetic soil types of China and their geographic distribution) Akademiya Nauk SSSR, Institut Geologicheskikh Nauk, Moscow. 86 pp.

GRANDET, C.
1955　Aspects de la morphologie dunaire dans la région de Beni-Abbès. Société Géologique de France, Bulletin sér. 6, 5:135-142.
1957　Sur la morphologie dunaire de la rive sud du Lac Faguibine. Institut de Recherches Sahariennes, Travaux 16:171-179.

GROVE, A. T.
1957　Patterned ground in northern Nigeria. Geographical Journal 123:271-274.
1958　The ancient erg of Hausaland and similar formations on the south side of the Sahara. Geographical Journal 124:528-533.
1959　A note on the former extent of Lake Chad. Geographical Journal 125:465-467.

HACK, J. T.
1941　Dunes of the western Navaho Country. Geographical Review 31(2):240-263.

HAMBLOCK, H.
1958　Das Alter einigen Dünen an der obren Ems. Erdkunde 12:128. (Review in Revue de Géomorphologie Dynamique, 1958, p. 95)

HANSEN, V.
1957　Sandflugten i Thy og dens indflydelse på kulturlandskabet. (The movement of sand dunes in Thy and its human consequences) Geografisk Tidsskrift 56:69-92.

HEFLEY, H. M., and R. SIDWELL
1945　Geological and ecological observations of some High Plains dunes. American Journal of Science 243:361-376.

HEMMING, C. F., and C. G. TRAPNELL
1957　A reconnaissance classification of the soils of the south Turkana desert. Journal of Soil Science 8(2):167-183.

HILLS, E. S.
1953　Regional geomorphic patterns in relation to climatic types in the dry areas of Australia. In Desert research, Proceedings, International Symposium held in Jerusalem, May 7-14, 1952, sponsored by the Research Council of Israel and Unesco. Research Council of Israel, Special Publication 2:355-364.

HOLM, D. A.
1953　Dome-shaped dunes of the Central Nejd, Saudi Arabia. International Geological Congress, 19th, Algiers, 1952, Compte Rendu, fasc. 7:107-112.
1957　Sigmoidal dunes, a transitional form. Geological Society of America, Bulletin 68(12:2):1746. (Abstr.)

HUFFINGTON, R. M., and C. C. ALBRITTON
1941　Quaternary sands of the High Plains. American Journal of Science 239:325-338.

HUFFMAN, C. G., and W. A. PRICE
1949　Clay dune formation near Corpus Christi, Texas. Journal of Sedimentary Petrology 19(3):118-127.

JAHN, A.
1956　Geomorphology and Quaternary history of the Lublin Plateau (translated title; English summary). Polska Akademia Nauk, Instytut Geografii, Prace Geograficzne 7, 453 p.

JONASSEN, H.
1954　Dating of sand drift east of Ulfborg. Botanisk Tidsskrift 51:134-140.

KAMEL, K.
1953　Sand dunes in the Kharga oasis. Société Royale de Géographie d'Egypte, Bulletin 25:77-80.

KARMAN, T. VON
1953　Considerations aerodynamiques sur la formation des ondulations du sable. In Actions éoliennes. Centre National de Recherches Scientifiques, Paris, Colloques Internationaux 35:103- .
1956　Collected work. Butterworth Sci. Publs., London.

KEPCZYNSKI, K.
1958　Flora und Geschichte des Moores Siwe Bagno in der Tucheler Heide (Polish with German summary). Zeszyty Naukowe Mniwersyletu M. Propernika. Torun.

KERR, R. C., and J. O. NIGRA
1952　Eolian sand control. American Association of Petroleum Geologists, Bulletin 36(8):1541-1573.

KIERSH, G. A.
1950　Small scale structure and other features of Navajo sandstone, northern part of San Rafael Swell, Utah. American Association of Petroleum Geologists, Bulletin 34:923-942.

KING, D.
1956　The Quaternary stratigraphic record at Lake Eyre North and the evolution of existing topographic forms. Royal Society of South Australia, Transactions 79:93-103.

KINZL, H.
1958　Die Dünen in der Küstenlandschaft von Peru. Geographische Gesellschaft, Wien, Mitteilungen 100(1-2):5-17.

KOBENDZA, J., and R. KOBENDZA
1958　Les dunes éparpillées de la Forêt de Kampinos (translated title; French summary). I:95-168. In R. Galon, ed., Wydmy Sródladowe Polski. PWN, Warszawa.

KOVDA, V. A.
1959　Ocherki prirody i pochv Kitaia. (Studies on the soils of China) Akademiia Nauk S.S.S.R. 455 p.

KRYGOWSKI, B.
1958　Quelques données sur les sables des dunes continentales (translated title; French summary). I:73-85. In R. Galon, ed., Wydmy Sródladowe Polski. PWN, Warszawa.

LANDSBERG, H.
1942　The structure of the wind over a sand dune. American Geophysical Union, Transactions 23:237-239.

LANDSBERG, H., and N. A. RILEY
1943　Wind influences on the transport of sand over

a Michigan sand dune. University of Iowa Studies in Engineering, Bulletin 27:342-352.

LANDSBERG, S. Y.
1956 The orientation of dunes in Britain and Denmark, with respect to the wind. Geographical Journal 122(2):176-189.

LE LUBRE, M.
1950 Une reconnaissance aerienne sur l'Edeyen de Mourzouk (Fezzân). Institut de Recherches Sahariennes, Travaux 5:219-221.
1952 Conditions structurales et formes de relief dans le Sahara. Institut de Recherches Sahariennes, Travaux 8:189-238.

LEONE, G.
1953 Origin and reclamation of the dunes in Tripolitania. In Desert research, Proceedings, International Symposium held in Jerusalem, May 7-14, 1952, sponsored by the Research Council of Israel and Unesco. Research Council of Israel, Special Publication 2:401-403.

MABBUTT, J. A.
1957 Physiographic evidence for the age of the Kalahari sands of the southwestern Kalahari. In Pan-African Congress on Prehistory, 3rd, Livingstone, 1955, pp. 123-126.

MADIGAN, C. T.
1945 Simpson Desert expedition, Scientific Reports: Introduction, narrative, physiography and meteorology. Royal Society of South Australia, Transactions 69(1):118-139.
1946 The sand formations. Simpson Desert Expedition, 1939, Scientific Report 6: Geology. Royal Society of South Australia, Transactions 70(1): 45-63.

MADJANOWSKI, S.
1958 Les problèmes climatiques des périodes des dunes (translated title; French summary). In R. Galon, ed., Wydmy Sródladowe Polski. PWN, Warszawa.

MALKOWSKI, S., and S. LENCEWICZ
1953 The inland dunes of Poland (translated title). Geol. Editions.

MARUSZCZAK, H.
1958 The dunes of the Lublin plateau and environs (translated title). II. In R. Galon, ed., Wydmy Sródladowe Polski. PWN, Warszawa.

MATSCHINSKI, M.
1952 Sur les formations sableuses des environs de Beni-Abbès. Société Géologique de France, Compte Rendu 9-10:171-174.
1954 Stabilita delle dune del Sahara. Servizio Geologico d'Italia, Rome, Bollettino 55:579-592.

MAULL, O.
1958 Handbuch der geomorphologie. Deuticke, Wien. 600 p.

McKEE, E. D.
1945 Small scale structures in the Coconino sandstone of northern Arizona. Journal of Geology 53:313-325.
1957 Primary structure in some recent sediments. American Association of Petroleum Geologists, Bulletin 41(8):1704-1747.

McKENZIE, L. A.
1952 Report on the Kalahari Expedition, 1949. South African Government Printer, Pretoria. 35 pp.

McLAUGHLIN, W. T., and R. L. BROWN
1942 Controlling coastal sand dunes in the Pacific northwest. U. S. Department of Agriculture, Circular 660:1-46.

MELTON, F. A.
1940 A tentative classification of sand dunes. Journal of Geology 48:113-174.

MICHEL, P.
1959 L'évolution géomorphologique des Bassins du Sénégal et du Haute Gambie et son connection avec la prospection minéral. Revue de Géomorphologie Dynamique 10:117-143.

MONOD, T.
1950a Autour du problème du dessèchement africain. Institut Français d'Afrique Noire, Bulletin 12(2): 514-523.
1950b Sur les conditions désertique anciennes au Sahara. Institut Français d'Afrique Noire, Bulletin 12(2):530-531.
1958 Majâbat-al-Koubrâ, contribution à l'étude de l'Empty-Quarter, ouest Saharien. Institut Français d'Afrique Noire, Mémoire, 52.

MONOD, T., and A. CAILLEUX
1945 Etude de quelques sables et grès du Sahara occidental. Institut Français d'Afrique Noire, Bulletin 7:174-190.

MOTT, R. J.
1959 Notes on the sand dunes near Prescott, Ontario. Revue Canadienne de Géographie 13:135-147.

MROZEK, W.
1958 The dunes of Tonin-Bydgosecz Valley (translated title). II. In R. Galon, ed., Wydmy Sródladowe Polski. PWN, Warszawa.

MURRAY, G. W.
1946 Possible changes in Arabian wind direction. Geographical Journal 108:127-128.

NORRIS, R. M.
1956 Crescentic beach cusps and barchans. American Association of Petroleum Geologists, Bulletin 40:1681-1686.

NOWICKA, I.
1958 Les dunes sur le sandr de Brda (translated title). Zeszyty Naukowe Uniw. M. Kopernika w Toruniu, 4(Geogr.):27-45.

OKOLOWICZ, W.
1952 Climatological criteria in geomorphological investigations in the north European lowland (translated title; English summary). Pol. Geol. Inst., Bulletin 65:121-136.

OLSON, J. S.
1958 Lake Michigan dune development. Journal of Geology 66:254-263, 345-351, 437-483.

PALAUSI, G.
1955 Au sujet du Niger fossile dans la région de Tombouctou. Revue de Géomorphologie Dynamique 6:217-218.

PASSARGE, S.
1940 Geomorphologische problem aus Algerien. Journal of Geomorphology 3:108-130, 227-243.

PERNAROWSKI, L.
1958 Les recherches sur les dunes de la Basse Silésie (translated title). I:171-198. In R. Galon, ed., Wydmy Sródladowe Polski. PWN, Warszawa.

PETROV, M. P.
1948 Le relief des barkhanes des déserts et les rapports de sa genèse avec la théorie. Akademiia Nauk SSSR, Institut Geografi, Vestnik 39: 184- .

PILARCZYK, L.
1958 Les dunes situées entre Warta et la Noteć (translated title). II:87-93. In R. Galon, ed., Wydmy Sródladowe Polski. PWN, Warszawa.

POLDERVAART, A.
1957 Kalahari sands. *In* Pan-African Congress on Prehistory, 3rd, Livingstone, 1955, Proceedings p. 106-114.

POLIANSKI, W.
1956 Pleistocene in the Vistula Gap across the southern uplands. Pol. Inst. Geol. Warszawa, Geol. Studies IX.

POSER, H.
1950 Zur Rekonstruktion der spätglazialen Luftdrukverhältnisse in Mittel- und Westeuropa auf Grand der vorzeitlichen Binnendünen. Erdkunde 4:81-88.

PRICE, W. A.
1950 Saharan sand dunes and the origin of the longitudinal dunes; a review (of Capot-Rey and Capot-Rey, 1948, *q.v.*). Geographical Review 40(3):462-465.
1959 The barchans of southern Peru; a review (of Finkel, 1959, *q.v.*). Geographical Review 50(4): 585-586.

QUENEY, P.
1953 Classification des rides de sable et théorie ondulatoire de leur formation. *In* Actions éoliennes. Centre National de Recherches Scientifiques, Paris, Colloques Internationaux 35:179-195.

QUENEY, P., and J. DUBIEF
1943 Action d'un obstacle ou d'un fossé sur un vent charge de sable. Institut de Recherches Sahariennes, Travaux 2:169-176.

RAVIKOVITCH, S.
1953 The aeolian soils of the northern Negev. *In* Desert research Proceedings, International Symposium held in Jerusalem, May 7-14, 1952, sponsored by the Research Council of Israel and Unesco. Research Council of Israel, Special Publication 2:404-433.

REIFENBERG, A.
1947 The soils of Palestine. 2nd ed. T. Munby and Co., London.
1950 Man-made dune encroachment on Israel's coast. International Congress of Soil Science, 4th, Amsterdam, 1950, 1:325-327.

RIM, M.
1948 The movement of dunes and the origin of red sand in Palestine from a physical point of view. Hebrew University, Jerusalem. (unpublished thesis)
1950 Sand and soil in the coastal plain of Israel. Israel Exploration Journal 1(1).
1951 The influence of geophysical processes on the stratification of sandy soils. Journal of Soil Science 2:188-195.
1953 Les classements des minéraux du sable par les agents naturels sur les dunes. *In* Action éoliennes. Centre National de Recherches Scientifiques, Paris, Colloques Internationaux 35:261-276.
1958 Simulation, by dynamical model, of sand tract morphologies occurring in Israel. Research Council of Israel, Bulletin 7-G(2/3):123- .

ROSBY, C. G.
1943 Introduction to the conference and some applications of boundary layer theory to the physical geography of the Middle West. New York Academy of Sciences, Annals 44:3-12.

ROSENAN, E.
1953 Discussion of Bagnold (1953a). *In* Desert Research, Proceedings, International symposium held in Jerusalem, May 7-14, 1952, sponsored by the Research Council of Israel and Unesco. Research Council of Israel, Special Publication 2:94.
1954 The direction of seif dunes and wind in Sinai and Negev (in Hebrew). 4 pp.

RUTTEN, M. G.
1954 Deposits of cover sand and loess in the Netherlands. Geologie en Mijnbouw, n.s. 16:127-129.

SANDFORD, K. S.
1953 Notes on sand-dunes and artesian water in Egypt and Sudan. Geographical Journal 119: 363-366.

SAWICKI, L.
1958 Le problème de l'âge des dunes (translated title; French summary). I:53-71. *In* R. Galon, ed., Wydmy Sródladowe Polski. PWN, Warszawa.

SCHELLING, J.
1957 Herkunft, Aufban und Bewertung der Flugs aude Binnelande. Erdkunde 11:129-135.

SCHOELLER, H.
1945 Le Quaternaire de la Saoura et du Grand Erg Occidental. Institut de Recherches Sahariennes, Travaux 3:57-

SCHOENEICH, K.
1958 Remarks on the morphology of dunes in the vicinity of Warsaw. Przeglad Geologiczny 6: 40-42.

SCHONHALS, E.
1953 Gesetzmässigkerten in Feinaufban von Talrand löesse mit Bemerkungen über die Entstehung des Lösses. Eiszeitalter und Gegenwart 3:19-36.

SEVENET, LIEUT.
1943 Etude sur le Djouf (Sahara occidental). Institut Français d'Afrique Noire, Bulletin 5:1-26.

SHOTTEN, F. W.
1956 Some aspects of the New Red desert in Britain. Liverpool and Manchester Geological Journal 1:450-465.

SIDORENKO, A. V.
1956 Differentiation eolienne de la matière dans les déserts. Akademiia Nauk S.S.S.R., Izvestiya, seriya Geograficheskaii 3:3-22. [*See* analysis by J. Tricart, Revue de Géomorphologie Dynamique, 1958, 9(1/2):29]

SIMMONET, D. S.
1949 Sand dunes near Castlereagh, New South Wales. Australian Geographer 5:3-10.
1951 On the grading of dune sands near Castlereagh, New South Wales. Royal Society of New South Wales, Journal and Proceedings 84:71-79.

SIMONS, F. S.
1956 A note on Pur-Pur dune, Virú Valley, Peru. Journal of Geology 64:517-521.

SIMONS, F. S., and G. E. ERICKSEN
1953 Some desert features of northwest central Peru. Sociedad Geológica del Peru, Boletín 26:229-246.

SMITH, H. T. U.
1940a Geological studies in southwestern Kansas. Kansas State Geological Survey, Bulletin 34: 153-168.
1940b Review of "A tentative classification of sand dunes" by F. A. Melton. Journal of Geomorphology 3:359- .
1942 Sand dune stratification. Geological Society of America, Bulletin 53:1852. (Abstr.)
1946 Sand dunes. New York Academy of Sciences, Transactions, ser. 2, 8:197-199.
1949a Dune forms in western Nebraska. Geological Society of America, Bulletin 60:1920. (Abstr.)

1949b Physical effects of Pleistocene climatic changes on non-glaciated areas—aeolian, frost and streams. Geological Society of America, Bulletin 60:1485-1516.

1951 Photo-interpretation studies in the sand hills of Nebraska. Naval Research Project NR089-016, Lawrence, Kansas. 49 pp.

1953 Classification of sand dunes. International Geological Congress, 19th, Algiers, 1952, fasc. 7:103. (Abstr.)

1954 Eolian sand on desert mountains. Geological Society of America, Bulletin 65:1036-1037. (Abstr.)

1956a Giant composite barchans of the northern Peruvian desert. Geological Society of America, Bulletin 67:1735. (Abstr.)

1956b Use of aerial photography for interpretation of dune history in Nebraska, U. S. A. Congrès International du Quaternaire, 4th, Rome-Pisa, 1953, Actes:152-158.

1959 Coastal sand dunes in Oregon and Washington. [A review of Cooper (1958), q.v.] Geographical Review 50:113-115.

SOARES DE CARVALLO, G.
1952 Les époques d'éolisation du Pleistocene dans la bordure Meso-cenozoique du Portugal. Universidade de Coimbra, Laboratorio Mineralogico Geologico, Mémorias e Notícias 33:53-58.

STANNARD, M. E.
1959 Wind studies in western New South Wales. New South Wales Soil Conservation Service, Journal 15(1):25- .

STEPHENS, C. G., and R. L. CROCKER
1946 Composition and genesis of lunettes. Royal Society of South Australia, Transactions 70: 302-312.

STRIEM, H. L.
1954 The seifs on the Israel-Sinai border and the correlation of their alignment. Research Council of Israel, Bulletin 4(2):195-198.

TAMHANE, V. A.
1952 Soils of the Rajputana and Sind deserts. National Institute of Sciences of India, Bulletin 1.

THESIGER, W.
1949 A further journey across the Empty Quarter. Geographical Journal 113:21-46.

TING, W. S.
1958 Geomorphology of the Tarim Basin. American Association of Geographers, Annals 48:293. (Abstr.)

TRICART, J.
1953 Géomorphologie dynamique de la steppe russe. Revue de Géomorphologie Dynamique 4:1-32.

1954 Une forme de relief climatique, les sebkhas. Revue de Géomorphologie Dynamique 5(3): 97-191.

1955a Aspects sédimentologiques du delta du Sénégal. Geologische Rundschau 43:384-397.

1955b Aspects sédimentologiques du delta du Sénégal. Revue de Géomorphologie Dynamique 6:145.

1955c Notes géomorphologiques sur les environs d'Atar (Mauritanie). Institut Français d'Afrique Noire, Bulletin, sér. A, 17:325-337.

1955d Présentation d'une carte géomorphologique du delta du Sénégal. Association de Geographes Français, Bulletin 251-252:98-117.

1956 Aspects géomorphologiques du delta du Sénégal. Revue de Géomorphologie Dynamique 5/6: 65-84.

1959 Géomorphologie dynamique de la moyenne

vallée du Niger. Annales de Géographie 68: 333-343.

TRICART, J., and M. BROCHU
1955 Le grand erg ancien du Trarza et du Cayor. Revue de Géomorphologie Dynamique 4: 145-176.

URVOY, Y.
1942 Les bassins du Niger, étude de géographie physique et paleogéographie. Institut Français d'Afrique Noire, Mémoire 4. 139 p.

VAN DER MERWE, C. R.
1954 Kalahari and Sahara sandy soils. International Congress of Soil Science, 5th, Leopoldville, Actes et Comptes Rendus 4:117-123.

VERLAQUE, C.
1958 Les dunes d'In Salah. Institut de Recherches Sahariennes, Travaux 17:12-58.

VINCENT-CUAZ, L.
1958 Les barkhanes de Mauritanie ont'elles toujours existe? Bulletin de Liaison Sahariennes 10(30): 141-147.

WALTHER, W.
1951 L'influence des facteurs physiques sur la morphologie des sables éoliens et des dunes. Revue de Géomorphologie Dynamique 2(6):242- . (A. Cailleux comments on this, Ibid. 1952, 3:99)

WAYLAND, E. J.
1953 More about the Kalahari. Geographical Journal 119(1):49-56.

WELLINGTON, J. H.
1955 The physical geography of South Africa. Cambridge University Press.

WHINCUP, S.
1944 Superficial sand deposits between Brighton and Franciston, Victoria. Royal Society of Victoria, Proceedings 52:315-332.

WIRTH, E.
1958 Morphologische und bodenkundliche Beobachtungen in der syrischirakischen wüste. Erdkunde 12:26-42. [Analysis and critique by J. Tricart in Revue de Géomorphologie Dynamique (1958), 9(1/2):29]

WRIGHT, H. E., JR.
1956 Origin of the Chuska sandstone, Arizona-New Mexico: a structural and petrographic study of a Tertiary eolian sediment. Geological Society of America, Bulletin 67:413-434.

WRIGHT, J. W.
1945 War-time exploration with the Sudan defence force in the Libyan Desert. Geographical Journal 105(3-4):100-111.

(d) 1960-1967

ABU BAKR, M.
1963 Physiography of the Changai-Kharan region, West Pakistan. Pakistan Geographical Review 18(2):1-12.

ALIMEN, H.
1965 The Quaternary Era in the northwest Sahara. In H. E. Wright, Jr. and D. G. Fray, eds., International studies on the Quaternary. Geological Society of America, Special Papers 84:273-291.

ALLIER, C.
1966 Formation et évolution d'une dune continentale au champ Minette (forêt de Fontainebleau). Revue de Géomorphologie Dynamique 16: 101-113.

ANONYMOUS
1963 Obituary: V. A. Doubiansky, sand desert geog-

rapher. Vsesoyuznogo Geografisheskogo Obshchestva, Izvestiya 2:191-   .

BAN, A., E. BRĂESCU and Z. GAFENOU
1964 The importance of fluvial and aeolian processes in forming relief in the Danube valley. International Geographical Congress, 20th, London, 1964, Abstracts of Papers, p. 86.

BAWDEN, M. G., and A. R. STOBBS
1963 The land resources of eastern Bechuanaland. Directorate of Overseas Surveys, Forestry and Land Use Section. 93 p.

BAYROCK, L. A., and G. M. HUGHES
1962 Surficial geology of the Edmonton District. Research Council of Alberta, Preliminary Report 62-6. 40 p.

BERTIN, E.
1964 Compte rendu d'une mission pour l'étude des phénomènes ondulatoires dans le sable au Sahara. Institut de Recherches Sahariennes, Travaux 23:181-185.

BETTENAY, E.
1962 The salt lake systems and their associated aeolian features in the semi-arid regions of Western Australia. Journal of Soil Science 13:11-17.

BIGARELLA, J. J., and R. SALAMUNI
1961 Early Mesozoic wind patterns as suggested by dune bedding in the Botucatú sandstone of Brazil and Uruguay. Geological Society of America, Bulletin 72:1089-1106.

BLACKBURN, G. et al.
1965 Soil development associated with stranded beach ridges in south-east South Australia. CSIRO, Soil Publication 22. 66 p.

BOSAZZA, V. L.
1962 The Kalahari system with particular reference to its occurence on the Macondes Plateau, northern Mozambique. Congrès Panafricain du Préhistoire et de l'Etude du Quaternaire, 4ᶜ, Leopoldville, 1959, Actes, p. 167-176.

BOWLER, J. M., and L. B. HARFORD
1963 Geomorphic sequence of the riverine plain near Echuca. Australian Journal of Science 26(3): 88-   .

BROGGI, J. A.
1961 Las ciclópeas dunas compuestas de la costa peruana, su origen y significación climática. Sociedad Geológica del Peru, Boletín 36:61-66.

BROWN, G. F.
1960 Geomorphology of western and central Saudi Arabia. International Geological Congress, 21st, Copenhagen, 1960, Report 21:150-159.

CAPOT-REY, R.
1963 Contribution à l'étude et à la représentation des barkhanes. Institut de Recherches Sahariennes, Travaux 22:37-60.

1965 Remarques sur quelques sables saharien. Institut de Recherches Sahariennes, Travaux 23:153-163.

CHENG JO-AI
1963 Studies on the characteristics of sweat in the sand-dunes of desert areas, Central Kansu (translated title). Acta Pedalogica Sinica 11(1): 84-91.

CHENOWITH, L., and M. E. COOLEY
1960 Pleistocene cinder dunes near Cameron, Arizona. Plateau 33:14-16.

CHEPIL, W. S.
1965 Function and significance of wind in sedimentology. U. S. Department of Agriculture, Miscellaneous Publication 970:89-94.

CHMIELEWSKA, M., and W. CHMIELEWSKI
1960 The stratigraphy and chronology of the Witow dune, Leczyca (translated title). Biuletyn Peryglacjalny 8:133-141.

CHMIELEWSKA, M., and K. WASYLIKOWA
1961 Witów: Late-Pleistocene dunes and peat-bogs. International Congress on Quaternary, 6th, Warsaw, 1961, Guide-Book of Excursion C, the Łódź region, p. 75-84.

CHURCHWARD, H. M.
1961 Soil studies at Swan Hill, Victoria, Australia. I: Soil layering. Journal of Soil Science 12:73-86.

1963 Soil studies at Swan Hill, Victoria, Australia. II: Dune moulding and parna formation. III: Some aspects of soil development on aeolian material. IV: Groundsurface history and its expression in the array of soils. Australian Journal of Soil Research 1:103-116, 117-128, 242-255.

CLAYTON, W. D.
1966 Vegetation ripples near Gummi, Nigeria. Journal of Ecology 54:415-417.

CLOS-ARCEDUC, A.
1967 La direction des dunes et ses rapports avec celle du vent. Académie des Sciences, Paris, Comptes Rendus, ser. D (11), 264:1393-1396.

COALDRAKE, J. E.
1963 The coastal sand-dunes of southern Queensland. Royal Society of Queensland, Proceedings 32: 101-116.

CORRELL, R. L., and R. T. LANGE
1963 Significant trends of surface lime in coastal dunes on Younghusband Peninsula. Australian Journal of Science 26(2):59-60.

DAVEAU, S.
1965 Dune ravinés et dépôts du Quaternaire Récent dans le Sahel Mauritanien. Revue de Géographie de l'Afrique Occidentale 1-2:7-47.

DINGMAN, R. J.
1962 Tertiary salt dunes near San Pedro de Atacama. U. S. Geological Survey, Professional Paper 450-D:92-94.

DRESCH, J.
1961 Observations sur le désert côtier du Pérou. Annales de Géographie 70:179-184.

DRESCH, J., and G. ROUGERIE
1960 Morphological observations in the Sahel of the Niger. Revue de Géomorphologie Dynamique 11:49-58.

DYLIKOWA, A.
1964 Les dunes de Pologne centrale et leur importance pour la stratigraphie du Pleistocene Tardif. International Congress on Quaternary, 6th, Warsaw, 1961, Report 4:67-80.

EARDLEY, A. J.
1962 Gypsum dunes and evaporite history of the Great Salt Lake Desert. Utah Geological and Mineralogical Survey, Special Studies 2. 27 p.

ERINC, S.
1962 On the relief features of blown sand at the Karapinar surroundings in the Interior Anatolia. University of Istanbul, Geographical Institute, Review 8:113-130.

EVANS, G. C.
1963 Geology and sedimentation along the lower Rio Salado in New Mexico (summary). New Mexico Geological Society, Guidebook to the Socorro Region, p. 209-216.

EVANS, J. R.
1962 Falling and climbing sand-dunes in the Cronese ("Cat") Mountain area, San Bernardino County, California. Journal of Geology 70:107-113.

EVERARD, C. E.
1964 Playas and dunes in the Estancia Basin, New Mexico. International Geographical Congress, 20th, London, Abstracts of Papers, pp. 89-90.

FEDOROVICH, B. A.
1963 Les types dynamiques du relief des sables comme fondement scientifique de la lutte contre les sables. Société Hellenique de Géographie, Bulletin 4:162-171.

FERNALD, A. T.
1964 Surficial geology of the Kobuk River Valley, Alaska. U. S. Geological Survey, Bulletin 1181-K. 31 p.

FINKEL, H. J.
1961 The movement of barchan dunes measured by aerial photogrammetry. Photogrammetric Engineering 27:439-444.

FRANKEL, J. J., and D. A. SCOGINGS
1960 The air photographic record of a destertina and dongas on the north coast, Natal, South Africa. Revue de Géomorphologie Dynamique 11:113-118.

FUJITA, T.
1967 Note on sand dunes. In U. S. Army Corps of Engineers / NASA, Earth resource surveys from spacecraft. v. 2, Appendix 1.

FÜRST, M.
1965 Hammada-Serir-Erg. Zeitschrift für Geomorphologie n.f. 9(4):385-421.

GABRIEL, A.
1965 Die Auswirkung vertikaler Luftströmungen und elektrischer spannungsfelder in kahlen sanden. Neue Gedanken zur Dünemorphologie als Diskussionsarbeitrag. Oesterreichische Geographische Gesellschaft, Mitteilungen 107(3):125-137.

GAY, P., JR.
1962 Origen, distribución y movimento de las arenas eólicas en el área de Yauca a Palpa. Sociedad Geológica del Peru, Boletín 37:37-58.

GAYELL, A. G., and A. A. TRISHKOVSKLY
1962 Soil age and classification of eolian sands of the steppe zone. Akademiia Nauk SSSR, Izvestiya, seriya Geograficheskaya 3:28-

GILE, L. H.
1966 Coppice dunes and the Rotura soil. Soil Science Society of America, Proceedings 30:657-660.

GREEN, F. E.
1961 The Monahans Dunes area. In F. Wendorf, ed., Paleoecology of the Llano Estacado. Fort Burgwin Research Center, Publication 1:22-47.

GRIPP, K.
1961 Uber Werden und Vergehen von Barchanen an der Nordseeküste Schleswig-Holsteins. Zeitschrift für Geomorphologie, n.f. 5:24-36.

GROMMELIN, R. D.
1965 Sediment-petrologie en herkomst van Jeng-Pleisteseen deksand in Nederland. Boor en Spade 14:138-150.

GROVE, A. T.
1960a Note following Prescott and White (1960), q.v.
1960b The geomorphology of the Tibesti Region. Geographical Journal 126:18-31.

GROVE, A. T., and R. A. PULLAN
1963 Some aspects of the Pleistocene paleo-geography of the Chad Basin. p. 230-245. In F. C. Howell and F. Bourlière, eds., African ecology and human evolution. Aldine Publishing Co., Chicago. 666 p.

HASTENRATH, S. L.
1967 The barchans of the Arequipa region, southern Peru. Zeitschrift für Geomorphologie 11:300-331.

HOLM, D. A.
1960 Desert geomorphology in the Arabian peninsula. Science 123:1369-1379.

HOYT, J. H.
1966 Air and sand movements in the lee of dunes. Sedimentology 7:137-144.

HSU CHUN-MIN
1965 The sources of dune sand in the region east of the Yellow River in Ninghsia (translated title). Acta Geographica Sinica 31:142-156.

INMAN, D. L., G. C. EWING, and J. B. CORLISS
1966 Coastal sand dunes of Guerero Negro, Baja California, Mexico. Geological Society of America, Bulletin 77:787-802.

JENNINGS, J. N.
1967 Cliff-top dunes. Australian Geographical Studies 5:40-49.

JOHNSON, R. B.
1967 The Great Sand Dunes of Colorado. U. S. Geological Survey, Professional Paper 575-C:177-183. Also in the Rocky Mountain Geologist, 1968.

JOHNSTONE, W. M., and J. C. WILKINSON
1960 Some geographical aspects of Qatar. Geographical Journal 126:442-450.

JORDAN, W. M.
1965 Prevelance of sand dune types in the Sahara. Geological Society of America, Special Publication 82:104-105. (Abstr.)

KADAR, L.
1966 Natural systems of eolian land forms (translated title). Földrajzi Ertesito 15:413-446.

KELLY, R. W.
1962 Michigan's sand dunes, a geologic sketch. Michigan Geological Survey Division. 22 p. (Expanded from Michigan Conservation, 1962, 31(4):10-16)

KING, D.
1960 The sand ridge deserts of South Australia and related aeolian landforms of Quaternary arid cycles. Royal Society of South Australia, Transactions 83:99-108.

KOBENOLZINA, J.
1961 Attempt to date dunes in the Kampinos primaeval forest (translated title). Przeglad Geograficzny 33:383-399. "Accompanying phenomena." Ibid. pp. 539-542.

LAMBERT, E. H., JR.
1964 Sand mounds of Livingstone and Tangipahoa parishes, Louisiana. Geological Society of America, Special Paper 82:303. (Abstr.)

LÁNG, S.
1964 Pleistocene climatic changes and evolution of relief. International Geographical Congress, 20th, London, Abstracts of Papers, pp. 66-67.

LEONTEV, O. K., and N. I. FOTEYEVA
1965 Proiskozhdeniye i vozrast Berovskikh bugrov. Akademiia Nauk S.S.R., Izvestiya, seriya Geograficheskaya 2:9-98.

LEWIS, P. F.
1960 Linear topography in the southwestern Palouse,

Washington-Oregon. Association of American Geographers, Annals 50(2):98-111.

LI HSIAO-FANG
1965 The genesis and development of stabilized sand dune soils in central eastern part of the Moyusu Desert of the Ordos plateau (translated title). Acta Pedologica Sinica 13(1):66-76.

LONG, J. T., and R. P. SHARP
1964 Barchan-dune movement in the Imperial Valley, California. Geological Society of America, Bulletin 75:149-156.

LOVELL, H. L.
1967 Geology of the Matachewan area. Ontario Department of Mines, Geological Report 51. 61p.

LUGN, A. L.
1962 The origin and sources of loess in the central plains and adjoining areas of the central Lowland. University of Nebraska Studies, n.s. 26. 105 p.

MAARLEVELD, G. C.
1960 Wind directions and cover sands in the Netherlands. Biuletyn Peryglacjalny 8:49-58.

MABBUTT, J. A.
1961 A stripped land surface in Western Australia. Institute of British Geographers, Transactions 29:101-114.
1962 Geomorphology of the Alice Springs area. In R. A. Perry et al, General report on lands of the Alice Springs area, Northern Territory, 1956-57. CSIRO Land Research Series 6:163-184.
1963 Wanderrie banks: micro-relief patterns in semi arid Australia. Geological Society of America, Bulletin 74:529-540.
1967 Denudation chronology in central Australia. p. 144-181. In J. N. Jennings and J. A. Mabbutt, eds., Land form studies in Australia and New Guinea. Cambridge University Press, Cambridge.

MABBUTT, J. A. et al
1963 General report on the lands of the Wiluna-Meekatharra area, Western Australia, 1958. CSIRO Land Research Series 7. 215 p.

MATHER, K. D., and G. S. MILLER
1966 Wind drainage of the High Plateau of Antarctica. Nature 209:281-284.

MACARTHUR, W. M.
1962 Development and distribution of soils of the Swan Coastal Plain (Western Australia). CSIRO, Soil Publication 16.

MCBRIDE, E. F., and M. O. HAYES
1962 Dune cross-bedding on Mustang Island, Texas. American Association of Petroleum Geologists, Bulletin 46:546-551.

MCCOY, F. W., JR., W. J. NOKLEBERG and R. M. NORRIS
1967 Speculations on the origin of the Algodones dunes southern California. Geological Society of America, Bulletin 78:1039-1044.

MCKEE, E. D.
1966 Structures of dunes at White Sands National Monument, New Mexico. Sedimentology 7:1-68.

MCKEE, E. D., and G. C. TIBBITTS, JR.
1964 Primary structures of a seif dune and associated deposits in Libya. Journal of Sedimentary Petrology 34:5-17.

MECKELEIN, W.
1960 Forschungen in der Zentralen Sahara: Fezzan. G. Westermann, Braunschweig.

MERK, G. P.
1960 Great sand dunes of Colorado. p. 127-129. In R. J. Weimer and J. D. Haun, eds., Guide to the geology of Colorado. Geological Society of America, N. Y.

MOLNAR, B.
1961 Die Verbreitung der Holischen Bildungen an der Oberflache und untertags im Swischenstromland von Danan und Theiss. Földtani Közlöny 91(3):300-315.

MONOD, T.
1961 Majâbat-al-Koubrâ (supplément). Institut Français d'Afrique Noire, Bulletin 23:591-637.
1962 Notes sur le Quaternaire de la region Tazzmout-El Bayyel (Adrar de Mauritanie). Congrès Panafricain du Préhistoire et de l'Etude du Quaternaire, 4e, Leopoldville, 1959, Actes, p. 172-188.

MORRISON, A., and M. C. CHOWN
1964 Photography of the western Sahara from the Mercury MA-4 spacecraft. NASA Contract No. NA Sr-140. McGill University. 125 p.

MOSELEY, F.
1965 Plateau calcrete, calcreted gravels, cemented dunes and related deposits of the Ma'allegh-Bomba region of Libya. Zeitschrift für Geomorphologie 9(2):167-185.

MYCIELSKA-DOWGIALLO, E.
1965 Mutual relation between loess and dune accumulation in southern Poland. Geographia Polonica 6:105-115.

MUKERJI, A. B.
1961 Morphogenetic nature of the Bhur. Indian Geographical Journal 36(2):53.

NEVYAZHSKIY, I. I., and R. A. BIOZHIEV
1960 Aeolian relief forms in Central Yakutia (translated title). Akademiia Nauk S.S.R., Izvestiya, ser. Geograficheskaya 3:90-

NIKIFOROV, L. G.
1960 The existence of an outlet of the Uzboy through the Abzhaib. Moskovskogo Universiteta, Vestnik, seriya Geograficheskaya May–June, p. 69- .

NICOLAEV, V. A.
1960 An analysis of the structure of steppe and semi-desert landscapes with aerial photographs. Akademiia Nauk S.S.R., Izvestiva, seriya Geograficheskaya, March–April, p. 82- .

NORRIS, R. M.
1966 Barchan dunes of Imperial Valley, California. Journal of Geology 74:292-306.

NORRIS, R. M., and K. S. NORRIS
1961 Algodones dunes of southeastern California. Geological Society of America, Bulletin 72(1): 605-620.

O'CONNOR, M. P.
1963 A summary of the Paleoclimates Conference, January 7–12, 1963, in Newcastle-upon-Tyne, England. Compass 40(3):155-160.

OPDYKE, N. D., and S. K. RUNCORN
1960 Wind direction in the western United States in the Late Paleozoic. Geological Society of America, Bulletin 71:959-972.

PEEL, R. F.
1960 Some aspects of desert geomorphology. Geography 45:241-262.
1966 The landscape in aridity. Institute of British Geographers, Transactions and Papers, Publication 38. 23 p.

PELISEK, J.
1963 Pleistozane Dünensande in der Tschekoslovakischen Sosialistischen Republik. Eiszeitalter und Gegenwart 14:216-223.

PERNAREWSKI, L.
1960 Application of statistical methods in investigating dune forms. Przeglad Geograficzny 32: 57-66. (Suppl.)
1962 O prosesach wydmocworczych w sweitte badan urtwalonych form wydmouch Delnego Slaska. Czasopismo Geograficzne 33:175-197.
1966 Glacjalna i postglacjalna cyrkulacja atmosfery w swietle kierunku wiatrow wy dmotworczych. (Glacial and postglacial atmospheric circulation in the light of directions of duneforming winds) Czaspismo Geograficzne 37(1):3-24.

PERRET, R.
1961 Images sahariennes. Acta Geographica, Paris, 39:3- .

PETROV, M, P.
1960 Geograficheskie issledovaniya v pustynyakh Tsentral'noi Azii (Geographic explorations in the deserts of Central Asia). Leningradskogo Universiteta, Vestnik 24, seriya Geologii i Geografii 4:118-130. Translation available as JPRS 9221.
1962 On grain size and mineral composition of sands in east Central Asian deserts (translated title). Leningradskogo Universiteta, Vestnik 12, seriya Geologii i Geografii 2:65- .

POOLE, F. G.
1962 Wind directions in late Paleozoic to Middle Mesozoic time on the Colorado Plateaus. U. S. Geological Survey, Professional Paper 450D:147-151.

POWERS, R. W. et al.
1966 Geology of the Arabian Peninsula: Sedimentary geology of Saudi Arabia. U. S. Geological Survey, Professional Paper 560-D. 147 p.

PRESCOTT, J. R. V., and H. P. WHITE
1960 Sand formations in the Niger Valley between Niamey and Bourem. Geographical Journal 126: 200-203.

PRICE, W. A.
1962 Stages of oxidation coloration in dune and barrier sands with age. Geological Society of America, Bulletin 73(10):1281-1283. (Abstr.) (Comment by R. M. Norris and S. K. S. Norris, Ibid., p. 1285)
1963 Physico-chemical and environmental factors in clay dune genesis. Journal of Sedimentary Petrology 33:766-778.
1964 The sand ridge deserts of Australia. Geographical Review 54:118-120. (Review)

PULLAN, R. A.
1962a A report on the reconnaissance soil survey of the Azare (Bauchi) area with special reference to the establishment of an experimental farm and the detailed soil survey of the N. A. farm, Azare. Ministry of Agriculture, Regional Research Station Samaru, Zaria, Nigeria, Soil Survey Section S, no. 19.
1962b A report on the reconnaissance soil survey of the Nguru-Hadejia-Gumel area with special reference to the establishment of an experimental farm. Ministry of Agriculture, Regional Research Station Samaru, Zaria, Nigeria, Soil Survey Section B, no. 19.

REEVES, C. C., JR.
1965 Chronology of west Texas pluvial lake dunes. Journal of Geology 73:504-508.

ROTH, E. S.
1960 The silt-clay dunes at Clark dry lake, California. Compass 38:18-27.

RÜHLE, E.
1961 Geomorphology (Poland). Review in Czwartorzed Europy Sridkowji i Wschodnief, p. 623-673.

SCHEIDEGGER, A. E.
1961 Theory of aeolian features. p. 287-291. In his Theoretical Geomorphology. Springer Verlag, Berlin. 333 p.

SEGERSTROM, K.
1962 Deflated marine terrace as a source of dune chains, Atacama Province, Chile. U. S. Geological Survey, Professional Paper 450-C:91-93.
1964 Quaternary geology of Chile, brief outline. Geological Society of America, Bulletin 75(3):157-170.

SELIVANOV, Y. I.
1961 Forms of eolian sand accumulations in the western part of Central Asia. Moskovskogo Universiteta, Vestnik (March–April).

SEN, A. K.
1967 Photo-interpretation to study arid zone geomorphology. In Symposium International de Photo-Interprétation, II, Groupe IV.1, Institut Français du Pétrole, Revue 21(12):1903-1906.

SHARP, R. P.
1962 Measurements on desert dunes – a testing of some concepts. Geological Society of America, Special Paper 73:238-239.
1966 Kelso dunes, Mojave Desert, California. Geological Society of America, Bulletin 77:1045-1074.

SIMONETT, D. S.
1960 Development and grading of dunes in western Kansas. Association of American Geographers, Annals 50:216-241.

SMITH, D. D., and R. E. SNEAD
1961 Thick eolian sand prism of probable Middle to Late Pleistocene age near Karachi, West Pakistan. Geological Society of America, Special Papers 68, p. 274. (Abstr.)

SMITH, H. T. U.
1963 Eolian geomorphology, wind direction, and climatic change in North Africa. U. S. Air Force, Cambridge Research Laboratories, Contract No. AF 19(628)-298. (Also cited as AD-405 144)
1964 Periglacial eolian phenomena in the U. S. International Congress on Quaternary, 6th, Warsaw, 1961, Report 4:177-186.

SNEAD, R. E.
1966 Physical geography reconnaissance, the Las Bela coastal plain, West Pakistan. Louisiana State University, Coastal Studies Series 13. 118 p.
1967 Note in U. S. Army Corps of Engineers / NASA, Earth resource studies from spacecraft, v. II.

SQUIRES, D. E.
1963 Carbon 14 dating of the fossil dune sequences Lord Howe Island. Australian Journal of Science 25(9):412-413.

STOKES, W. L.
1964 Incised, wind-aligned stream patterns of the Colorado Plateau. American Journal of Science 262(6):808-816.

STONE, R. O.
1967 A desert glossary. Earth Science Reviews 3:211-268.

STRAW, A.
1963 Some observations on the 'cover sands' of north Lincolnshire. Lincolnshire Naturalists Union, Transactions 15(4):260-269.

SUSLOV, S. P.
1961   The desert regions of Central Asia. *In* S. P. Suslov, The physical geography of Asiatic Russia. W. H. Freeman and Co., San Francisco, London. 594 p.

SWAN, L. W.
1962   Eolian zone. Science 140:77-79.

TADA, F.
1963   Geomorphological study of sand dunes in the Kunshan desert, Inner Mongolia. Société Hellénique de Géographie, Bulletin 4:172-173. (Abstr.)

TRAINER, F. W.
1961   Eolian deposits of the Matanuska Valley agricultural area, Alaska. U. S. Geological Survey, Bulletin 1121-C. 34 p.

TRICART, J.
1961   Le modelé du Quadulatero Ferifero, Sud de Belo Horizonte, Brésil. Annales de Géographie 70: 255-272.

1965   Reconnaisance géomorphologie de la moyenne valle du Niger. Institut Français d'Afrique Noire, Mémoire 72, 196 p.

1966   Un chott dans le désert chilien: la pampa del Tamarugal. Revue de Géomorphologie Dynamique 16:12-22.

TRICART, J., and A. CAILLEUX
1962-   Le modelé des régions séches. Centre de Docu-
1963   mentation Universitaire, Paris. 2 vols. 129, 179 p.

TRICART, J., T. CARDOSO DA SILVA, and M. BROCHU
1960   Etude géomorphologique du projet d'aménagement du lac Faguibine (République du Mali). Sols Africains 5(3):207-289.

TRICART, J., and M. MAINGUET
1965   Caractéristiques granulometriques de quelques sables eoliens du desert Péruvien; aspects de la dynamique des barkanes. Revue de Geomorphologie Dynamique 15:110-121.

URBANIAK, U.
1962   The structure of a dune in Goren Duzy (translated title; English summary). Przeglad Geograficzny 34:743-758.

1966   Sklad mineralny piaskow wydmowych w Kotlinie Plockiej. Przeglad Geograficzny 38:435-453.

VEJISOV, S.
1966   The mechanics of the formation of barchan chains, from experimental material (translated title). Akademiia Nauk SSR, Isvestiya, seriya Geograficheskaya 3:66-70.

WALKER, H. J.
1967   Riverbank dunes in the Colville delta, Alaska. Louisiana State University Coastal Studies Institute, Coastal Studies Bulletin 1:7-14. (Also cited as *its* Technical Report 36)

WANG AN-CHI
1960   Quaternary research in China (in Russian, translated from the Chinese). Akademiia Nauk S.S.R., Izvestiya, seriya Geograficheskaya 2:123-126.

WARREN, A.
1964   The dunes of Kordofan. Hunting Group Review 3:5-9.

1966   The Qoz region of Kordofan. Cambridge University Library (unpublished Ph.D. thesis)

WEBB, B. P., and H. WOPFNER
1961   Plio-Pleistocene dunes north-west of Lake Torrens, South Australia, and their influence on the erosional pattern. Australian Journal of Science 23:379-381.

WEIR, J. E.
1962   Large ripple marks caused by the wind near Coyote Lake (dry), California. Geological Society of America, Special Paper 73:72. (Abstr.)

WILCOXON, J. A.
1962   Relationship between sand ripples and wind velocity in a dune area. Compass 39:65-76.

WILLIAMS, M. A. J., and D. N. HALL
1965   Recent expeditions to Libya from the Royal Military Academy, Sandhurst. Geographical Journal 131:482-501.

WILSON, I.
1967   The nature and development of sand seas. University of Reading (unpublished M.Sc. thesis)

WOPFNER, H., and C. R. TWIDALE
1967   Geomorphological history of the Lake Eyre Basin. p. 118-143. *In* J. N. Jennings and J. A. Mabbutt, eds., Landform studies in Australia and New Guinea. Cambridge University Press, Cambridge.

## A BIBLIOGRAPHY OF SAND MOVEMENT BY WIND (until 1967)

AMERICAN SOCIETY OF CIVIL ENGINEERS, HYDRAULICS DIVISION, COMMITTEE ON SEDIMENTATION, TASK COMMITTEE ON PREPARATION OF SEDIMENTATION MANUAL
1965   Sediment transportation mechanics: wind erosion and transportation; progress report. Journal 91, HY 2(1):267-287.

BAGNOLD, R. A.
1935   The movement of desert sand. Geographical Journal 85:343-

1936   The movement of desert sand. Royal Society, London, Proceedings, ser. A, 157:594.

1937a  The transport of sand by wind. Royal Society, London, Proceedings, ser. A, 163:250-264.

1937b  The transport of sand by wind. Geographical Journal 89:409-438.

1938   The measurement of sand storms. Royal Society of London, Proceedings, ser. A, 167:282-291.

1941   The physics of blown sand and desert dunes. Methuen, London. (reprinted 1954, 1960)

BELLY, P.-Y.
1964   Sand movement by wind. U. S. Army Corps of Engineers, Coastal Engineering Research Center, Technical Memorandum 1.

CARROL, E.
1939   The movement of sand by wind. Geological Magazine 76:6-22.

CHEPIL, W. S.
1941a  Relation of wind erosion to the dry aggregate structure of a soil. Scientific Agriculture 21(17): 488-507.

1941b  Wind erosion of soil in relation to roughness of surface. Soil Science 52:417-431.

1942   Measurement of wind erosiveness of soils by the dry sieving procedure. Scientific Agriculture 25:154-160.

1943   Relation of wind erosion to the water-stable and dry clod structure of soil. Soil Science 55:275-287.

1945-   Dynamics of wind erosion. Soil Science 60:397-
1946   411; 61:167-177, 257-263, 331-340.

1950-   Properties of soil which influence wind erosion.
1952   Soil Science 69:149-162, 403-414; 71:141-153; 72:387-401, 465-478.

1956   Influence of moisture on erodability of soil by wind. Soil Science Society of America, Proceedings 20:288-292.

1957 Sedimentary characteristics of dust storms. American Journal of Science 225:12-22, 104-114, 206-213.

1958 Soil conditions that influence wind erosion. U. S. Department of Agriculture, Technical Bulletin 1185. 40 p.

1965 Function and significance of wind in sedimentology. U. S. Department of Agriculture, Miscellaneous Publication 970:89-94.

CHEPIL, W. S., and R. A. MILNE
1939 Comparative study of soil drifting in the field and in a wind tunnel. Scientific Agriculture 19:249- .

1941 Wind erosion of soils in relation to size and nature of the exposed area. Scientific Agriculture 21(7):479-487.

CHEPIL, W. S., and N. P. WOODRUFF
1963 The physics of wind erosion and its control. Advances in Agronomy 15:211-302.

CHEPIL, W. S., N. P. WOODRUFF, and A. W. ZINGG
1955 Field study of wind erosion in West Texas. U. S. Department of Agriculture, SCS-TP-125. 60 p.

CLEMENTS, T. et al.
1963 A study of windborne sand and dust in desert areas. U. S. Army Natick Laboratories, Technical Report ES-8. 61 p.

DUBIEF, J.
1943 Les vents de sable dans le Sahara français. Institut de Recherches Sahariennes, Travaux 2:11-35.

1952 Le vent et le déplacement du sable au Sahara. Institut de Recherches Sahariennes, Travaux 8:123-164.

1953 Les vents de sable dans le Sahara français. In Actions éoliennes. Centre National de Recherches Scientifiques, Paris, Colloques Internationaux 35: .

FÉLICE, P. DE
1956 Processus du soulèvement des grains de sable par le vent. Académie des Sciences, Paris, Comptes Rendus 242:920-923.

FORD, E. F.
1957 The transport of sand by wind. American Geophysical Union, Transactions 38:171-174. [Discussion by Bagnold, op. cit., 1958, 39(1):127-128]

FREE, E. C.
1911 The movement of soil material by wind. U. S. Bureau of Soils, Bulletin 68. 263 p.

GIDELIS, V., and V. MINKEVICIUS
1963 Lithodynamic spectra of sand drift in the coastal dunes of Lithuania (translated title). Baltica 1:211-232.

GILL, E. W. B.
1948 Frictional electrification of sand. Nature 162:568-569.

HEWITT, B. R.
1954 Coastal sand drift investigations in New South Wales. Soil Conservation Service of New South Wales, Journal 10:45- , 90- .

HOPKINS, E. S.
1935 Soil drifting in Canada. International Congress of Soil Science, 3rd, Oxford, 1935, Transactions 1:403-405.

HORIKAWA, K., and H. W. SHEN
1960 Sand movement by wind. U. S. Army Corps of Engineers, Beach Erosion Board, Technical Memorandum 119. 51 p.

JOHNSON, J. W.
1965 Sand movement on coastal dunes. U. S. Department of Agriculture, Miscellaneous Publication 970:747-755.

KALINSKE, A. A.
1943 Turbulence and the transport of sand and silt by wind. New York Academy of Science, Annals 44:41-54.

KAWADA, S.
1953 Quelques expériences sur l'entrainement du sable par le vent. In Actions éoliennes. Centre National de Recherches Scientifiques, Paris, Colloques Internationaux 35:109- .

KAWAMURA, R.
1951 Study on sand movement by wind. Institute of Science and Technology, Tokyo, Report 5 (314).

1953 Movement du sable sous l'effect du vent. In Actions éoliennes. Centre National de Recherches Scientifiques, Paris, Colloques Internationaux 35:117-151.

KREUTZ, W., and W. WALTER
1956 Stream flow, erosion processes and snow deposition at artificial wind breaks on the basis of studies in a wind tunnel: a contribution to the wind protection problem (translated title). Deutsche Wetterdienst, Berichte 4(24):25- .

KUENEN, P. H.
1960 Experimental abrasion. IV: Eolian action. Journal of Geology 68:427-449.

MacCARTHY, G. R., and J. W. HUDDLE
1938 Shape sorting of sand grains by wind action. American Journal of Science 35:64-73.

MARSLAND, J.
1937 A study of the effects of wind transport on several minerals. Journal of Sedimentary Petrology 13:18- .

O'BRIEN, M. P., and B. D. RINDLAUB
1936 The transportation of sand by wind. Civil Engineering 6:325-327.

PETITJEAN, L.
1937a Carte mensuelles de la répartition des vents de sables et des pluis au Sahara. Office National Météorologique de France, Mémoire 27.

1937b Généralités sur les vents de sable et pluies de Boue. Office National Météorologique de France, Mémoire 27.

RIM, M.
1951 The influence of geophysical processes on the stratification of sandy soils. Journal of Soil Science 2:188-195.

1953 Le classements des minéraux du sable par les agents naturels sur les dunes. In Actions éoliennes. Centre National de Recherches Scientifiques, Paris, Colloques Internationaux 35:259- .

ROUSE, H.
1940 Criteria for similarity in the transportation of sediment. University of Iowa Studies, Studies in Engineering, Bulletin 20:22-49.

SHARP, R. P.
1964 Wind driven sand in Coachella Valley, California. Geological Society of America, Bulletin 75:785-804.

STUNTZ, S. C.
1911 Bibliography of eolian geology. In E. E. Free, The movement of soil material by wind. U. S. Bureau of Soils, Bulletin 68:174-263.

WALTHER, W.
1951 L'influence des facteurs physiques sur la morphologie des sables éoliens et des dunes. Revue de Géomorphologie Dynamique 2(6):242-258. (A. Cailleux comments, Ibid., 1952, 3:99)

WILLIAMS, G.
1964 Some aspects of the eolian saltation load. Sedimentology 3(4):257-287.

ZINGG, A. W.
1949 A study of the movement of surface wind. Agricultural Engineering 30:11-13, 19- .
1953a Some characteristics of eolian sand movement by saltation process. *In* Actions éoliennes. Centre National de Recherches Scientifiques, Paris, Colloques Internationaux 35:197-208.
1953b Wind-tunnel studies of the movement of sedimentary material. University of Iowa, Studies in Engineering, Bulletin 34:111-135.

# A BIBLIOGRAPHY OF STUDIES OF AEOLIAN SAND (until 1967)

ALIMEN, H.
1940 Traces de l'action éolienne dans les sables auversiens du Bassin de Paris. Société Géologique de France, Bulletin sér. 5, 10(7/9):178-186.
1953a Caractéres granulométriques d'un dépôt effectué par le vent de sable à Beni-Abbès (Algérie). Societé Géologique dé France, Bulletin ser.6,t.3, Compte Rendu Sommaire 12:234-237.
1953b Variation granulométrique et morphoscopiques du sable le long de profil dunaires au Sahara occidental. *In* Actions éoliennes. Centre National de Recherches Scientifiques, Paris, Colloques Internationaux 35:217-225.

ALIMEN, H., and M. MERCIER
1951 Remaniements eoliens d'âge bartonien dans les sables de la partie orientale du Bassin de Paris. International Sedimentological Congress, 3rd, Groningen-Wageningen, 1951, Proceedings, pp. 25-42.

ALIMEN, H., and A. VATAN
1937 Contribution à l'étude pétrographique des sables stampiens. Société Géologique de France, Bulletin sér. 5, 7:141-162.

ALIMEN, H. *et al.*
1957 Sables quaternaire du Sahara nord-occidental (Saoura-Ougarta). Service de la Carte Géologique de l'Algérie, Bulletin n.s. 15. 207 p. (Critique *in* Revue de Géomorphologie Dynamique, 1958, p. 180- )

ALMEIDA, F. F. M. DE
1953 Botucatú, a Triassic desert of South America. International Geological Congress, 19th, Algiers, 1952, fasc. 7:9-24.

AMSTUTZ, G. C., and R. CHICO
1958 Sand size fractions of southern Peruvian barchans and a brief review of the genetic grain shape function. Vereinigung Schweizerischen Petroleum-Geologen-und-Ingenieure, Bulletin 24:47-52.

BARRET, W. H.
1930 The grading of dune sand by wind. Geological Magazine 67:159-162.

BEAL, M. A., and F. P. SHEPARD
1950 A use of roundness for determining depositional environments. Journal of Sedimentary Petrology 26:49-60.

BELLAIR, P.
1938 Les éléments lourds dans les sables désertiques. Académie des Sciences, Paris, Comptes Rendus 207:1054-1055.
1939 Sur la composition minéralogique des sables du Grand Erg Occidental. Société Géologique de France, Comptes Rendus 14:212-213.
1940a Les sables de la dorsale saharienne et du bassin de l'Oued Rhir. Service de la Carte Géologique de l'Algérie, Bulletin sér. 5, no. 5.
1940b Les sables du Souf (Algérie). Société Géologique de France, Bulletin sér. 5, v. 10, Compte Rendu Sommaire 7:75.
1941 Etude granulométrique de quelques formations arénacées du Gourara et la Saoura. Société d'Histoire Naturelle de l'Afrique du Nord, Bulletin 32:191-196.
1943 Les éléments lourds de quelques sables sahariens. Société d'Histoire Naturelle de l'Afrique du Nord, Bulletin 34.
1945 Les éléments lourds dans les sables de l'erg d'Oubari (Fezzan). Société Géologique de France, Bulletin ser. 5, t. 5, Compte Rendu Sommaire de Séance du 23 avril, 8:95-97.
1952 Sables désertiques et morphologie éolienne. International Geological Congress, 19th, Algiers, 1952, fasc. 7:113-118.

BONATTI, E., and G. P. S. ARRHENIUS
1965 Eolian sedimentation in the Pacific off northern Mexico. Marine Geology 3(5):337-348.

BOND, G.
1948 The direction of origin of the Kalahari sands of southern Rhodesia. Geological Magazine 85: 305-313.
1954 Surface textures of sand grains from the Victoria Falls area. Journal of Sedimentary Petrology 24:191-195.
1957 Quaternary sands at the Victoria Falls. *In* Pan-African Congress on Prehistory, 3rd, Livingstone, 1955, p. 115-122.

BOSAZZA, U. L.
1957 The Kalahari system in southern Africa and its importance in relationship to the evolution of man. *In* Pan-African Congress on Prehistory, 3rd, Livingstone, 1955, p. 127-132.
1962 The Kalahari system with particular reference to its occurrence on the Macondes Plateau, Northern Mozambique. Congrès Panafricain du Préhistoire et de l'Etude du Quaternaire, 4e, Leopoldville, 1959, Actes, p. 167-176.

BOURCART, J., and V. MALYCHEFF
1926 Premiers résultats de recherches sur les sables du Sahara. Société Géologique de France, Bulletin sér. 4, 26(4):191-208.

CAILLEUX, A.
1951 Interprétation climatique des éolisations pliocènes et quaternaires en France. Société Géologique de France, Bulletin sér. 6, Compte Rendu Sommaire 3:44-46.
1952a Observations à l'article de M. Walther sur les sables éoliens. Revue de Géomorphologie Dynamique 2:99.
1952b L'indice d'émousée des grains de sable et grès. Revue de Géomorphologie Dynamique 2:78-87.
1953 Les limons et loess éoliens de France. Service de la Carte Géologique, Paris, Bulletin 51:437-460.

CAPOT-REY, R.
1965 Remarques sur quelque sables sahariens. Institut de Recherches Sahariennes, Travaux 23:153-163.

CAPOT-REY, R., and F. CAPOT-REY
1948 Le déplacement des sables éoliens et la formation des dunes désertiques, d'après R. A. Bagnold.

Institut de Recherches Sahariennes, Travaux 5:47-80.

CARROL, D.
1944    Desert sands. Simpson Desert Expedition, 1939, Scientific Report 2: Geology. Royal Society of South Australia, Transactions and Proceedings 68(1):49-59.

CHAKRABARTI, A.
1965    Selective removal of sand in dune sediments. Geological, Mining and Metallurgical Society of India, Quarterly Journal 37:189-190.

CLEMENTS, T. et al
1963    A study of windborne sand and dust in desert areas. U. S. Army Natick Laboratories, Technical Report ES-8. 61 p.

CROMMELIN, R. D.
1964    A contribution to the sedimentary petrology and provenance of young Pleistocene cover sands in the Netherlands. Geologie en Mijnbouw 43:389-402.

DAPPLES, E. C.
1941    Surficial deposits of the deserts of Syria, Transjordan and Iraq. Journal of Sedimentary Petrology 11:124-141.

DEVILLERS, C.
1948    Les dépôts quaternaires de l'Erg Tihodaine (Sahara Central). Société Géologique de France, Bulletin sér. 5, t. 18, Compte Rendu Sommaire 10:189-191.

DUNHAM, K. C.
1952    Red coloration in desert formations of Permian and Triassic age in Britain. International Geological Congress, 19th, Algiers, 1952, fasc. 7: 25-32.

EMERY, K. O.
1954    Some characteristics of southern California sediments. Journal of Sedimentary Petrology 24(1):50-59.

EVANS, O. F.
1944    Some structural differences between wind-laid and water-laid deposits on the west shore of Lake Michigan. Journal of Sedimentary Petrology 14:94-96.

FRIEDMAN, G. M.
1961    Distinction between dune, beach and river sands from their textural characteristics. Journal of Sedimentary Petrology 31:514-529.

GALLOWAY, J. J.
1922    Value of the physical characteristics of sand grains in interpreting the origin of sandstones. Geological Society of America, Bulletin 33:104-105 (Abstr.)

GERVAIS, D.
1954    Etude morphoscopique de divers sables, application au transport éolien sur une dune maritime. Centre d'Etudes et Documentation Palaéontologiques, Paris, Annales 5. 64 p.

GIBSON, E. S. H.
1946    Singing sands. Royal Society of South Australia, Transactions 70:35-44.

HAILS, G.
1967a    Heavy mineral concentrations in coastal sediments and Pleistocene cliff-top dunes: an evaluation of eolian activity. Preprints of the 7th International Sedimentological Congress, Reading and Edinburgh.
1967b    Significance of statistical parameters for distinguishing sedimentary environments in New South Wales, Australia. Preprints of the 7th

International Sedimentological Congress, Reading and Edinburgh.

HAMDAM, R. A.
1965    The size and shape characteristics of some modern desert sands. University of Sheffield (unpublished Ph.D. thesis)

HARRIS, S. A.
1955    The mechanical composition of certain recent and fossil beach and desert sand deposits. University of London (unpublished M.Sc. thesis)
1957    The mechanical constitution of certain present day Egyptian dune sands. Journal of Sedimentary Petrology 27:421-434.
1958a    Differential analysis of aeolian sand. Journal of Sedimentary Petrology 28:164-174.
1958b    Probability curves and the recognition of adjustment to depositional environment. Journal of Sedimentary Petrology 28:151-163.

HAYES, M. O.
1964    Grain modes in Padre Island sands. Gulf Coast Association of Geological Societies, Annual Meeting, Field Guidebook, Austin, p. 121-126.

HEFLEY, H. M., and R. SIDWELL
1945    Geological and ecological observations of some High Plains dunes. American Journal of Science 243:361-376.

HUFFINGTON, R. M., and G. C. ALBRITTON
1941    Quaternary sands of the High Plains. American Journal of Science 239:325-

HUMPHRIES, D. W.
1966    A comparison of the booming sand of Korizi (Sahara) with the squeaking sand of the Gower (South Wales). Sedimentology 6:135-153.

JONASSEN, H.
1954    Dating of sand drift east of Ulfborg. Botanisk Tidsskrift 51:134-140.

KOLBUSZEWSKI, J.
1950    Notes on the deposition of sands. Research (London) 3(10):478-483.
1953    Porosity of wind deposited sands. Geological Magazine 90:48-56.

KOLBUSZEWSKI, J., L. NADOLSKI, and Z. DYDACKI
1950    Porosity of wind deposited sands. Geological Magazine 87:433-435.

KRINSLEY, D., and T. TAKAHASHI
1962    Electromicroscopy of the surface textures of natural and artificial sand grains. Geological Society of America, Special Paper 68:213-214. (Abstr.)

KRUMBEIN, W. C.
1941    Measurement and geologic significance of shape and roundness in sedimentary particles. Journal of Sedimentary Petrology 11:64-72.

KRUMBEIN, W. C., and L. L. SLOSS
1963    Stratigraphy and sedimentation. 2nd ed. W. H. Freeman and Company, San Francisco. 660 p.

KRYGOWSKI, B.
1958    Quelques données sur les sables des dunes continentales (translated title; French summary). In R. Galon, ed., Wydmy śródladowe Polski. PWN, Warszawa.

KUENAN, P. H., and W. G. PERDOK
1961    Frosting of sand grains. Koninklijke Nederlandse Akademie van Watenschappen, Proceedings 64:343-345.
1962a    Experimental abrasion. 4: Eolian action. Journal of Geology 68:427-449.
1962b    Experimental abrasion. 5: Frosting and defrosting of quartz grains. Journal of Geology 70:648-658.

LEWIS, A. D.
1936 Roaring sands from the Kalahari. South African Geographical Journal 19:33-49.

MacCARTHY, G. R.
1935 Eolien sands, a comparison. American Journal of Science 30:81-95.

MacCARTHY, G. R., and J. W. HUDDLE
1938 Shape sorting of sand grains by wind action. American Journal of Science 35:64-73.

MARSLAND, P. S., and J. G. WOODRUFF
1937 A study of the effects of wind transportation on grains of several minerals. Journal of Sedimentary Petrology 7:18-30.

MASON, C. C., and R. L. FOLK
1958 Differentiation of beach, dune and eolian flat environments by size analysis, Mustang Island, Texas. Journal of Sedimentary Petrology 28:211-226.

MATTOX, R. B.
1955 Aeolian shape sorting. Journal of Sedimentary Petrology 25:111-114.

McCRONE, A. W.
1962 Classification of the winnowing concept in geology. Geological Society of America, Bulletin 73(4):517-

McKEE, E. D.
1933 The Coconino sandstone—its history and origin. Carnegie Institution of Washington, Publication 440:78-125.

1945 Small scale structures in the Coconino sandstone of northern Arizona. Journal of Geology 53:313-325.

1957 Primary structures in some recent sediments. American Association of Petroleum Geologists, Bulletin 41:1704-1742.

1966 [see column 1, p. 92]

McKEE, E. D., and G. C. TIBBITTS, JR.
1964 [see column 1, p. 92]

McKIE, W.
1897 On the laws that govern the rounding of grains of sand. Edinburgh Geological Society, Transactions 7:298-311.

1899 The sands and sandstones of eastern Moray. Edinburgh Geological Society, Transactions 9:148-

METTLER, D. E.
1955 Dune sands of the Syracuse area in Kansas. University of Kansas (unpublished M.S. thesis)

MONOD, T.
1958 Majâbat-al-Koubrá, contribution à l'étude de l'Empty-Quarter, ouest Saharien. Institut Français d'Afrique Noire, Mémoire 52.

1961 Majâbât-al-Koubra. Institut Français d'Afrique Noire, Bulletin 23:591-637. (Supplément)

MONOD, T., and A. CAILLEUX
1945 Etude de quelques sables et grès du Sahara occidental. Institut Français d'Afrique Noire, Bulletin 7:174-190.

MOSS, A. J.
1962 The physical nature of common sandy and pebbly deposits. American Journal of Science 260:337-373.

NEWELL, N. D., and D. W. BOYD
1955 Extraordinarily coarse eolian sand of the Ica Desert, Peru. Journal of Sedimentary Petrology 25(3):226-228.

NORRIS, R. M., and K. S. NORRIS
1961 Algodones dunes of southeastern California. Geological Society of America, Bulletin 72:605-620.

PETROV, M. P.
1961 The mineralogical and granulometric composition of the eolian sands of the Ordos, the eastern Atashau and the middle Yellow River (translated title). Leningradskogo Universiteta, Vestnik 6, seriya Geologii i Geografii 1.

1962 On grain size and mineral composition of sands in East Central Asian deserts (translated title). Leningradskogo Universiteta, Vestnik 12, seriya Geologii i Geografii 2:65-

PHILLIPS, J. A.
1882 The red sands of the Arabian Desert. Geological Society of London, Quarterly Journal 38:110-113.

POLDERVAART, A.
1957 Kalahari sands. In Pan-African Congress on Prehistory, 3rd, Livingstone, 1955, Proceedings, p. 106-114.

PRESCOTT, J. A., and C. S. PIPER
1932 The soils of the South Australian mallee. Royal Society of South Australia, Transactions 56:118-146.

REED, R. D.
1930 Recent sands of California. Journal of Geology 38:223-245.

RETGERS, J. W.
1895 Ueber die mineralogische und chemische zusamensetung der dünensand Hollands und ueber die wichtigkeit Fluss- und Meeressanduntersuchungen in Allgemeinen. Neues Jahrbuch für Mineralogie, Geologie und Palaeontologie 1:16-74.

RIM, M.
1948 The movement of dunes and the origin of red sand in Palestine from a physical point of view. Hebrew University, Jerusalem (Unpublished thesis)

1950 Sand and soil in the coastal plain of Israel. Israel Exploration Journal 1(1).

1951 The influence of geophysical processes on the stratification of sandy soils. Journal of Soil Science 2:188-195.

1953 Le classements des minéraux du sable par les agents naturels sur les dunes. In Actions éoliennes. Centre National de Recherches Scientifiques, Paris, Colloques Internationaux 35:259.

ROSFELDER, A.
1960 Contribution à l'analyse texturale des sédiments. Université d'Alger (Thèse) 360 p.

SEVON, W. D.
1966 Distinction of New Zealand beach, dune and river sands by their grain size distribution characteristics. New Zealand Journal of Geology and Geophysics 9(3):212-223.

SHARP, R. P.
1963 Wind ripples. Journal of Geology 71:617-636.

1964 Wind driven sand in Coachella Valley, California. Geological Society of America, Bulletin 75:785-804.

SHEPARD, F. P., and R. YOUNG
1961 Distinguishing between beach and dune sands. Journal of Sedimentary Petrology 31:196-214.

SHERZER, W. H., and A. W. GRABAU
1909 The Sylvania sandstone and its distribution, nature and origin. Michigan Geological and Biological Survey, Geology Ser. 1:61-86.

SIDORENKO, A. V.
1956 Differentiation éolienne de la matière dans le désert. Akademiia Nauk SSSR, Izvestiya, Seriya Geograficheskaya 3. [See analysis by J. Tricart,

1958, Revue de Géomorphologie Dynamique 9(1/2):29].

SIDWELL, R., and W. F. TANNER
1939   Sand grain patterns of west Texas dunes. American Journal of Science 239:181-187.

SIMONETT, D. S.
1951   On the grading of dune sands near Castlereagh, New South Wales. Royal Society of New South Wales, Journal and Proceedings 84:71-79.

1960   Development and grading of dunes in western Kansas. Association of American Geographers, Annals 50:216-241.

SIMONS, F. S.
1956   A note on Pur-Pur dune, Virú valley, Peru. Journal of Geology 64:517-521.

SMITH, D. D., and R. E. SNEAD
1961   Eolian sand beds near Karachi. Geological Society of America, Bulletin 68:274. (Abstr.)

STUART, A.
1924   The petrology of the dune sands of South Wales. Geologists' Association, Proceedings 35:316-331.

1927   Notes on the South Wales dune sands. Swansea Scientific and Field Naturalists' Society, Proceedings 1:16-

TALMAGE, S. B.
1932   The origin of the gypsum sands of the Tala Rossa River. Geological Society of America, Bulletin 43:185-186.

THOULET, J.
1908a   De l'influence du vent dans le remplissage du lit de l'océan. Académie des Sciences, Paris, Comptes Rendus 146:1184-1186.

1908b   Origine éolienne des minéraux fins contenus dans fonds marins. Académie des Sciences, Paris, Comptes Rendus 146:1346-1348.

TRAINER, F. W.
1961   Eolian deposits of the Matanuska Valley agricultural area, Alaska. U. S. Geological Survey, Bulletin 1121-C. 34 p.

TREMBACZOWSKI, J.
1948   Origin of beach- and dune-sands in Pulawy (translated title; English summary). Universitatis Mariae Curie-Sklodowska, Lublin, Annales ser. B, 3.

TRICART, J., and M. MAINGUET
1965   Caractéristiques granulométriques de quelques sables éoliens du désert Péruvien, aspects de la dynamique des barkanes. Revue de Géomorphologie Dynamique 15(7-9):110-121.

TWENHOFL, W. H. et al.
1932   Treatise on sedimentation. 2nd ed. Williams and Wilkins Co., Baltimore. 926 p.

1943   Origin of the Black sands of southwestern Oregon. Oregon Department of Geology and Mineral Industries, Bulletin 24. 25 p.

1945   The rounding of sand grains. Journal of Sedimentary Petrology 15(2):59-71.

1946   Mineralogical and physical composition of the sands of the Oregon coast from Coos Bay to the mouth of the Colorado River. Oregon Department of Geology and Mineral Industries, Bulletin 30. 64 p.

UDDEN, J. A.
1894   Erosion, transportation and sedimentation by the atmosphere. Journal of Geology 2:318-331.

1896   Dust and sand storms in the West. Popular Science Monthly 49:655-664.

1898   The mechanical composition of wind deposits. Augustana Library, Publication 1. 69 p.

URBANIAK, U.
1962   The structure of a dune in Goren Duzy (translated title; English summary). Przeglad Geograficzny 34:749-758.

1966   Mineral composition of dune sands from the Plock Basin (translated title). Przeglad Geograficzny 38(3):435-453.

VAN DER MERWE, C. R.
1954   The soils of the desert and arid regions of South Africa. Interafrican Soil Conference, 2nd, Leopoldville, Proceedings 2:827-834.

VERLAQUE, C.
1958   Les dunes d'In Salah. Institut de Recherches Sahariennes, Travaux 17:12-58.

VIGNERON, H.
1938   Les effects du vent sur le sable. Nature 3028: 14-17.

WALTHER, J.
1924   Das Gesetz der Wüstenbildung in Gegenwart und Vorziet. Verlag von Quelle und Heyer, Leipzig. 421 p.

1951   L'influence des facteurs physiques sur la morphologie des sables éoliens et des dunes. Revue de Géomorphologie Dynamique 2(6):242- (A. Cailleux comments on this, Ibid, 1952, 3:99)

WHINCUP, S.
1944   Superficial sand deposits between Brighton and Franciston, Victoria. Royal Society of Victoria, Proceedings 52:315-332.

WILSON, C. CARUS-
1897   Correspondence on drifting sands. Geographical Journal 9:570.

WRIGHT, H. E., JR.
1956   Origin of the Chuska sandstone, Arizona–New Mexico: a structural and petrographic study of a Tertiary eolian sediment. Geological Society of America, Bulletin 67:413-434.

YAALON, D. H., and D. GINZBOURG
1966   Sedimentary characteristics and climatic analysis of easterly dust storms in the Negev (Israel). Sedimentology 6:315-332.

# FUTURE USE OF DESERT SEACOASTS

PEVERIL MEIGS

147 Pelham Island Road
Wayland, Massachusetts, U. S. A.

# ABSTRACT

The forecast for the next 10 or 20 years is that there will be no natural permanent change in rainfall, river flow, or groundwater level, and that men will not by artificial means increase rainfall in desert seacoasts. Desalinated sea water will be used more and more for domestic water supplies and for manufacturing but not for field farming. Cities and towns will continue to grow more than rural lands; suburbs, dairying, and truck farming will increase just outside the cities. Recreation and retirement facilities will grow within the sphere of cities and far from cities, aided by the mild temperature of winter and summer and the natural attractions of beaches, sea sports, scenery, and antiquities, and aided by desalination of sea water. Agriculture will grow sporadically, in places where men develop ground-water and river water previously unused, and, locally, surface water. Hydroponics will be increasingly used where water is expensive. Fishing will continue to develop slowly, and fish will be protected increasingly by governments working in accord with conservationists to prevent undue exploitation. There may be national or provincial parks created to preserve wildlife or landscapes of particular interest. Scientists will continue to work in the coastal deserts. The making of salt from sea water will continue to increase, as will harvesting of seaweed. Mining of oil, iron, and other minerals will grow and will make possible improved social conditions in the coastal deserts. The use of boats at desert ports will grow, and the use of airplanes will grow faster. No part of the desert seacoasts can any longer be considered inaccessible.

# FUTURE USE OF DESERT SEACOASTS
## Peveril Meigs

Deserts extend down to the seacoasts of the world for about 32,000 kilometers, 37,000 if we count the inland seas of the Caspian and Aral. If we assume the coastal deserts to be 40 kilometers wide on the average (1, p. 3), we get an area of 1,500,000 square kilometers. Most of this vast stretch of land is unused, or poorly used, but small stretches of its support rich agriculture and large cities. The future use of any desert presents interesting possibilities for the expanding population of the world. The coastal deserts have some extra endowments that are lacking in the inland deserts and therefore are worthy of special consideration.

Desert seacoasts have the usual problems of all deserts. Rainfall is as low as anywhere on earth, in spite of the bordering ocean. The ocean makes the air humid, but in order for rain to fall, there must be an uplift of humid air, and this, for a number of reasons, depending on the locality, is lacking in coastal deserts. The low rainfall causes scarcity of water on the land, unless water comes from humid places by rivers or underground flow. Much of the land is rough and soilless, as a result either of the washing away of land during the few rains, or of the work of the wind that blows sand into dunes or blows away the fine particles and leaves rocks. The soil is poor or lacking in the upper slopes of the land, and where the soil is rich along the stream beds or deltas it is likely to be highly alkaline. The conduct of agriculture must be carefully practiced or the alkalinity may increase. There is sparse vegetation for livestock, and that, together with the scarcity of water, makes travel by land difficult.

## ADVANTAGES OF DESERT SEACOASTS

The sea is itself an advantage which the coastal deserts have over the inland deserts. Since the earliest days, man has gathered the food of the sea to support villages along the coast. Immense shell mounds from shellfish mark the sites of primitive cultures around the world. Gradually the people went farther from land to catch fish, and at the present day there are large commercial fisheries in the desert seacoasts. In carrying on agriculture, too, the sea helped. The eastern American Indians used fish for fertilizer, and the colonists followed their example. Seaweed is used today for fertilizer by farmers.

Living near the sea a man never need lack salt.

Of late years, he never need lack enough water to drink, if he knows the principle of evaporation of sea water and has the minimum equipment necessary to put it into practice. Solar power can be used for the extraction of both fresh water and salt from the ocean.

The sea provides for travel by boats in the desert seacoasts, and sea routes have been used before historic times along the Old World deserts of the northern Sahara, the deserts bordering the Red Sea and the Persian Gulf, and the Peruvian and Chilean deserts. In modern times, the sea provides for the export of products produced by the coast and the import of fuels and other things from the non-desert world. In addition, many coastal deserts have wave-cut terraces or flat deltas on which it is easy to build airports.

Wind blows more strongly and constantly along the coast than in many places inland; it is a potential source of power that has been little used. Tidal power is another source of power that inland deserts lack, but it has been used even less than wind power in the desert seacoasts. The coasts of Ecuador, Northwest India, Western Australia, and the Gulf of California have tides of 3 to 8 meters in height and may some day use the power of the tides with tide mills.

One great advantage of many coastal deserts over those farther inland is their temperature. The temperature of the sea changes less in a day or a season than that of land. In the summers the comparatively low temperatures of the sea during the daytime pass over to the land bordering the sea, often impelled by a sea breeze. Thus the climate is pleasant during the summer. Alexandria on the coast, for example, has a mean July maximum temperature 6°C below that of Cairo, 150 kilometers inland. On the west coastal deserts the difference is even more striking. In southwest Africa, for example in the vicinity of Port Nolloth, the mean daily maximum of the coast in January (summer) is 20°C lower than in a place 140 kilometers inland (1, Figs. 5 & 6).

Some coastal deserts have the most oppressive summers of the world. Many places on the west side of the Red Sea or along the south shore of the Persian Gulf lack cool water off their coasts, and with high humidities such locations are unbearably hot in summer.

So much importance do the climates have on the future use of the land, and so widely do the climates

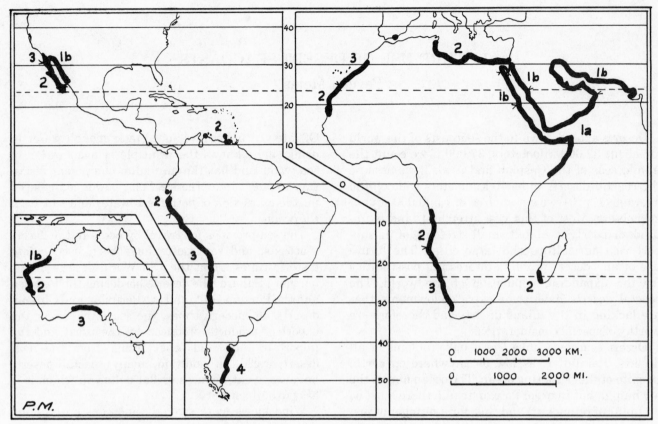

Fig. 1. Desert seacoast climates. *1a:* hot, tropical; *1b:* hot, subtropical; *2:* warm; *3:* cool; *4:* cold.

of the desert coasts differ, that the coastal deserts can be classified on the basis of climate. Figure 1 is a simplification of previously published maps (*2*). Briefly, the coastal deserts are classified as *hot* (tropical and subtropical), *warm, cool, cold.* The west coast deserts are cool to warm. Actually, each of the desert seacoasts differs in both natural and cultural landscape and deserves to be treated by itself, as has been done elsewhere for thirty-one coastal deserts (*1*), especially with regard to climate and culture. Schreiber (*3*) deals especially with an inventory of maps and of geology.

## GENERAL ASSUMPTIONS

It is dangerous to rely upon long-time forecasts of the future, in arid lands or anywhere else. Yet the forecasters have always had a good following, from the scapulimancers of the primitive tribes of central Asia, to the oracle of the prophetic tripod of Delphi, to the modern forecasters of weather by sunspots and of stock market reports by any number of ingenious devices. One cannot live without making forecasts. Generally the forecast is made based on what has gone on before: the historical forecast.

When one sets an alarm clock on going to bed at night, he assumes from history that the new day will dawn as usual after the night. As science pursues its discoveries, more and more is learned of the effect that certain acts of nature and of man will

have. The problem consists of getting scientific findings to man in a way that he can and will apply them, and discovering how the politics of nations will affect the use of the land. Stamp's *History of Land Use in Arid Regions* (*4*) and White's *The Future of Arid Lands* (*5*) are essentially histories of the distant and recent past. Yet both are essential to forecasting the coming events, and if we follow the experience of the past, we will do better in the desert in the future (we hope).

The first assumption we make as to the future use of desert seacoasts is that man has enough intelligence and foresight to use increasingly the lessons learned from past experience (*4* & *5*): that over-irrigation causes the soil to become alkaline, that groundwater level will diminish if the groundwater is overused, that the range will be overgrazed and erosion encouraged if too many animals are allowed to graze, that reservoirs behind dams will fill up with silt, and that cooperation is needed in the desert between regions and between countries in many cases to make the best use of water.

The locust and grasshopper control program is a good example of international cooperation to increase the productivity of deserts. Locusts are well adapted to the patchy conditions of deserts, where vegetationless land is juxtaposed with intensely vegetated rivers and oases. They lay their eggs in open land and must migrate to vegetated land to eat. Their eating of crops is one of the great scourges of the desert. If the wind is in their favor, they can spread to a cropland in another country many kilometers from their breeding place and eat the crop in a matter of hours (Fig. 2). It takes an international organization to bring the locusts under control (6). The Food and Agriculture Organization of the United Nations is conducting an anti-locust campaign that has resulted in studies and the spreading of insecticides in breeding places, both by airplane and on the ground.

Most deserts need more water, for man, beasts, and crops (Fig. 3). The assumption is here made that nature will not make additional water available: that man must help nature. Rainfall fluctuates over a period of variable time, but there is nothing to indicate that it is increasing (7). Nor is there any way to forecast future trends (4, p. 54). Men have tried to force clouds to contribute more rain, by seeding; the evidence is confusing, and it may be that they can produce more rain from a rainstorm that is falling already, but so far they have not produced any rain, or enough rain to make any difference with coastal deserts.

There has been talk of flooding some interior basins with seawater, as the Qattara Depression in the Libyan Desert, and causing more rain in the vicinity of the artificial lake. But there are deserts along the Red Sea and the Caspian Sea, and one of

Fig. 2. A plague of locusts. These locusts, on a doorstep in Iran, may have come hundreds of miles from a breeding place in the deserts of Asia or Africa. The Food and Agriculture Organization of the U.N. coordinates an international campaign in an anti-locust drive to prevent the crops of the deserts and steppes from being eaten. (Courtesy Unesco)

Fig. 3. The need for water in deserts. Parents and children wait for water tanks to bring their weekly supply of water, in Parahibo, Brazil, in the semi-desert that extends to the coast in the eastern part of Rio Grande do Norte. (Courtesy Unesco)

the driest deserts in the world (the coast of Peru) is along the largest body of water in the world, the Pacific Ocean; it is unlikely that a local lake in the hot desert would do anything but increase the relative humidity unbearably.

## POSITIVE ASSUMPTIONS FOR INCREASING WATER SUPPLIES

The assumption is justified that practically all the water to be used in the coastal deserts for irrigation in the future will be developed from natural rainfall, as in the past. It will be used either directly or from dams from through-flowing rivers which have risen in rainy climates; from intermittent streams, which will be dammed or arranged to flow into the groundwater supplies; from flood runoffs; or from the ground-water direct from wells. These supplies have been developed only partially in many parts of the desert seacoasts.

For other than irrigation, there will be an increasing use of precipitating dew or fog by mechanical means. While that will not bulk large in the total amount of water, it will be used for drinking by man or animal where water is scarce. There have been many reports of dew mounds in the coastal deserts and of fog reducers in the west coast deserts. In the foggy *lomas* of the Peruvian coastal deserts there is evidence that in pre-Columbian time the people produced actual springs of water from heaps of rock which condensed the fog and clouds that swept in from the sea (8). A little farther south, in the province of Antofagasta, Chile, modern men have built in the passes heavy wooden frames with vertical nylon strings through which the wind blows the mountain mists. The mists condense on the strings, and the water runs down the strings into pipes (9). The obtaining and the use of dew is well-treated by Masson in an article prepared for Unesco (10).

The most important assumption of all for the future use of the desert seacoasts is that the desalination of seawater will be used increasingly. There is no reason why the *higher* forms of freshwater use (domestic, stock-raising, most manufacturing) should be limited near the seacoast by the absence of fresh water. In the last ten years the use of desalination plants to make fresh water out of seawater has been increasing rapidly. The U. S. Department of the Interior has established an Office of Saline Water to conduct experimental research on the best methods of desalinating sea water and brackish water; it has set up plants for desalinating water at the rate of a million gallons per day by various systems in several parts of the United States and abroad. The United States is cooperating with various foreign countries (U. S. S. R., Israel, Mexico, Saudi Arabia, and others) to exchange research and set up operating stations.

The First International Symposium on Water Desalination, held in Washington in 1965, shows the extent of world coverage on this subject. The three thick volumes of *Proceedings* go into all major techniques of water desalination and give many examples (11). There are several principal methods:

(a) Evaporation of water by boiling, including the use of multi-effect distillation. This is the principal method used to date. It requires a cheap source of heat (generally petroleum or natural gas, but may be atomic power).

(b) Ion exchange by electrodialysis. This is much used where a small amount of water is needed, from brackish water from lagoons or from the ground.

(c) The hydrate process, which is a chemical process.

(d) Freezing of salt water. Theoretically this is a cheap method, but so far has not been used successfully.

(e) Solar distillation. The simplest form, simple covered diaphragms exposed to the sun, has been used, for example in the little island of Symi, Greece (12). This method is suitable for small towns in coastal deserts. A more sophisticated sort of solar distillation was the plant set up at Puerto Peñasco on the Gulf of California, Mexico. There a complex and ingenious plant treated the seawater heated in a large solar collector. The condenser of

the water vapor was cooled by the seawater when first pumped in from the Gulf (13). The experimental plant produced 5,000 gallons of distilled water per day, and it is figured that at the rate of one million gallons per day it could produce water at $1.00 per 1,000 gallons, which compares with a cost of water trucked in from normal supplies of $6.00 per 1,000 gallons (11, pp. 125, 429-459).

The section of this book by Carl Hodges points out that the Puerto Peñasco plant has been changed to a different kind of experimental system since 1965. The salt water is now desalinated by the waste energy of a diesel-electric plant instead of by solar energy. The crop plants are grown in controlled-environment plastic greenhouses instead of in the open. The humidity of the air in the greenhouses is kept high by direct evaporation from seawater; enough water condenses on the enclosure to water the plants, and desalinated water from the diesel engine can be put to other uses. Perhaps this system can be combined with hydroponics (see below under "Growth of Agriculture"). Some authorities have high hopes set on this system, but it is still experimental. In the present stage of experiment in water desalination, there are many slips between the hopes and the results. For example, the Bolsa Island project of Southern California, for which the contracts were signed on November 20, 1967, and which had been studied well in advance, has had to be *restudied* as the costs were too high, according to a news release by the Office of Saline Water on July 25, 1968. The project was to have produced 150,000,000 gallons of fresh water per day. It may still be carried out under some revised form.

The use of desalination of seawater is proceeding at an accelerating pace. In 1962 the United Nations made a study of how many desalting stations there were in developing countries and issued a large book describing the stations and telling how they fitted into the economy of the countries (14). It lists 61 large desalination plants and 19 plants under construction or planned, many in desert seacoasts. A summary of this and other studies is made in "The geographical status of water desalination" (15).

Hodges (16) reports that the plants built in 1967 alone had a total capacity of 15,000,000 gallons per day. Hodges and most other commentators who have studied desalination warn that the increase of desalting water does not mean that the distilled water is going to be used for agriculture: it is too expensive, now and in the indefinite future. Hodges, on the basis of his experience, gives a tentative limit of about $0.20 per 1,000 gallons of water in the year 2000, of the most economical desalting plant operating on seawater: way below the cost obtainable today. That makes the price still way above what the farmer pays for irrigation water today. And to transport it away from the seacoast costs roughly about $5.00 per acre-foot mile in addition. In other words, the assumption can be made that desalinated seawater will not be used for field farmers for the indefinite future. But for domestic use and manufacture, the cost of desalinating water is already below the price of natural water in many places, and the realization of this has caused the desalting plants to burgeon.

I wish we could assume that no more wars would interrupt the future of the coastal deserts. War has caused the disruption of old cultures based on irrigation at least in the desert seacoasts of Irak, southern Israel and Jordan, and pre-Inca Peru, and, more recently, Libya. These cultures have not yet come all the way back from the disruption of irrigation canals, the displacement of populations, and the other things that accompany war. Andrianov and Murzaev place war as the most important social-political factor leading to the disappearance of oasis culture in the desert (17).

I am pessimistic as to the future, based on existing conditions, where so many leaders in government still will go to war if they are not able to convince other nations peaceably to think as they do. Many of the coastal deserts are in remote places, but some are in the center of potentially conflicting nations. And even remote places may be drawn into conflict, so close has the world grown together. It is to be hoped that the agreement between nations as to the use of rivers in large-scale irrigation, or as to locust control, will make war less likely.

Epidemics also cause the depopulation of deserts. They may be introduced by war. Baja California, Mexico, for example, was depopulated for a century or more when diseases, especially smallpox and syphilis, introduced by white conquerors, virtually wiped out the native Indian population (18). In this day, public and private medicine, it is hoped, would not allow infectious disease to wipe out the population of any part of the world.

## FUTURE TRENDS

The future trends of desert seacoasts will be classified for this chapter as: (a) growth of cities, (b) growth of agriculture, (c) growth of mining, (d) growth of sea industries, (e) growth of recreation, and (f) growth of science and conservation.

### Growth of Cities

Big cities are growing larger on the desert seacoasts, little cities are turning into big cities, and towns are turning into cities, the same as in other parts of the earth. For the present, this is likely to continue, though population trends are notoriously undependable. Wilson (19), who calls attention to this trend among desert cities, says that in the few

cases he knows, manufacturing and tourism are among the causes for this growth of desert cities. There are three cities in the desert seacoasts with populations over 1,000,000 (not counting Cairo, which is farther inland). The latest populations, taken from 1961 to 1967, are: Lima, Peru, 2,093,434; Karachi, Pakistan, 1,912,598; and Alexandria, United Arab Republic, 1,587,700 (Fig. 4). The population of each city has increased from 5 to 11 per cent a year for the last 10 or 15 years. Two are capitals, and capitals are important in today's highly centralized government. In addition, the operation of large desert irrigation projects needs state planning.

The lesser cities are growing about as fast as the big cities. There are four cities in the Nile Delta with populations of about 200,000 and they are growing as fast as Alexandria. Tijuana (Baja California, Mexico), which had a population of about 6,000 in 1930 (20), in 1963 had a population of 222,534 (19). Lying on the United States border, it is attractive to tourists from the north. Mexicali, across the peninsula from Tijuana on the delta of the Colorado River, grew from about 15,000 in 1930 to 290,000 in 1967. It is an administrative and commercial center. Kuwait grew 27 per cent in the four years 1957 to 1961 because of development encouraged by oil and gas (including the world's largest desalination plants). Jiddah (Saudi Arabia) grew 85 per cent in the 10 years between 1952 and 1962 because of religion: it is visited by pilgrims of the *hadj* to Mecca. The occasions for growth are varied, but all the cities are growing. Overall, there are attractions to living in a city or near a city, in the form of services, medical care, and opportunities.

Where there is limited water, a city uses it much more economically than does agriculture. Koenig (21) has pointed out that the water to support one worker in arid land agriculture will support 60 workers in manufacturing. Instead of raising crops in arid regions, it would be more profitable to manufacture radios, clothing, and cameras to ship out, and then ship in food from more humid regions. In the services (retail, medical, insurance, museums, and similar occupations) there is still less dependence on water than in manufacturing. The trend toward growth of cities in arid regions in comparison with farm lands is in line with the economics of water.

In the coastal deserts, there is even more likelihood of the growth of cities as desalinated water comes into greater use. Desalinated water is too expensive for agriculture, but not too expensive for domestic purposes and many kinds of manufacturing. As to what kind of manufactures have come in and are coming into the coastal deserts, there have been few studies made: there should be more. We know, at least, that the products of the coast are giving rise to manufactures (as in the fish meal of

Fig. 4. Alexandria, a city of one and a half million, where the Nile delta meets the Mediterranean Sea. A view in the heart of the city and a beach in the city. (Courtesy United Arab Republic Tourist Office)

Peru and petroleum refineries in Saudi Arabia). We can assume that the big cities are devoting increasing attention to small industries that don't take much water.

### Growth of Agriculture

Although desalinated water will not affect field agriculture for some time, if ever, there is every prospect that farming is not dead in the desert seacoasts. It will expand, although not as fast as manufacturing, as long as the world needs more food.

Agriculture will expand in the more conventional ways first, using natural water in streams and in the ground. There are still some dams being built on the great rivers of the world, which will make more lands of the coastal deserts usable. The most famous is the dam at Aswan on the Nile, which is designed to irrigate the coastal delta. Much water of the Nile now escapes to the sea in time of flood. That is true of most of the large rivers flowing across the coastal deserts. The small rivers in the Near East could be used to irrigate crops, too. Many of the rivers that cross the coastal desert of Peru have been used for millenia, but Peru is now going through a period of new development, in the northern seacoast desert, for example.

Ground-water is usually pumped from vertical wells. Ground-water is still being found with modern portable drills in coastal deserts. In some arid places, mostly away from the coasts, ground-water has been over-used to the point where the depth is such that water cannot profitably be lifted. Ground-water has the advantage that it can be used by the individual without much capital, whereas river development generally requires government control to build the dams and distribute the water. On the other hand, it is hard to regulate digging of wells, and as a result the water level may be lowered to disastrous proportions.

In the deserts from the northern Sahara to China, and in the Atacama Desert of Chile, a specialized system is used to get ground-water, called *qanat* in Asia, *foggara* in Arabic lands, and *socavónes* in Chile. A hole is dug from 10 to over 100 meters down to ground-water in an alluvial fan, and a tunnel about the height of a man is dug at a slight downgradient to the land to be served. Every 25 meters or so there is a hole to the tunnel, to be used in excavating and repairing the tunnel. An almost constant flow of water comes out of the tunnel by gravity. These systems cost much to build and maintain, and they can only be used where there is abundant labor. Unless there is worked out some way to cheapen the cost, they are not likely to be built much in the future. There is a good summary of the system in a chapter by Dixey in a book edited by Hills (22).

Another way of using ground-water is trench farming. A trench $1\frac{1}{2}$ to 2 meters wide is dug to a depth generally one or two meters down to capillary water. Plants can be raised in the bottom of the trench. Vegetables, apples, pears, currants, and other plants are raised this way on the borders of the Caspian and Aral Seas. On the northern coastal desert of Egypt, grapes and figs have been raised (23). In about the middle of the sixteenth century the Spaniards noticed with amazement maize, vegetables, and fruit trees growing without irrigation in a desert valley 65 kilometers south of Lima and found that they were in deep basins dug to where the earth was watered by ground-water. There also have been found remains of basins fronting the ocean floor, dug to a depth of about 1 meter and up to 100 meters long by 50 meters wide, in the lower Virú Valley. These basins, locally known as *pukios*, enabled the plants to get down to ground-water (24, p. 270). Probably there are many areas in the coastal deserts where the pukio system would work, and it is likely to be tried more and more in areas of shallow ground-water.

Another way that agricultural production can be increased in the desert seacoasts is by using plants best suited to little water and salty water. Collin-Delavaud has shown that one reason less land is irrigated today in the coastal deserts of Peru than in pre-Columbian times is that since the Spanish conquest, and particularly today, the policy has been to raise sugar cane for export, rice, annual cotton, and alfalfa, which require much more water than did the maize, beans, and tree cotton raised by the ancients. A crop of sugar can, he says, takes four times as much water to raise as a crop of maize (25). This analysis agrees with what Armillas has said, that in the early years of the Christian era there was at least 40 per cent more land under irrigation in the Virú Valley in Peru than at present. He says it may be due to the higher water requirements of the irrigated crops today and not to any reduction of the volume of water in the river (24).

Wars with Tiahuanacan groups from the interior, with the Incas, and with the Spaniards, temporarily upset the cultures of the coasts of Peru and damaged their systems of irrigation. But even if there is as much water used today as in the best of pre-Columbian times, there would be much less land used because the social, political, and economic objectives lead to a use of plants that consume large amounts of water. If the policy is changed to raising of crops that consume less water in Peru and other developments elsewhere, there could be a substantial increase in the land cultivated in coastal deserts. Geographic and economic studies should be made to show whether that would be an advantage or not.

Many scientists studying soil and plant life have been studying the use of salt and alkaline water to grow crops. Unesco has financed some of these studies, notably in Tunisia, and has brought groups of scientists together to discuss the problem (26). More and more attention will be paid to this subject, which is particularly appropriate to coastal deserts with salt water along their coasts. The use of salt marsh hay has long been used to feed livestock. Only a few of our cultivated plants are salt tolerant: dates, barley, beets, onions, celery, and radishes, among others (25, p. 130). As more and more discoveries are made of the relations of plants to salt water, the crops will be pushed slightly closer to the sea.

Reuse of water that has already been used will become more and more accepted. Even sewage water is being refined and used for drinking. This applies especially to water used in cities. But irrigation water can also be used after it has drained off the field if it is not too alkaline.

Irrigation by the water that runs off the surface of the ground has been used in some districts for thousands of years. Now the modern users of desert water are experimenting with all systems for the use of runoff, and have begun to put some of them into practice. In many desert coasts the practice consisted of allowing runoff from slopes from an occasional rain to be concentrated in a limited acreage by means of dikes. The rainfall that fell on 10 hectares of slopes could be concentrated, for example, on 1 hectare of good land at the foot of the slopes.

Another system, widely used in the Near East from southern Tunis to Irak, was to lead the surface water to deep cisterns. The cisterns, cut in the rock, would hold the water until it was drawn out by a bucket on the rope for domestic use or livestock use (22, pp. 232-235). Arroyos flood in any kind of coastal desert that has even occasional rain. Many articles have been written on the use of water resulting from arroyo floods in ancient times. Evenari and Koller have described how the ancients controlled floods by building a series of light dams in the minor arroyos, building fields lined by stones along the larger wadis, and deflecting the waters by means of partial dikes (27). Surface water can be used today by the methods discovered by the ancients, and gradually they will be so used in the coastal deserts.

Hydroponics (the raising of plants in fertilized tanks of water) will be more and more used as the cities grow in the desert seacoasts. All the water can be used by the plant, and no water is evaporated by the soil. Even desalinated water can be used economically by this method. The outstanding example of the use of hydroponics in the desert seacoast is in the island of Aruba, Netherlands Antilles. Aruba had 20,000 square meters (216,000 square feet) of hydroponic beds in operation in 1962 and made a substantial profit raising cucumbers,

tomatoes, beans, melons, and other crops (14, pp. 67-68). With proper training, this system could go far in the coastal deserts lacking natural water.

Less livestock is raised in coastal deserts, with one exception. Gradually the herdsmen are going into some other industry, in cities or elsewhere, that rewards them better. As the cities grow, they have more requirement for fresh milk and other dairy products, so there is a revival of the dairying industry near, and there will probably be towns like Pucusana (see below, under "Growth of Sea Industries") near the cities.

### Growth of Mining

As the result of the development of desalination, by large installations and by portable sets, there is no part of the desert seacoasts that need hold up mineral prospecting by lack of water. The increase of mining need depend merely upon the discovery of minerals in adequate amounts and the world demand for minerals.

There have been many minerals found in coastal deserts, from the gold along the Red Sea coast and Baja California to the oil of the present day. The world's greatest oil deposits have been found in the coastal deserts, in the eastern part of the Arabian Peninsula. The oil has made the average wealth of the desert nations, such as Kuwait, Bahrain, and Qatar, the greatest in the world. The wealth falls mostly into the hands of the rulers, and how the people fare depends largely on the wishes of the rulers. In Kuwait, the people are not faring badly, with plenty of food, air conditioning, and other needs supplied. Saudi Arabia, Iraq, and Iran have other great oilfields. Libya has begun to produce more oil lately, and is still trying to pass her prosperity to the average citizen.

At first, the oil industry does not offer many jobs to the people. Then as refineries are built there are more jobs. The government uses parts of its revenues from oil to encourage agriculture, building, and public health. Entirely new towns grow up, and certain agricultural areas are at first decimated by the families going to the towns. There is no way of knowing which deserts will find oil in the future, but it is likely that the coastal deserts that have found it (northern Peru as well as the Near East) will continue to prosper (1).

Next to oil, iron ore is likely to affect coastal deserts in the near future. Two great deposits have been found in northern Mauritania in northwest Africa and the west coast of Australia. Although both deposits are too far inland to be in the coastal deserts, their shipment will be through ports in the desert seacoast. Port Etienne, in Mauritania, is growing from a small fishing village into a major port, and Port Hedland, which used to be a small gold-field port, has potentiality, too (1).

## Growth of Sea Industries

Fish, seaweed, and salt are resources of desert seacoasts but not of the interior deserts. These resources have been used for thousands of years by the seacoast dwellers, but all are under-utilized at present.

Shellfish, gathered in great quantities by the pre-historic settlers, are still gathered in some places on the desert coasts. But increasingly the fishing has become larger scale, and is now carried on in boats many miles from the coast. Many of the boats come from outside the deserts. There are two main types of fishing: demersal fishing, which is carried out in comparatively shallow waters; and pelagic fishing, which is generally conducted in deep waters.

Demersal fish were the first kinds caught, chiefly with hand-lines and baited hooks; however, power-ful trawlers with nets are being used more and more. Shallow waters, in which this kind of fishing is done, lie off the coastal deserts of Patagonia, the Red Sea, and the Gulfs of Persia, Cutch, Aden, Gabes (Tunisia), and California. Efficient boats and equipment would result in a much greater catch on all of these coasts, and international agencies are already providing the modern tools and education in some countries, where fish will provide much-needed protein to the insufficiently-fed peo-ple. Some of these areas are sought by sport fisher-men looking for abundant fish.

The most spectacular large-scale fisheries today are the pelagic type. This type of fishing is conducted off the west coasts of continents, where the water is well-fed by plankton from the upwelling of water from the deeps. The last fields to be developed were the western South American coast of Peru and Chile and the coast of South West Africa. The pilchard, or sardine—a member of the herring family—swims in schools or shoals, some of which have been reported to be 3 to 5 miles wide and 15 miles long off South West Africa. These fish are scooped up in large power-driven nets and appear to be inexhaustible; but the industry has been going on in the African field only since 1947. In Peru, where the industry has been going on much longer, the government has found it necessary to take conservation meas-ures to prevent an undue decrease in the numbers of fish.

Only the pelagic fisheries of the west coast of South America are fished largely from desert ports along the coast. The fisheries of South Africa, the Canaries, and Baja California are fished from boats often hundreds of miles from the desert. Even here the desert is affected by the fish processing plants set up on the nearby coasts (1, pp. 18-19).

In Peru, many ports carry on a fishing industry, as they have long before the Columbian or even the Christian era. The fish go through cycles of abun-dance and scarcity, in part dependent on oceano-graphic developments. Since shortly before 1960 the fish meal industry for export has increased mark-edly, somewhat making up for the decline of the nitrate mining. Salinas Messina gives an account of the "boom" industry of fish meal in Iquique, its chief center in Peru (28). Orbegoso Rodríguez tells of a much smaller fishing village (Pucusana), just south of Lima, and how its nearness to Lima, a good road, electric power (making possible the develop-ment of wells), and organization of cooperatives have affected its present well being (29). Its citizens, while not giving up fishing, have turned their at-tention to raising dairy cattle and providing places of recreation for Lima people on weekends and summers, favored by good beaches. More and more as the cities of the coastal desert develop, there will be Pucusanas in the suburbs.

As for pelagic fishing, probably it has developed about as far as it will, except in Australia.

Seaweed is a potential resource of the desert coasts that merits further study. It has been used little in deserts, but more in densely-settled humid countries, such as Japan, China, and northwestern Europe. There it is used somewhat as food. Most seaweeds are not digestible by man, but some are digestible by animals. Camels, donkeys, and horses eat seaweed cast upon the beach in Tunisia, and there is a vast reservoir that could be harvested from the sea. Northern European farmers who live near the seacoast use the crops for fertilizing their farmland. Seaweed is used for making chemi-cal preparations: potash, iodine, or agar, when prices are high enough. There has been much scien-tific work done on the green, brown, and red sea-weeds, but there needs to be more to make this a contribution to coastal deserts (1, pp. 19-20).

Salt is made in most coastal deserts. Seawater is allowed to evaporate in shallow basins along the coast, and the salt can be scraped into piles (Fig. 5). The dry atmosphere and the cheap lands give deserts an advantage in making salt. The salt is used for local industries, such as the salting and drying of fish and meat, and some is exported. The sea-salt industry is increasing at a faster rate than the land-salt industry around the Sahara. Salt is likely to show a gradual increase in coastal deserts in the future.

## Growth of Recreation

Of all economic industries of the desert seacoasts, none is likely to grow more than recreation. The cli-mate of many coastal deserts is incomparable for tourists. The gentle sea breezes of summer in areas 2 and 3 on the map of Figure 1 are a relief to anyone, especially the people who live in the hot interior. These areas, and area 1b on the map, also make good winter resorts for the people of northern Europe and North America.

Fig. 5. Salt from sea water in the desert seacoast of Somalia. Scenes like this occur in most coastal deserts. (Photo Zöhrer, courtesy Unesco)

Fig. 6. Cliffs and a little bird island at picturesque Playa de la Herradura, Lima. (Courtesy Consulado General del Peru, New York, N. Y.)

The desert seacoasts have fine swimming beaches; waters for boating and fishing; picturesque cliffs, islands, animals, and plants; underwater animal and plant life, and some coral in the tropical deserts, for skin divers to observe; some fine examples of ancient works of man, both above and underwater, for archeologists and tourists to study, especially on the Mediterranean coast but to some extent in all deserts; and an abundant supply of fresh water from the ocean, now that the cost of desalination has decreased.

Cities of the desert coasts use the nearby recreation features of the coasts for the daily, weekend, and vacation use of their citizens; hotels and other features of recreational areas are multiplying. The beaches of Alexandria have been famous since the days of the ancient Greek dwellers (Fig. 4). Fashionable resorts are advantageous to the beaches and picturesque cliffs on the coast near Lima (Fig. 6). As the cities increase in population and in wealth, more and more use will be made of the surrounding lands for recreation and retirement.

Away from the cities, places are developing for the vacationist to spend his time. Libya and southern Tunisia have their Phoenician and Roman antiquities to show. The island of Djerba has hotels supplied with desalinated water.

Eilat, in Israel, at the head of the Gulf of Aqaba of the Red Sea, is a good example of a town that has decided to make recreation its major industry. It is steadily attracting more tourists, both in winter and in summer. Eilat in 1949 had only a lonely desert police post. Today it is a town of 10,000 people. A dual-purpose electric power and distillation plant was installed in 1965; in 1966 distillation supplied 60 per cent of the water of the town, and ground water supplied 40 per cent. There were 11 hotels in 1966; new ones and enlargements to existing ones are steadily being made (30). Besides the climate, the beaches, the water, and the picturesque nearby mountains, Eilat supplies tourists with an artificial small harbor and boats (Fig. 7).

In the western hemisphere, Baja California, a province of Mexico, has developed recreation more than any of the other coastal deserts. This is due not only to the attractions which it shares with other coastal deserts but to its nearness to the United States. Tijuana, next to the border, has grown to a large city. The small seaport of Ensenada, 65 miles south of the border, has grown to a city of 50,000 people, with 60 hotels, motels, and trailer parks in or around it. The five-mile-long beautiful beach, the fishing, and the yacht harbor are main attractions for tourists. Generally, when a place has become frequented by tourists, many people move in to serve the tourists, and gradually other industries grow up. It has been this way with

Ensenada, which now has light manufacturing, in part for tourists.

Farther afield in Baja California, plenty of wild and picturesque land attracts nature lovers and campers (Fig. 8). About a dozen hotels in various coves in the southern part of the peninsula are served by airplane as well as by boat. More and more sportsmen fish the rich waters of the Gulf of California. Recently a book has been written about the beaches of the northern 400 kilometers of coast (31). Of the ancient works of man, in the northern part of the peninsula, only the trace remains of Indian settlements and the ruins of the old missions built by the Dominican fathers from 200 to 150 years ago. A proposal has been made to reconstruct the mission buildings to increase the public respect towards Baja California (especially towards Tijuana) and to attract tourists, as tourists have been attracted to similar mission buildings in "Alta" California in the United States (32). Remarkable cave paintings about halfway down Baja California, on the Gulf side, should be preserved (33). With or without archeological remains, Baja California will continue to develop a recreational industry.

Retirement is a special, permanent form of recreation. Most retirees like to live in a pleasant climate with other environmental attractions, in or near a town with good medical services and recreational facilities. Some coastal deserts have the natural attractions and are getting the cultural amenities.

## Growth of Science and Conservation

Since Unesco began the study of arid zones in 1949, there has been a surge of activity in this field. Part of this surge was due to Unesco grants, education, and international meetings of experts. Part has been due to the organizations and individuals already conducting scientific studies in the desert. Many centers devoted to scientific study of the desert have sprung up in the past 10 years, in universities, scientific societies, and elsewhere. This trend shows no signs of abating. Recently more attention has been given to studying the coastal deserts as such. Unesco published the first book on coastal deserts as a whole in 1966 (1), and the University of Arizona formally published its bibliography on desert coasts (3) in 1968, although it was available earlier in preprint form. To the few research institutions on desert coasts, such as the Namib Desert Research Association and the Puerto Peñasco Research Station (on the Gulf of California), others will probably be added as time goes on.

Scientific studies of coastal deserts have shown a need for greater conservation of animal life, as in deserts anywhere. The gray whales of Baja California are a case in point. After spending part of their life in the Arctic Sea, the gray whales come

Fig. 7. Eilat, Israel, primarily a recreational town at the head of the Gulf of Aqaba, Red Sea. The general view of town and desert mountains shows the little harbor for pleasure boats. The close-up of the harbor shows a new hotel. (Courtesy Israel Government Tourist Office)

south to Baja California to three large lagoons (of which Scammon's Lagoon is the most frequented) to bear their young. When they were first discovered and whaling began on them in 1857, their number was estimated at 30,000; they were reduced in a few years of whaling to only a few dozen. The International Convention for the Regulation of Whaling in 1938 forbade the killing of gray whales, except in certain limited cases. Now they number about 6,000 and are fascinating to watch and study (34).

As roads improve, Scammon's Lagoon could become a world-famous observatory for the gray whale. There has lately been talk of killing these whales for meat and oil, and in 1967 the Mexican Government proposed contracts for killing them. Furthermore, a private company, affiliated with a United States firm, is extracting salt from seawater on large areas about five miles north from Scammon's Lagoon, around small Guerrero Harbor. The firm would like to build a canal to Scammon's Lagoon to

Fig. 8. Basalt cliff at Cape Colnett, northwest Baja California, Mexico, that drops 100 meters (350 feet) to the edge of the sea. Back of the cliff is
       Colnett Mesa, a table land of about 130 square kilometers (50 square miles) without permanent surface water, a zone of fog-fed vegetation
       and wilderness. (Photo by Meigs)

permit entrance of larger ships for salt, thus breaking up the grounds where the whale breeds, and probably driving the whales away (35). Biological and conservation experts should be consulted before a step is taken that may result in destruction of animal life and the attraction of coastal deserts. With science exploring and people getting into coastal deserts more and more, incipient acts of destruction of conservation and the remedies will more and more come into play.

Eventually, parks probably will be set up on the desert seacoasts in places like Scammon's Lagoon or the rock paintings, where great wonders of nature and man should be preserved for the future. It will take a combination of scientific knowledge and popular pressure to bring this about.

## REFERENCES AND NOTES

1. MEIGS, P.
   1966 Geography of coastal deserts. Unesco, Paris. Arid Zone Research 28.
2. MEIGS, P.
   1965 Coastal deserts: prime customers of desalination. International Symposium on Water Desalination, 1st, Washington, D. C., 1965, Proceedings 3:721-736.
3. SCHREIBER, J. F., JR.
   1968 Inventory of research on desert coastal zones. pp. 649-724. In W. G. McGinnies, B. J. Goldman and P. Paylore, eds., Deserts of the world: an appraisal of research into their physical and biological environments. University of Arizona Press, Tucson.
4. STAMP, L. D. (ed.)
   1961 A history of land use in arid regions. Unesco, Paris. Arid Zone Research 17. 388 pp., maps.
5. WHITE, G. F. (ed.)
   1956 The future of arid lands. American Association for the Advancement of Science, Washington, D. C., Publication 43.
6. UVAROV, B. P.
   1956 The locust and grasshopper problem in relation to the development of arid lands. pp. 383-389. In G. F. White, ed., The future of arid lands. American Association for the Advancement of Science, Washington, D. C., Publication 43.
7. UNESCO
   1963 Changes of climate with special reference to the arid zones. Unesco, Paris. Arid Zone Research 20. 488 pp.
8. ENGELL, F.
   1967 Notas referentes a la adaptación de los pueblos precolumbianos del Peru a la vida en tierras áridas. Symposium on Coastal Deserts, Lima, 1967, Proceedings. (Mimeographed)
9. NEW YORK TIMES
   1966 Jersey tests highway fog "broom." February 18, 1966.
10. MASSON, H.
    1954 La rosée et les possibilités de son utilisation. Institut des Hautes Etudes, L'Ecole Supérieure des Sciences, Annales. Dakar, Senegal. 44 pp.
11. INTERNATIONAL SYMPOSIUM ON WATER DESALINATION
    1965 Proceedings, 1st, Washington, D. C., 1965. U. S. Department of the Interior. 2254 pp., 3 vols.
12. NATIONAL COUNCIL OF THE CHURCHES OF CHRIST IN THE UNITED STATES OF AMERICA, CHURCH WORLD SERVICE

    1964 Water, man's primary need. New York. 6 pp.
13. U. S. DEPARTMENT OF THE INTERIOR, OFFICE OF SALINE WATER
    1964 Puerto Peñasco plant for solar desalting. pp. 197-200. In Saline Conversion Report for 1963. Washington, D. C.
    HODGES, C. N., J. E. GROH, and T. L. THOMPSON
    1965 Solar powered humidification cycle desalination, a report on the Puerto Peñasco pilot desalting plant. International Symposium on Water Desalination, 1st, Washington, D. C., 1965, Proceedings 2:429-459.
14. UNITED NATIONS
    1964 Water desalination in developing countries. United Nations, New York. 325 pp., maps.
15. MEIGS, P.
    1965 The geographical status of water desalination. Geographical Review 55:596-599.
16. HODGES, C. N.
    1968 Desalting the oceans: real or fantasy technology? American Association for the Advancement of Sciences, Committee on Desert and Arid Lands Research, Arid Lands Research Newsletter 28:3-5.
17. ANDRIANOV, B. V., and E. M. MURZAEV
    1967 Certain ethnographic problems of the arid zones. Symposium on Coastal Deserts, 1967, Proceedings. (Mimeographed)
18. MEIGS, P.
    1935 The Dominican mission frontier of Lower California. University of California Press, Berkeley. 192 pp., maps.
19. WILSON, A. W.
    1967 Trends in urban, industrial, and recreational use of the arid zone. Symposium on Coastal Deserts, Lima, 1967, Proceedings. (Mimeographed)
20. DEASY, G. F., and P. GERHARD
    1944 Settlements in Baja California: 1768-1930. Geographical Review 34. Fig. 10.
21. KOENIG, L.
    1956 The economics of water sources. pp. 320-328. In G. F. White, ed., The future of arid lands. American Association for the Advancement of Science, Washington, D. C., Publication 43.
22. HILLS, E. S. (ed.)
    1966 Arid lands: A geographical appraisal. Methuen, London; Unesco, Paris.
23. RODIN, L. E.
    1967 Coastal deserts of the old world and their reclamation. Symposium on Coastal Deserts, Lima, 1967, Proceedings. (Mimeographed)

24. Armillas, P.
    1961   Land use in pre-Columbian America. pp.
           255-276. *In* L. D. Stamp, ed., A history of
           land use in arid regions. Unesco, Paris.
           Arid Zone Research 17.
25. Collin-Delavaud, C.
    1967   Les variations de l'ager dans les oasis de
           piemont du Lambayeque et de la Libertad
           depuis l'époque Chimu. Symposium on
           Coastal Deserts, Lima, 1967, Proceedings.
           (Mimeographed)
26. Unesco
    1961   Salinity problems in the arid zones, pro-
           ceedings of the Teheran Symposium. Unesco,
           Paris. Arid Zone Research 14. 395 pp., maps.
27. Evenari, M., and D. Koller
    1956   Desert agriculture: problems and results
           in Israel. pp. 402-408. *In* G. F. White, ed.,
           The future of arid lands. American Associa-
           tion for the Advancement of Science, Wash-
           ington, D. C., Publication 43.
28. Salinas Messina, R.
    1967   Un ejemplo de valoración pesquera del
           litoral norte de Chile: La industria de la
           harina de pescado en Iquique. Symposium
           on Coastal Deserts, Lima, 1967, Proceedings.
           (Mimeographed)

29. Orbegoso Rodríguez, E.
    1967   Pucusana: distrito de tierras aridas. Sym-
           posium on Coastal Deserts, Lima, 1967,
           Proceedings. (Mimeographed)
30. Amiran, D. H. K.
    1967   Eilat: seaside town in the desert of Israel.
           Symposium on Coastal Deserts, Lima, 1967,
           Proceedings. (Mimeographed)
31. Wheelock, W.
    1968   The beaches of Baja. La Siesta Press, Glen-
           dale. 72 pp.
32. Zavala Abascal, A.
    1964   Las misiones Domincas, el turismo y la
           leyenda negra de Tijuana y de la Baja Cali-
           fornia, Mexico. 43 pp.
33. Gardner, E. S.
    1962   The hidden heart of Baja. Morrow, New
           York. 256 pp., map.
    Meighan, C. W.
    1966   Prehistoric rock paintings in Baja Cali-
           fornia. American Antiquity 31:372-392.
34. Gardner, E. S.
    1960   Hunting the desert whale. Morrow, New
           York. 208 pp.
35. Marx, W.
    1966   Eviction of whales; threat to breeding
           ground of the great gray whale. Atlantic
           Monthly 213:91-95.

# A DESERT SEACOAST PROJECT AND ITS FUTURE

CARL N. HODGES

Environmental Research Laboratory

The University of Arizona

Tucson, Arizona, U. S. A.

# ABSTRACT

With the financial support of the Rockefeller Foundation, the University of Sonora and the University of Arizona are developing an integrated facility for the production of power, water, and food for coastal desert areas. Power is supplied by a relatively conventional diesel-electric facility. The waste heat from the diesel engine provides the thermal energy for a desalination plant. The waste seawater from the desalination plant plus additional seawater is used to modulate the environment inside inflated plastic greenhouses. Since plants growing within the greenhouses are exposed to a humid environment, their demand for the relatively expensive desalted seawater is small. It is hoped that the cost-sharing provided by this integrated approach will make it possible to provide power, water, and food at many coastal desert areas at a cost that is less than more conventional alternatives.

# A DESERT SEACOAST PROJECT AND ITS FUTURE
## Carl N. Hodges

As indicated in the section of this book by Peveril Meigs, and elsewhere, about 20,000 miles of coastal desert is potentially available for human habitation, for the most part. Further, it is well demonstrated that the coastal desert areas provide one of the most desirable regions for human habitation, if the basic amenities of life can be supplied. Ample evidence of this fact is provided at the desert coasts of southern California (U. S. A.) and Baja California (Mexico) to which people are migrating rapidly. For the case of southern California, to supply the water necessary for this influx, aqueducts have already been built 340 miles long to the Colorado River. Other aqueducts over 400 miles long are under construction to the Feather River (1), and it has even been proposed that some day water may be brought from Alaska, thousands of miles away (2).

Recently, however, there has been growing interest in the world in the possibility of the utilization of desalted seawater as a resource alternative to importing fresh water from distant sources for use at coastal locations. Probably the most ambitious investigation into this possibility has been conducted by the Oak Ridge National Laboratories into the feasibility of utilizing large nuclear energy centers to support industrial and agro-industrial complexes (3). Their study evaluated the potentialities of a central facility that could produce 2,000 megawatts of electricity and a billion gallons per day of distilled water from the sea. This quantity of water would be adequate to support an industrial complex as well as a 300,000-acre farm.

The authors of the Oak Ridge report are optimistic that such a facility could be constructed in perhaps 10 to 30 years and be economically sound. They do point out, however, the dangers of projecting economic analyses that far into the future; they suggest that their report be taken strictly as a preliminary feasibility report and that certainly any actual construction in the future would require a detailed analysis for a specific site and location.

If such a facility were ever to be constructed, due to the size it undoubtedly would have to be done on a national or possibly international scale. Further, the capital investment would be so great that few developing countries of the world (which contain much of the coastal desert areas) could afford such a venture without great assistance from the developed world.

In 1961, the University of Arizona began an investigation into the possibility of utilizing solar energy as a thermal source to provide a means of economically supplying desalted seawater for relatively small coastal desert communities. Desalting may be a more attractive alternative to importation for small communities than for larger communities. This situation comes about because long aqueducts are generally justified only if relatively large quantities of water are required.

With the support of the Office of Saline Water, U. S. Department of Interior, in 1963, a pilot facility was constructed in cooperation with the University of Sonora (4) at Puerto Peñasco, Sonora, Mexico. This facility is located approximately 220 miles southwest of Tucson, Arizona, on the Gulf of California. The plant was operated for a period of two years, utilizing solar energy as the primary energy source. During this time the nature of the research program began to change.

As potential sites for future solar distillation plants were investigated, it became obvious that in almost all coastal communities of the world (and probably in all that could ever support any significant desalting plant) there would be available some waste thermal energy from either a municipal power center or individual private power-generation facilities. In general, the waste thermal energy would be more economical than solar energy when the capital investment necessary to construct the solar collector devices was legitimately amortized. Therefore, the Puerto Peñasco desalting plant was modified to operate from the waste thermal energy of the diesel-electric plant that generates the power for the experimental complex, and research on solar collectors was discontinued.

As further investigations were conducted into the actual requirements of coastal desert areas and the economics of desalting seawater, it also became clear that for both sociological and economic reasons it would be desirable to produce at least part of the food consumption of the community within the community. It was also evident that for anything but gigantic desalting plants, it is unlikely that in the foreseeable future desalted seawater would be inexpensive enough for conventional-type agriculture. Even for high-value items such as vegetables, the production of crops with irrigation water that costs $0.50 to $3.00 per thousand gallons would not be feasible.*

*The real cost of water from a desalting plant is very difficult to specify; however, most efficient plants now operating or projected for the near future produce water within this range.

A plant growing in an open-field condition will require water amounting to 100 to 1,000 times its final weight. Most of this water serves no important physiological function except the cooling of the plant by means of transpiration. If the plant could be encased in a very humid environment in which the potential for transpiration could be greatly reduced, and at the same time a mechanism besides transpiration substituted for the cooling requirements of the plant, it is possible that plants could be grown on one to ten per cent of the fresh water that they would require in the open field. With a relatively small requirement for water, the cost of the water would become much less important in the total crop economics, and food production could be accomplished in connection with present-day desalting technology. In an attempt to do this, with financial support from the Rockefeller Foundation, a system of controlled-environment structures for plant growth was added to the ongoing research program at Puerto Peñasco. They completed the central facility as envisioned for the production of power, water, and food for coastal desert areas (Figs. 1 & 2).

## POWER PRODUCTION

Power for the Puerto Peñasco experimental facility is generated in a manner similar to that utilized by many small communities of the world. A Caterpillar D330T, 80-horsepower, diesel engine turns a 60-kilowatt electric generator. The 60 kilowatts of electrical power is used for pumps and blowers in the desalting plant, for pumps and fans in the greenhouses, and for lighting and powering the experimental equipment throughout the facility. In any internal combustion engine, such as the diesel engine, approximately two-thirds of the fuel energy supplied to the engine is normally rejected as waste heat. Half of this waste heat is usually thrown away in the hot exhaust gases, and half is rejected through the water cooling system of the

Fig. 1. University of Arizona–University of Sonora experimental facility, Puerto Peñasco, Sonora, Mexico.

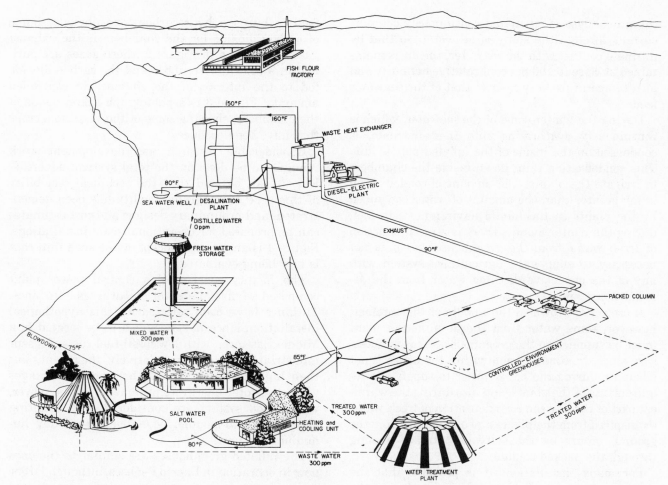

Fig. 2. Schematic of the controlled-environment community shown in Figure 1.

engine. However, the Puerto Peñasco engine is equipped with waste heat recovery equipment which is cooled by the salt water feed to the desalting plant. The waste heat from the engine warms the salt water to 160°F.

## WATER PRODUCTION

The Puerto Peñasco desalination plant is a 6,000-gallon-per-day humidification type distillation unit. The humidification process was developed at the University of Arizona particularly for use with low-temperature energy sources. By operating the plant at temperatures below 160°F, no scaling from the seawater occurs and no pretreatment of the seawater is necessary.

The 160°F stream of seawater supplied from the power facility is pumped to a packed column evaporator (the round tower in Figure 2) where it flows countercurrent to a high-velocity stream of air. This stream of air is transported to an extended surface condenser (the square tower in Figure 2) which is cooled by 78°F seawater taken from a salt-water well. It is the water from the condenser (*preheated* by the latent heat of condensation) that is fed to the power plant where it is heated to 160°F. The water vapor in the airstream condenses on the fins of the

surface condenser and falls to the bottom of the condenser, where it is collected to be pumped to storage.

Only 10 per cent of the salt water that is fed to the desalting plant is actually evaporated and collected as product. The other 90 per cent of the seawater is rejected from the plant at approximately 88°F, still containing all the original salts. The 6,000-gallon-per-day productivity of the pilot plant is more than adequate to meet the requirements of the 20,000 square feet of experimental greenhouse area and supply the drinking-water bottling facility operated by the University of Sonora for the residents of the city.

## FOOD PRODUCTION

As shown in Figure 1, there are four experimental inflated greenhouses at the Puerto Peñasco facility. Each of the four greenhouses consists of two chambers connected by underground tunnels. The tunnel at one end serves as the entryway into the chambers as well as an air passage between the two halves. The tunnel at the opposite end contains a packed column heat exchanger for environmental control. The waste salt water from the desalting plant, as well as additional seawater, is continuously pumped into the greenhouse heat exchangers. The air within the greenhouse is circulated continuously counter-current through the spray of seawater, so that by intimate contact with the seawater, the air is maintained at close to 100 per cent relative humidity and at a temperature very near to that of the seawater feed.

During the winter some of the seawater, which is continuously evaporating into the airstream, is condensed on the inside of the inflated plastic film. This condensation rains down inside the chambers to irrigate the plants. The amount of condensation is far greater than the amount of water consumed by the plants in the humid environment; in fact, during the winter months there is a net production of fresh water from the greenhouses; so it is not necessary to supply the agricultural system with any of the expensive distilled water from the desalination plant.

If crops are grown in the middle of the summer, however, some water from the desalination plant must be supplied to the greenhouses. This is necessary since the condensation within the house is not adequate during the hot weather to supply the requirements of the plants, and also during the warmest part of the day some condensation of fresh water (transpired from the plants and evaporated from the ground) occurs on the saltwater stream passing through the packed column.

For many installations it is probable that the growing of crops will not be desirable during the two warmest months of the year. Since the sizing of the mechanical equipment (pumps, fans, packed columns) is almost completely a function of the maximum temperature allowed in the house, it may be more economical to simply set aside the two warmest months of the year for renovation of the greenhouses. If an expendable plastic is utilized, as is now envisioned, these two nongrowing months may be used for such things as removing and discarding the plastic film, cleaning the greenhouses, renovating cultural systems, installing new plastic film, sterilizing, and replanting. If permanent plastic film is utilized, then possibly larger mechanical equipment for the additional cooling load will be justified, or a shading program initiated, to allow year around food production.

It has been determined that for some crops grown in these chambers, where many of the environmental factors can be controlled, the limiting factor on growth is the carbon dioxide level of the atmosphere surrounding the plants. By elevating the carbon dioxide level (5), significant increases in productivity can be obtained.*

Since the engine that provides power for the

facility burns a hydrocarbon, one obvious source of carbon dioxide for the chambers is the exhaust gases. As shown in Figure 2, these gases are purified in a seawater scrubber and the carbon dioxide fed to the interior of the chamber in controlled amounts. Figure 3 is a photograph taken inside of the chambers, showing some of the vegetable crops that have been produced.

Considerable research and development work remains to be done on the total system; nevertheless, it can be said that the technical feasibility of this type of integrated facility has been demonstrated, and preliminary designs and cost estimates can be prepared for operational-scale installations. Figure 4 is an artist's concept of one such unit that is now being evaluated.

The facility consists of a central power plant composed of five diesel or natural gas turbines. Turbines have been selected for this hypothetical installation, since it is assumed to be located in a Middle East area with low fossil-fuel costs. Present industrial turbines are relatively inefficient, but their high heat-rejection rate has some advantages when the total energy requirements of the power, water, food system are considered. The desalting plant is a two-module, 64,000-gallon-per-day humidification unit.

The inflated greenhouses are similar to the ones now in operation in Puerto Peñasco, although larger and with one central service tunnel. The anticipated productivity from the greenhouses is over a thousand tons per year of high-quality vegetables. Such an installation then not only has the potential to supply the local community, but it might also export to interior regions.

To specify productivity costs from such a unit is difficult, due to the extreme effect on the cost of the amortization schedule selected, labor costs used, and miscellaneous other costs. One specific analysis was made for Puerto Peñasco, Sonora, Mexico, however, in which a 15-year amortization schedule for permanent equipment was utilized and actual local labor costs were considered; the projected production costs of power, water, and vegetables would then be slightly less than 50 per cent of the cost of present alternative sources.

It is the opinion of the author that systems such as that of Figure 4 will be utilized in some coastal areas of the world within the next few years. Undoubtedly such installations will be limited for some time to specific areas where economic factors most favor the expenditure of the capital necessary to construct such a facility. It is encouraging to note, although not likely to happen, that if it were necessary, with the development of 5 per cent of the world's desert coasts (925 miles) to a depth of 20 miles using such a system producing food at productivity rates already obtained, one billion people could be fed.

---

*The most responsive plant evaluated to date is IR-8 rice, which showed a 100 per cent increase when the carbon dioxide level was increased from 300 ppm, the normal atmospheric level, to 1200 ppm.

Fig. 3. Vegetable crops inside Puerto Peñasco greenhouses.

Fig. 4. Proposed power-water-food facility for coastal desert areas.

## REFERENCES

1. MONROE, L. E. (ed.)
   1967  Metropolitan Water District of Southern California report for the fiscal year July 1, 1966, to June 30, 1967.
2. NORTH AMERICAN WATER AND POWER ALLIANCE (NAWAPA)
   1967  Brochure 606-2934-19. The Ralph M. Parsons Company, Los Angeles.
3. OAK RIDGE NATIONAL LABORATORY REPORT
   1968  Nuclear Energy Centers, Industrial and Agro-Industrial Complexes, Summary Report. ORNL-4291. 30 pp.

4. HODGES, C. N., *et al.*
   1966  Solar distillation utilizing multiple-effect humidification. Department of the Interior, Office of Saline Water Research and Development, Report 194.
5. HODGES, C. N.
   1967  Controlled-environment agriculture for coastal desert areas. University of Arizona, Institute of Atmospheric Physics, Environmental Research Laboratory, Report. 32 pp.

# GEOGRAPHY OF THE CENTRAL NAMIB DESERT

RICHARD F. LOGAN
Professor of Geography
University of California
Los Angeles, California, U. S. A.

# ABSTRACT

The Namib Desert of the South West African coast is representative of the cool coastal desert type. Little affected by seasonality, its climate is cool and extremely damp all year at the coast, but markedly warmer and drier inland. The very scanty vegetation is strongly succulent near the coast, with grasses and nonsucculent bushes dominating the interior. Large areas are totally barren.

Three regional units are recognized: the Coastal Namib, the area of strongly marine climate; the Namib Platform, an erosional surface (pediment) of extreme flatness cut by a few stream valleys and interrupted by widely scattered *inselberge;* and the Dune Namib, a vast sand-sea.

Aside from primitive herders of goats in riparian situations and very marginal commercial grazing in the interior, economic development is based entirely on fisheries and port activities and a rapidly expanding seaside resort activity.

# GEOGRAPHY OF THE CENTRAL NAMIB DESERT
## Richard F. Logan

## LOCATION AND GENERAL NATURE

The Namib Desert of South West Africa is a part of one of the two great belts of desert that encircle the world in the vicinity of the tropics. The northern hemisphere belt is composed of the Sahara and the deserts of Arabia, Iran, Afghanistan, and Pakistan in the Old World, and of Mexico and the southwestern United States in the New World. In the southern hemisphere, the arid belt includes much of Australia and the desert areas of Chile and southern Africa. While the land areas in the southern-hemisphere arid belt are much smaller than those of the northern-hemisphere arid zone, it must be realized that tremendous expanses of ocean within tropical latitudes actually have arid climates, and despite the presence of water everywhere, are as climatically dry as the Sahara.

The landward portions of the tropical deserts suffer from extreme aridity, great summer heat, and very low humidities, except in certain extraordinary, anomalous areas: the cool coastal deserts.

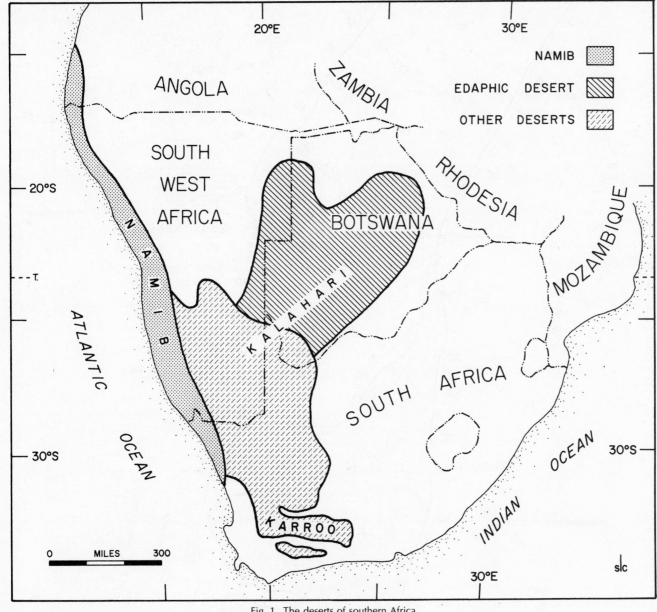

Fig. 1. The deserts of southern Africa.

Three such deserts exist: the western coastal desert of Baja California in Mexico, the Atacama desert of Chile, and the Namib of South West Africa. All occur in about the same latitudes on the western coasts of continents, where the water alongshore is abnormally cold for the latitude. All have the same climatic characteristics, quite far removed from the popular concept of deserts. While virtually rainless, they have very high humidities and are frequently swathed in fog; their summers are cool,

their winters mild, and their rare hot spells come only in winter. The Namib is in every way representative of this cool coastal desert type.

The Namib is actually the western coastal phase of the much larger desert area of southern Africa, with which it merges almost imperceptibly (Fig. 1). But the desert of southern Africa is by no means simple or uniform, and so the transitions are multiple and varied in natures. Southern Botswana, the southeastern part of South West Africa, and

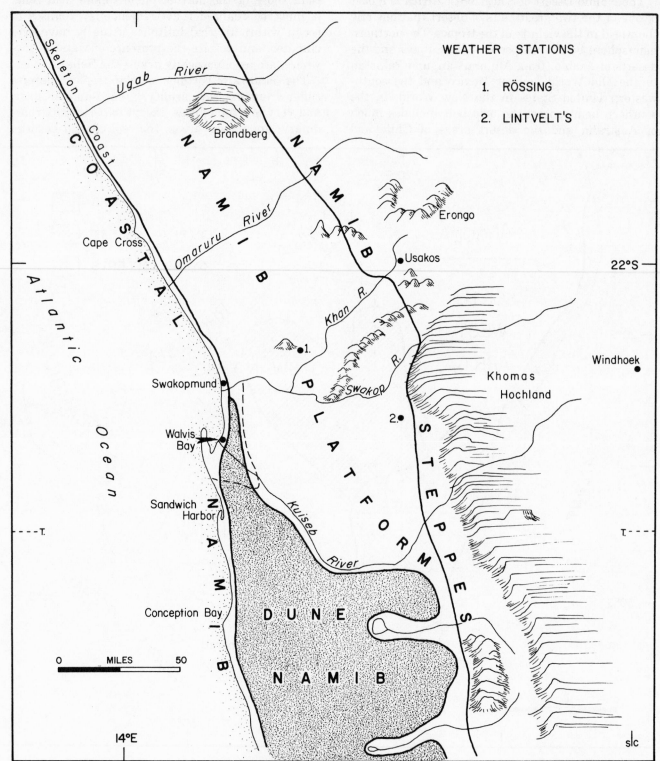

Fig. 2. Regional divisions of the central Namib Desert.

part of the Cape Province of South Africa lie within the Kalahari Desert—an area of very little rain and quite devoid of vegetation. The Kalahari is generally considered to extend much farther northward, to include most of northern Botswana and much of eastern South West Africa. Climatically, however, this area does not qualify as a desert, for its annual average rainfall is relatively high. It is considered desert because its rainfall is absorbed directly by the thick layers of sand that cover its surface, making surface water unavailable. On the other hand, deep-rooted trees and shrubs tap this water, and an open woodland and/or a brush cover exists over much of the so-called desert. This then is a desert of edaphic, rather than climatic, origin.

Within both parts of the Kalahari, such rain as does fall comes in the form of summer convectional showers, and the winters are dry, clear, and cool. Farther south, within the Cape Province, lies the Karroo, wherein desert conditions also prevail but where the precipitation occurs in winter, sometimes in the form of snow. The southernmost part of the Namib exhibits tendencies in this direction and represents a transition between the Namib proper and the Karroo.

Southern Africa thus has four desert types—the Namib: mild, rainless, but with high humidity; the Kalahari proper: with hot summers, cool, dry winters, and summer precipitation; the *edaphic* Kalahari: with summer precipitation of non-desert proportions, hot in summer and cool in winter, and desertic because of the porosity of its surface materials; and the Karroo: with cold winters, warm summers and winter precipitation.

The Namib occupies an elongated tract on the western (Atlantic) coast of southern Africa. It extends from the vicinity of Mossamedes in Angola across the breadth of South West Africa to the mouth of the Olifants River in the Cape Province of South Africa—a distance of about 1,300 miles. From the coastline, it reaches inland to the foot of the Great Western Escarpment of southern Africa—a distance of 50 to 90 miles. From the purely climatic viewpoint, the same desertic condition also extends over the neighboring ocean—northward almost to the equator and northwestward almost to the bulge of Brazil.

This paper will concern itself with the core area of the Namib, from the sea to the foot of the Escarpment between the parallels of 22°30′S and 23°30′S (the Tropic of Capricorn). Within that area (Fig. 2) it is possible to study representative samples of most aspects of the Namib and to observe contrasts between its coastal and interior parts.

## CLIMATE

Throughout the year, the semi-permanent high-pressure cell occupying the center of the South Atlantic exerts a great influence on the weather of the Namib. Within the cell, cool, dry air subsides from aloft. With descent, this air is warmed adiabatically and comes to the surface as warm dry air. A similar cell is normally developed over the high plateau of southern Africa. During the winter its descending air usually reaches the surface, producing warm, dry, clear air over all of the interior. During the summer, however, surface heating is strong, and the resultant convectionality not only prevents the descending air from reaching the surface but develops a fairly strong, low-pressure circulation pattern on the surface, although the high-pressure zone still exists in the upper air.

In response to the South Atlantic pressure cell, the air movement off the South West African coast is from the south-south-east throughout the year. Since at any season, the temperatures of the interior Namib are warmer than the sea, a thermal pressure gradient develops which deflects some of the moving air across the coast and inland, creating a sea breeze; this is strictly a diurnal phenomenon, from late morning until sunset. The direction of air movement is usually from the south-south-west, and it blows with moderate vigor (force 3 to 4 Beaufort). Blowing across the whole width of the Namib, it rises in irregular streams up the face of the Escarpment before dispersing.

The Benguela Current flows northward in the eastern Atlantic, bringing water from the Southern Ocean far equatorward. This already cold water is made even colder by the admixture of water upwelled from the depths. As a result, water temperatures along the Namib coast are far below the normal for the latitude, being in the middle fifties (Fahrenheit) at all seasons. The air in contact with this water is chilled thereby and is normally at or near the saturation point. It is the daily movement of this air into the desert that makes the Namib so different from the usual concept of a desert.

This regular importation of air from the cool sea produces a strong temperature inversion over a wide zone back of the coast. At the coast, the layer of cool air is normally some two-thousand feet thick (1). Within it, the surface temperature is that of the sea offshore, and humidities approach the saturation point. Fog is common; and when fog is not present, a stratus layer often exists, with its base less than a thousand feet above the coastal surface. In its passage across the desert, this air warms considerably, evaporating the fog and cloud and lowering the relative humidity. But as far inland as the top of the Escarpment, the air is distinctively different from the interior and the superior air.

Throughout the year, the air above the inversion is devoid of moisture, having descended from high aloft in one of the high-pressure cells; hence it cannot be looked upon as a source for precipitation. During the summer, however, the surface low-pressure draws in moisture-bearing air from the Indian Ocean to the interior, and sometimes as far west as the Namib. This air is the source of the summer rains of the interior of the continent, where strong surface-heating produces strong convectional currents, which in turn produce thunderstorms. Over the Namib, however, this chain of events is stopped before it starts by the interposition of the temperature inversion between the ground and the moist air aloft; the cool sea air precludes any convectional activity in the moist upper air. The lower layer is moist also, of course, but too cool and too shallow to contain a great volume of moisture, and too cool for convectionality. Thus we have the anomaly of a virtually rainless area whose air is almost permanently saturated with moisture. Were this layer of moist air thicker, or were the ocean off shore warmer, conditions potentially productive of precipitation would prevail and the desert would disappear.

No area is totally without rainfall, however, and the Namib does receive some. On the rare occasions when moist air is present aloft and not sealed off from terrestrial heating by the inversion, rain may occur. Such conditions develop usually only at the beginning and the end of the summer. Two conditions are necessary: absence of the inversion and the sea breeze, and presence of moist Indian Ocean air aloft. The chances of the coincidence of these conditions are very slight, and hence rains occur very infrequently.

Rain is least plentiful in the coastal sector. Swakopmund, situated on the coast, has an annual average of 0.65 inches (2). In one year (1934), however, 6.13 inches were received—2 inches of it in a single day. If this very abnormal year is omitted from the record, the average annual rainfall is only 0.52 inches. Many years are completely rainless.

Rainfall increases with the distance inland. At Goanikontes (787 feet, 20 miles from the coast), the average is 1.35 inches. At Khan Mine (925 feet, 32 miles inland), it is 1.71. At Donkerhuk, in a valley between massive outliers of the Escarpment (4,200 feet, 90 miles from the sea), it is 6.8 inches, or 5.8 if the exceptional year of 1934 (when 29.67 inches fell) is omitted.

The nature of weather conditions can be best described by a detailed study of representative stations on representative days. For this purpose, three stations at which the writer maintained weather instruments have been selected: Swakopmund, at the coast; Rossing, 26 miles inland and 1,394 feet above sea level; and Lintvelt's, a gap in the first outliers of the Escarpment 76 miles from the coast and 3,750 feet in elevation. Hygrothermograph traces for these stations on representative midsummer and midwinter days are presented here.

Meteorologically, conditions at Swakopmund in summer are monotonously regular. The trace for Christmas Day, 1956, is representative (Fig. 3). During the first half of the day, from midnight until noon, the humidity remained consistently at 100 per cent, and the temperature fluctuated only between 58° and 56°F. The temperature curve crossed 60°F about 1 p.m., at which time the humidity first dropped below the saturation point. During most of this period, however, there was no fog; but a very low stratus cover persisted, at times shrouding the top of the lighthouse. Through the remainder of the afternoon the sun occasionally broke through, and the temperature rose to a maximum of 63°F at 5 p.m., while the humidity dropped as low as 92 per cent. At 6 p.m., the humidity went once more to 100 per cent and the stratus cover was resumed; these conditions prevailed the rest of the day.

During the summer, the humidity is normally at 100 per cent for 19 hours per day, and the mean minimum is 90 per cent. Summer days have a range of 8.6°F, from a minimum of 58.5° to a maximum of 67.1°. In all respects, the climate of Swakopmund is the epitome of oceanity.

Along all of the Namib coast, most days dawn slowly, with gray fog shrouding the area. There is no sunrise—only a gradual graying, an enlightening of the darkness of the night. The air is calm, or wafting lightly from the north, and feels very damp. The few objects that can be seen seem all out of proportion. One is conscious of his whereabouts chiefly from the sounds about him—the dull boom of the surf or the crying of the sea birds on the invisible banks.

By 9:00 a.m. or so, the fog lifts. Actually there is no *lifting;* the air stays still while droplets in the lower levels are evaporated. Short wavelengths of insolational energy have penetrated the fog blanket and warmed the soil of the marsh and the sand of the dunes. Long wavelengths radiated by these now-warmer bodies heat the air above, and the fog particles go rapidly from the liquid into the gaseous state. And so, suddenly, one can look for miles up and down the coast under the fog; but the ceiling is at 100, or 50, or even 25 feet.

About this same time, the sea breeze begins to blow gently. About 10:00 or 11:00 a.m., the sun becomes visible intermittently as a pale gray disc, and soon thereafter, following several premature bursts, it comes out to stay. The fog has now succumbed to evaporative attacks from both above and below and has gone for the day. But banks of it will persist just offshore, and wisps will scud by all day, just overhead.

The sea breeze gradually increases its vigor until by noon it is blowing at force 4 or 5 Beaufort. At first soil and sand are held in place by the dampness of the preceding night, but eventually they dry out and the finer particles are released, filling the air

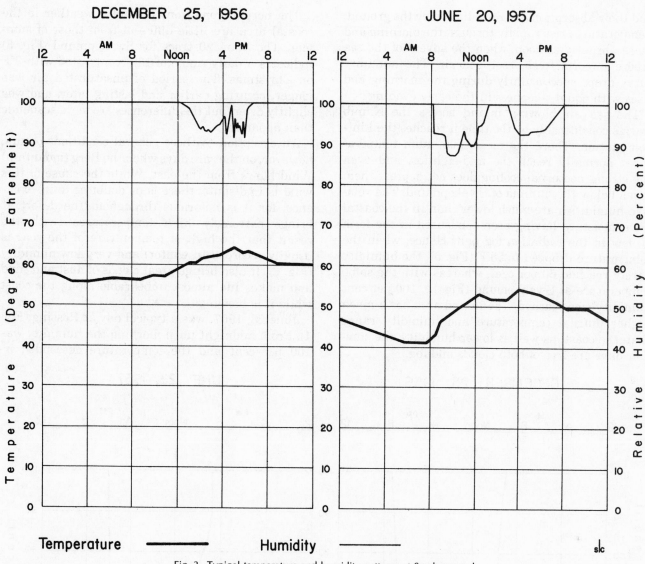

Fig. 3. Typical temperature and humidity patterns at Swakopmund.

with dust. Small ripples of sand blow across the ground and collect behind every obstacle.

With late afternoon, the insolational heating gradually decreases, the evaporative attack on the offshore fogbank lessens, and the wall of fog begins to creep shoreward, cutting off the sun from the shore. And so, just as there was no sunrise, so there will be no sunset: the sun merely disappears into the fogbanks, and much later, darkness comes on imperceptibly.

But the fog does not come in at sea level: the radiating warmth of lagoon, marsh, and sand keeps the air warm, and therefore clear, for a few feet above the surface. Once again one can look for miles up and down the coast under a gray roof of fog. Above the lagoons, the air chills rapidly. The shallow water, warm from the day's sun, gives off moisture more rapidly than the chilled air can absorb it, and so the lagoon *steams,* forming little waftings of fog that float off eastward on the declining sea breeze. Eventually, stored heat of pool and land is exhausted;

with nothing left to keep the fog aloft, it settles down and blankets everything.

The Christmas Day trace for Rössing (Fig. 4) shows strikingly greater variation in both temperature and humidity than was recorded for Swakopmund. The air was saturated for fourteen hours but was markedly drier (to 37 per cent) in midday. The cause of the saturated condition is different from that at Swakopmund. The fog at the coast is advectional sea fog, blown in from the neighboring ocean. That at Rössing appears to be radiational fog resulting from pronounced nocturnal cooling. It will be noticed that the night temperatures at Rössing are well below those of the coast. Sea air drawn inland is warmed greatly during the daytime; but as night comes on, radiational cooling is strong, temperatures drop below the dewpoint, and condensation occurs, forming fog.

The rise in temperature at Rössing starts a little before the fog clears away, due to the penetration of shorter wavelengths of insolation through the fog and their absorption and re-radiation by the ground. Temperatures rise rapidly through the morning and cease abruptly at noon, when the advent of the sea breeze imports fresh, cool, sea air. The humidity curve drops spectacularly during the morning and rises with equal steepness in the early evening.

The sea air blowing inland across the Namib warms considerably by the time it reaches the Lintvelt station. There the mean maximum temperatures normally reach the mid-eighties, and even the strong nocturnal cooling does not depress them much below the minima of Swakopmund. The relative humidities are much lower than in the coastal area, moisture having been condensed in the form of dew in the radiation fog belt. Hence, when the temperature dropped to 58°F (Fig. 5), the humidity barely reached 60 per cent, whereas with the same temperature at Swakopmund (Fig. 3), 100 per cent humidity was reached. The inland area has a much higher diurnal temperature and humidity range than the coast, as well as lower humidity; the summer days are free of both clouds and fog.

The normal conditions of winter weather in the coastal area are little different from those of summer. The June 20 trace for Swakopmund (Fig. 3) indicates a daily range of 13°F contrasted with 9°F on Christmas. The period of unsaturated air was longer, occuring earlier and lasting later, and was slightly drier; but the differences are more academic than apparent.

An exception to these general conditions occurs, however, on the rare days when the Berg (mountain) Wind blows from the east. While the cause of this wind is in dispute, there is no doubt as to its existence, for it is notorious throughout the desert. It brings extremely high temperatures (with the result that the highest temperature of the year is usually recorded in winter) and very low humidity (Fig. 6). It also brings great clouds of dust and sand and makes life almost unbearable along the coast while it is blowing.

June 23, 1957, was a typical day at Rössing (Fig. 4). From midnight until morning the humidity was 100 per cent and the temperature descended ir-

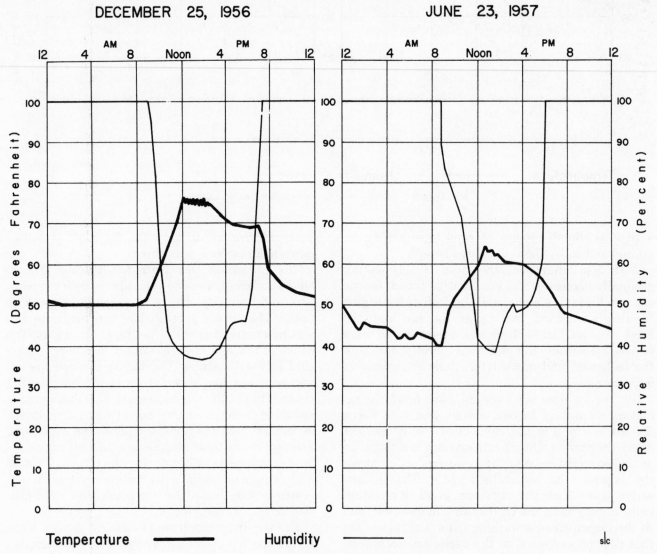

Fig. 4. Typical temperature and humidity patterns at Rössing.

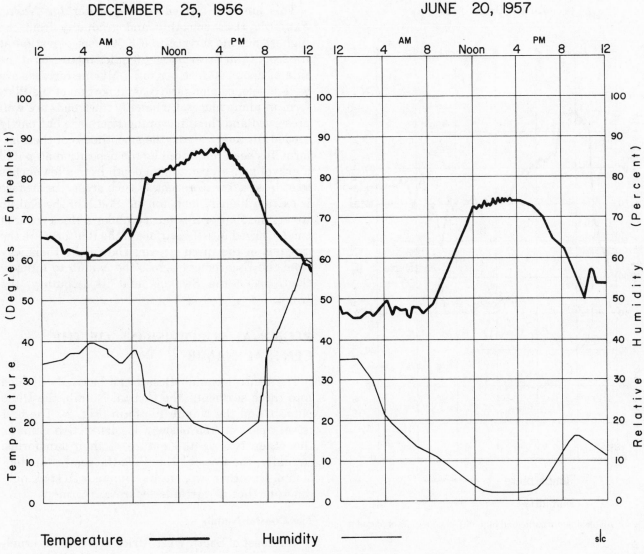

Fig. 5. Typical temperature and humidity patterns at Lintvelt.

regularly from 50° to 40°F. At 8:30 a.m., the humidity dropped spectacularly, reaching a minimum of 38 per cent at 1 p.m., and the temperature soared to 62°F. With sundown, the temperature dropped rapidly and the humidity rushed back to the saturation point. The fog here is, as in summer, radiation fog. But in contrast to the summer situation, there is no midday effect from the sea breeze, which does not penetrate strongly into the interior in the winter months.

June 20, 1957, provided a typical midwinter trace (Fig. 5) for the Lintvelt station. Thermometrically, the night hours showed a serrate trace, the result of local winds upsetting an inversion developed from terrestrial radiational cooling. Temperatures averaged in the middle forties. Temperatures rose strongly from 8 a.m. until noon, cresting in the middle seventies. Humidity was low throughout the period, ranging from a high of 35 per cent to a minimum of 2 per cent. Fog is almost unknown in the Inner Namib, where the winter is characterized by

cool, brilliantly clear nights and sunny, very dry, warm, clear days.

## GEOMORPHOLOGY

In the area under consideration here, the Namib is underlain by rocks of highly metamorphosed nature and of very great age, probably largely pre-Cambrian (3). The most widespread and the most characteristic, although not the most conspicuous, formation is a great mass of mica schist. Grading from almost pure biotite-muscovite schist to micaceous quartzite, these beds are found from sea level to the top of the Great Western Escarpment at an altitude of well over a mile. Although silvery-gray on close inspection when fresh, they commonly appear darker on weathering and make a dark and foreboding landscape. Over much of the Namib, they strike in a northeast-southwest direction. Certain more arenaceous members resist weathering and stand as ridges above troughs eroded in their weaker

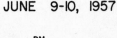

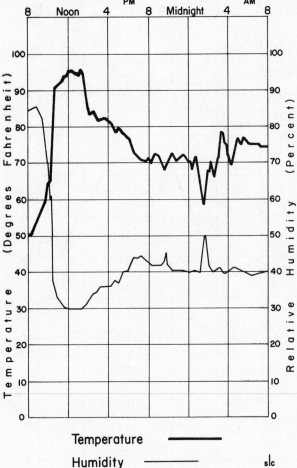

Fig. 6. Atypical temperature and humidity pattern at Swakopmund.

This surface, termed by the writer *the Namib Platform,* rises smoothly and gradually from the highest marine terrace to a 3,000-foot elevation at the foot of the Escarpment 80 miles inland, or at the rate of about 40 feet per mile. Marine terraces are well-developed along the coastal portion of the Platform in many places. Commonly the treads are well preserved and the intervening risers are but partly eroded. In many areas, the smoothness of the Platform has been reinforced by the deposition atop it of a gravel veneer, ranging in depth from a few inches to at least a few score feet. In such areas, the surface is extraordinarily monotonous. South of the Kuiseb River, the Platform is buried under a thick cover of sand. Carved into the surface of the Platform are the canyons of the three streams that make their way from the Escarpment across the Namib to the sea: the Kuiseb, the Swakop, and its tributary, the Khan.

## REGIONAL SUBDIVISIONS OF THE CENTRAL NAMIB

The central Namib Desert can be easily divided into three segments: the Coastal Namib, the Dune Namib, and the Namib Platform (Fig. 2). The first is set apart on the basis of its distinctive climate, the other two on the nature of their landforms. Actually, the Coastal Namib is the seaward extension of the other two, but its climate is so strikingly different that it warrants separate treatment.

### The Coastal Namib

The coast of South West Africa is one of the cruelest shorelines imaginable. It is devoid of drinkable water, uninhabited, bleak and barren, and endlessly pounded by a wild surf. It is a land of paradox. Rain almost never falls, yet the area is drenched in moisture. There is nothing to eat, yet the surface reveals myriads of tracks: of beetles, mice and lizards, jackals and antelopes, flamingoes and penguins. Here the hottest days come in winter, rain seldom falls but floods are violent, fog shrouds a parched land, and sterile sands conceal gravels rich in diamonds.

Within the area under study here, the coast consists of two major indentations closed or partly closed by sand spits, and a series of long beach ridges and barrier beaches separating elongated lagoons from the sea. The oblique breaking of the rollers from the southwest, combined with the steady southwest wind, tends to move water and sand northward along the coast. The effect has been to construct spits across the indentations, including a large one that has closed Sandwich Harbor almost completely since 1889, and another (Pelican Point) that forms the present (1968) harbor of Walvis Bay and is steadily building forward (northward).

neighbors—giving the country a corrugated aspect of low relief. Other formations include bands of marbles, pristinely white and standing as conspicuous ridges, and quartzites, including pre-Cambrian tillites (*4*). Intruded into all of these are granites of indefinite age (*5*).

All of these formations have been beveled by an erosional surface whose origin has not been definitely determined. Earlier writers (*6*) pronounced it a surface of marine planation; later authorities have leaned toward sub-aerial agencies as the major causative factors. The present writer believes it to be a peneplain or a pediplain, basing that opinion on the presence of scattered unreduced remnants (monadnocks or inselberge) which rise above the general level, are more frequent toward the Escarpment, and bear no sign of marine erosion; the face of the Escarpment, which is a product of normal stream erosion, with the re-entrant valleys graded accordingly with the peneplain; and the fact that where marine erosion has occurred, in the last few miles seaward, its evidence is clear and unmistakable.

While the foreshore is totally devoid of vegetation (aside from seaweed cast up by the surf), it is particularly rich in avifauna, a reflection of the abundance of fish in the adjacent sea. Flamingoes line the shore, fishing standing on spindly legs at the surf-edge. With them are snowy-white pelicans, with their great yellow beaks. Small strandlopers run in and out of the surf, stopping nervously to probe in the sand with their long, thin bills. Overhead fly flocks of cormorants, duikers, gannets, mews, gulls, and a variety of small sea birds. Small groups of penguins walk along the beaches in great solemnity. Jackals live in the dunes back of the beach and prey on the flamingoes, pulling them down at night as they sleep standing on one leg at the water's edge. Hyenas scavenge the beach at night, and seals bask on it in the sunny hours of the afternoon.

Back of the beach, a sand ridge has been accumulated by the waves and built higher by the wind into irregular dunes. Behind small mat-like bushes (notably *Arthraerua leubnitziae* and *Mesembryanthemum salicornioides*), elephant-head dunes up to 6 feet in height are seen. Where the orientation of the coast is favorable, larger dunes (to 40 feet) have developed. On them, four additional plants occur: two shrubby perennial grasses (*Aristida sabulicola* and *Eragrostis spinescens*) and two halophytic shrubs (*Salsola aphylla* and *S. nollothensis*). Amidst the dunes live the jackals, a little mouse (*Gerbillus swakopensis*), some lizards and scorpions.

The lagoon at Sandwich Harbor consists of a large expanse of open water with a level rising and falling with the tide, broad expanses of sand flats and bars, small areas of mud flats, and a large development of salt marsh. The latter results from deposition of fine silts by the tidal currents, sand blown from the adjacent beaches and dunes, decay of plants, and shells of mollusks, crustaceans, and other shellfish. Thus the marsh ranges from slick black organic mud to beds of sand and banks of shell. It is cut by a maze of channels that alternately flood and drain the marsh. Very low natural levees border the channels, preventing the interior areas between channels from draining, and thus creating stagnant pools with slimy mud bottoms. The vegetation is that of the typical coastal salt marsh the world over: thick-leaved, thick-stemmed, recumbent and sprawling succulents. The marsh swarms with birds and crustaceans, but mammalian fauna is virtually absent.

The mud flats of Sandwich and Walvis Bay are usually inundated during the higher high tides. While they have very low relief and are cut by only very shallow channels, they are pock-marked by low, ring-shaped mounds, from 1 to 10 feet in diameter and up to a foot in height. These are explosion craters from which hydrogen sulphide gas has erupted. Such eruptions occur along the central Namib coast at irregular intervals during the summer. They apparently result from the generation of gas through biochemical action in the thick mud deposits underlying the area, although the exact mechanism is not thoroughly understood (7). During such eruptions, great numbers of fish are killed, paint and silverware is discolored, and the odor is smelled for miles inland. The mud flats are totally devoid of vegetation but are extremely popular with many forms of bird life.

Between Sandwich Harbor and Walvis Bay, discontinuous chains of basins lie inland from the beach ridge, separated from each other by older, abandoned beach ridges. These basins are occupied by pans, which may take the forms of shallow semipermanent salt lakes, or basins flooded at high tide, or salt flats dry at all seasons. Some can be traversed by a car at any time, while others are always morasses of slimy mud. Most are completely barren and have no regular animal inhabitants. Salt is commercially extracted on a fairly large scale from certain pans north of Swakopmund.

In the stretch of coast between Walvis Bay and Swakopmund, the waves beat against the feet of sand dunes that rise to heights of more than a hundred feet. The dunes (like those to be described in a later section) are vigorously active and pose a serious and endless problem to the railway line and highway that parallel the coast in this section. The position of the coastline here is apparently determined by the presence of bedrock that outcrops in a number of places and sometimes creates a rock-defended bench between the dunes and the sea.

The Kuiseb River, which rises in the plateau east of the Escarpment and flows across the Namib to the sea at Walvis Bay, has constructed an estuarine delta completely filling a former embayment of the coast. The stream flows on the surface only once in several years, but when it does, it flows with surprising volume and violence. Since its source and most of its course are in areas of mica-schist, it carries a great load of mud, and its delta consists chiefly of a clay plain. Across it the daily sea breeze blows an endless succession of small dunes, some of them barchans. These obstruct the flow of surface waters and create temporary stagnant ponds in flood years. Normally, the delta has a scanty cover of *Salsola* species, a few *Tamarix austroafricana,* and the import from South America, *Nicotiana glauca.*

The city of Walvis Bay is situated on the seaward edge of these delta flats. Few places in the world have a more dismal and unprepossessing location than it: on a mud-flat in danger of inundation, plagued by drifting sand and sulphur eruptions, in a windy, foggy, rainless environment. Yet it is today a relatively attractive, prosperous, fast-growing, and forward-looking community.

Being vulnerable to flooding, the early homes were built on stilts. Today, an earth dike prevents inundation by deflecting the waters. A wind-obstruction fence has been built, creating a *municipal sand dune,* which halts the dunes that formerly migrated right through the town. The soils are heavily impregnated with salts, but homeowners grow attractive lawns and gardens in soil brought from the uplands over a hundred miles to the east. Water of excellent quality is brought by pipeline from the underflow of the Kuiseb. Some of the streets are tarred. Others are surfaced with a mixture of clay and salt from the nearby pans; at night the deliquescent salts absorb moisture from the fog, which binds the clay through the ensuing day.

Walvis Bay is the only harbor in South West Africa where full-sized freighters can come alongside a wharf. Before improvement, the bottom shoaled so gradually that the 18-foot depth curve was a mile from shore. The port handles most of the export-import trade of the Territory and is a major base for fisheries in the plankton-rich Benguela Current. Canneries line the shore-front; large tanks store fish oil; and acres of fish-meal stand out-of-doors, awaiting shipment.

Swakopmund, the only other town of the Central Namib, is situated on the coast just north of the mouth of the Swakop River. It is an old German town, the former port for the former German colony; and today it is the chief shore resort and summer capital of the Territory. From early December through February its hotels flourish, and a tent colony springs up on the beach as great numbers of people flock from the hot interior down to the cool coast. While formerly provided with highly saline water from the Swakop River, the city is now connected by pipeline to Walvis Bay and the Kuiseb River supply.

The town exhibits remarkable examples of German architecture: towers and turrets, half-timbering, balconies, and steep roofs. In its language, culture, architecture and ethnic composition, it is still thoroughly German.

The coast north of Swakopmund has undergone remarkable changes in the past decade. Long completely isolated and uninhabited, famed only for its shipwrecks (which earned it the name *Skeleton Coast*), it is today much in use. A road up the coast is graded and surfaced with salt and clay well beyond the limits of this study, and another road runs directly from Usakos west to the coast. Three communities have developed: Vineta, Wlotzka's Bakken, and Henty's Bay. The latter, near the mouth of the Omaruru River, has a good water supply from wells, and a store and a motel. All three communities have scattered buildings ranging from shacks to quite pretentious houses, much used in summer and on weekends by residents of the interior who come here to escape the heat, to rest, and to fish.

While there is no road along it, the coast south of Walvis Bay is much used by sports fishermen, many of them going all the way to Sandwich Harbor for their angling. The area is accessible with difficulty by Land Rover and other four-wheel-drive vehicles as far as Sandwich, but there are no habitations or occupants, other than one Coloured family living at Sandwich Harbor. Travel south of Sandwich is completely impossible both by edict (it is closed to entry because of possible diamond-bearing formations) and by trafficability (dunes that are practically impassable).

The main line of the South African Railways from Windhoek passes through Usakos to Swakopmund and on to its terminus at Walvis Bay. It uses diesel-electric engines and operates on a 3.5-foot gauge. A paved highway parallels the rail line, and graded roads run inland from Walvis Bay directly east to Windhoek, and southeast to the southern part of the Territory. A large airfield a few miles inland serves Walvis Bay, and there is regular air traffic to Windhoek.

## The Dune Namib

Stretching as an unbroken sea from the Kuiseb River to Luderitz (250 miles) and from the surf of the Atlantic nearly to the Great Western Escarpment (50 to 80 miles) are the Great Sand Dunes. They rank among the great sand dune accumulations of the world, both in area and in height (in excess of 800 feet in some places). The dunes lie atop the Namib Platform, which shows through in a few places amidst the dunes.

The sand has been derived from the erosion of the plateau east of the Escarpment. It has been carried westward by streams that emerge from the Escarpment only to dwindle away and be lost by evaporation on the Namib Platform. The sand, left behind as the water disappears, becomes the plaything of the winds. The daily sea breeze carries it obliquely inland in a northeasterly direction; the rare but violent Berg Wind blows it directly seaward. The dunes lie in great irregular waves, aligned chiefly at right angles to the sea breeze, but the regularity of their crests is much interrupted, chiefly by breakthroughs caused by the Berg Winds.

The Dune Namib is bounded abruptly on the north by the Kuiseb River, which flows across the Namib Platform in a sharply incised gorge. Over a period of years, sand cascades down the steep south wall and encroaches on the channel. Then, in one of its violent floods, the river sweeps the sand away, and the process begins again. Only at its mouth, where the stream spreads out across the flats and loses its carrying power, are the dunes able to cross the Kuiseb and extend themselves northward to the Swakop River.

The dunes near the coast are yellow-white; inland they are deeply red. This would seem to indicate a greater degree of oxidation inland, and hence a greater age.

Near the coast the dunes are almost totally barren, with only an occasional narra bush *(Acanthosicyos horrida)* growing on small hummocks in hollows amidst the larger dunes. The narra is a leafless, thorny bush growing in a scraggly manner, looking more like a tangle of green barbed wire than a plant. Its melons bear an edible, almond-like seed, formerly an important food of the local Nama tribesmen, and today much used in the confectionery industry of the Cape. The red dunes of the interior have a scattered vegetation of bushes such as *Galenia* and a couple of shrub-like grasses: *Aristida sabulicola* and *Aristida namaquensis.*

The Dune Namib is totally uninhabited and totally unused.

## The Namib Platform

The Namib Platform can be subdivided into a number of typical areas; inasmuch as there is a marked variation in climate between the Outer (more coastal) and Inner Namib, two variations for the same landscape type may sometimes be described.

### Gravel Flats, Outer Namib

Great expanses of the Outer Namib are unbelievably flat, totally barren, and surfaced entirely with gravels. While the extent of individual units may be measured in tens of miles, their relief is measurable in inches per acre. While apparently flat, their surfaces actually slope up inland at the rate of about 40 feet per mile.

Normally, the surfaces have a slight to moderate development of desert pavement, with pebbles covering up to 50 per cent of the surface. The color of the landscape varies from gray-white to yellow. The surface is never cemented, but layers of lime or gypsum cementation occur at varying depths, and all areas investigated had some cementation in the upper two feet of soil, although it was universally weak. Gypsum is most frequent in the seaward portion. Its presence, in place of the more common calcium carbonate, is unusual. It has been proposed by Martin (8) that the carbonate radical has been replaced by the sulphate radical derived from the hydrogen sulphide of the submarine explosions, borne landward by the sea breeze and condensed out with the dew.

Vegetation consists of very widely scattered bushes of *Arthraerua leubnitziae* and *Zygophyllum stapffii*, none over two feet in height and each with a small elongated dune on its southwest side—product of the Berg Winds. The larger stones have coatings of lichens on their seaward faces but are bare on the eastern side.

It has often been stated that the curious plant, *Welwitschia mirabilis,* grows only on the gravel flats of the Outer Namib. Field work by the writer indicates that while the plant does grow in such areas, it is also to be found on rocky benches in the Swakop gorge, on very stony relict fans west of the Brandberg, and in the mopane *(Copaifera mopane)* steppes of the Petrified Forest area near Franzfontein. Often termed a *living fossil,* the *Welwitschia* is a gymnosperm. It has only two leaves, often over six feet in length, split longitudinally into numerous strips, and exhibiting in their leathery texture their extreme xeromorphism. The leaves grow out of the top of a huge woody trunk that protrudes only a few inches above the ground and has a depth of only two or three feet, yet has a diameter of at least two feet. Each plant is host to a number of brilliantly red-orange beetles that seem peculiar to only one plant species.

### Gravel Flats, Inner Namib

The increased precipitation of the more inland portions of the Namib alters both the vegetation and the morphology of the flats. The flats apparently represent vast amounts of alluvium fluvially deposited in the past and since partially dissected. Over great areas the gravels have been strongly cemented by calcium carbonate. This cementation may represent the present situation in which occasional heavy rains saturate the gravels, dissolving the calcareous material, which is redeposited as a cementing agent upon the evaporation of the ground-water. But the writer feels that the present climate is not sufficiently moist for this; he believes that the present cementation reflects a former period of greater moisture.

In many areas the cementation occurs at the surface. Such areas are virtually devoid of vegetation except for widely scattered grasses on exceptionally rainy years. Lichens are totally absent, due to the absence of dew. Areas underlain by uncemented gravels support noticeably heavier growths of grass, especially *Aristida ciliata* and *Aristida obtusa,* which become heavier still in areas of sandy soils. But even the *heaviest* grass covers are very thin by humid land standards, shading less than 20 per cent of the total area in the best years and seldom exceeding two feet in height.

Lines of heavier vegetation, chiefly bushes such as *Lycium* sp. and *Acacia hebeclada,* growing to 10 feet in height, and widely scattered camelthorn trees, *Acacia giraffae,* reaching heights of 30 feet, mark the position of *omurambas,* or stream courses incised a few feet below the level of the gravel flats and carrying water from the face of the Escarpment or the plateaus at its top. Such streams run only a few hours a decade, but they apparently carry a moderate underflow at all seasons.

### Granite Pediments

Over broad tracts, mass downwastage of granite has produced an almost featureless landscape with bedrock at the surface everywhere. Low mounds of bedrock, covered with exfoliating scales, protrude from beneath a thin mantle of *grus* (disintegrated granite), which masks the presence of bedrock in the intervening pockets. Unreduced remnants of various sizes rise a few feet above the common surface; most of them seem to be lithologic in origin: aplite or pegmatite dikes and quartz veins.

Most of the surface is undergoing continuous granular disintegration. On all exposed surfaces, individual grains and exfoliation sheets are only lightly attached and fall off at the touch of a finger, joining the adjacent accumulation of grus. Endless attack of the more exposed surfaces by granular disintegration and protection of other surfaces by a layer of grus results in the eventual attainment of a common level of surface—the pediment. Considering the immensity of such surfaces, the lack of precipitation and run-off, and the omnipresent evidence of the role of weathering and disintegration, it is quite apparent that lateral corrasion by streams, once theorized as the cause of pediments, could never produce such features, and that mass wastage through weathering is obviously the chief agent involved.

In the Outer Namib, vegetation is very sparse on the pediments—yet actually more plentiful than on the adjacent gravel surfaces. This is probably due to the facts that dew is more important than rain as a source of water, that dew condenses more readily on rock than on gravel, and that this water is stored in crevices in the rocks and remains available to shallow-rooted plants, but is lost through percolation in the gravels. Common shrubs are *Arthraerua leubnitziae*, *Zygophyllum stapffii*, and *Salsola aphylla*. Grass is completely lacking. In the Inner Namib, similar pediments support a thin phase of bush steppe, with *Barleria* spp., *Petalidium* sp., and *Boscia foetida* as the chief components, together with a thin growth of grass. The bushes are dwarfed (under one foot in height). In other areas, erect, 6-foot *Euphorbia gregaria* and recumbent *Aloe asperifolia* are major components.

### Granite Inselberge

A few isolated knobs rise conspicuously above the pediments. Those of the Outer Namib have an abrupt angular contact with the pediment and are exfoliating, dome-shaped, granite masses. On their seaward faces, they are pitted with large cavernous recesses and myriads of small pits. The inland faces are not pitted; hence the pits are not sand-blast phenomena, since the chief abrasive aeolian action would be associated with the Berg Wind. It would appear that the pitting results from chemical action involving, perhaps, dew, sea-fog, salt and/or hydrogen sulphide from the eruptions. These inselberge are devoid of vegetation, except for the lichens that grow plentifully on the seaward faces.

### Mica-schist Pediments

Over large areas of the Central Namib, steeply dipping beds of mica-schist have been truncated by erosion, producing what appears to be a pediment surface. The method of formation of such a surface is not easily explained. Granular disintegration is not possible, for the schist is not granular; hence mass wastage is unlikely. The writer has no ready explanation for these developments.

In some areas the schist is apparently homogeneous, and the surface is monotonously flat and featureless. Elsewhere, beds of ordinary mica schist alternate with a more arenaceous and hence more resistant variety. The latter forms low ridges, aligned in a northeast-southwest direction, making a corrugated country. The relief is small (in the neighborhood of 20 to 50 feet), but the ridges culminate in jagged crests and are steep enough to make travel even by jeep or Land Rover difficult and uncomfortable. In the coastwise portions, lichens grow on the seaward sides of rock outcrops. In the central portion, they are joined by a yellow matlike succulent, but together they cover less than one per cent of the surface. The lower areas between the ridges are veneered with materials washed and blown from the higher parts, especially quartz from inclusions in the schist. Such areas are totally devoid of vegetation. Some cementation by calcium carbonate occurs.

### Rocky Ridges of the Outer and Central Namib

Occasional elongated rocky ridges rise conspicuously above the monotony of the Namid Platform. They break sharply with the surrounding flats, with very little slope wash or talus accumulation at their bases. Sandblasting is common on their lower northeastern sides, the result of Berg Winds, but sand accumulations are negligible.

The ridges of the more coastal section have a cover of lichens plastered over most of their southwestern faces. The ridges, composed of white limestone, usually have a number of species of plants growing on them. While actually sparse, this vegetation is relatively rich compared to the surrounding gravel flats. Included are such surprising plants as certain *Lithops* species, which look almost exactly like the pebbles amidst which they grow. Other succulents include *Trichocaulon dinteri*, *Aloe asperifolia*, *Hoodia curori*, and *Senecio longiflorum*. As mentioned before, the rocky areas are better suited to water accumulation and retention than are the surrounding areas; the white limestone ridges are perhaps especially so, since the light color makes for less heat absorption, a cooler condition, and hence a longer period of temperature below the condensation point of the surrounding air.

### Rocky Ridges of the Inner Namib

The greater precipitation of the Inner Namib permits the development there of a much richer vegetation, both in number and sizes of individuals and in numbers of species. Most conspicuous are the sentinel-like *Aloe dichotoma*, standing erect and tall (to 8 or 10 feet), large cactiform *Euphorbia dinteri* (to 6 feet), *Croton* sp. (6 feet), and *Boscia albitrunca* (3 feet), together with two species of *Commiphora*, which while only 3 feet in height, have trunks 2 feet in diameter. Succulence is not as common a feature here as in the Outer Namib. Curiously interesting is *Myrothamnus flabellifolia*, known as Resurrection Plant because its seemingly dead foliage becomes green with reactivated chlorophyll soon after being moistened.

### Exotic Stream Courses

Incised well below the level of the Namib Platform are the canyons of three streams (the Kuiseb, the Swakop, and the Khan) which rise in the rainier plateaus above the Escarpment and hence occasionally carry flood waters all the way to the sea. While the streams flow through an essentially rainless area, their sides are nonetheless dissected by streams of purely local origin – the result of once-in-a-decade downpours. In areas of softer and impervious sediments, the valley sides are eroded into an intense dissected belt with an intricate stream pattern, extremely rugged relief, and strikingly rugged landforms. Were the precipitation greater locally, the rejuvenation of these tributaries would have reduced the whole country to a dissected hill land. As it is, the degree of dissection that does exist is remarkable, considering the extreme aridity of the region. These dissected belts are totally devoid of vegetation.

On the other hand, the underflow of the major streams supports a relatively rich riparian growth, characterized by several trees and a number of bushes. The most common trees are *Acacia giraffae*, (camelthorn) growing to a height of 30 feet, and *Acacia albida* (anaboom), which grows even taller. *Euclea pseudebenis* and *Ficus damarensis* grow to heights of 20 feet. Bushes include *Lycium* sp. and *Tamarix austroafricana* and a brushlike grass, *Eragrostis spinescens*. This vegetation is totally different from that of the surrounding desert; it is composed of nonsucculent trees and bushes, growing far taller, larger, and more densely than anywhere

in the surroundings – a response to the relatively humid edaphic environment.

## VEGETATIVE SUMMARY

The Outer Namib is almost totally devoid of vegetation (9), particularly on the gravel flats, mica schist, and granite pediment areas. Rocky ridges condense and store some water from the fog, and thus support a slightly richer vegetation. Nearly all vegetation of the Outer Namib is succulent, whereas very few succulents grow in the Inner Namib. Lichens are common in the Outer Namib but not in the Inner, whereas grass is restricted to the Inner Namib. Arborescent riparian developments are common along the through-flowing streams.

## ANIMAL LIFE

Both the Dune Namib and the Outer Namib Platform are quite devoid of higher forms of animal life. Certain lower forms are common, however: several types of beetles live in the dunes, lizards are surprisingly plentiful, a dune-dwelling snake (*Bitis peringueyi*) has developed a locomotion similar to that of the American sidewinder, and several types of smaller rodents are to be found (10).

The grassy plains of the Inner Namib are the home of great herds of herbivorous grazers – especially several species of antelope: gemsbok, springbok, steenbok, duiker and klipspringer. The first two normally travel in herds – up to a score for the first, and hundreds for the second. The others are more solitary. Zebras occur in large troops. Ostriches roam in flocks numbering in the scores. Several burrowing herbivores are found. Carnivores include several members of the cat family (including cheetah, leopard, and until recently, the lion), jackals, hyenas, the termite-eating aardvark, the mongoose, and a small fox.

The ease of accessibility of the area, especially since the development of the Land Rover and jeep, plus the role of hunting in the philosophy of the white South African, has resulted in a sharp reduction in the number of the larger animals through both unrestricted and illegal hunting. In an effort to stem this trend and to preserve the animals, most of the Central Namib has been proclaimed a game reserve, but there remains the major problem of effectively controlling so large an area.

## HUMAN OCCUPANCE

Until a few decades ago, small wandering bands of Bushmen known as *Saan* hunted and gathered food over the wide expanses of the Namib. None remain today, they having been either exterminated by other groups (11) or merged into the almost unidentifiable mass of natives on certain reserves or within urban communities.

Along the banks of the Swakop and Kuiseb Rivers, small groups of Topnaar Hottentots (Namas) graze their flocks of goats on the riparian vegetation and gather narra seeds to sell to dealers in Walvis Bay. No attempt is made to irrigate, and no crops are raised, even though water is just below the surface and is lifted by hand to water the goats, and even though plenty of good land is available. Unfortunately, cultivation of the soil is not inherent in the philosophy of these people. The Namas have always been graziers and gatherers, and it would be difficult to change their way of life. Date palms, planted on the Kuiseb by an early German missionary, grow unattended; the Topnaars have made no effort to care for them or to plant them elsewhere, although it is quite likely that they would grow well along the Kuiseb and the Swakop across the full width of the Namib.

Several farms have been developed by Europeans along the lower Swakop River, with irrigation by pump from shallow wells. They are chiefly engaged in a small-scale garden type of production.

Despite the restricted size of their areas, it would seem that the stream valleys of the Namib afford greater potentialities than are now realized. Much more agriculture could be carried on, producing fresh vegetables and milk for the coastal towns and for ships calling at Walvis Bay, as well as dates, citrus, and other subtropical fruits. An overall plan for allocation of the rather limited water supply between the urban and port uses and the agricultural requirements would be necessary. Should agriculture be developed, the problem of providing for the Topnaars would also have to be solved.

The surfaces of the Outer Namib Platform, whether gravel, pediment or other, and of the Dune Namib, defy the imagination in regard to future developmental potentialities. Even if water from some inexpensive source (such as desalinization of sea water by solar or atomic power) were available, almost none of these would be useful agriculturally, due to their extreme porosity (sand dunes), gypsum or salt content (gravel flats), or lack of soil (pediments).

The Inner Namib alternates between being a sea of highly nutritious grasses and a vast barren waste, depending on the rainfall in the preceding several months. Until recently, the area was totally unoccupied, and only the game animals suffered from droughts. In the past two decades, however, the land in the more-humid interior of South West Africa has largely been settled by European graziers and their native herdsmen, and many would-be ranchers have looked with interest at the Namib borderlands.

Seeing the lands in good years, covered with grass and game, the outsider fails to understand the total picture: that the game, being mobile, can range far afield in search of food in drought years, whereas his own domestic stock, confined by arbitrary property boundaries, will starve on his own farm. As a consequence, graziers during the late 1940s and early '50s settled too far westward into the desert borderlands and were badly hit by the droughts of the early '60s. Today, the last row or two of farms (12) should be, by administrative edict, taken back by the government and their areas added to the Game Reserve. No ranching activity should be permitted west of the foot of the Escarpment.

## FUTURE OF THE NAMIB

Within the portion of the Namib under consideration here, extensive future mining developments seem relatively unlikely. Copper, tin, and other ores are produced sporadically at several places, but their production is not likely on a large and continuing scale. Diamonds are found both north and south of the area and conceivably might be found in the local gravels. Marble may be exploited for cement in the future. A search for petroleum goes on offshore, but there is little likelihood of its discovery on the mainland.

Grazing should be prohibited entirely, and agriculture will be restricted by nature to small-scale operations along the major stream valleys.

On the other hand, the coastal area will continue to develop in the same ways as at present. Assuming continuation of existing conditions, Walvis Bay will flourish and expand as the port for the Territory, as a fisheries and canning center, and as a growing industrial community. Its water supply is adequate for considerable growth and can be augmented by desalinized sea water if necessary.

Swakopmund will continue its role as summer capital and seaside resort of the Territory. Further development of recreation along the coast is to be expected. Touristic developments are not to be anticipated on any large scale; the area simply does not have anything that will attract visitors from afar.

While other desert areas of the world will no doubt develop industry based on the harnessing of solar energy, the fogginess of the coastal Namib will preclude or sharply limit such developments.

In summary, the Namib, in the foreseeable future and with the assumption that political situations remain relatively stable, will probably remain much as it is today, with a modest development based more on its position relative to the interior of the Territory than on any inherent basic resource (13).

## REFERENCES AND NOTES

1. JACKSON, S. P.
   1941  A note on the climate of Walvis Bay. South African Geographical Journal 23:49, 52.
2. METEOROLOGICAL OFFICE, Official records, Windhoek.
3. WAGNER, P. A.
   1916  Geology and mineral industries of South West Africa. Geological Survey Memoir 7. Pretoria.
4. GEVERS, T. W.
   1931  An ancient tillite in South West Africa. Geological Society of South Africa, Transactions 34:1-17.
5. REUNING, E.
   1923  Der Intrusionsverband der Granite des mittleren Hereroland und des angrezenden Küstengebietes in Südwest-Afrika. Geologische Rundschau 14:232-239.
6. GEVERS, T. W.
   1936  The morphology of western Damaraland and the adjoining Namib Desert of South West Africa. South African Geographical Journal 19:78.

   Presents a good résumé of theories.
7. COPENHAGEN, W. J.
   1953  Periodic mortality of fish in the Walvis region: a phenomenon of the Benguela Current. Union of South Africa, Division of Fisheries, Investigational Report 14.
8. MARTIN, HENNO (formerly Geologist, Water Resources Branch, South West Africa Administration, currently Professor of Geology, University of Göttingen, Germany), personal communication.

9. WALTER, HEINRICH
   1936  Die Ökologischen Verhältnisse der Namib-Nebelwüste (Südwest-Afrika). Pringheims Jahrbuch für Wissenschaftliche Botanie 84:58-222.

   This contains an excellent study of the plant ecology of the Outer Namib.
10. KOCH, C.
    1960  The Tenebrionid Beetles of South West Africa. Bulletin of the South African Museum Association 7:73-85.
    1960-  Some aspects of abundant life in the vegeta-
    1961   tionless sand of the Namib Desert Dunes. South West African Scientific Society Journal 15:9-92.
11. VEDDER, H.
    1938  South West Africa in early times. Oxford University Press, p. 174.
12. In South West Africa, the term *farm* is used indiscriminately both for units that till the soil and produce crops, and those that commercially herd cattle, sheep, or other animals. The term *ranch* is not used.
13. LOGAN, R. F.
    1960  The Central Namib Desert. National Academy of Sciences, National Research Council, Publication 758. 162 pp.

    Most aspects covered in the present paper have been discussed previously in greater detail in this monograph by the same writer.

# FORMATION OF DESERTS
# OF THE NEAR EAST AND NORTH AFRICA
## Climatic, Tectonic, Biotic, and Human Factors

ROBERT L. RAIKES
Raikes and Partners, Consulting Engineers and Hydrologists
Rome, Italy

## ABSTRACT

The concept of weather fluctuations of relatively short duration is basically more sound in many respects than is the more widely accepted climate-change explanation as the cause of the formation of deserts. The irreversibility of drought effects in low-rainfall areas is both inevitable and more desert-creative, because of its greater severity, than the generally postulated secular climatic change. Weather-induced ecological change is the fundamental factor in the creation of deserts; it is the ineluctable background against which the abuse of land and its vegetation by man and animals acts only as a possible accelerator. The effect of tectonics is relatively minor and localized. The chronology of deserts is considered only for the period—the Holocene—approximately datable by the normal archeological methods of relative chronologies fixed at intervals by historical or isotope dating.

# FORMATION OF DESERTS OF THE NEAR EAST AND NORTH AFRICA
## Climatic, Tectonic, Biotic, and Human Factors
### Robert L. Raikes

Let it be admitted immediately that much of what is expressed in this paper is contentions: some of it is entirely contrary to widely held views and beliefs. As I have been invited to contribute to this book in the dual capacity of hydrologist and archeologist, I can only interpret my brief in this dual sense. Climatic change, therefore, for me means primarily secular climatic change. As appears later, the effects of alleged secular climatic change and of known short-term climatic oscillations are often practically indistinguishable, and even more often not distinguished, the one from the other.

Climate has become, for some archeologists and for a few followers of other disciplines who work with them, a convenient whipping boy. Civilizations disappear without evidence of invasion or other obvious reason—obvious in the archeological sense—and all too often the disappearance is attributed to climate change. I was delighted to see in the London *Times* that the downfall of Mr. Khrushchev was ascribed indirectly to a *weather* change in Russia! The protagonists of climate change seem, all too seldom, to be concerned with investigating whether what they are invoking is really climate change—secular and, therefore, inevitably worldwide—or what might perhaps be better and more understandably described as weather fluctuation.

## CLIMATIC CHANGE OR WEATHER FLUCTUATION

There is plenty of evidence for and against my views. If anyone doubts this he may read the published reports in "Changes of Climate . . ." (1) and "World Climate from 8000 B.C. to 0. B.C." (2).

After carefully reviewing the evidence from a hydrological point of view (3), I cannot find any convincing worldwide evidence for secular climate change during the greater part of the Holocene. Let us say, in order not to trespass onto marginal ground where the archeological evidence is at present scarce, that there has been continuity of the general, inherently variable, conditions of today from about 7000 B.C., or perhaps a little earlier. The evidence for this view has already been published and is not directly relevant to this paper. It is relevant only in that the process of desert formation has to be considered, in my opinion, against a background of *no* secular climate change.

Dr. Robert Stephenson, formerly of the University of Nevada, has suggested to me orally—and I am extremely grateful to him for it—that what we are really concerned with is *weather* change and not climate change.

Climate is a worldwide phenomenon depending, in the long term, on the amount of global energy supplied by the sun, how this energy is converted into kinetic energy as atmospheric mass circulation, and how it is spatially distributed. Weather concerns the localized manifestation of various combinations of meteorological parameters against the background of worldwide circulation patterns. Localized manifestations on the causative side can include such things (themselves dependent variables) as differential heating of land and sea areas or of adjacent but different land areas, sometimes as a result of localized cloud cover or insolation*; local *wave* instabilities between tropical and polar air masses; the effect of changes in albedo (reflected radiation) caused by vegetational changes, in turn due to rainfall which perhaps occurred several weeks previously; the dynamic succession of changes in pressure fields; and similar matters.

Weather is subject to a great amplitude of variation in all its aspects due to the complex—and at the moment only partially understood—interaction of many factors, of which only some can rightly be described as parameters. Climate, on the other hand, to the limited extent to which it has been quantified, can be no more than a generalized description of the mean of weather parameters.

It seems to me to be of relatively little consequence whether or not there have been *climatic* changes, in the sense implied by archeologists or paleoclimatologists. For the *quantified* range of variation from the mean of various climatic or *weather* parameters postulated by the proponents of *climate change* are of a different order of magnitude, or orders of magnitude, less than the amplitudes of variation of weather parameters.

### Weather Fluctuations as Agents of Desert Formation

It seems to me that it is against this background—of weather variations rather than of climate change—that desert formation must be considered from the *climatic* point of view. At a given place in the Middle East, or North Africa, the year-to-year

---

*Insolation: not to be confused with insulation. Insolation is a rather unscientific way of describing solar radiation.

variation of monthly mean temperature can be two or three, or occasionally more, times the difference, estimated by paleoclimatologist, of mean annual temperatures between then and now. At the same given place the year-to-year variation of rainfall, about the short-term mean, can be between 30 per cent of the mean and 300 per cent of it.

Quite reputable workers in this field have come out with statements to the effect that the mean rainfall of a certain place is one or two millimeters per annum; statements of this kind can only be described as immortal. Such estimates, and the statements based on them, are either deliberately misleading, or a measure of the inability of the worker concerned to realize the futility of applying the statistical concept of the *mean* to any place where, first the records (if they exist at all, or for more than a few years) are of doubtful validity, and, secondly, include for what they are worth tens of successive years of zero rainfall. On this kind of basis one rather exceptional year yielding 20 millimeters of rain and another slightly less exceptional year yielding 10 millimeters of rain, in 30 years of records, give a mean annual rainfall of 1 millimeter. It is difficult to think of a more flagrant abuse of statistical methods; or, indeed alternatively a more abysmal misunderstanding of the purely statistical value of the otherwise meaningless mean.

### The Effect of True Secular Climate Change

Deserts could be created or intensified by secular climate change, if it were reasonable to postulate a steady change in the course of time from, let us say 200 millimeters mean seasonal rainfall to 50 millimeters notional mean seasonal rainfall; however, all the evidence from areas which today enjoy—if one can use the word in this context—200 millimeters indicate that anything in the nature of steady trending change is effectively masked by the range of short-term change.

Few people would seriously argue that genuine climatic difference did not exist in the northern hemisphere between the glacial periods and the interglacials; genuine, that is to say, in that global air-mass circulation patterns were different and that all the main climatic parameters such as temperature and rainfall were significantly different. The words *significant* and *significantly* are exceedingly vague and difficult to define. What I mean in this case, in using them, is that the difference was ecologically determinant. Few workers, however, agree as to the extent of northward or southward movement of such genuine differences. I would guess (what else can one do in areas where, even today, the ecological and pedological differences between notional means of 200 and 50 millimeters rainfall are distinguishable only, if at all, by an expert?) that the glacially induced climatic changes were not ecologically significant more than 300 kilometers south of the present day rainfall and temperature zones where 200 millimeters of mean rainfall have been more or less reliably determined.

### Holocene Weather or Climate Change

In considering the Holocene, is climatic change significant? The answer must be "No." If we change the question to: "Is weather change significant . . . to desert formation?" then the answer must, I think, be an emphatic "Yes." Even the comparatively short period of reliably observed weather parameters indicates vividly how the variability of the main parameter—rainfall, as the ultimate source of water—increases with decreasing mean rainfall. In this paper we are concerned with areas prone to desert formation, either from sheer lack of rainfall or from unfortunate distribution of what might otherwise be barely adequate rainfall.

When lack of rainfall is the cause, the high variability associated with low rainfall is the agency. When unfortunate distribution is the cause, the agency is the high variability of that part of the rainfall that is ecologically effective: that which falls during the season of plant growth or that has survived as soil moisture from an immediately preceding period when growth was not possible. For instance, winter snow may suffer such high sublimation losses that the resultant effective depth as rainfall equivalent may be inadequate. To quote another example, high-intensity summer rainfall producing rapid runoff, particularly if it occurs after fruiting of the vegetation, contributes little directly to plant growth and often a negligible amount to soil moisture.

## HYDROMETEOROLOGY AND ECOLOGY

### How Meaningful is the Mean?

Let us take the example of a place that now has an authenticated mean rainfall of 350 millimeters, based on (say) 50 years of record, and try to quantify what I mean by the difference between climate change and weather variability. If it were possible to have accurate records of the rainfall of that place for, let us say, 8,000 years, I do not believe that the long-term mean rainfall would differ from any selected 50-year mean by more than about 10 per cent. At the same time, those same hypothetically accurate records might show an absolute maximum that could easily be 300 per cent of the recent mean and an absolute minimum that might approach zero.

I am trying, all too inadequately, to demonstrate two facts that are of vital ecological significance. The first fact is the comparative unimportance of the statistical mean and of any departures from this—whether in a trending sense or as regards any one period of so many years, compared with other periods of similar length (provided that the periods concerned are of, say, 50 years). The second fact is the vastly greater range of departure from the mean, in one direction or another, of single isolated years of record. Between the maximum and minimum of any parameter over a long period—probably occurring at a long interval of time the one from the other—and the range of variation of the means for 50-year periods or for the arbitrarily selected so-called standard periods of 30 years, there is a variability of almost certainly far greater ecological significance; that is, the extent of variation associated with varying numbers of consecutive years expressed as a mean of the rainfall parameter for those consecutive years.

I am not referring to any particular number of years but will refer to them as short periods without trying to introduce directly or by implication any concept of frequency. I am simply concerned, in a general way, with the approximate magnitude of the means for varying *short* periods and their difference from the long-term mean, however defined.

If we tabulate this algebraically, perhaps a clearer picture will emerge. Let us select a hypothetical place in a desert-margin area having a mean annual rainfall of 300 millimeters, based on a completely hypothetical record of, say, 10,000 years. We can give values of $R$, for maximum, and $r$, for minimum, with suffixes denoting the number of years concerned, to indicate the possible maximum and minimum values of mean rainfall for those numbers of years during the period of record. $R_2$ and $r_2$ will then represent, respectively, the maximum and minimum means for two successive years; $R_3$ and $r_3$, the maximum and minimum means for three successive years; and so on.

All we can say with certainty is that $R_2$, $R_3$, $R_{10}$, and $R_{20}$, and $r_2$, $r_3$, $r_{10}$, and $r_{20}$, are diminishing series. A given year may have a value of $R = 900$ millimeters and value of $r = 0$ millimeters. The values of $R_2$, $R_3$, etc., and $r_2$, $r_3$, etc., are not known, simply because nowhere, in desert margin countries, is the period of reliable record long enough; but even the short-period records available to us suggest that $r_5$ (the minimum mean rainfall of five successive years of lower than normal rainfall) may be as low as 175 millimeters in the case cited. I have not attempted estimates of the corresponding maximums, as they are largely (as will be seen later) irrelevant.

## The Effect of Minimal Conditions

It is the minimum with which we are chiefly concerned from the point of view of desert formation.

A definition of desert formation, however imprecise, is overdue. What we really mean by desert formation is the reduction of vegetation to a point where only a few highly adapted species can survive and where, as a consequence, only a limited and highly adapted fauna can be supported. The availability of water for drinking under such circumstances has only a minor significance. For to vary my usual approach to this matter, which is that life of all kinds is ultimately dependent on water, it must be emphasised that water alone is not sufficient.

In practice, it is seldom that, *in a natural state*, water for drinking exists where food for animals and men is nonexistent. Man, in recent years, has often with the best intentions upset this rule. Examples could be cited of the storing of occasional floods in near-desert regions that have created completely desertic conditions locally. The rain that caused the flood, intercepted and stored by man, gave rise in such conditions to a more-or-less abundant local flush of vegetation of the species already available and established. The stored water attracted flocks and herds and their attendant men in numbers related to the amount of water and not to the amount of vegetation. In a short time the vegetation disappears almost totally from the surrounding area to a radius determined by the distance that grazing animals can travel, there and back, without drinking.

This illustration, of what man is now doing on a scale made possible only by modern techniques, has slipped in out of turn, as it really belongs (later in this contribution) to the list of desert-forming factors. On a smaller and less immediately destructive scale it has probably been going on for millennia.

## The Effect of Maximal Conditions

Before examining the nonclimatic factors in desert formation and those inherent in weather fluctuation, the effect of higher than normal rainfall must be described in order that its significance can be assessed and virtually eliminated.

Studies of rainfall, evapo-transpiration, and infiltration indicate that above-normal rainfall may be expected to behave in one or more of the following ways, depending on the characteristics of the rainfall and surface:

(*i*) If the rainfall is of high intensity—a characteristic of much desert-margin rainfall—all that which falls at an intensity higher than the infiltration rate of the surface on which it falls will run off the surface and create stream flow in whatever drainage system exists. According to the characteristics of the drainage system (e.g., sandy or gravelly wadis or relatively impervious channels leading to depressions, lakes, or the sea or perennial river) the runoff will, to a varying extent, recharge the shallow or deep aquifers of the region. Whatever runs off however (whatever its ultimate fate—recharge or loss to evaporation or some external system) is not available to encourage plant growth where the rain falls.

(*ii*) Whether the rainfall is of high intensity or not, it will, if sufficient, first raise the water content of the soil to *field capacity* and subsequently pass through the surface *field-capacity* zone as a *wetting front* to recharge the underlying aquifer. The depth of the surface zone liable to re-evaporation or to use by plant-life depends on the soil and its temperature gradient but is generally very shallow (less than a meter), and the soil temperature gradient generally inhibits upward movement (and loss) of water except for brief periods in the winter when evaporation and transpiration rates are low anyhow.

As the vegetation is already conditioned by normally low rainfall to being of one of three general types, the amount of use that can be made of additional rainfall is limited. One type of vegetation is a shallow-rooted plant enjoying a brief existence; with this type higher than normal rainfall can at most, if consisting of repeated events, result in multiple flushes of the same ephemeral plants. Another type employs a tap root to make a slightly extended (in time) use of the available moisture by following the descending water; the result of higher than normal rainfall is much the same as in the first case. The third type, almost exclusively represented by trees, follows the water down to considerable depths and can, and apparently does, make use of the mean rainfall rather than of any brief period of higher rainfall, because of the modifying influence of groundwater storage.

Evidently higher than normal rainfall can only increase temporarily the quantity of types of vegetation already adapted to low rainfall conditions. If it could even temporarily establish other less-adapted types these would not survive the next dry spell.

I have kept phreatophytes out of this discussion, for they are limited to those restricted wadi or stream-bed locations where the level of the water table does not reflect immediately or to any large degree fluctuation in the rainfall that is the ultimate source of the ground-water.

## The Ecological Effect of Drought Conditions

It seems fairly clear then that, from the weather or *climate* point of view, it is drought and its extent and duration that controls vegetation and desert formation. It would follow that weather fluctuation alone in desert margin areas—let us say those having anything from 300-millimeter *mean* rainfall downward, with the effect more pronounced in hot areas than in cold ones—must inevitably have the

effect of gradually eliminating less-resistant or less-adapted species. In other words, if man and his domesticated animals and all wild fauna, from locusts to gazelles and goats, that make use of desert margin vegetation were suddenly to be eliminated, it is probable that the vegetational association would still deteriorate from natural causes to a limited number of *species*. The density of these species would almost certainly vary, but elimination of the eaters of it would not result in the establishment or re-establishment of a new species. Exceptions to this general statement can certainly be cited in limited and favored areas where a source of population (or repopulation) of less drought-adapted species exists, both upstream and uprain. In the absence of a source of population or repopulation, the vegetational spectrum becomes inevitably and progressively impoverished.

## Hydrological Background for Other Factors

This is the background against which the effect of other factors has to be considered. These factors are many, but it must be emphasised that in general their effect is to reduce the density of a weather-determined association and, because of varying degrees of palatability, the proportions of species within it. The Anatolian plateau, although its arid parts have a rather higher rainfall (mostly between 300 and 400 millimeters) can be cited as an example. The combined efforts of man, domesticated animals, and all other biotic factors have not entirely eliminated the various associations of grassland and oak/pistachio, or of grassland and pines. Cutting of trees for firewood and heavy grazing have reduced the density and changed the proportion, but the species have mostly survived. Where land has been cleared and successively cultivated, the whole picture has of course been radically altered, as man has made it his business to eliminate as far as possible everything that competes with his crops.

The nonclimatic factors in desert formation are not normally as irreversible as the fundamental weather fluctuation factor, because almost always a nearby (possibly relict) source of the impoverished population exists. In certain isolated localities, however, they *have* been irreversible; these are those localities where a source of repopulation of plant species simply does not exist. I will come back to this later.

The principal nonclimatic factors are:

(*i*) Locust infestations: these tend, in fact, to be a self-adjusting factor. In years of higher than normal rainfall denser vegetation encourages locust proliferation to the swarming stage, and scarce vegetation inhibits swarm formation. This factor will be ignored in what follows.

(*ii*) Fire.

(*iii*) Overgrazing by wild animals to a small

extent and by domesticated animals to an immensely greater extent.

(*iv*) Browsing to a limited extent, and wood-cutting by man to a vastly greater extent.

(*v*) Cultivation in marginal areas during periods of higher than normal rainfall.

Of these factors (*iii*), (*iv*), and (*v*) are by far the most grave, but the geomorphological effect of (*iii*) and (*iv*) are often over-estimated. With the natural density of vegetation associated with desert margin conditions, and the fact that most of the vegetation has withered (even if not grazed) before the onset of the next season's rain, protection afforded to the soil by annual vegetation is negligible, while that afforded by open woodland is virtually so. The apparently natural density of woodland, now limited to isolated small patches of semiarid areas of Turkey, Jordan, and Pakistan, for example, can have little effect on runoff, and the attendant low herbaceous growth has none whatsoever at the outset of the rainy season. But the emphasis should here be on the word *semiarid,* which implies, according to the temperature and insolation regime (both dependent on altitude), anything from 250 millimeters mean annual rainfall in northerly latitudes to some 350 millimeters and up in more southerly latitudes bordering on the desert belt.

### Hydrology and Ecology

In low rainfall areas the concept of *woodland* virtually disappears. Either there are a few isolated trees in ecologically favorable niches, or there may be, at low altitudes, scattered thorn scrub, mostly acacia. This may have approximately the same density, in terms of numbers of trees in a given area, as the oak or pine woodland of higher rainfall areas, but there the resemblance ceases. Arid-area thorn scrub or trees have a survival mechanism that involves shedding their leaves during periods—either very cold or very hot—unsuitable to growth. In very hot times the loss of leaves stops transpiration losses. Consequently for much of the year the only obstacle to erosion by intense rain is bare twigs and trunks—virtually negligible. The trees themselves, being deep-rooted, are little affected by loss of whatever soil there is or may once have been. Even if such trees survive the axes of man or the browsing activities of a limited semidesert fauna, their presence does not make an area visibly less desertic, except during the brief period of leafing. The removal of such trees by man or animals does not make their habitat any more desertic, for they are largely an irrelevance, as is their occasional and localized reestablishment.

Much the same can be said of fire, with the caveat that fire is not normally a serious risk in desert areas, because of the dispersed nature of trees and other vegetation.

## BIOTIC AND HUMAN FACTORS

### Land Abuse: Overgrazing

As regards overgrazing, it is important to distinguish between semiarid areas, where its effects can simulate the appearance of truly arid or desert areas, and true desert areas. In arid or desert areas the chances of nomad-controlled herds and flocks being in the right place at the right time to benefit from the vegetational effects of widely dispersed and very localized rainfall are generally very low; only in areas such as Saudi Arabia, where truck-borne water solves the problem of access to areas without wells, is there a serious risk of depletion of desert vegetation.

Whether or not the vegetation survives, and irrespective of its density when flourishing, it has practically no inhibiting influence on geomorphological change. The reasons behind this statement: firstly, the vegetation does not germinate until *after* the occurrence of any geomorphologically effective rains that cause erosion (those of high intensity and runoff) and so it cannot afford even limited protection against such rain; secondly, the annual vegetation is ephemeral in the extreme so that such little protective effect as it may have on the soil endures for short periods only. Perennial grasses or low-growing herbs are generally found only in desert areas where there is a high water table; much the same applies to the rare trees, except that they can make use of deeper water.

Let us go back, however, to the effects of overgrazing in semiarid areas. Here the natural vegetation is somewhat more dense, and the effects of its over-exploitation are correspondingly more apparent. It must be borne in mind that overgrazing may often have relatively short-term effects on vegetation due to a form of the law of diminishing returns. It may be possible, in theory, for grazing animals to destroy every vestige of the more palatable species, but the results of the exclusion of animals from range lands suggest that even the palatable species are remarkably tenacious, for they reappear. The proportions of species may be so changed that the vegetation is no longer of value for range purposes, but here we are concerned with desert formation rather than range deterioration.

The removal of palatable species provides more living space for those that are not grazed so that in time a vegetation association, of reduced variety but often of the normal density associated with the available moisture, becomes established and affords some protection. The area may have become a desert from the subjective viewpoint of the rancher, but this does not mean that it has become vegetationally a desert. I do not wish to give the impression of ignoring the disastrous *economic* effects of overgrazing but simply to put these into perspective as an agency of desert formation.

An overgrazed area can be converted, at least temporarily, into desert when practically all the low-growing vegetation is palatable, but this is a fairly rare situation. I have seen it happen in the short-grass savannah of the Northern Sudan in the vicinity of watering-points.

### Land Abuse: Improvident Cultivation

The most serious, nonweather culprit in desert formation is cultivation in areas where no cultivation should be practiced. It should not be practiced in semi-desert areas without irrigation, simply because of the vagaries of rainfall. A succession of a few years may yield a worthwhile crop on rainfall alone, but in the ensuing dry period (which is statistically inevitable) plowed soil, not even partially fixed by moisture, is at the mercy of every wind that blows and is rapidly deflated.

### The Main Causes of Desert Formation, with Examples

The evidence of archeology suggests that in semi-desert areas cultivation may have an irreversible effect, changing steppe to desert, simply because the normal processes of soil formation cannot—without at least limited protection by plants and their roots—keep pace with deflation.

We appear, therefore, to be left with two main causes of desert formation, neither of which involve either secular climate change or that much-maligned and useful animal the goat. In both semiarid and arid areas, rainfall fluctuations have a gradually irreversible effect on vegetation with the inexorable elimination of all but long-term drought-resistant species and impoverishment of the floral spectrum. In semiarid areas, marginal to arid areas, the farmer (or should one call him soil miner?) who wants to turn a fast buck during rare periods of sufficient rainfall can very rapidly indeed simulate the same virtually irreversible desert formation by acting as the agent for soil removal. A few examples, described briefly, illustrate the foregoing:

*a*) Earlier I mentioned localities where a source of repopulation of plant species does not exist. The most striking examples are to be found in the mountain oases of the Sahara. These are isolated by hundreds of kilometers of true desert from each other and from either the Mediterranean littoral or the *monsoonal* belt south of the Sahara. Any plant species not dependent on windborne dispersal of its seeds could not possibly establish or re-establish itself in such localities, so that cutting, browsing, and the effects of prolonged drought must inevitably eliminate such species, particularly if they are palatable to animals or efficient as fuel.

The rock paintings and graffiti of some 4,000 years B.C. do *not* mean that, at that time, the climate or weather conditions were more favorable than they are now, for species of which the only evidence is their pollen. They mean rather that the semi-nomadic herdsmen who painted or scratched them cut the trees to keep warm, and their flocks and herds did the rest. Occupation of these mountain oases was probably seasonal on a transhumance pattern. In the winter the sporadic and dispersed desert winter vegetation was probably used (as now), while in the hot summers men and animals moved up—from all around—to the limited high areas where, even today, rainfall of 300 to 400 millimeters mean is experienced.

*b*) In Jordan desert-formation was probably already well advanced through the agency of man and his animals in very early times. In more recent times the fuel-hungry locomotives of the Hejaz railway were fed by the Turks—in the First World War—by denuding the East Bank Highlands of trees. What these Highlands could be like, without man's wholesale destruction, can be seen today around Hai and Wadi Musa where small patches of oak and pistachio survive on the higher and rainier parts, and juniper on the lower steep slopes. Their survival is probably due to the engineering difficulties of constructing roads or light railways for their extraction. Even so, they have fairly obviously suffered from the normal depredations of man and his animals and have completely disappeared from the cultivable high plateau.

In even more recent times, attempts to settle the Bedu and the availability of tractor-drawn plows have resulted in the plowing of marginal lands that should not be cultivated.

*c*) In Turkey one cannot write of desert formation, but, in the autumn for instance, the high plateau country has a desertic appearance mainly because cultivable slopes at altitudes below about 1200 meters have been deforested to make way for the plow.

*d*) The North African littoral is not, despite popular belief, a desert except locally (where it has been so for millenia). It was at one time called "the granary of Rome," and it could still ecologically produce considerable grain crops if the soil and water conservation practices of the Romans were maintained. It has the rainfall necessary, and, locally (where malpractice has not ensured its disappearance), it has the soil. It was not the "granary of Rome" because of its more favorable climate but because labor along the "Colonial" lands bordering the *limes* was cheaper than that of Puglia.

Many other cases could be cited.

## THE EFFECT OF TECTONICS

I have not so far mentioned tectonics. This is partly because they are mainly the concern of geologists (which I am not) and partly because generally the effect of tectonic movements make themselves felt along valley lands which occupy only a small fraction of so-called desert lands. There is little doubt that in certain cases, in Baluchistan for example, areas previously irrigable and irrigated have now become barren because earth movements have changed drainage gradients and drainage datums. The most common immediate result of such movements has been either down-cutting, which has put formerly irrigable lands out of reach of water, or alluviation, which has buried formerly cultivable lands. Such effects have, however, seldom if ever been widespread, although they are archeologically of much interest.

As an example of the archeological interest, one can cite the central part of the Baluchistan (Pakistan) Highlands, known as Jhalawan, where chalcolithic sites (of which the lowest probable dates are around 1800 B.C.) have been obviously cut off from their irrigation livelihood; while iron age sites, with the highest dates around 1100 B.C., have not similarly suffered. Sometime between these dates — and possibly coinciding with, and one of several reasons for, the end of the chalcolithic settlements — major earth movements have occurred. These did not result in desert formation however; indeed, the abandonment of calcolithic sites may well have permitted regeneration of the flora.

## CHRONOLOGY OF DESERTS AS AFFECTING HUMAN SETTLEMENT

Chronological evidence is a very tricky subject in the context of desert formation. If one is seeking the date from which weather fluctuation, man and his animals, and, to a limited extent, tectonics have been the desert-forming factors, I think one must settle for somewhere around 8000 B.C. along the present southern desert margin and a little later in more northerly situations such as the Anatolian plateau. The evidence is of two kinds: first there is the fact that the most ancient sites known, such as Çatal Höyük in Anatolia, Jericho in Jordan, and Beidha in Jordan are all (as are most of the more recent sites down to about 2000 B.C.) situated precisely where a spring or rivulet makes life possible for the people of today. In the case of Beidha, several thousand years of erosion of valley fill that almost certainly once supported either a small stream or spring have left the little residual mound about 3 kilometers away from the modern springs. In Anatolia, visited recently, no single prehistoric site (early bronze about 2000 B.C. or earlier) of the hundreds seen was more than a few meters from surface water. This argues a continuity of the present regime, for if there had been a general drying of the climate, expressible only in terms of lower rainfall, infiltration, and spring and stream flow, some sites would be found remote from the available water of today.

Secondly there is the question of the grains, pulses, fruits, and so on from the collection of which man proceeded to cultivation. The sites of the neolithic farming revolution are found, almost without exception, in situations within an hour's walk of the natural habitats (today) of the wild barley, wheat, pulses, pistachio and so on, with the emphasis on barley and wheat (which have become properly cultivated whereas pistachio remains a more-or-less wild *orchard* crop). Such incontrovertible matters as the law of gravity — for heavy-seeded cereals — preclude secular climate change of ecological significance. For while cereals could migrate downstream in response to rainier conditions, they would be hard put to it to make the return journey.

## CONCLUSION

To summarize briefly, it seems to me that desert formation is the result of man's improvidence, aided here and there by tectonic movements, against a background of vegetational impoverishment inherent in the variability of weather.

## REFERENCES

1. UNESCO
   1963    Changes of climate, proceedings of the Rome Symposium organized by Unesco and the World Meteorological Organization. Arid Zone Research 20. 488 pp.
2. ROYAL METEOROLOGICAL SOCIETY
   1966    World climate from 8000 to 0 B.C. International Symposium, Proceedings, Imperial College, London. Royal Meteorological Society, London.

3. RAIKES, R.(L.)
   1967    Water, weather and prehistory. John Baker, London; Humanities Press, New York. 208 pp.
   *Some other works on this subject by the author.*
   RAIKES, R. L., and R. H. DYSON, JR.
   1961    The prehistoric climate of Baluchistan and the Indus Valley. American Anthropologist 63:268-281.

RAIKES, R.
1965    Sites in Wadi Shu'eib and Kufrein, Jordan. Palestine Exploration Quarterly, July-December 1965, pp. 161-168.
RAIKES, R. L.
1966    Prehistoric climate and water supply, a preliminary report. *In* D. Kirkbride, Five seasons at the pre-pottery Neolithic village of Beidha, in Jordan, Appendix C. Palestine Exploration Quarterly, January-June 1966, 68-72.
RAIKES, R.
1965    Physical environment and human settlement in prehistoric times in the Near and Middle East, a hydrological approach. East and West, n.s. 15(3/4): 179-193.

# VEGETATION CHANGES IN SOUTHERN NEW MEXICO DURING THE PAST HUNDRED YEARS

JOHN C. YORK

and

WILLIAM A. DICK-PEDDIE
Department of Biology
New Mexico State University
Las Cruces, New Mexico, U. S. A.

# ABSTRACT

Vegetation patterns have changed in southern New Mexico during the past hundred years. A comparison of the results of a statewide transect and detailed studies of several additional areas with original territorial survey records (1858 and later) showed that mesas of southern New Mexico, now virtually devoid of grass and occupied mostly by mesquite or creosotebush, had been largely covered with a good stand of grass at the time of the territorial survey.

Small pockets of mesquite and creosotebush grow here and there on the mesas. Mesquite pockets invariably coincide with past Indian campsites. These pockets suggest an explanation for the rapid expansion of mesquite: that grazing by cattle can be considered the cause for the reduction of the grass and the subsequent occupation by shrubs.

Many of the sites have been so modified by the succession of species and erosion that they should no longer be considered potential grassland areas.

# VEGETATION CHANGES IN SOUTHERN NEW MEXICO DURING THE PAST HUNDRED YEARS

## John C. York and William A. Dick-Peddie

## INTRODUCTION

Man constantly strives to increase the productivity of his lands. This is such a universal problem today that the planners of the International Biological Program (IBP) have selected *productivity* as a primary goal for the initial stage of their program.

In order to assess or improve long-term productivity potential, it is desirable to have a standard vegetation condition or frame of reference to use for comparison and as a guide for possible manipulations. This standard should be relatively stable. The ideal situation would be to use a *climax* condition as a base. Eyre (*1*) points out that climate is the primary controlling factor for continental vegetation patterns. Climax vegetation is said to be in dynamic equilibrium with the climate and the soil. Since most lands of the world have been used to varying degrees by man, however, much vegetation is not in *climax* condition but is instead in a state of imbalance (secondary succession) or in a temporarily stable state due to man-caused factors (plagiosere or disclimax). Consequently, it is often difficult to determine climax conditions, let alone find climax vegetation. Such situations make it necessary to determine what the vegetation was before disturbance in the case of an unchanged climate or to determine what the vegetation might be if disturbance were to be removed.

These disturbance conditions have often persisted for long periods of time. Also, in areas where annual biomass production is small, such as arid lands, the succession may be so slow as to virtually escape detection. Under these conditions the climax situation is difficult to ascertain.

### The Problem

Southern New Mexico is such an area. It is 1,200 meters above sea level; it has an annual evaporation rate of 2,400 millimeters, and as much as one-third of the 200 millimeters of annual precipitation may fall outside the 180-day growing season. These conditions constitute an extremely demanding environment.

Most early ecologists called the vegetation of this portion of New Mexico *desert grassland* (*2*). This vegetation, thought to be an ecotone between the grassland and desert formations, then presumably constituted the edge of the grassland on the xeric side. The vegetation patterns were thought to be as follows: Black grama (*Bouteloua eriopoda* Torr.) dominated all of the more mesic well-drained sites such as mesas and rolling hills. Low sites (swales) were dominated by tobosa grass [*Hilaria mutica (Buckl.)* Benth.], often surrounded on their sandy margins by a belt of yucca (*Yucca elata* Engelm.). Shallow soil sites such as gravelly hill tops were dominated by creosotebush [*Larrea tridentata* (DC.) Coville]. Mesquite [*Prosopis juliflora* (Swartz) DC.] and tarbush (*Flourensia cernua* DC.) were also to be found on sites unsuitable for grass. As the available moisture conditions improved with an increase in elevation or latitude, blue grama [*Bouteloua grasilis* (H.B.K.) Lag.] tended to dominate on mesa sites. As this trend continued, the grassland gave way to forest in a savannah-like zone called a piñon (*Pinus edulis* Engelm.) / juniper (*Juniperus* spp.) belt.

Today the mesas and rolling hills of southern New Mexico are dominated by creosotebush, tarbush, and mesquite. There are many tobosa swales, and occasionally large stands of yucca occur in swales and on the surrounding mesas. Juniper stands often meet and merge with the desert shrubs. In almost all cases grama grass is merely a subdominant, if it is present at all.

Because the existing vegetation patterns appear to be stable, there is an increasing tendency for workers to consider southern New Mexico as a desert climax rather than a desert grassland climax (*3*). There is even speculation that the area never was dominated by grass and that it was incorrectly classified as desert grassland. Others suggest that if it was desert grassland, it is now a desert shrub disclimax or plagiosere. Still others maintain that it is truly a desert shrub climax now because there has been a climatic change. It is also possible that the pristine vegetation of southern New Mexico was desert grassland and that the climate is still virtually the same; however, disturbance such as grazing or the suppression of fires may have allowed the sites to deteriorate to the extent that they would not now return to grassland if left undisturbed (*4*).

### The Approach

We will attempt to answer the following questions.

*a*) Was southern New Mexico a desert grassland

157

prior to extreme disturbance? If so, to what extent was grass dominant? Considerable controversy exists as to the exact ratio of shrubs to grass in the past. This difficulty occurs because most of the early information has been obtained from explorers' accounts, army diaries, and settlers' records (5) which are usually subjective, often contradictive, and are seldom detailed enough to permit accurate reconstruction of vegetation patterns.

b) Is the vegetation of southern New Mexico in a successional stage, a disclimax, or a climax condition at the present time?

The answers to a number of additional questions must be found before this one can be handled. These are: Has the climate changed? Could the area have carried fire? If there have been changes in the vegetation have these changes coincided with disturbances such as grazing?

The source for these answers will come from determining pristine vegetation and comparing it with today's vegetation. Workers in other parts of the United States have found that the Territorial Survey records provide a relatively unbiased and objective account of the vegetation prior to extensive settlement (6). Thirty years ago Kenoyer (7) said: "The writer is convinced that the most accurate picture of plant distribution before settlement of the country can be obtained from the study of survey records as indicated." This approach has been used in New Mexico by Buffington and Herbal (8) to correlate vegetation and soils in a relatively limited area.

## PROCEDURES

Systematic New Mexico land surveys were begun in 1858 and continued sporadically for sixty years. These early surveys were the ones used for this study. High mountain areas and Spanish Land Grants were not included in these early surveys.

The land was evaluated by the surveyors in terms of dry land farming and statements such as ". . . broken area is destitute of vegetation except for grama grass. It is unfit generally for cultivation and adapted only for grazing cattle."—A. P. Wibar, October 1, 1858. The phrase "good grama grass" was often used after stating that the land was "poor." Grama grass refers to a number of species in the genus *Bouteloua*.

In the West, presettlement vegetation patterns cannot be reconstructed with the same precision as areas where there were forests. The use of trees to *witness* a section corner constituted a one mile square grid sample. A *witness tree* was selected in each quadrant as near the corner as possible, and the species, trunk diameter, and distance and azimuth from the corner were recorded.

In the New Mexico surveys the surveyor noted the vegetation on each line of all sections in every township. He then also summarized the vegetative and topographic features of the entire township.

The surveyors were quite consistent in their use of common names for plants (9); however, some names received varied spelling through time. Table 1 gives a list of often-used surveyors' plant names, the present common name, and the scientific name.

TABLE 1

### PLANT NAMES USED BY EARLY SURVEYORS IN NEW MEXICO

| Surveyors' Plant Names | Present Common Name | Scientific Name |
| --- | --- | --- |
| Cedar | Juniper | *Juniperus* sp. |
| Chamisal | Chamisa | *Atriplex canescens* |
| Edeondilla | Creosotebush | *Larrea tridentata* (DC.) Coville |
| Gatuño | Catclaw | *Acacia greggii* Gray |
| Greasewood | Creosotebush | *Larrea tridentata* (DC.) Coville |
| Hediondilla | Creosotebush | *Larrea tridentata* (DC.) Coville |
| Hediondo | Creosotebush | *Larrea tridentata* (DC.) Coville |
| Largoncillo | Whitethorn | *Acacia constrica* Benth. |
| Mimbres | Desert willow | *Chilopsis linearis* (Cav.) Sweet |
| Palmilla | Soaptree yucca | *Yucca elata* Engelm. |
| Palmira | Soaptree yucca | *Yucca elata* Engelm. |
| Poñil | Apache plume | *Falugia paradoxa* (D. Don) Endl. |

We were given free access to the survey records which are kept in the Bureau of Land Management office in Santa Fe, New Mexico. All information for the areas studied was copied from the original survey *handbooks*.

Current vegetation estimates were made using a modified wandering quarter method as described by Catana (10).

The choice of study sites had to be highly selective for two reasons. First, many areas were not surveyed at all, as already stated. Secondly, present military reservations prevented the use of other areas because present vegetation could not be assessed for comparison with the early survey data.

The sites studied are shown (shaded) in Figure 1. Each shaded area is one township or larger. Some of the sites have been grouped for special comparisons, and these groups (A-D) are referred to as *areas*.

A. Corralitos Area: Used to indicate a possible source for the rapid expansion of mesquite.

B. Tortugas Area: Used to illustrate creosotebush expansion and mesquite and grassland reduction.

C. Bingham Area· Used to illustrate past and present boundaries of juniper-grassland and brush-grassland.

D. Shinnery Area: This was and is an area of shin oak [*Quercus havardi* (Rydb.)] in southeastern New Mexico. The past and present stand boundaries are compared.

We obtained a partial transect from west to east

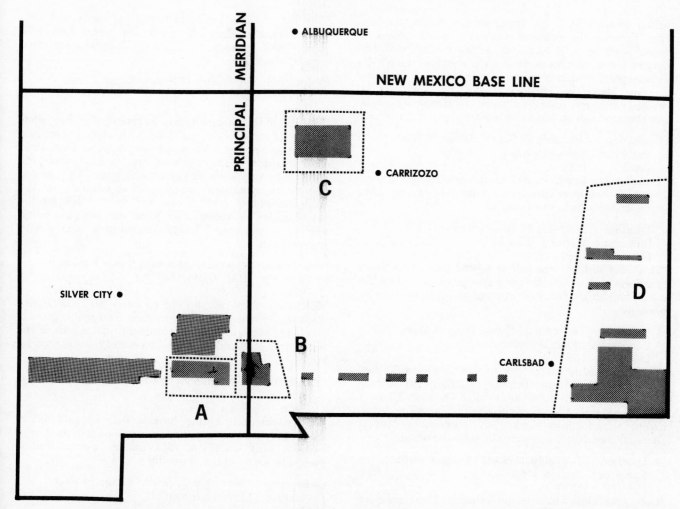

Fig. 1. Sites where past and present vegetation patterns are compared. Special comparison areas are: *A*, Corralitos Area; *B*, Tortugas Area; *C*, Bingham Area; *D*, Shinnery Area.

across the state. As stated previously, this could not be an unbroken transect. All studied areas were at least one township (36 square miles) wide and one or more townships long (Fig. 1).

## OBSERVATIONS AND DISCUSSION

The survey of southern New Mexico was begun at about the same time or slightly after the large cattle buildups began. During and after the survey, hundreds of thousands of cattle were grazed over most of the study area.

References to grama grass by the surveyors evidently referred mostly to black grama. The remnants found today on the study sites are invariably black grama. The topographic relief of southern New Mexico is very erratic as indicated by the broken nature of the original survey. These two observations plus the low litter accumulation of black grama on long-term study plots on the college ranch and the Jornada Experimental Range make fire a questionable factor in past vegetation patterns in southern New Mexico. It is doubtful if the area could ever carry a fire. Even if a fire could be

carried, topographic features would likely prevent an extensive burn at any given time.

### Extent of Past and Present Grass Cover

The 31 townships in the transect were approximately 75 per cent covered with grass at the time of the surveys. The same townships today have less than 5 per cent grass cover. Following are some examples of the township "general descriptions" followed by a summary of today's vegetation. These examples are from west to east on the transect.

*a)* Location:  Township 23 South, Range 21 West
    Surveyor:  Clarance Goddard
    Date:      1904
"In general the township is rolling with third rate soil. Greasewood (creosotebush), mesquite, and gatuno brush with mimbres found only along the dry runs. There is a fair to good grass cover otherwise over the entire township." Today: Virtually no grass, area dominated by creosotebush, mesquite, and catclaw.

*b)* Location:  Township 23 South, Range 16 West
    Surveyor:  Walter Davies
    Date:      1886
"Township is in general gently rolling or level. The soil is of a sandy character except in the extreme northeastern

corner where it is rocky. The Southern Pacific Railroad runs through the extreme southwest corner. There is dense brush on a great portion of the township. Water may be found near the surface in the arroyo in the northeastern part. No timber except a few willows, hackberries, growing along the arroyos." Today: The township is covered with mesquite and yucca on the sandy areas and creosotebush on the rocky foothills.

c) Location:    Township 22 South, Range 17 West
   Surveyor:   Walter Davies
   Date:       1886

"There is no brush except in the southwest corner." Today: Township is covered by a mixture of mesquite, yucca, creosotebush, and chamisa.

d) Location:    Township 23 South, Range 11 West
   Surveyor:   Walter I. Rumple
   Date:       1886

"This land is level, the soil is second rate, there is no brush, no timber, and there is a good grass cover over the entire township." Today: Yucca and mesquite over entire township.

e) Location:    Township 23 South, Range 9 West
   Surveyor:   Albert Robinson
   Date:       1881

"In this township a tent city has just been established at the Nutt (New Mexico Territory) Southern Pacific Railroad junction. This junction is called Deming. There is fairly good grass throughout the township and no brush except in the foothills." Today: All that is not the town of Deming, New Mexico, is virtually solid mesquite.

f) Location:    Township 23 South, Range 6 West
   Surveyor:   Andrew Moore
   Date:       1883

"Good grass covers most of the township. There are a few mesquite in the very southern part of the township. Land is rolling, soil is third rate and sandy." Today: Township is covered by mesquite and yucca.

g) Location:    Township 23 South, Range 5 West
   Surveyor:   Andrew Moore
   Date:       1883

"Good grass, some brush in southern portion of township (Palmillan and mesquite). Land is rolling, sandy and third rate." Today: Township is covered by mesquite and yucca.

h) Location:    Township 23 South, Range 3 West
   Surveyor:   Moore Fisher
   Date:       1882

"The land of this township is level open prairie. The soil is sandy, second rate and is covered with good grass, in some places a few mesquite. No timber grows on the prairie. L. M. Black Ranch is in the northwest corner of section 1." Today: Township is covered by mesquite and yucca.

i) Location:    Township 23 South, Range 1 West
   Surveyor:   A. P. Wilbar
   Date:       1882

"The surface in the southwestern part of this township is high rolling prairie. The eastern part is rolling broken, thickly covered with mesquite. The Mesilla Civil Colony Grant takes a portion of the northeast corner. The surface of the prairie is covered with a fair growth of grama grass." Today: Township is covered by mesquite and yucca.

j) Location:    Township 23 South, Range 2 East
   Surveyor:   Joseph Lively
   Date:       1885

"The land in this township is rolling, sandy and with good grazing in places. No permanent water. Lime kiln in northeast 1/4 section 23." Today: Township is covered with creosotebush.

k) Location:    Township 23 South, Range 3 East
   Surveyor:   Joseph Lively
   Date:       1858

"The soil of this township may be classified rather below average of second rate land. Its principle [sic] value consists in grazing grounds which are valuable from their close proximity to the settlements on the Rio Grande, the herds from which are driven here to graze. The springs also in the mountains afford good opportunity for the establishment of stock farms and were it not for the apprehension of molestation from the Indians, would soon be settled upon." Today: Township is covered with creosotebush.

l) Location:    Township 23 South, Range 5 East
   Surveyor:   R. E. Clements
   Date:       1858

"Soil sandy, third rate, rolling, no water, no timber, there is a very light scattering of mesquite. The grama grass is excellent." Today: All of the township not a part of the White Sands Missile Range Community is covered by mesquite.

m) Location:    Township 22 South, Range 8 East
   Surveyor:   Samuel R. Biggs
   Date:       1884

"This township is rolling, broken, with a small hill in the center. There is excellent grass, some scattered mesquite, a rank species of sage, and no water." Today: Mesquite covers entire township.

n) Location:    Township 22 South, Range 11 East
   Surveyor:   Davis Laupton
   Date:       1884

"The extreme western part of this township is mountainous, however, the entire township is covered with a heavy growth of fine grama grass. No brush and no timber." Today: The township is covered by mesquite.

o) Location:    Township 22 South, Range 2 East
   Surveyor:   Joseph Lively
   Date:       1858

"This township is now much resorted to by herdsmen from the river settlements with their herds and flocks to graze upon the abundant pasturage which it affords. Hay is also made here to supply Dona Ana and Las Cruces. Much of the land is of good quality, well adapted to the production of all serials [cereals]." Today: Township is covered by mesquite and creosotebush.

These comparisons indicate a rapid and drastic change in the vegetation over a period of 110 years. Actually, all but three of the citations are from surveys conducted after 1881 (only 87 years ago). When the three 1858 descriptions (k, l, and m) are compared to the others, it appears that grass cover may have already begun to diminish by the end of the century.

## Speed of Brush Occupation

The diminution of grass cover appears to have been so extensive and rapid that factors other than climate must have been responsible. The Corralitos area (Fig. 1) yields information which could account for the speed with which mesquite has occupied many sites.

Following are five township summaries from this area. The townships summarized in items *h* and *i* of the preceding discussion are also in the Corralitos area.

*a*) Location:   Township 22 South, Range 1 West
   Surveyor:   William Sanders
   Date:       1881

"The surface in the northern portion of the township is mountainous. The south and west portions are level and rolling. The surface is covered with chamisal (chimiso) brush except the southwest part which is covered with a good growth of grama grass on third rate soil." Today: Township covered with yucca, chamisa and mesquite complex.

*b*) Location:   Township 22 South, Range 2 West
   Surveyor:   Andrew Moore
   Date:       1882

"The land on this township is rough and broken. The soil is second rate. Very good grama grass, a few cedar [juniper]. There is very little brush. Pedro Pachecos house and corral is in Section 15." Today: Township is covered with mesquite at varying densities.

*c*) Location:   Township 23 South, Range 2 West
   Surveyor:   Moore Fisher
   Date:       1882

"The land of this township is rolling, broken, with a few hills. The soil is second and third rate with a very good growth of grass and a few scatterings of mesquite. The

township contains no major shrubs or timber." Today: Township is covered with mesquite and yucca.

*d*) Location:   Township 22 South, Range 3 West
   Surveyor:   Andrew Moore
   Date:       1882

"The land of this township is generally rolling prairie. The soil is sandy, and second rate. No shrubs, no timber grows in this township except on the tops of the few hills. There is good grass of grama or prairie type covering the entire township." Today: The township is covered with mesquite and creosotebush at varying densities.

*e*) Location:   Township 22 South, Range 4 West
   Surveyor:   Andrew Moore
   Date:       1882

"The land of this township is level, open prairie. Soil is sandy and second rate. There are no shrubs or timber." Today: Tobosa flats, creosotebush, mesquite at varying densities.

Figure 2, Corralitos Area, is a detailed map of the vegetation in 1882 as reconstructed from the survey. This area west of Las Cruces is a mesa type formation that is very extensive. The land is eroded and broken where the mesa bounds the Rio Grande Valley. Mesquite and creosotebush abound in all sections of the area. When the early surveys were made, mesquite was found only along the mesa edge and in a few isolated pockets. Juniper was found occasionally on the tops of the hills in the area.

The pockets of mesquite found in the area eighty

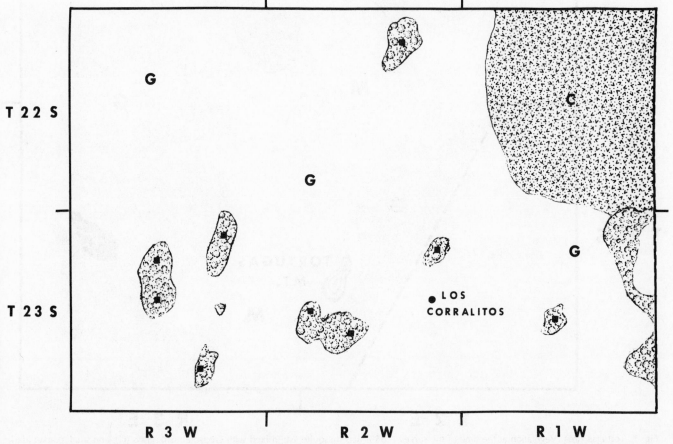

Fig. 2. Corralitos Area. Vegetation at the time of survey (1881) included grass (G) and chamisa (C) in northeast corner. Black squares indicate Indian campsites. Stippled "pockets" associated with the campsites are mesquite, which also appears in the lower right-hand portion of the map.

years ago were at first puzzling; however, upon inspection it has been determined that each of these small locations has an Indian campsite near its center. These sites had evidently been occupied by Indians long enough to allow the mesquite to become established around the camp. The Indians used mesquite extensively for their food and later as food for their horses.

The soil of this mesa country is sandy and underlain with a caliche hardpan. Local Soil Conservation Service scientists have concluded that this entire area is one of the oldest undisturbed land formations in the state. Grama grass was growing on this area, but it was a marginal grama grass range. After grazing started, rapid deterioration of the range took place.

As increased grazing pressure reduced the competitive advantage of grass, this *grid* of mesquite pockets provided an ideal source for rapid and complete expansion. The presence of cattle undoubtedly speeded the process even more, as indicated by Buffington and Herbel (8).

### Secondary "Invasion"

Certain sites were found to have mesquite even 100 years ago. These sites were found on shallow sandy soils, *breaks* on the edges of water courses, or around Indian camp sites. As indicated previously, in most cases we found that these mesquite areas have become contiguous over the intervening years.

One extreme exception was found and deserves

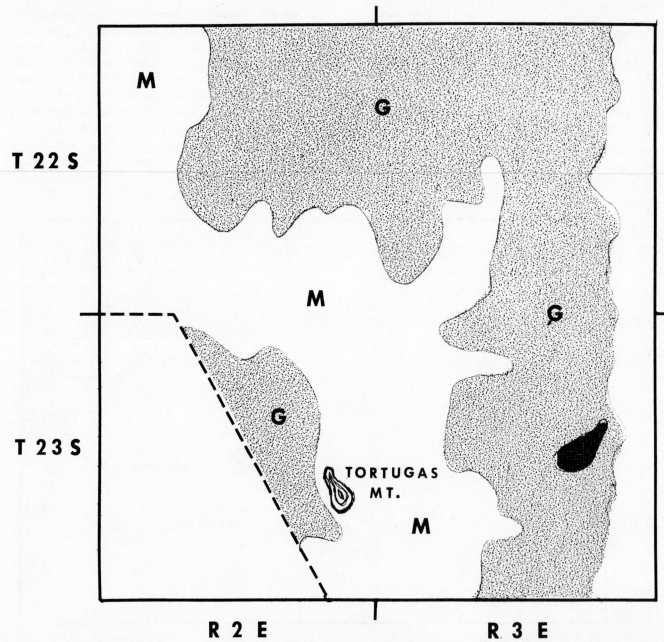

Fig. 3. Tortugas Area. Vegetation at the time of the survey (1858) was mesquite (M) (mixed with creosotebush), grass (G), and small pocket (dark patch in southeastern portion) of creosotebush. Checked line indicates boundaries of Dona Ana Bend Colony Grant (not surveyed)

more detailed consideration. This exception occurs on sites east of the Rio Grande River referred to as area *B*, the Tortugas Area, in Figure 1. The area extends from the river valley ridge east into the foothills of the Organ Mountains. Most of the area is on a 10 to 15 per cent slope and today is highly dissected by arroyos.

Figure 3 illustrates the detailed vegetation patterns of this area at the time of the survey. Summary descriptions *h* and *j* included in the section on grass cover were of townships included in this Tortugas Area.

At the time of the survey, grama grass covered the northern and eastern (foothills) portions of the area. Except for the southeast quarter, the rest of the area was covered by mesquite. On the surveyor's map the words "dense mesquite underbrush" were written through three sections east of Tortugas Mountain. Previous to survey times there had been a large Indian village in much of the central mesquite portion. Creosotebush was mentioned on about one-half of the sections in the mesquite portion. It was apparently growing on well-drained, gravelly

hilltops. One small stand of creosotebush was in the southeast corner of the grass-covered portion (Fig. 3).

Today, except for a narrow band of mesquite up in the foothills, the mesa portion of the entire area is completely dominated by creosotebush! The grass replacement is to be expected, but we initially considered the mesquite replacement to be quite unexpected. Here then is an instance where mesquite, which had replaced grass due to Indian camps, being itself almost completely replaced by creosotebush in a period of 70 or 80 years.

The explanation would appear to be that the presence of mesquite on a 10 to 15 per cent slope allowed severe and rapid sheet erosion. The summer rains in the area are often of great intensity for short periods of time. The surface was quickly washed into the arroyos, leaving the gravelly subsurface exposed and setting the stage for occupation by creosotebush.

## Variations in Changing Patterns

The evidence we have presented to this point has

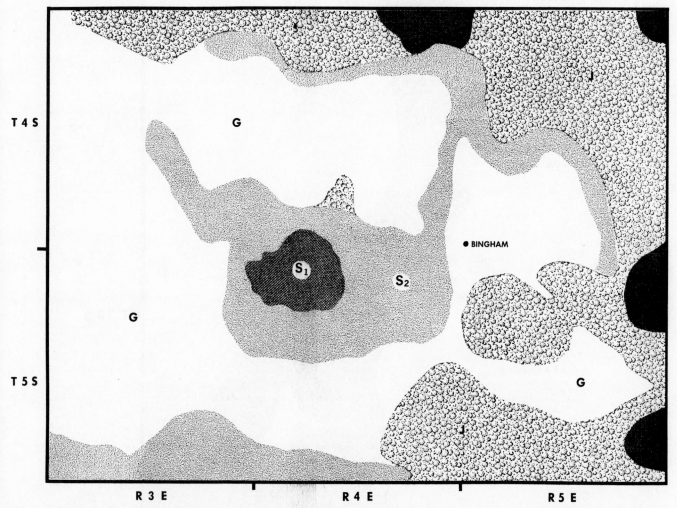

Fig. 4. Bingham Area. Vegetation patterns past and present are compared. Dark areas to the north and east represent the extent of juniper at the time of the survey; lighter patterned contiguous area (J) is the boundary of juniper today. Area in the center (S₁) was occupied by sagebrush at the time of the survey; lighter area (S₂) shows sagebrush coverage today.

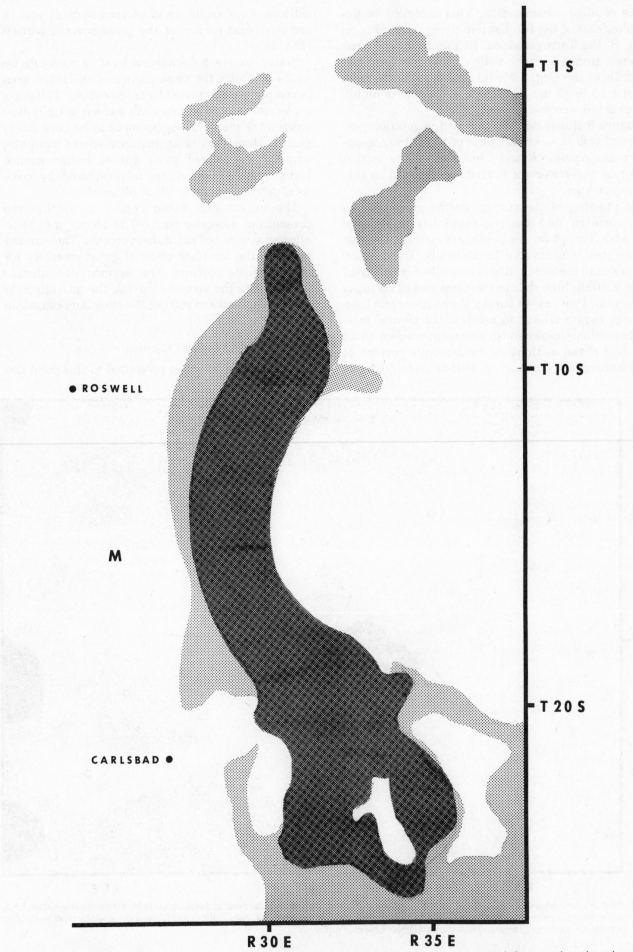

Fig. 5. Shinnery Area. Shin oak patterns past and present are indicated. Dark portion is pattern at time of survey; lighter area shows boundary today. Mesquite (M) meets shinnery on the west.

emphasized the great reduction in grass cover and the speed with which the reduction has occurred. In much of the study area grass has been completely replaced, and usually by two dominants—mesquite and creosotebush. Other shrubs or trees have also encroached on areas previously in grass. Also, some sites still have grass on them.

The Bingham Area and the Shinnery Area (Fig. 1) are used to compare other shrub cover, past and present, and also to demonstrate that the changes have not always meant the complete disappearance of grass. Figure 4 shows the advance of juniper and sagebrush (*Artemesia filifolia* Torr.) since the survey. The blank area (G) is still covered by grass. This condition was also found in the Nutt Area (not included in a figure). These areas have subsequently been found to have been lightly stocked over the past 100 years.

This evidence of grass still occupying sites after 100 years supports the evidence of Linney and García (*11*) and of Martin (*12*), which indicates that slight, if any, climatic change has taken place in the past 100 years. The fact that grazing pressure has been less where grass remains than in areas where it has virtually disappeared is strong indication that grazing is the factor responsible for the increase in desert shrubs.

The juniper areas (J) at the time of the survey were largely restricted to foothills. Since that time they have moved down and occupied the flats. The sagebrush (S) was restricted to a small sandy area and has moved out in a rather strange pattern. Some of it now occupies an ecotone between the juniper and grass. Further investigation will be necessary to arrive at an explanation of this sagebrush expansion. The Bingham Area has more rainfall than farther south and juniper appears to have replaced creosotebush as the "invader" from the foothills.

The Shinnery Area is detailed in Figure 5. This area is unique to the state and provides an example of a shrub-tree, shin oak (*Quercus havardii* Rydb.), replacing grass. This shrub oak type extends into Texas where you find its best expression. The recently occupied areas to the north appear to be an expansion out of river valleys. The cap-rock (a geological formation) has understandably prevented any expansion of the shinnery belt to the east; however, the rather modest expansion to the west is puzzling. The shinnery appears to be growing on the same sandy sites as the mesquite which bounds the belt on the west (Fig. 5, M). More information is necessary before we can propose an explanation

for the portion of the shinnery belt which has apparently been held in check in its westward expansion.

## SUMMARY AND CONCLUSIONS

The mesas of southern New Mexico were covered by grass in the middle of the last century. This grass was primarily grama grass (*Bouteloua* sp.).

Mesquite occurred in limited areas on shallow sandy soil, but more importantly it occurred around Indian campsites. Creosotebush was restricted to well-drained gravelly hilltops and narrow patches in the foothills of mountains. Juniper stands were on mountain foothills usually higher or further north than creosotebush. The area was correctly classified desert grassland.

All the other species have greatly expanded their ranges during the past hundred years, usually in response to a reduction in grass cover. The grass has not completely disappeared or even changed its boundaries in all cases. In one instance a secondary replacement of mesquite by creosotebush has taken place. The speed of the recent occupation by mesquite may be attributed to the effects of cattle in the presence of ideal source pockets of mesquite around old Indian campsites.

The speed with which this almost complete replacement of grass has taken place in southern New Mexico coupled with the fact that isolated areas are unchanged indicates that climate has not been a factor. The topography and biomass potential of southern New Mexico make it highly unlikely that the area could ever have carried a fire. The appearance of the grazing industry is the only factor which coincides with the time of this spectacular change.

Although the grasslands of southern New Mexico were extensive and dominated the area, they were on the xeric edge of the continental Grassland Formation. A single factor such as grazing was evidently enough to set in motion a series of relatively rapid events which culminated in a desert shrub vegetation.

Even though the climate is virtually unchanged, the surface horizon which 100 years ago supported grassland is undoubtedly long gone down the arroyos or formed into dunes. Until there is a climatic change, therefore, most of southern New Mexico can be considered to be a desert climax rather than a desert grassland climax as it was 100 years ago.

## REFERENCES

1. EYRE, S. R.
    1963 Vegetation and soils, a world picture. Aldine Publishing Company, pp. 1-24.

2. GARDNER, J. L.
    1951 Vegetation of the creosotebush area of the Rio Grande Valley in New Mexico. Ecological Monographs 21:379-403.

3. KUCHLER, A. W.
   1964  Potential natural vegetation of the conter-
         minous United States. American Geographi-
         cal Society, Special Publication 36. 116 p.
4. BRYAN, K.
   1925  Date of channel trenching (arroyo cutting)
         in the arid Southwest. Science, n.s. 62(1607):
         338-344.
   HASTINGS, J. R.
   1958- Vegetation change and arroyo cutting in
   1959  southeastern Arizona during the past cen-
         tury: an historical review, pp. 24-39. *In*
         Arid Lands Colloquia. University of Ari-
         zona, Tucson.
   LEOPOLD, L. B.
   1951  Vegetation of southwestern watersheds in
         the nineteenth century. Geographical Re-
         view 41:295-316.
5. FASSETT, N. C.
   1904  Vegetation of Brule Basin, past and present.
         Wisconsin Academy of Sciences, Arts, and
         Letters, Transactions 35:33-56.
6. BLEWETT, M. B., and J. E. POTZGER
   1951  The forest primeval of Marion and Johnson
         Counties, Indiana, in 1819. Butler Univer-
         sity, Botanical Studies 10:40-52.
   DICK, W. B.
   1936  A study of the original vegetation of Wayne
         County, Michigan. Michigan Academy of
         Science, Papers 22:329-334.
   DICK-PEDDIE, W. A.
   1955  Presettlement forest types in Iowa. Iowa
         State College, Ames (Ph.D. dissertation)

   MASON, L.
   1963  Using historical records to determine climax
         vegetation. Journal of Soil and Water Con-
         servation 18:190-194.
   POTZGER, J. E., and M. POTZGER
   1950  Composition of the forest primeval from
         Hendricks County southward to Lawrence
         County, Indiana. Indiana Academy of sci-
         ences, Proceedings 60:109-113.
   SEARS, P. B.
   1925  Vegetation of Ohio. Ohio Journal of Science
         25:139-149.
7. KENOYER, L. A.
   1933  Forest distribution in southwestern Michi-
         gan as interpreted from the original land
         survey. Michigan Academy of Science,
         Papers 19:107-111.
8. BUFFINGTON, L. C., and C. H. HERBEL
   1965  Vegetational changes on a semidesert grass-
         land range. Ecological Monographs 35:
         139-164.
9. POTZGER and POTZGER, ibid.
10. CATANA, A. J., JR.
    1964  The wandering quarter method of estima-
          ting population density. Ecology 44:349-360.
11. LINNEY, C. E., and F. GARCÍA
    1918  Climate in relation to crop adaptation in
          New Mexico. New Mexico Agricultural Ex-
          periment Station, Bulletin 113.
12. MARTIN, P. S.
    1963  The last 10,000 years; a fossil pollen record
          of the American Southwest. University of
          Arizona Press, Tucson. 87 p.

# COLD DESERT CHARACTERISTICS AND PROBLEMS RELEVANT TO OTHER ARID LANDS

ROY E. CAMERON
Bioscience Section
Jet Propulsion Laboratory
California Institute of Technology
Pasadena, California, U. S. A.

# ABSTRACT

Cold deserts are the last deserts to be investigated on this planet. They closely resemble polar and high-mountain deserts in many respects and share many problems in common with other arid regions. Antarctica contains typical cold deserts in its dry valley areas; various aspects of the geomorphology, hydrology, climate, soils, vegetation, and microbiology show similarities and differences compared to those of high Arctic and other desert areas. The harsh environment of cold deserts limits the extant biota. Colonization and exploitation of cold deserts are difficult because of geographical position, remoteness, harsh environment, and the large logistic burden. Problems and adaptation of man in cold deserts are similar to those of an outpost society. Antarctica is a valuable test area prior to extraterrestrial exploration and colonization.

# COLD DESERT CHARACTERISTICS AND PROBLEMS RELEVANT TO OTHER ARID LANDS

Roy E. Cameron

Cold deserts receive little or no attention in considering studies on arid and semiarid regions on this planet, and it is a common misconception that "arid lands are hot regions" (1). A discussion of deserts usually includes only those regions of temperate or tropical climates, soils, and vegetation below a given elevation (2). Desert geographical areas are generally considered as those outlined in maps by Peveril Meigs (3). The classification of these areas is based on moisture, season of precipitation, and temperature. The most temperate of these deserts have mean temperatures of below 0°C for the coldest month. These deserts include parts of the Turkestan, Takla-Makan, Gobi, Pamir-Karakoram, Tsaidam, and North American Great Basin; all are in the northern hemisphere, approximately between 35° and 55°N latitude. Mean summer temperatures for these same deserts are approximately between +10° and +20°C. There is no marked season of precipitation except for the Great Basin Desert, which receives most of its precipitation as winter snow. In the southern hemisphere only the Chilean Puna de Atacama (2,000 to 4,000 meters elevation) and eastern Patagonian Desert in Argentina can approach the austral winter coldness of the northern deserts.

Cold deserts, to be considered within the category of other desert regions, must have some of the basic characteristics of warmer arid lands. For this purpose, the primary characteristic is aridity. This includes irregular or low-frequency distribution, quality, and total quantity of available water, and high actual or potential losses through evaporation, sublimation, percolation, run-off, and so on. Other useful characteristics for desert classification include a complex of interacting dependent factors such as wind-molded topography, salty mineral soils, large radiation (energy) influxes and outfluxes, wide vapor-pressure differences, and certain common geographical characteristics. Finally, in terms of occupation by organisms and in similarity with other deserts, a cold desert must present a harsh environment for life. This type of environment is indicated by a paucity of endemic life-forms or organisms which have developed adaptive mechanisms for surviving, growing, and reproducing in cold as well as aridity. In terms of human inhabitation and agricultural productivity, these values would be extremely low or nil.

Cold deserts on this planet are limited to polar regions or to high mountain areas (generally > 4,000-5,000 meters) which simulate high-latitude regions, since an increase in elevation results in climatic conditions approximately equivalent to those found by an increase in latitude. Similarities have been noted between cold Antarctic deserts and periglacial high mountain deserts (4). The high altitude Pamir "Roof of the World," at 3,700 to 4,300 meters elevation and with 25 to 30 millimeters annual precipitation, is a glacial valley with undulating desert plains and is remindful of a polar desert (5), as are the great ranges of Tien Shan, Kunlun, Karakoram and Hindu Kush, and associated parts of Sinkiang.

The main geographical regions for cold deserts are in Antarctica; polar deserts are characteristic of the high Arctic. In the North, these areas are principally between 70° and 90°N latitude, except for Brooks Range, Alaska, which is between the Arctic Circle and 70°N latitude (Fig. 1). In the south there is only the Antarctic continent, which is essentially within the Antarctic Circle, 67°63' latitude (Fig. 2). The high Arctic would include parts of Alaska (northern Brooks Range), Canadian Archipelago (Banks Island, Prince Patrick Island, Cornwallis Island, Axel Heiberg Island, Northern Ellesmere Island), Northern Greenland, Spitzbergen, and Northern Eurasia (areas of Franz Josef Land, Novaya Zemlya, Nicholas Land, Taimyr Peninsula, New Siberian Islands, and Wrangel Island (6). If only exposed land areas and soils are considered, rather than ice and snow and partially vegetated areas, then only a very small per cent of the Antarctic would fall entirely within the category of a cold desert (7). When moisture becomes available for a sufficient time-period to support growth of vegetation, there is a noticeable change from cold desert to polar desert (partially vegetated) and tundra and bog (heavily vegetated with considerable deposits of organic matter) (8).

## THE ANTARCTIC CONTINENT AND ARCTIC POLAR DESERT

For the purpose of considering characteristics and problems of a cold desert, it is necessary to become familiar with Antarctica. Antarctica is the fifth largest of our seven terrestrial continents. It

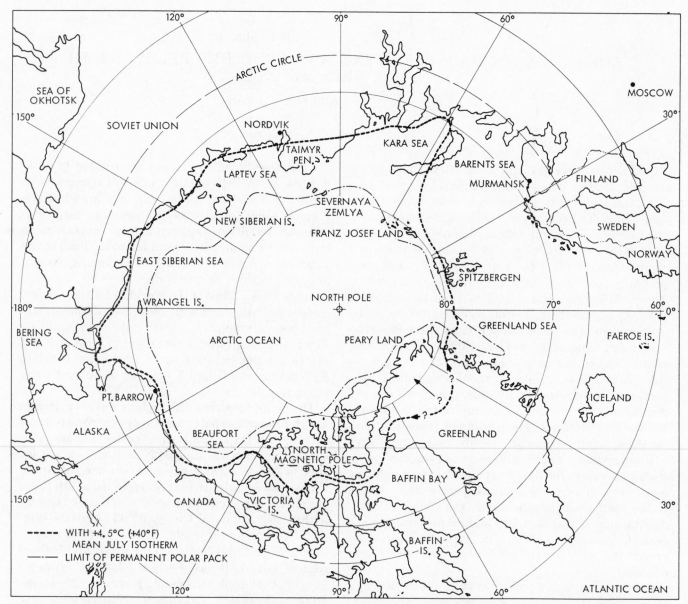

Fig. 1. Arctic region, showing limit of permanent polar pack and +4.5°C (+40°F) mean July isotherm for the polar desert region.

is a true polar continent surrounded by the Southern Ocean, the largest unbroken expanse of sea in the world, which extends from about 40°S latitude to the Antarctic Circle. During austral summer, Antarctica contains an area of $14 \times 10^6$ km², but this area is greatly increased and almost doubled during the winter when the sea freezes beyond the periphery of the continent proper, reaching toward the 55th parallel, as shown by Nimbus I satellite views (9). The glacial cold of Antarctica results in a discernible climatic and oceanic boundary, the Antarctic Convergence, which reaches as far north as the 50th parallel (Fig. 2), a latitude comparable to the southern tip of the Patagonian Desert or the northern boundaries of the Gobi, Turkestan, and North American Great Basin Deserts.

Antarctica has the highest average elevation of all continents, about 2 kilometers, with 55 per cent of its area higher than 2 kilometers and 43 per cent

higher than 2.5 kilometers (10). It is principally a continental ice sheet, which contains about 90 per cent of the world's ice. The average ice thickness is about 2,100 meters, with a volume of approximately $33 \times 10^6$ km³. The significance of this can be appreciated when it is realized that if all the ice were to melt, the sea level would rise at least 75 meters! In the past, during colder climates, glaciated mountain tops indicate that the former continental ice level was 300 to 600 meters above the present level. The Antarctic ice sheet is neither increasing nor decreasing at this time, although the Greenland ice sheet, which is of considerably less average thickness and less extensive, has been receding inland since 1930 (11).

The Antarctic continent as a whole does not have a typical climate, and its climate is not paralleled by the Arctic, because of its geographical position, distribution of sea and land mass, high elevation, and areal extent. The Arctic climate is less harsh, and the Arctic Ocean, covered by 125 to 3,000 centimeters of floating ice, is a source of heat, which escapes through the ice from the water below. By contrast, the Antarctic functions as a tremendous heat sink, releasing immense amounts of heat into space, especially during the dark austral winter. Blizzards are a common phenomenon of both polar regions and both regions also give rise to storms, but off the coasts near the Antarctic Circle, storms are much more frequent, intense, and far reaching in influence. Blizzard conditions result whenever the surface winds exceed about 10 meters per second (12). Winds reach and exceed hurricane force in the Antarctic.

Intensity of insolation is less at the poles, but it is increased at higher altitudes. Daily temperature ranges are therefore much higher at high elevations in middle and low latitudes, as indicated by temperatures above freezing in the aeolian zone (13). The temperature of the Arctic regions and Antarctica remain substantially the same from year to year. Temperatures in the Antarctic fall rapidly from summer maxima to near minimum levels in autumn and reach an absolute minimum in late winter (14).

Winter temperatures at Antarctic coastal stations are seldom below −50°C (15). At McMurdo Station, across from southern Victoria Land, the mean monthly temperature fluctuates between −3.5°C in January to −29°C in August (16). Slightly warmer means are reported for Mirnyy (17). Colder temperatures have been reported for the Antarctic dry valleys and Bunger Hill Oasis (18). Tempera-

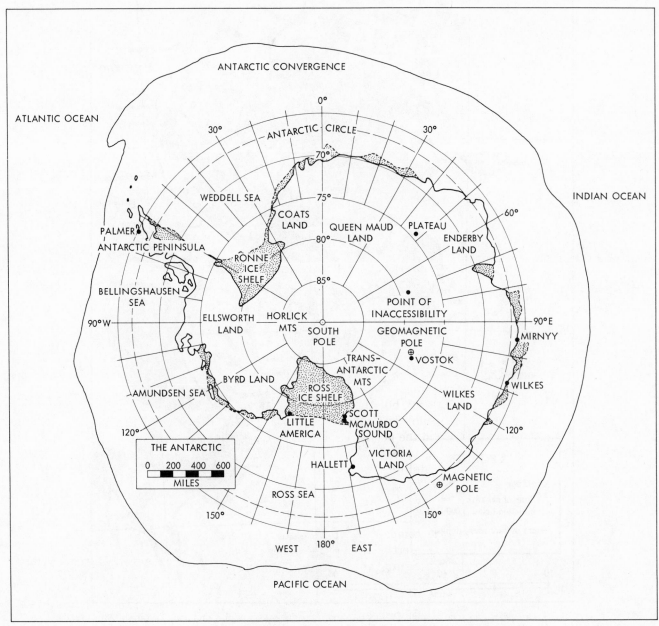

Fig. 2. Antarctic continent and Antarctic Convergence.

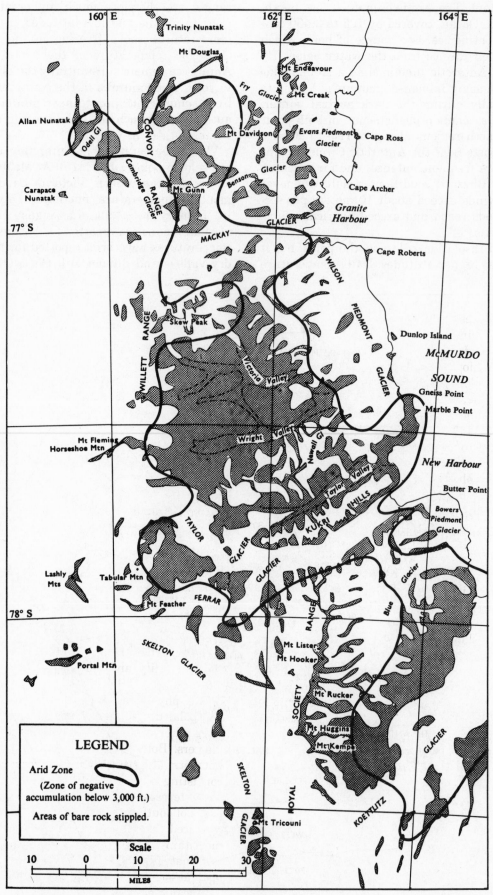

Fig. 3. Dry valley arid zone of southern Victoria Land, Antarctica. Map by B. Gunn and G. Warren (*28*).

ture differences between the Antarctic plateau (1,500 to 2,000 meters) of 55°C are not uncommon (19). The coldest temperature, −88.3°C, was recorded at Vostok in August 20, 1960.

In the Arctic temperatures are not as low as in the Antarctic, and the mean temperature is at least 3°C lower in southern latitudes than at the corresponding northern ones (20). In Peary Land, Greenland, at Jørgen Brønlunds Fjord, the July mean is +8°C and the January mean −9°C (21). Interior temperatures are always low. In Spitzbergen, for latitudes corresponding to Greenland, the mean for March, the coldest month, is −15°C, and for July, +6.5°C (22). Although not as far North as Spitzbergen, at Novaya Zemlya, for February (the mean coldest month) the temperature is −17°C, and for July, +7°C (23). In the Canadian Archipelago at Isachsen, comparable in latitude to Spitzbergen, the average low for February (the mean coldest month) is −36.5°C and the average high for the mean warmest month, July, is +2.5°C (24).

Meteoric precipitation in Antarctica varies with temperature and elevation. The coldest and highest inland stations, Amundsen-Scott (South Pole) (90°00′S latitude, elevation 2,800 meters), Plateau (79°15′S latitude, elevation 3675 meters) and Vostok (78°20′S latitude, elevation 3,488 meters) show probable annual precipitation as snow of less than 7 centimeters to less than 2.5 centimeters, with increasing precipitation on the coast (25). There is very little melt of this snow in the interior from year to year. The snow facies of Antarctica and Greenland have been compared, and for Antarctica it has been found that ablation occurs in less than one per cent of the region and dry snow facies cover 90 per cent of the continent, with the dry snow line generally above 100 meters elevation and limited by annual isotherms of −15°C to −25°C (26). In terms of precipitation alone, most of Antarctica would qualify as a desert. The annual rainfall equivalent in polar deserts is about 5 to 7 centimeters (27). In fact, Antarctica and other extensive ice-covered regions in the Arctic and Greenland are commonly referred to as "ice or white deserts."

In regard to the cold deserts of Antarctica, these are found in certain coastal areas where glaciers have receded. The most extensive areas are in southern Victoria Land, east of McMurdo Sound and Scott Stations (Fig. 2), with a few other exposed areas scattered around the continent, that is, the Bunger Hills area and Knox Coast in the vicinity of Mirnyy and Wilkes Stations (Fig. 2). This does not include Hallett or the Antarctic Peninsula, which are more moist than the dry valley areas. The total exposed area in Antarctica is small compared to its overall area and constitutes 5,600 km² (28). Investigations of cold deserts in the Antarctic

dry valley areas have provided information which can be used for comparisons with warmer arid lands.

## PHYSIOGRAPHICAL SETTING

The "arid zone" of the Antarctic dry valleys is indicated in Figure 3 (29). This is the approximate area in which there is low precipitation and high summer ablation at elevations generally up to 900 meters. Parts of valleys at higher elevations, 900 to 1,500 meters (depending on orientation and slope) may also be free of snow during the summer months. The existence of these valleys is due primarily to (a) presently reduced influx of glaciers from the polar ice-sheet, (b) favorable topographic and climatic conditions of the Transantarctic Mountain Range, which extends from 70°S to about 83°S latitude, and has limited the seaward flow of the icecap and reduced precipitation, (c) greatly increased ablation on bare rock and typical desert pebble or lag pavement, with a much lower albedo compared to snow or ice, and (d) comparatively warmer and drier katabatic winds which flow from a northerly or westerly direction through the valleys. There is also some indication that the carbon dioxide concentration may be higher than average in some of the dry valleys.

Within a given valley these factors are interdependent, since the bare rocks and pebble pavement absorb heat, cause a local rise in temperature, reduce the relative humidity, increase ablation, and again expose more rock and soil. These conditions are operative to some extent in the high Arctic where retreating glaciers and low precipitation have resulted in the formation of polar desert topography and soils (Fig. 4). Glaciers, glacial moraines and some of the Antarctic dry valley topography are shown in Figure 5. It is believed that nondestruction of landscape in the high altitude Pamirs, which resemble a polar desert, was accomplished by the slow wastage of ice through sublimation, in turn attributed to factors of low precipitation, humidity, and temperatures at high elevations (30).

The physiographic processes operative in cold and polar deserts give them some geomorphic unity with other arid lands, except for the influence of glaciers. Both Inglefield Land and Peary Land in northern Greenland show a predominant terrain consisting of dissected plateaus or tablelands and some terraces of rolling relief (Fig. 6). The arid areas of southern Victoria Land show a topography produced by glacial dissection of block-faulted mountain ranges resulting in deep troughs, each terminated at the head by a deep cirque and either open at the foot or blocked by a glacier.

Volcanic activity is negligible in hot deserts (31),

Fig. 4(a). Polar desert terrain, southwestern flank of Prince Patrick Island, Canadian Arctic Archipelago. Photo by J. C. F. Tedrow (27).

Fig. 4(b). Polar desert terrain, vicinity of "Main River," Inglefield Land, Greenland. Photo by D. Gaskin, Interpretation Research Division, Department of the Army.

Fig. 5. Viewing West over Taylor Glacier, south Victoria Land, toward McMurdo Sound and Ross Island, Antarctica. Photography flown by U. S. Navy for U. S. Geological Survey. TMA 541, F33, exposure 69.

Fig. 6. Aerial view of dissected plateaus and rolling relief in vicinity of "Main River," inland from Cape Ingersoll and Rensselaer Bay, Inglefield Land, Greenland. Photo by U. S. Department of the Army.

176

and this is also true of cold deserts. Mt. Erebus on Ross Island is active, as has been Deception Island, in the Antarctic Peninsula, in 1966 and 1969. Extinct volcanic mountains are found in some high altitude deserts (Fig. 7). Warm springs occur in Woodbay, Spitzbergen, and there is evidence of thermal warming in Antarctica.

Local relief in cold deserts and polar deserts differs from that in hot deserts as indicated by the presence of receding glaciers and morainic debris, occasional frozen lakes, solifluction, soil creep, heaved and patterned ground (Fig. 8). Forms of patterned ground are related to cycles of freezing and thawing of rocks and moist soil and especially with activity and depth to permafrost (Figs. 8 & 9). The most desiccated of the Antarctic dry valleys lack frost crack polygons (Fig. 10), whereas some of the most recent valleys have high rubble-rimmed polygons (32). Influence of moisture is shown in Fig. 11.

Solifluction features are not favored by aridity. They are therefore better developed in the moister coastal areas around McMurdo Sound (33) as well as the wetter and warmer areas of the Arctic. Solifluction in Peary Land was probably due to a more humid climate than occurs at present (34),

Fig. 7. Talus slope of Mt. Aucanquilcha, Chile; elevation 6,000 meters. An extinct volcano is mined for sulfur.

Fig. 8. Aerial view of Wheeler Valley, Antarctica. A relatively favorable dry valley for microorganisms exists at 1,000 to 1,500 meters elevation. U. S. Navy photograph.

Fig. 9. Patterned ground, Prince Patrick Island, Canadian Arctic Archipelago. Darker areas are more stable lichen-covered rocks; light areas show recent frost-heaving. Photo by J. C. F. Tedrow (27).

Fig. 10. McKelvey Valley. This dry valley lacks frost crack polygons. Helicopter transport is used to bring men, supplies and equipment to the field. The elevation is 800 meters. U. S. Navy photograph.

Fig. 11. Polar desert-tundra interjacence as a result of differences in drainage and moisture relationships. Dark, vegetated tundra is on the left. Inglefield Land, Greenland. Photograph by J. C. F. Tedrow 6, (1968a).

and in Antarctica, probably occurred soon after glaciers retreated (35). Large solifluction lobes in cold deserts are primarily relics (36), although there is presently some activity (37). Hot deserts may present conditions for earth movement, especially as gravel, sand, boulder and mud flows, following runoff from short periods of intense precipitation.

Weathering, as in other deserts, is physical (mechanical) as well as chemical and is promoted especially during diurnal temperature fluctuations of moist rocks and soils. Felsenmeers are found in

nearly all known deserts—including cold and polar deserts—and in high mountains (Fig. 12). Wind-sculptured rocks are found in many deserts, and ventifacts are a common feature in some Antarctic areas, for example, Don Juan Pond and Wright Valley (Fig. 13). Pitted and fretted surfaces, exfoliation and cavernous weathering can be observed. In the high Arctic conspicuous ventifacts are poorly developed or absent (38). Wind-polishing is stronger in the high Arctic deserts than in most hot deserts (39), and the same is true for exposed areas in Ant-

Fig. 12. Rocky summit and research laboratory at White Mountains, Inyo National Forest, California; elevation 4,343 meters. This high altitude area simulates some polar conditions.

Fig. 13. Weathered ventifacts and salt crusts of antarcticite at Don Juan Pond, upper Wright Valley, Antarctica. The unique $CaCl_2$ pond freezes at $-57°C$ (45). Elevation 118 meters.

arctica. Sand and pebbles as well as ice needles and snow grains may contribute to this intense polishing, where winds are of greater frequency and intensity than in hotter deserts. As is found in hotter deserts, crusts of desert varnish coat many of the exposed rocks in cold deserts (40). McKelvey

Fig. 14. Barchan sand dunes, lower Victoria Valley. Dunes were salty and hard frozen beneath the surface. Lower Victoria Glacier is in the distance. Elevation 430 meters.

Valley, Antarctica, one of the older dry valleys, shows a high degree of varnished desert pavement (Fig. 10). Aeolian deposits such as dunes are common features of both hot and cold deserts. Dunes, except along coastal areas, are outstanding evidence of deserts (41). There are vast sand deposits and dunes in southwest Peary Land (42), and some deposits occur in a few dry valley areas of Antarctica (Fig. 14). In lower Victoria Valley dunes are more than 6.5 kilometers long and 1.6 kilometers wide, with barchans up to 15 meters high and 90 meters long (43). In contrast to hot desert dunes, these sand deposits are ice- and salt-cemented beneath the surface.*

Salt lakes and ponds are other features held in common with all deserts where evaporation of water increases salt concentration. Large salt lakes are found in Peary Land (44), and salt lakes and ponds also occur in coastal areas of Antarctica (Fig. 15). Don Juan Pond is an unique calcium chloride accumulation in the South Fork of Wright Valley, which freezes at $-57°C$ (45) (Fig. 13). Evaporite deposits resulting from increasing aridity in cold deserts provide many minerals similar to those found in hot deserts, including sodium iodate and nitrate (46), which are also found in the Atacama. A recent summary of hot desert geomorphology and surface hydrology has been given by Lustig (47).

The moisture problems of cold deserts exceed those of hot deserts, since water is entirely unavailable during the period of the dark, cold, austral winter. During the short summer season, low temperatures, low humidities, and desiccating winds also limit the availability of water. There may be some glacial melt that traverses an otherwise arid area, but in Antarctic dry valleys, lakes and ponds are primarily saline and also frozen, except at the margins (Fig. 16) (48), and may remain entirely

*See footnote, p. 182.

Fig. 15. Lake Vanda, upper Wright Valley; elevation 123 meters. This frozen, saline lake shows previous bench marks resulting from increasing aridity. U. S. Navy photograph TMA 489, exposure 57.

Fig. 16. Island camp site in western arm of Lake Bonney, Taylor Valley; elevation 98 meters. During midsummer, edges of the lake are unfrozen.

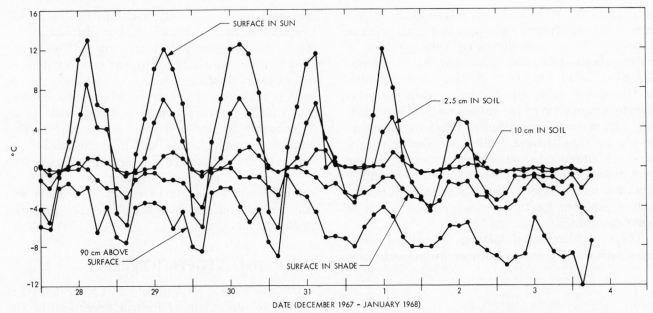

Fig. 17. Diurnal temperature fluctuations for soil and air temperatures at Wheeler Valley camp site; elevation 1236 meters. (51).

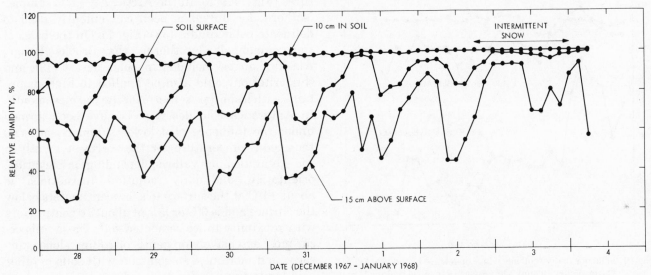

Fig. 18. Diurnal relative humidity fluctuations for soil and air at Wheeler Valley site.

frozen during the summer (Fig. 8). The vast reservoir of permafrost does not provide moisture except when favorable temperature relationships promote upward movement of vapor through the soil to the atmosphere.

The local climate of the Antarctic dry valleys east of McMurdo Sound or the "oases," such as Bunger Hills, is quite different from that of the snow- and ice-covered areas (49). Principal dry-valley characteristics are greater warmth, drier air, lower precipitation, and almost continuous wind. Complete data on the dry-valley surface meteorology are not available, although some measurements have been made during previous austral summers (50).*

The thermal regime is greatly influenced by the albedo of the rock and soil surfaces, which give a

*See footnote, p. 182.

positive surface radiation budget. With an albedo of about 20 per cent, the exposed surfaces can retain more than four times as much solar radiation (51); soil surface temperatures have momentarily reached +18°C in some of the dry valleys, and at Oazis and Mirnyy, rock temperature highs of +15 to 30°C have been measured (52). In the Victoria Valley complex, a mean minimun is reported of −62°C (53), and at Bunger Hills of −44°C (54).

Air and soil temperature and relative humidities show diurnal fluctuations in the valleys, particularly during midsummer, as shown for Wheeler Valley (Figs. 17-18). Although soil-surface temperature in the sun is above freezing, air temperature can be below freezing. The "active zone" in the soil is quite limited, and decreases with depth from the air-surface interface to depth of permafrost; however, the soil relative humidity increases with depth to permafrost. Wind direction and velocity have a considerable influence on temperature and relative humidity. Wind direction (and velocity) have a noticeable effect on relative humidity as determined for King-David Valleys (Fig. 19). Wind direction in the valleys is determined to a considerable extent by orientation of the valleys and the surrounding mountainous topography. Previous measurements in the Victoria Valley complex during a six-week summer period showed a mean of 45 per cent relative humidity (55) and are much higher than the usual 80 to 90 per cent relative humidity values reported for the summer season in Arctic polar deserts (56).

These meteorological factors, as well as other topographic and climatic features, are important in

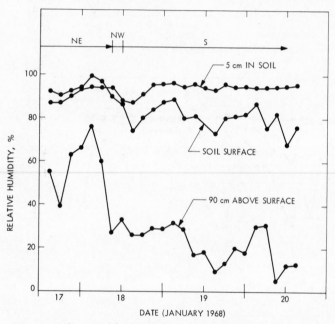

Fig. 19. Influence of dry southerly winds on relative humidity at King-David Valleys camp site, Asgard Range, elevation 1335 meters. (51).

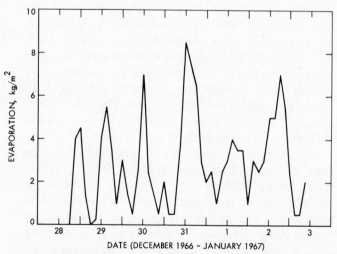

Fig. 20. Diurnal evaporation rate during midsummer at Victoria Valley site near Lake Vida, elevation 395 meters.

any arid region and vary with altitude. A typical evaporation rate at relatively low altitude near a frozen lake in Victoria Valley is shown in Figure 20.* Cloud cover, which is negligible in most deserts, can at times be of considerable extent and duration in Antarctic dry-valley areas. White-outs or very low rapidly moving clouds and fog also occur. Year-round studies and more extensive summer studies of macroclimate and microclimate of the Antarctic dry valleys are still lacking and necessary to understand ecologic conditions in these cold, barren desert areas. An inventory of weather and climate research on hot deserts has been prepared by Reitan and Green (57).

## SOILS AND VEGETATION

The presence of soils in cold and polar deserts is limited to the extent of ice-free areas and by the influence of radiation gains despite low temperatures (Figs. 1 & 3). In the Arctic, a +4.5°C summer isotherm for July has been conveniently used to delineate polar tundra (58) (Fig. 1). On the basis of this circumpolar isotherm, the Brooks Range, Alaska, can be included in polar desert (59), and this criterion could also be applied to high mountains with a similar isotherm for the warmest month. In Antarctica cold desert soils have been formed under the influence of a lower summer isotherm; the mean temperature for the warmest month in the Antarctica dry valleys, depending on elevation, orientation, and daily radiation fluctuations, is about +1°C at the surface to a few centimeters below the surface and is 0°C or less at about 10 centimeters with proximity to ice-cemented soil. These soils occur most extensively on gentle or rolling slopes, terraces and moraines. In common with soils of other arid lands, the paucity of liquid water and its short period of availability would greatly limit the presence and activity of biota (especially vegetation) and subsequent soil development.

Most of the Antarctic dry-valley soils show the effects of desertic weathering processes: wind, moisture, and glacial erosion and deposition. Removal of fines, except in the dune areas, has produced surfaces of residual lag gravels, pebbles, and boulders—desert pavement (Fig. 10). Many of the soils found beneath desert pavements are poorly developed, shallow, saline or alkaline, yellowish or brownish sands and loams. Older soils are brownish, whereas younger soils are more greyish (60). Changes in texture or color are not common except with approach to or within ice-cemented soil, when there is an increase in lighter colored fines. Soils

*Measured in cooperation with Prof. Robert E. Benoit, Virginia Polytechnic Institute, Blacksburg, Virginia.

may show concentrations of $Na^+$, $Ca^{++}$, $Mg^{++}$, $Cl^-$, $SO_4^=$, $CO_3^=$ ($HCO_3^-$), and $NO_3^-$ ions, sometimes in the form of gravel- or pebble-sized crystals. In older soils, salts may accumulate at the surface as well as below the surface. Electrical conductivity, buffer capacity, and cation exchange capacities are variable but are within the ranges of other coarse-textured desert soils (61). While low in organic matter, they are not ahumic, although reported as such (62), and most of them contain about 0.3 to 0.01 per cent organic carbon, which may be old carbon (63). Low organic nitrogen content corresponds to the lowest levels found in other subsurface desert soils (64). Although not entirely resolved, the organic carbon and nitrogen content in Antarctic dry valley soils is due to growth and/or windblown dislodgment and settling of plant debris (especially algae but also lichens and mosses), possibly some animal contamination (especially penguins and other sea birds), fixation processes in the aurora australis, or windborne products resulting from geologic sediments, such as nearby coal deposits (65).

In the drier Antarctic valleys, the surface moisture content is usually low, about 0.1 to less than 2 per cent. Summer snowfall in the dry valleys occurs intermittently, and when the snow cover is exposed to solar radiation it usually sublimates with little consequent wetting of the soil. With depth of soil to the hard, icy, cemented permafrost, and with proximity to surface moisture supply, for example, melt-water streams, ponds, and lakes, the moisture content can be considerably higher. A hard, icy permafrost layer occurs at about 10 to 30 centimeters in many of the dry valley soils, with moisture contents approaching 50 per cent or more.

The influence of salts versus moisture and depth at a McKelvey Valley site is shown in Figure 21. The subsurface soil was decreasingly friable with increasing moisture to a depth of about 45 centimeters where the temperature did not vary appreciably from −11°C. Most dry valley soils do not show appreciable subsurface freeze-thaw cycles unless there is sufficient available water, and some soils are notably dry with depth to ice-cemented soil (the "ice table"). Stony and gravelly constituents within the soils usually show very little indication of layering by frost action.

Cold desert soils and polar soils may be as fertile as other arid land soils, but the short summer season, low temperatures, shallow depth to permafrost, desert pavement, unfavorable permeability and drainage relationships, salinity and other factors limit their productivity. A comparison of some Antarctic soil properties with other desert soil properties is shown in Table 1.

Polar desert soils, as indicated previously, are predominant in the high Arctic (Fig. 22). In polar desert regions there can be a gradation from polar desert soils to polar desert interjacence (Fig. 11), Arctic brown, tundra and bog (66). In southern Victoria Land there are a few small isolated or protected areas with favorable environmental characteristics where some Arctic soils may be approximated. These may occur under moss and lichen cover near glacial drainage, as "algal peat" deposits such as at Marble Point; even a spongy, wet bog-like relict soil has been encountered in a depression of Balham Valley.*

Polar desert soils are generally sands and loams that are drier than other Arctic soils, are usually unfrozen for 2 or 3 months of each year, and have good internal drainage (67). Polar desert areas (Fig. 4) may be compared with some of the areas in southern Victoria Land, showing desert landscape (Fig. 15), effects of wind erosion (Figs. 13 & 23), patterned ground (Figs. 8 & 9) desiccation cracks and salt efflorescence (Figs. 24 & 25), and a weakly developed soil overlain by desert pavement (68) (Figs. 21, 22 & 26). In polar deserts, however, the presence of some available moisture and warmer temperatures provides a more favorable environment than in cold deserts, and there may be a few, low, scattered vascular plants (69) (Fig. 24). Soils of the Antarctic Peninsula and at Cape Hallett have been insufficiently studied at present, but on the basis of climatic, vegetation, and ecologic studies (70) it appears that some soils resemble those found in the high Arctic (Table 1). For a general comparison of characteristics of soils in hotter deserts, see the report prepared by Dregne (71).

Vegetation in Antarctica is sparse, unevenly distributed, and possesses a low growth form. Plants not only have the ability to resist desiccation for long periods but can also endure long periods of below freezing temperatures and months of darkness. Only in Graham Land and its offshore islands are there any phanerogams: two grasses and a pink (endemic grass, *Deschampsia antarctica,* introduced grass, *Poa annua,* and the pink, *Colobanthus crassifolius*) (72). South of latitude 68°50'S the flora is restricted to cryptogams and bacteria. The Arctic polar desert shares some physiognomic similarity with Antarctic vegetation in favorable habitats, although phanerogams are much more abundant in the Arctic (73).

It is generally now well recognized that the number of plant species in Antarctica is much less than at comparable latitudes in the Arctic. The property of a limited number of species in cold and polar deserts is a similarity shared with some other desert vegetation (74).

*See footnote, p. 182.

TABLE 1

## PHYSICAL AND CHEMICAL PROPERTIES, SURFACE DESERT SOILS

| Desert Region and Location | Texture | Air-dry Soil Color and Munsell Notation | In situ Moisture wt. % | pH Saturated paste | Electrical Conductivity EC × 10^6 mhos cm^2 @ 25°C 1:5 extract | Ions ppm in 1:5, soil:$H_2O$ extract | | | | | | | | | Organic C % | Organic N % | Cation Exchange Capacity meq/100 gm | Remarks |
|---|---|---|---|---|---|---|---|---|---|---|---|---|---|---|---|---|---|---|
| | | | | | | $Na^+$ | $K^+$ | $Ca^{++}$ | $Mg^{++}$ | $Cl^-$ | $SO_4^=$ | $HCO_3^-$ | $NO_3^-$ | $PO_4^{\equiv}$ | | | | |
| 1. Cold desert, McKelvey Valley, Antarctica | loamy sand | 10YR 6/2 reddish brown | 1.4 | 7.9 | 4950 | 650 | 5 | 190 | 71 | 665 | 510 | 24 | 780 | 0.1 | 0.09 | 0.007 | 8 | JPL #500 |
| 2. Cold desert, Victoria Valley, Antarctica | sand | 5Y 7/3 pale yellow | 0.24 | 8.9 | 88 | 8 | 1 | 8 | 1 | 5 | 8 | 24 | 2 | 0.2 | 0.02 | 0.002 | 4 | JPL #537 |
| 3. Cold desert, Asgard Range, Antarctica | sandy loam | 2.5Y 5/2 greyish brown | 5.0 | 7.4 | 8400 | 1150 | 245 | 37 | 100 | 2340 | 450 | 61 | 130 | 0.8 | 0.04 | 0.007 | 11 | JPL #664 |
| 4. Cold desert, Wheeler Valley, Antarctica | sand | 2.5Y 5/4 light olive brown | 8.9 | 8.1 | 380 | 21 | 8 | 20 | 5 | 41 | 15 | 30 | 7 | 0.2 | 0.17 | 0.024 | 2.5 | JPL #615 |
| 5. Polar desert, Inglefield Land, Greenland | gravelly sand | 7.5YR 5/4 brown | – | 8.8 | – | 2.3 | 11.7 | 15 | 3.7 | – | – | – | – | – | 0.42 (organic matter) | – | 0.9 | Tedrow 1968 (6), site 2, A horizon |
| 6. Polar desert, Prince Patrick Island, N. W. T., Canada | sand | 10YR 5/6 yellowish-brown | – | 4.4 | 79000 (1:2 extract) | – | – | – | – | – | – | – | – | – | 1.70 (organic matter) | – | – | Tedrow et al. 1968a (59). Profile 1, loose and dry sand. |
| 7. Polar desert, Jørgen Brønlunds Fjord, Pearyland, Greenland | – | – | – | 7.5 | – | – | – | – | – | – | – | – | weak pos. | – | 35.6 (humus) | 1.05 (total) | – | Jensen, 1951 (96), #4 somewhat dry ground; stonefield |
| 8. High Mts., White Mt. Summit, Inyo National Forest, California | stony sandy loam | 10YR 6/3 pale brown | 9.2 | 6.3 | 16 | 5 | 3 | 9 | 1.1 | 4 | 11 | 15 | 28 | 1.1 | 0.55 | 0.070 | 10.0 | JPL #14 stony peak, elev. 4,350 m |
| 9. High Mts., Mt. Aucanquilcha, near Amincha, Chile | angular cobbly sand | 2.5Y 8/2 white | 1.2 | 2.6 | 17600 | 35 | 10 | 350 | 45 | 105 | 2700 | 0 | 0 | 2 | 0.06 | 0.006 | 0.0 | JPL #256 extinct volcano, elev. 7,800 m |
| 10. Hot desert, Atacama, near Prosperidad, Chile | sandy loam | 7.5YR 7/2 pinkish grey | 0.90 | 8.2 | 39200 | 1450 | 155 | 4250 | 75 | 530 | 10,000 | 0 | 2500 | 1 | 0.04 | 0.003 | 0.5 | JPL #288 |
| 11. Hot desert, eastern Sahara near Abu Simbel, Egypt, U. A. R. | loamy sand | 7.5YR 6/4 light brown | 0.78 | 8.0 | 1230 | 55 | 40 | 40 | 15 | 5 | 10 | 43 | 3 | 0.5 | 0.17 | 0.006 | 7.0 | JPL #295 |
| 12. Hot desert, Negev, Dead Sea, near Chamai Zohar, Israel | sandy loam | 10YR 6/4 light yellowish brown | 2.2 | 7.6 | 2750 | 270 | 50 | 150 | 54 | 940 | 12 | 24 | 1 | 0.5 | 1.45 | 0.012 | 14 | JPL #220 |
| 13. Hot desert, Patagonian, west of San Julian, Argentina | sandy loam | 10YR 6/2 light brownish grey | – | 7.4 | 153 | 11 | 18 | 100 | 1 | 40 | 225 | 18 | 20 | 0.2 | 0.44 | 0.052 | 6 | JPL #237. Dr. L. Halperin #605/35, Instituto de Suelos y Agrotechnica, Buenos Aires, Argentina |
| 14. Hot desert, Great Basin Oregon Desert, near Denton, Oregon | loamy sand | 10YR 6/2 light brownish grey | 35.5 | 6.7 | 151 | 10 | 16 | 9 | 7 | 30 | 300 | 24 | 10 | 1 | 0.58 | 0.056 | 10 | JPL #154 snow cover on soil at time of collection |

| No. | Location | Soil | Munsell color | Color | | | | | | | | | | | | | | | | JPL # |
|---|---|---|---|---|---|---|---|---|---|---|---|---|---|---|---|---|---|---|---|---|---|
| | Wyoming Red Desert, near Thermopolis, Wyoming | loam | | yellow | | | | | | | | | | | | | | | | |
| 16. | Hot desert, Mohave Desert, Eureka Valley, California | sandy loam | 2.5Y N5/ | grey | 8.1 | 2.4 | 3100 | 270 | 10 | 375 | 11 | 2 | 1500 | 0 | 0.0 | 0.0 | 0.70 | 0.190 | 2.0 | JPL #47 |
| 17. | Hot desert, Sonoran Arizona Upland Desert, near Mammoth, Arizona | loamy sand | 10YR 5/3 | brown | 7.6 | 1.7 | 143 | 1 | 15 | 100 | 45 | 2 | 30 | 31 | 1.0 | 3.5 | 0.46 | 0.045 | 6.5 | JPL #99 |
| 18. | Hot desert, Sonoran Colorado Desert, near Thermal, California | clay | 5YR 7/4 | pink | 7.6 | 3.3 | 3700 | 550 | 47 | 125 | 16 | 586 | 1150 | 43 | 14 | 0.1 | 0.27 | 0.026 | 16.5 | JPL #6-1 (#6) #6-1 collected 1 year after #6 at same site |
| 19. | Hot desert, Sonoran Yuma Desert, Algodones Dunes, near West Cactus, California | sand | 10YR 6/4 | light yellowish brown | 8.3 | 0.16 | 41 | 2 | 2 | 2.5 | 0.6 | 2 | 0 | 21 | 1 | 0.4 | 0.04 | 0.002 | 0.0 | JPL #88 |
| 20. | Hot desert, Sonoran Viscaino-Magdalena Desert, near Punta Prieta, Baja California, Mexico | loamy sand | 10YR 6/3 | pale brown | 8.2 | 0.74 | 1850 | 8 | 21 | 29 | 8 | 40 | 300 | 60 | 1 | 1.5 | 0.29 | 0.021 | 5 | JPL #188 |
| 21. | Hot desert, Sonoran Gulf Coast Desert, near Navojoa, Mexico | loamy sand | 5YR 5/1 | grey | 8.0 | 4.3 | 81 | 1.5 | 1 | 3.4 | 1.8 | 4 | 10 | 24 | 3 | 0.5 | 0.84 | 0.074 | 18.5 | JPL #250 |
| 22. | Hot desert, Chihuahuan Desert, near Las Cruces, New Mexico | sandy loam | 10YR 5/3 | brown | 8.0 | 3.1 | 115 | 2 | 10 | 3.5 | 1 | 3.5 | 10 | 51 | 0 | 1.05 | 0.39 | 0.023 | 6 | JPL #395 |
| 23. | Volcanic desert (within Mohave Desert) near Little Lake, California | coarse sand | 10R 4/3 | weak red | 9.1 | 0.05 | 1550 | 17 | 5 | 12 | 5 | 35 | 300 | 6 | 0 | 2 | 0.03 | 0.005 | 1 | JPL #196 |
| 24. | Volcanic desert, Valley of 10,000 Smokes, Katmai Nat. Mon., Alaska | loam | 5YR 7/3 | pink | 4.5 | 20.0 | 16 | 7.8 | <5 | <40 | <2 | 13 | <150 | 0 | 1.3 | <0.5 | 0.07 | 0.004 | 3.0 | JPL #116 |
| 25. | Volcanic desert, Kau Desert, Hawaii Nat. Parks, Hawaii | sandy loam | 2.5YR 3/2 | dusky red | 5.2 | 6.3 | 145 | 50 | 34 | 6 | 5 | 20 | 150 | 0 | 0 | 0.5 | 0.06 | 0.005 | 3.5 | JPL #34 |
| 26. | Volcanic desert, Surtsey, Iceland | sand | 5Y 4/2 | olive grey | 6.6 | 0.04 | 730 | 52 | 60 | 220 | 87 | — | 68 | — | 0 | 1 | 0 | 0.0003 | — | Ames #SC-A. Collected by Dr. R. Young, Ames Research Lab., Moffett Field, Calif. (Analyses provided by Ames). |

Fig. 21.(a) Soil pit at McKelvey Valley site. Salty soil was not frozen below 0°C. Soil samples contained only 100 to 1000 bacteria per gram of soil. Photo by G. B. Blank. (b) Soil moisture curve for soils 500 thru 503, McKelvey Valley, Antarctica [see (a)]. Influence of salts is shown by displacement of actual hygroscopic coefficient values from the expected values. *In situ* moisture values indicate field moisture status relative to other moisture constants.

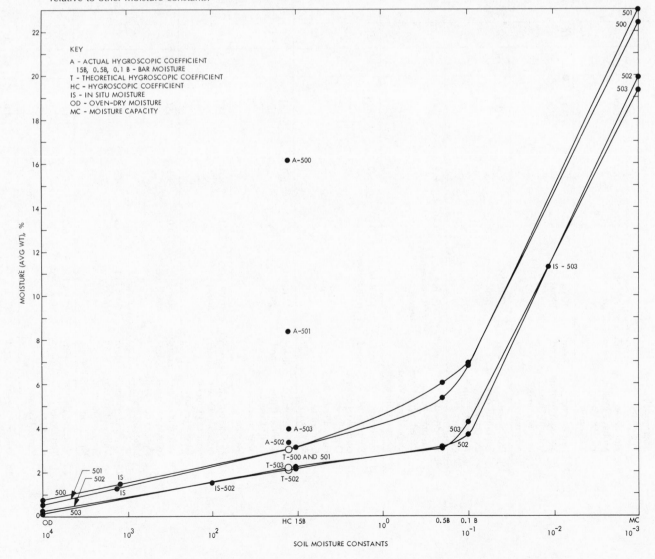

In the high Arctic, for the Canadian Archipelago and northern Greenland, a detailed map has been made showing the distribution of vascular plants (75). For areas designated as polar desert (Fig. 1), there are about 75 phanerogams representing 180 species. With at least 5 species each per genus, these include the following: *Carex* spp., *Saxifraga* spp., *Draba* spp., and *Potentilla* spp. In the Canadian Archipelago, 150 of these species occur north of the 75th parallel (76). In the Eurasian Arctic islands, Wrangel Island was never glaciated and has a rich and ancient flora, with 160 vascular plants in the polar desert (77), whereas Novaya Zemlya has been glaciated several times and has a youthful flora (78), including only three endemic vascular plants, all *Taraxacum* spp., which have the potential for producing new forms rapidly (78). Vegetation on Franz Josef Land includes 14 vascular plants, and the New Siberian Islands have a dwarf willow about 2.5 centimeters high (79).

In contrast to the Arctic, as indicated previously, south of the 68th parallel there are no vascular

Fig. 22. Barren polar desert soil beneath pavement, central Prince Patrick Island, Canadian Arctic Archipelago. Similar profile morphology is observed in cold desert soils of south Victoria Land. Photograph by J. C. F. Tedrow (27).

Fig. 23. Lichen-incrusted rocks on wind eroded surface of polar desert, central portion of Prince Patrick Island, Canadian Arctic Archipelago. Photograph by J. C. F. Tedrow. (27).

plants—only scattered mosses, lichens, algae, and some hepatics. Mosses, lichens, and algae are, of course, also present in polar desert soils, and the algae can contribute significantly to the soil organic matter as in other arid lands. In terms of biomass, the algae are the most abundant in Antarctica, having the ability to grow not only in some lakes, ponds, and melt-water streams, but within and on ice and snow, under translucent stones, and on moist soil (80).

Taxonomic discrepancies make it difficult to determine the species abundance of cryptogams in the Antarctic. There are less than a dozen species of mosses (81) and 70 species of lichens, although only about 25 are found on the coast of Victoria Land (82). The blue-green and green algae are the most predominant terrestrial algal flora, but these do not constitute nearly as many as the approximately 100 species indicated previously (83) as shown by the extensive studies of Drouet (84). In the Antarctic dry valleys, when soil moisture is sufficiently available during the summer months, the blue-green algae grow and become active; they are usually found to be coccoid or oscillatorioid forms such as *Coccochloris* spp. and *Schizothrix* spp. (85). The

green algae resemble the coccoid forms of *Protococcus* sp. (more commonly identified as *Chlorococcum* sp.). Similar species occur in other desert soils (86). A crustose lichen, *Rinodina frigida,* containing a coccoid green algal symbiont, is shown in Fig. 27.

When environmental factors are favorable, lichens and mosses are present, but there may not be any detectable biota or else only a limited number of bacteria (87) (Fig. 28). In ecological studies of Antarctic dry valleys it has been shown that the period of available soil moisture is important in the development of soil bacteria (88), and this is also important in the Arctic (89).

In the Antarctic dry valleys there is a wide variation in abundance of microorganisms found in the various valleys, ranging from essentially zero to about $10^7$ per gram of soil, usually with relatively low abundances compared to previously investigated desert soils (90). As many (if not more) microorganisms are found in subsurface soils as at the surface, especially at the level of hard, ice-cemented permafrost (91). In more favorable Arctic soils, however, the opposite has been found (92). The subsurface bacteria are predominantly white, opaque, translucent, or nonpigmented colonies, whereas

Fig. 24. Desiccation cracks, salt efflorescence (thenardite), and xeric vegetation on polar desert soil, Prince Patrick Island, Canadian Arctic Archipelago. Photograph by J. C. F. Tedrow (27).

the chromogenic bacteria are most abundant at the surface. This has been observed for other desert soils as well; however, the Antarctic subsurface microorganisms may represent an ancient "freeze-dried" microflora. A microscopic examination of buried algal peat shows microflora similar to that now extant in valley soils.

The microflora populations are composed primarily of bacteria which constitute the following major groups: (*a*) gram positive cocci, *Micrococcus* and *Mycococcus* spp., (*b*) soil diphtheroids, *Corynebacterium, Brevibacterium, Arthrobacter* and related spp., (*c*) gram positive and negative rods, *Bacillus* and *Pseudomonas* spp., and (*d*) actinomycetes, primarily *Streptomyces* spp.* The opaque or white colonies are usually soil diphtheroids or micro-

*Bollen, W. B., K. Byers, and S. Nishikawa. Microorganism Study of Bacteria and Actinomycetes from Harsh Environments, Jet Propulsion Laboratory Contract No. 950783.

Fig. 25. Desiccated salt and algal soil crusts in protected seepage area, King-David Valleys, Asgard Range, elevation 1451 meters.

Fig. 26. Desert pavement and underlying soil on a glacio-fluvial deposit. Inglefield Land, Greenland. Note similarities with soils in Figures 21 and 22. Photograph by J. C. F. Tedrow (23).

Fig. 27. Blackish crustose lichen, *Rinodina frigida* in protected cracks of sandstone (light stone above ruler). Wheeler Valley, Antarctica, elevation 1236 meters.

cocci, and the translucent or nonpigmented colonies are gram positive or negative rods. Pigmented colonies are generally micrococci. Algal populations are composed of coccoid blue-green algae—*Anacystis* and *Coccochloris* spp., oscillatorioid blue-blue-green algae—*Schizothrix, Microcoleus,* and *Oscillatoria* spp., and coccoid green algae such as *Protococcus grevillei* resembling *Chlorococcum* spp. Fungi consist of various ascomycetous molds, e.g. *Penicillium* spp., a common inhabitant of desert soils, and a few yeasts, e.g. *Sporotrichum, Pullularia,* and *Cryptococcus* spp. (93). Fungi most frequently occur in association with algae (Fig. 28). No bacteriophages were found by means of the bacterial plaque technique.

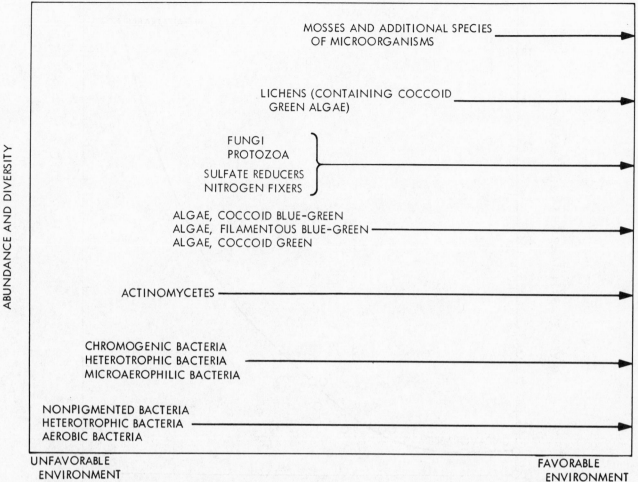

Fig. 28. Variability of population density with variability of ecological factors in Antarctic dry valleys (85).

For other desert soils, neither coccoid blue-green algae nor coccoid bacteria have been found as abundant or as predominant in the soil microbial community. Characteristics of the Antarctic soil bacterial species are not easily resolved, but they appear to be most similar to those isolated from soils of the Chilean Atacama Desert. *Mycococcus* spp., found in Antarctica as well as Chile, also occur in high mountain soils (*94*). They exhibit pleomorphism, which may possibly aid in survival in harsh environments.

In general, a favorable complex of interacting environmental (microclimatic and edaphic) factors is necessary to obtain an abundance of mixed populations of microorganisms (Fig. 29). Regardless of elevation, a north-south valley orientation is extremely important, as are factors of favorable slope, drainage, and exposure so as to obtain maximum duration, frequency, and quantity of insolation and available moisture and protection from wind (Table 2). It was found, however, that an otherwise favorable environment can be limiting for microorganisms by one or more indigenous soil properties, such as unfavorable mineralogy, texture, structure, salts, pH, or moisture relationships—for example, soil samples with increasing concentrations of salts as indicated by conductivity measurements, generally had lower abundances of microorganisms,

although salt-tolerant forms were frequently found (*95*). In the Antarctic dry valleys the diurnal cycling of moisture (Figs. 18 & 19) and heat (Fig. 17) are not advantageous for growth and reproduction of soil microorganisms as a favorable steady-state condition. As shown by laboratory and other desert field studies, a relative humidity above approximately 80 per cent and temperatures above approximately 15°C for extended time periods are more favorable for most microorganisms.

With progression from extremely harsh to more favorable environments, especially with increase in quality, quantity, duration, and frequency of available moisture, it was found that there is a sequential increase in abundance, kinds, and complexity of organisms. In general, the following sequence was observed: (*a*) heterotrophic, aerobic, non-pigmented, white, translucent or opaque bacteria, (*b*) heterotrophic microaerophilic and chromogenic bacteria, (*c*) actinomycetes, (*d*) coccoid blue-green and green algae, and oscillatorioid blue-green algae (*e*) fungi—molds and yeasts (and protozoa), (*f*) lichens containing coccoid green algae, and (*g*) mosses and other algae (filamentous green, nitrogen-fixing blue-green, and diatoms) (Fig. 28).

More specialized microflora, for example, sulfate reducers and nitrogen-fixing bacteria, were gen-

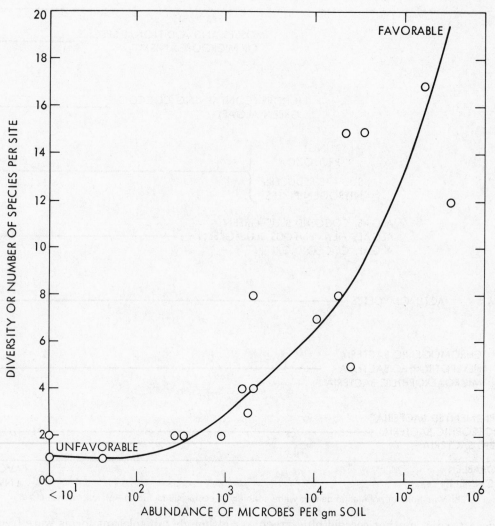

Fig. 29. Diversity versus abundance of populations for surface samples at each soil site, Asgard Range, Antarctica. Population diversity and abundance increases from unfavorable to favorable habitats (85).

TABLE 2

## ECOLOGICAL FACTORS DETERMINING DISTRIBUTION OF LIFE IN ANTARCTIC DRY VALLEYS*

| Favorable | Unfavorable |
|---|---|
| N-S ORIENTATION | E-W ORIENTATION |
| NORTHERN EXPOSURE | SOUTHERN EXPOSURE |
| GENTLE, NORTH-FACING SLOPES | FLAT OR SOUTH-FACING SLOPES |
| HIGH SOLAR RADIATION | LOW SOLAR RADIATION |
| MICROCLIMATE ABOVE FREEZING | MICROCLIMATE BELOW FREEZING |
| ABSENCE OF WIND | HIGH WINDS |
| NORTHERLY WINDS | SOUTHERLY WINDS |
| HIGH HUMIDITIES | LOW HUMIDITIES |
| SLOW OR IMPEDED DRAINAGE | RAPID DRAINAGE |
| LENGTHY DURATION OF AVAILABLE $H_2O$ (PRESENCE OF GLACIERS, LAKES, STREAMS, SNOW AND ICE FIELDS) | SHORT DURATION OF AVAILABLE $H_2O$ (ABSENCE OF GLACIERS, LAKES, STREAMS, SNOW AND ICE FIELDS) |
| TRANSLUCENT PEBBLES | OPAQUE PEBBLES |
| NON-SALTY SOILS, BALANCED IONIC COMPOSITION | SALTY SOILS, UNBALANCED IONIC COMPOSITION |
| APPROX NEUTRAL pH | HIGH (OR LOW) pH |
| ORGANIC CONTAMINATION (SKUAS, SEALS, ETC) | NO ORGANIC CONTAMINATION (NO LARGE INCREMENTS OF ORGANIC MATTER) |

*After Cameron, King, and David

erally not found unless algae were also present. Anaerobes, obligate psychrophiles, thermophiles, obligate halophiles, photosynthetic bacteria, and coliforms were rarely detected. The absence of

anaerobes and photosynthetic bacteria is especially significant, since this observation also has been made in investigations of other harsh desert soils, for example, the Atacama and parts of the Sahara.

On the basis of one study in Northern Greenland, microorganisms found in soils from Jørgen Brønlunds Fjord, Peary Land (95), compare favorably in abundance with the higher numbers in some Antarctic dry valleys (96). Some species similarities are also evident, for example, *Micrococcus* spp., and *Streptomyces* spp (97). A comparison of abundance of soil microflora in various desert regions is given in Table 3.

Compared to most other arid lands, except where there are barren, expansionless wastes, the terrestrial fauna of polar zones is quite limited. In the Antarctic, only the arthropods are endemic. In comparing abundances of species in the Arctic with those in the Antarctic, there are about 300 species of insects (not including *Collembola*) in the Arctic (98), but considerably fewer in the Antarctic, where the total arthropod fauna is probably fewer than 100 species (99). In southern Victoria Land there are about 10 species of tardigrades, 6 springtails, and 5 mites, as well as undetermined numbers of nematodes, rotifers, and protozoans (100). The arthropod fauna can be found where there are lichens and mosses and in other favorable environments of damp soil and at edges of ponds (101). Protozoa, amoeboid, and flagellate forms are sometimes found when algae and fungi are present (102). Nematode skeletons have also been observed. A short developmental period and low temperatures are limiting factors (103), similar to those which restrict the fauna in the aeolian zone (104). The presence of favorable microenvironments is extremely important for growth and activities of terrestrial fauna, as well as vegetation, in cold deserts. A resume of hot desert fauna has been prepared by Lowe (105).

## MAN AND THE FUTURE OF COLD DESERTS

Polar and cold deserts, because of their harsh environments, remoteness, and apparent low economic value, have been the very last deserts to be invaded and investigated by man. They constitute the "last frontiers" on Earth; until recently there have been comparatively few attempts to establish and maintain settlements in these areas.

In polar desert areas there have been small settlements on Novaya Zemlya by the Samoyedes, and Eskimos have ranged into northwestern Greenland where there is a settlement at Etah, 78°20′N latitude. Etah has served as a departure point for polar expeditions, as have Spitzbergen, Franz Josef Land, and other islands in the Eurasian Archipelago. There have not been any native populations in Spitzbergen, Franz Josef Land, Wrangel Island, Nicholas Land, the New Siberian Islands, or Taimyr

Peninsula (106). There is currently some coal mining in Spitzbergen, and the Soviets established a small colony which has become permanent—Chuckhee on Wrangel Island. Over 500 people inhabit northern Greenland (107). Transient coastal hunters have visited various areas in search of shrimp, fish, seal, wolves, lemming, bear, whale, and other animals, and some outposts have been established for military, geological, meterological, and other scientific purposes. Some of the activities at these stations include routine weather and storm observations, permafrost studies, topographical mapping, installation of attack warning systems, testing of cold-weather gear, equipment and shelter designs, testing of thermonuclear weapons, and maintenance of airbases and submarine depots. Although some minerals (for example, iron, copper, and nickel) are known to be present, there have been few explorations for mineral resources or few mining developments and operations. There are indications of vast coal and oil deposits, but they have not been extensively exploited (108).

Among other activities, the Arctic polar "Great Circle" air route is now well established, but submarine transport, although feasible, is not economical. Oceanic research by satellite, plane, ship, and drifting ice-floe continues, and along with land observations shows that the Arctic region is becoming warmer. During the International Geophysical Year (IGY) more than 300 Arctic stations were established. The Arctic Institute of North America is prominent in sponsoring studies of Arctic areas.

In the Antarctic there have been no endemic populations by man, and prior to the IGY coastal hunters, explorers, and scientists were the main visitors. During the IGY, 12 countries made a large effort to study the continent and establish stations. In contrast to the Arctic, following a treaty proposed in 1959 and ratified by the 12 nations, Antarctica is currently an international continent. Each country has agreed to give up existing territorial claims, eliminate weapons testing, exclude disposal of nuclear wastes, refrain from warfare, and preserve the area for scientific research. The treaty also contains a provision for inspection of bases and promotion of cooperative scientific research by exchange of scientists and data (109).

In Antarctica, as in the Arctic, scientific objectives, logistic difficulties (such as transportation, food, water, fuel, power, waste disposal, shelter, communication), and the *pioneer spirit* necessary to endure environmental and other hardships are important factors in the establishment of stations. A tremendous and costly effort is necessary to maintain and supply the various bases. Williams Field (Fig. 30) is located on sea ice and is used to bring in men, supplies, and equipment for McMurdo Station (Fig. 31) and Scott Base; it is a jumping-off place

TABLE 3

## MICROBIOLOGICAL DETERMINATIONS, SURFACE DESERT SOILS
### (Microorganisms per gram of soil)

| Desert Region and Location | Aerobic Bacteria and Actinomycetes | Microaerophiles (positives at highest dilution) | Anaerobic Bacteria | Fungi molds | Fungi yeasts | Algae (positives at highest dilution) | Remarks |
|---|---|---|---|---|---|---|---|
| 1. Cold desert, McKelvey Valley, Antarctica | 25 | $10^0$ | 0 | 0 | 0 | 0 | JPL #500 |
| 2. Cold desert, Victoria Valley, Antarctica | $8 \times 10^3$ | $10^4$ | 0 | 0 | 0 | $10^3$ | JPL #537 |
| 3. Cold desert, Asgard Range, Antarctica | 0 | 0 | 0 | 0 | 0 | 0 | JPL #664 |
| 4. Cold desert, Wheeler Valley, Antarctica | $10^5$ | $10^6$ | 0 | 200 | 0 | $6 \times 10^6$ | JPL #615 |
| 5. Polar desert, Franz Josef Land, Hooker Island | $2.4 \times 10^6$ | | — | $11 \times 10^3$ | — | — | Mishustin & Mirzoeva 1964 (97) gravel and marshy soil (mean values) |
| 6. Polar desert, New Siberian Islands (Kotelnyi Island) | $2.9 \times 10^6$ | — | — | $8.3 \times 10^3$ | — | — | Mishustin & Mirzoeva 1964 (97) peat moss turf and bare ground (mean values) |
| 7. Polar desert, Jørgen Brønlunds Fjord, Peary Land, Greenland | $14.1 \times 10^6$ | Present | 10 | Present | Present | Present | Jensen, 1951 (95), #4 somewhat dry ground; stonefield |
| 8. High Mts., White Mt. Summit, Inyo National Forest, California | $2 \times 10^6$ | $10^6$ | $15 \times 10^4$ | $3.2 \times 10^3$ | — | $10^3$ | JPL #14-1 |
| 9. High Mts., Mt. Aucanquilcha, near Amincha, Chile | 600 | $10^6$ | 0 | 0 | — | 0 | JPL #256 |
| 10. Hot desert, Atacama, near Prosperidad, Chile | <10 | 100 | 0 | 0 | — | 0 | JPL #288 |
| 11. Hot desert, eastern Sahara, near Abu Simbel, Egypt, U. A. R. | $160 \times 10^3$ | $10^3$ | 5 | 15 | — | 0 | JPL #295 |
| 12. Hot desert, Negev, Dead Sea, near Chamai Zohar, Israel | $132 \times 10^5$ | $10^4$ | 0 | 260 | — | $10^3$ | JPL #220 |
| 13. Hot desert, Patagonian, west of San Julian, Argentina | $71 \times 10^4$ | $10^6$ | $53 \times 10^3$ | 370 | — | 10 | JPL #237. Dr. L. Halperin #605/35, Instituto de Suelos y Agrotechnica, Buenos Aires, Argentina |
| 14. Hot desert, Great Basin Oregon Desert near Brothers, Oregon | $72 \times 10^5$ | $10^7$ | $16 \times 10^4$ | 0 | — | $10^3$ | JPL #154 |
| 15. Hot desert, Wyoming Red Desert, near Thermopolis, Wyoming | $11 \times 10^5$ | $10^7$ | 520 | $23 \times 10^4$ | — | $10^5$ | JPL #311 |
| 16. Hot desert, Mohave Desert, Eureka Valley, California | $9 \times 10^4$ | $10^7$ | 0 | $34 \times 10^4$ | — | 100 | JPL #47 |

TABLE 3 (cont)

| Desert Region and Location | Aerobic Bacteria and Actinomycetes | Microaerophiles (positives at highest dilution) | Anaerobic Bacteria | Fungi molds | Fungi yeasts | Algae (positives at highest dilution) | Remarks |
|---|---|---|---|---|---|---|---|
| 17. Hot desert, Sonoran Arizona Upland Desert, near Mammoth, Arizona | $15.3 \times 10^5$ | $10^8$ | $8 \times 10^5$ | 170 | — | $10^3$ | JPL #99 |
| 18. Hot desert, Sonoran Colorado Desert, near Thermal, California | $87 \times 10^3$ | $10^4$ | $1.5 \times 10^3$ | 400 | — | $10^3$ | JPL #6-1 |
| 19. Hot desert, Sonoran Yuma Desert, Algodones Dunes, near West Cactus, California | $19.5 \times 10^3$ | $10^5$ | 525 | 10 | — | 0 | JPL #88 |
| 20. Hot desert, Sonoran Viscaino-Magdalena Desert, near Punta Prieta, Baja Calif., Mexico | $25 \times 10^5$ | $10^7$ | 0 | $4.9 \times 10^3$ | — | $10^4$ | JPL #188 |
| 21. Hot desert, Sonoran Gulf Coast Desert, near Navojoa, Mexico | $17.2 \times 10^5$ | $10^7$ | $1.8 \times 10^3$ | $1.3 \times 10^3$ | — | $10^6$ | JPL #250 |
| 22. Hot desert, Chihuahuan Desert, near Las Cruces, New Mexico | $17 \times 10^4$ | $10^3$ | $11 \times 10^3$ | $2 \times 10^3$ | — | 100 | JPL #395 |
| 23. Volcanic desert (in Mohave Desert) near Little Lake, California | $75 \times 10^3$ | $10^8$ | 0 | 30 | — | 0 | JPL #196 |
| 24. Volcanic desert, Valley of 10,000 Smokes, Katmai Nat. Mon., Alaska | 10 | 10 | 0 | 10 | 0 | $10^4$ | JPL #116 |
| 25. Volcanic desert, Kau Desert, Hawaii Nat. Parks, Hawaii | $13.2 \times 10^3$ | $10^6$ | 50 | $3.3 \times 10^3$ | — | $10^5$ | JPL #34 |
| 26. Volcanic desert, Surtsey, Iceland | 0 | 0 | 0 | 0 | 0 | 0 | Ames #SC-A Collected by Dr. R. Young. Ames Research Lab., Moffett Field, California (JPL analysis) |
| Media for JPL soils | Trypticase soy agar | Fluid thioglycollate | Trypticase soy agar in $CO_2$ | Rose bengal agar | diMenna yeast agar | Pochon's medium or Thornton's medium less organics | All incubations at "room" temperature |

for further inland flights, such as to Byrd or Pole Station (Fig. 32). Propeller aircraft and helicopters are used because jets are not usually practical in the Antarctic environment. During the short summer season it is often possible to use sea transport, following ice-breaker activity (Fig. 33), to bring in the bulk of needed supplies and equipment (Fig. 34). Local transportation is by surface vehicles or helicopters. Sudden and severe inclement weather can inhibit or halt operations at any time. Haste is oftentimes necessary to take advantage of good weather (Fig. 10).

As in any polar area, construction problems on land are tremendously increased by the necessity to establish roads and buildings on permafrost. Heating, ventilation, insulation, and installment of facilities, and protection against high winds and fire are major problems. Fire is recognized as the greatest single hazard in Antarctica, and once started, may be difficult or impossible to control. Water is always in short supply and must be continually replenished (Fig. 35). Inadequate ventilation can subject occupants to resultant low level performances, or even to the risk of carbon monoxide poisoning. At McMurdo, a nuclear power plant now furnishes 1500 kilowatts electrical power for heat, light, and operations of equipment for 90 per cent of the time (Fig. 36), and a nearby salt distillation plant furnishes 55,000 liters of fresh water daily. Other world desert and populated areas also need nuclear power plants and distillation facilities for fresh water.

In the field, away from established stations, scientific teams carry out their investigations using tents (Fig. 10) or crude huts for bases. It is in these isolated situations that teamwork and cooperation are absolutely essential for survival and accomplishment of duties. For this purpose, it has been found advisable to issue a booklet to incoming personnel and to advise them of the hazards of Antarctica and the precautions necessary for existence there (110). As in other desert areas, attempts must be made to maintain personal hygiene and health and avoid sunburn, windburn and dehydration, but in addition, care must be taken to avoid carbon monoxide poisoning, snow blindness, frostbite, and windchill. The combined effects of cold and wind give an equivalent lower temperature expressed as windchill, which indicates the increased danger of frostbite (Table 4). Nearly all field workers have had experience with frostbite.

As in other harsh desert areas, clothing is specialized to meet the needs of personnel exposed to the environment. Solar and reflected radiation are important; for example, the temperature of a person's clothing facing the sun could be +61°C, and his back +12°C (111). Insulation and windproofing are necessary, and the resulting heavy and bulky clothing fatigue the wearer and make him clumsy and prone to accidents from lack of agility and visibility.

Physiological problems are increased, since sweating is dangerous and overheating leads to freezing. Vigorous activity is sometimes unavoidable, such as in movement of field teams and their equipment. Frequency of urination (cold diuresis) is increased from lack of ventilation, cold temperatures, and excessive coffee intake, and constipation is not uncommon. Runny noses, watery eyes, and cracked lips are normal responses to the dry, cold air. At higher altitudes, acclimatization is advisable, and effects from altitude sickness are pronounced at cold temperatures (112). Oxygen deprivation of the extremities and resulting numbness are experienced by some personnel working at higher altitudes (and this was also the author's experience).

Fig. 30. Aerial view of Williams Field, located on sea ice near Ross Island. This is an active embarkation point in Antarctica. U. S. Navy photograph 822716.

Fig. 31. October view of McMurdo Station from Observation Hill, Antarctica. U. S. Navy photograph K-34095.

TABLE 4

## WIND CHILL CHART INDICATING EQUIVALENT TEMPERATURE DURING WIND CHANGES AND THE DANGER OF FROSTBITE (110)

| Estimated Wind Speed MPH | Actual Thermometer Reading °F. | | | | | | | | | | | |
|---|---|---|---|---|---|---|---|---|---|---|---|---|
| | 50 | 40 | 30 | 20 | 10 | 0 | −10 | −20 | −30 | −40 | −50 | −60 |
| | Equivalent Temperature °F | | | | | | | | | | | |
| Calm | 50 | 40 | 30 | 20 | 10 | 0 | −10 | −20 | −30 | −40 | −50 | −60 |
| 5 | 48 | 37 | 27 | 16 | 6 | −5 | −15 | −26 | −36 | −47 | −57 | −68 |
| 10 | 40 | 28 | 16 | 4 | −9 | −21 | −33 | −46 | −58 | −70 | −83 | −95 |
| 15 | 36 | 22 | 9 | −5 | −18 | −36 | −45 | −58 | −72 | −85 | −99 | −112 |
| 20 | 32 | 18 | 4 | −10 | −25 | −39 | −53 | −67 | −82 | −96 | −110 | −124 |
| 25 | 50 | 16 | 0 | −15 | −29 | −44 | −59 | −74 | −88 | −104 | −118 | −133 |
| 30 | 28 | 13 | −2 | −18 | −33 | −48 | −63 | −79 | −94 | −109 | −125 | −140 |
| 35 | 27 | 11 | −4 | −20 | −35 | −49 | −67 | −82 | −98 | −113 | −129 | −145 |
| 40 | 26 | 10 | −6 | −21 | −37 | −53 | −69 | −85 | −100 | −116 | −132 | −148 |

| Wind speeds greater than 40 MPH have little additional effect | LITTLE DANGER FOR PROPERLY CLOTHED PERSON | INCREASING DANGER | GREAT DANGER |
|---|---|---|---|
| | | DANGER FROM FREEZING OF EXPOSED FLESH | |

To use the chart, find the estimated or actual wind speed in the left-hand column and the actual temperature in degrees F. in the top row. The equivalent temperature is found where these two intersect. For example, with a wind speed of 10 mph and a temperature of −10°F., the equivalent temperature is −33°F. This lies within the zone of increasing danger of frostbite, and protective measures should be taken.

Fig. 32. December view of South Pole, Antarctica. Tunnel entrance, forward scatter antennas, and "barber pole" marker. U. S. Navy photograph K-35396.

Fig. 33. December view of icebreakers heading into Winter Quarters Bay, McMurdo Station, Antarctica. U. S. Navy photograph XAM 71083.

Fig. 34. January view of unloading cargo for McMurdo Station, Antarctica. U. S. Navy photograph 826582.

Fig. 35. Providing snow for snowmelter at McMurdo Station, Antarctica. Available water is in short supply and must be conserved at all times. U. S. Navy photograph XAM-40598.

Because of the unique situation of working in the Antarctic environment, an orientation course is conducted each field season by the National Science Foundation, Division of Environmental Sciences, Office of Antarctic Programs. This organization is also responsible for all U. S. scientific efforts in Antarctica. Physical and psychological tests are given to personnel before leaving for the Antarctic. Not only must personnel be physically fit, but they must be able to cope psychologically with a number of problems. On this basis, Antarctica is both a medical and sociological laboratory.

The environmental features of isolation and man-to-man relationships affect personnel far more than the cold or other physical stresses. Behavioral adjustment is necessary for existence (*113*). In this regard, there is some similarity with sparsely manned outposts or small tribal units in other isolated areas, as well as similarities to living on a ship (*114*). All is not adventure in Antarctica. The individual must be tuned to the region emotionally and be able to cope with small groups, lack of privacy, monotony, and sexual deprivation (*115*). Men with personality problems do not adjust readily to outpost society. In Antarctica it is generally recognized that the smaller the station the more the likelihood of interpersonal problems, and it is suggested that similar problems will arise when small teams explore space (116). In fact, a Soviet scientist has reported that scientists and technicians at their Antarctic bases have a great interest in establishment of lunar bases and regard their Antarctic experience as preparation for this event!

Length of tenure of a man's duty in Antarctica also affects his adjustment to isolation. With increasing tenure, strong feelings of interdependence develop (and are necessary), and in the absence of families, family-like relationships become evident. Strong personalities are eventually apparent, and conflict situations are avoided insofar as possible by full-time occupation in work, bull-sessions, sports, religion, hobbies, personal and educational correspondence, and other activities. It has been found that the carrying of offensive weapons, such as sheath knives, is not advisable at the relatively large McMurdo base, but is allowable out of necessity in the field.

Psychological adjustment is necessary in Antarctica and includes adaptation to the physical environment of barren landscape, losing fear of the cold, accepting continual sunshine or darkness, and accepting one's companions. In both the Arctic and the Antarctic, in contrast to other deserts, the extended periods of continuous sunshine or darkness require added physiological as well as psychological adjustment. Some of the maladjustments which may be manifested with time include over-eating, insomnia, depression, inability to concentrate, as well as loss of interest and absent-mindedness (fugue state), hallucinations, repressed hostility, regressive behavior, and "Antarctic headache." Extreme rigidity is undesirable in Antarctica, and "demanding, sensitive, narcissistic individuals usually become a source of friction and a demoral-

izing influence" (117). Personal autonomy is a characteristic of Antarctic scientists as well as of astronauts.

Although more permanent facilities are now being built at some of the Antarctic stations, without women and families it is still an outpost society and "permanent" colonization cannot take place until such conditions are present. Their presence would probably reduce some of the effects and accompanying problems of isolation and long tenure.

At present Antarctica is useful mainly for scientific purposes, but tourism has recently been introduced and may increase (118). Although there are some mineral deposits, the nature of the environment, the remoteness, and tremendous problems of logistics make exploitation financially impractical at present. As the "world's ice-box," Antarctica's greatest resource is its tremendous ice deposit; however it has not been determined what practical use could be made of this material, although it is so desperately needed in available form in all deserts.

In some ways, the geographic position and remoteness of Antarctica have been of value to increasing knowledge in certain scientific studies. These have included such fields as cosmic ray, ionosphere and magnetic-field studies, meteorology, satellite monitoring, geology, glaciology, paleontology, pedology, ecology, and various aspects of biology, ranging from psychology to microbiology. Studies on penguins, skuas, seals, and other marine life, especially within the Antarctic Convergence, have shown that the seas surrounding Antarctica are the richest in the world. As a potential food source for increasing world populations, the Antarctic seas may help to alleviate the problem. It is imperative, however, that the ecology be thoroughly studied before any

Fig. 36. Winter view of nuclear power plant, McMurdo Station, Antarctica. U. S. Navy photograph.

attempt is made to interfere with marine populations, their turnover rate and productivity.

At present, the dry valleys are the only extensive colonizable land area, and this area is quite limited compared to most other desert regions. Life in the valleys would have to be of a protective or "greenhouse" nature, and it is not likely at present that introduced vegetation will survive, even though soil nutrients are adequate, as shown by studies of Rudolph (119) (Fig. 37). In combination with an unfavorable climate, unfavorable substrate properties of soil-salt-moisture-permafrost relationships would put agriculture on the subsistence level and it would be highly uneconomical with the present state of technology. Among other things, nuclear power or the entrapment and storage of solar energy during the period of continuous daylight would undoubtedly be necessary for the establishment of "permanent" self-subsisting colonies.

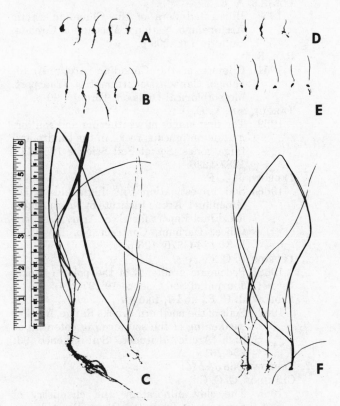

Fig. 37. Length of imported grass following growth during summer season, Hallett Station, Antarctica. Upper two rows (A,B,D,E) indicate *in situ* growth; same species and soil under warm laboratory conditions, lower row (C,F). Photograph by E. D. Rudolph (1966).

Antarctica has also been considered as a testing area prior to space exploration (120). Antarctic soil sites have been investigated and the samples tested before looking for life in extraterrestrial soils, such as on Mars. Similarities of small team operations in the hostile Antarctic environment and those expected in extraterrestrial environments would make Antarctica a valuable training ground for astronauts. It could also be a valuable test area for future extraterrestrial colonists who would live and work together before facing the more extreme environment of the moon and outer space. Antarctica, the last continent and desert to be investigated on this planet, can serve as a less expensive and less hazardous simulated extraterrestrial base before actual extraterrestrial manned exploration and colonization take place.

## ACKNOWLEDGMENT

During the past 12 years of research on hot and cold deserts, especially, for the Jet Propulsion Laboratory desert microflora program, a number of individuals have provided assistance in various ways. Dr. N. Horowitz, Bioscience Section Manager, first suggested and encouraged studies in the Antarctic dry valleys because of their value to extraterrestrial life detection. Dr. G. Llano, Program Director, Antarctic Biology, first suggested that I write on the comparison of cold and hot deserts. In Antarctica, associate investigators included R. Benoit, G. Blank, H. Conrow, C. David, C. J. Hall, J. King and, H. Lowman III. Measurements at Taylor and Victoria dry valleys were made in cooperation with Prof. R. Benoit and his team members from Virginia Polytechnic Institute.

Some of the others who have provided assistance in various ways, especially in terms of suggestions, technical information, photographs, and soil analysis for the Arctic and Antarctic, include the following: S. Babcock, A. Bauman, W. Becker, K. Bentley, W. Bollen, E. Bollin, V. Boothe, W. Boyd, M. Brewer, C. Bull, K. Byers, M. Cameron, G. Chapel, L. Chiang, A. Cherry, W. Crawford, W. Davies, G. Denton, M. diMenna, C. Dodge, B. Dowling, F. Drouet, B. Fairbanks, M. Frech, M. Frias, M. Gadsen, P. Geiger, D. Gensel, B. Goldman, E. Goodale, R. Gorny, R. Haack, L. Hall, R. Haugen, L. Helperin, C. Heuser, G. Hobby, J. Hubbard, W. Hovis, J. Huffman, H. Kruger, H. Jaffe, D. Jenkins, W. Massell, C. McDanald, E. Merek, A. Miller, F. Morelli, R. Nichols, S. Nishikawa, O. Reynolds, M. Rubin, E. Rudolph, K. Sandved, E. Schofield, G. Shulman, P. Simmonds, P. Smith, O. Soule, E. Staffeldt, E. Stuhlinger, L. Swan, L. Taylor, J. Tedrow, J. Turnock, F. Ugolini, W. von Braun, W. Weyant, R. Young, and U. S. Navy Task Force 43.

This paper presents the results of one phase of research carried out at the Jet Propulsion Laboratory, California Institute of Technology, under the Contract No. NAS-7-100, sponsored by the National Aeronautics and Space Administration. Logistic support and facilities for this study in Antarctica were provided by the Office of Antarctic Programs, National Science Foundation, Washington, D. C., U. S. A.

# REFERENCES AND NOTES

1. HILLS, E. S., O. S. OLIVER, and C. R. TWIDALE
   1966 Geomorphology. p. 53-76. *In* E. S. Hills, ed., Arid lands. Methuen and Company, London; Unesco, Paris.

2. BROWN, S.
   1963 World of the desert. Bobbs-Merrill Company, Inc., Indianapolis. 224 p.
   RESEARCH COUNCIL OF ISRAEL/UNESCO
   1953 Desert Research Proceedings, International Symposium held in Jerusalem, May 7-14, 1952. Research Council of Israel, Special Publication 2. 641 p.
   HILLS, E. S. ed.
   1966 Arid lands. Methuen and Company, London; Unesco, Paris. 461 p.
   LEOPOLD, A. S., and the Editors of *Life*
   1961 The desert. Life Nature Library. Time Inc., New York. 192 p.
   McGINNIES, W. G., B. J. GOLDMAN, and P. PAYLORE (eds.)
   1968 Deserts of the world: an appraisal of research into their physical and biological environments. University of Arizona Press, Tucson. 788 p.
   UNESCO
   1953- Arid Zone Research, reviews of research.
   HODGE, C., and P. C. DUISBERG (eds.)
   1963 Aridity and man. American Association for the Advancement of Science, Washington, D. C., Publication 74. 584 p.

3. MEIGS, P.
   1953 World distribution of arid and semi-arid homoclimates. *In* Reviews of Research on Arid Zone Hydrology, Unesco, Paris, Arid Zone Programme 1:203-210.

4. GLAZOVSKAIA, M. A.
   1958 Weathering and primary soil formation in Antarctica. Institute of Geology and Geography, Scientific Paper 1:63-76. (Translated by Rutgers, The State University).

5. GERASIMOV, I. P., and R. P. ZIMINA
   1968 Recent natural landscapes and ancient glaciation of the Pamir. p. 267-269. *In* H. E. Wright, Jr., and W. H. Osburn, eds., Arctic and alpine environments. Indiana University Press, Bloomington.

6. ALEKSANDROVA, V. D.
   1960 Some regularities in the distribution of the vegetation in the Arctic tundra. Arctic 13: 146-162.
   ENCYCLOPAEDIA BRITANNICA
   1962 Various pages on the Arctic, Antarctic, Greenland, Northwest Territories, Spitzbergen, and Eurasian Archipelago. William Benton, Chicago.
   GORODKOV, B. N.
   1939 Pecularities of the Arctic top soil (cover). Izvestiia Gosudarstvennego Geograficheskago Obshchestva 71:1516-1532. (Translated by Rutgers, The State University).
   GORODKOV, B. N.
   1943 Polar deserts of Wrangel Island (translated title; French summary). Botanicheskii Zhurnal SSSR 28:127-143.
   ACADEMY OF SCIENCES OF THE USSR
   1962 Soil geographical zoning of the USSR.

Translated by Israel Program for Scientific Translations, Jerusalem. 494 p.
DAVIES, W. E.
1961 Surface features of permafrost in arid areas. *In* G. O. Raasch, ed., Geology of the Arctic, 2:981-987. University of Toronto Press, Toronto.
FRISTRUP, B.
1952 Physical geography of Peary Land. I: Meteorological observations for Jørgen Brønlunds Fjord. Meddelelser om Gronland 127: 1-143.
FRISTRUP, B.
1953 High Arctic deserts. International Geological Congress, 19th, Algiers, 1952, fasc. 7:91-99.
MIKHAYLOV, I. S.
1962 The soils of polar wastes and the role of B. N. Gorodkov in their study. Vseyouznogo Geograficheskogo Obshchestva, Izvestiya 94:520-523. Translated by Jet Propulsion Laboratory, California Institute of Technology.
PORSILD, A. E.
1957 Illustrated flora of the Canadian Arctic Archipelago. National Museum of Canada, Bulletin 146. 209 p.
RAE, R. W.
1951 Climate of the Canadian Arctic Archipelago. Canada Department of Transport, Meteorological Division, Toronto. 90 p.
TARGUL'YAN, V. O.
1959 The first stages of weathering and soil formation on igneous rocks in the tundra and taiga zones. Soviet Soil Science 1957(11): 1287-1296.
TEDROW, J. C. F.
1968*a* Soil investigations in Inglefield Land, Greenland. Arctic Institute of North America, Final Report for U. S. Army Research Office, Durham, Contract No. DA-ARO-D-31-124-G820. 126 p.
TEDROW, J. C. F.
1968*b* Pedogenic gradients of the polar regions. Journal of Soil Science 19:197-204.
TEDROW, J. C. F., and J. BROWN
1962 Soils of the northern Brooks Range, Alaska: weakening of the soil-forming potential at high Arctic altitudes. Soil Science 93: 254-261.

7. TEDROW 1968 *a&b* (6)
   CLARIDGE, G. G. C.
   1965 The clay mineralogy and chemistry of some soils from the Ross Dependency, Antarctica. New Zealand Journal of Geology and Geophysics 8:186-220.
   MARKOV, K. K.
   1956 Some facts concerning Periglacial phenomena in Antarctica (preliminary report) (translated title). Moscow State University, Herald, Geography 1:139-148.
   TEDROW, J. C. F., and F. C. UGOLINI
   1966 Antarctic soils. *In* J. C. F. Tedrow, ed., Antarctic soils and soil forming processes. Antarctic Research Series 8:161-177.

8. TEDROW 1968*b* (6).

9. GODDARD SPACE FLIGHT CENTER
   1965 Nimbus I High Resolution radiation data. Greenbelt, Maryland. 1:1-226.

10. DALRYMPLE, P. C.
    1966 A physical climatology of the Antarctic Plateau. *In* M. J. Rubin, ed., Studies in Antarctic meteorology. Antarctic Research Series 9:195-231.

11. ENCYCLOPAEDIA BRITANNICA (6).

12. WEYANT, W. S.
    1966 The Antarctic climate. *In* J. C. F. Tedrow, ed., Antarctic soils and soil forming processes. Antarctic Research Series 8:47-59.

13. SWAN, L. W.
    1967 Alpine and aeolian regions of the world. p. 29 ff. *In* H. E. Wright, Jr., and W. H. Osburn, eds., Arctic and alpine environments. Indiana University Press, Bloomington.

14. WEYANT (12).

15. RUBIN, M. J.
    1965 Antarctic climatology. p. 72-96. *In* J. Van Mieghem and P. Van Oye, eds., Biogeography and ecology in Antarctica. W. Junk, The Hague.

16. U. S. ENVIRONMENTAL SCIENCE SERVICES ADMINISTRATION
    1957- Environmental data service: Climatologi-
    1966 cal data for Antarctic stations 1-9.

17. MEYER, G. H., M. B. MORROW, and O. WYSS
    1967 Bacteria, fungi, and other biota in the vicinity of Mirnyy Observatory. Antarctic Journal of the U. S. 2:248-251.

18. RUBIN (15).
    BULL, C.
    1966 Climatological observations in ice-free areas of southern Victoria Land, Antarctica. *In* M. J. Rubin, ed., Studies in Antarctic meteorology. Antarctic Research Series 9:177-194.
    RUDOLPH, E. D.
    1968 Environmental factors, climate. *In* V. C. Bushnell, ed., Terrestrial life of Antarctica. Antarctic Map Folio Series, Folio 5:5.

19. U. S. ENVIRONMENTAL SCIENCE SERVICES ADMINISTRATION (16)

20. ENCYCLOPAEDIA BRITANNICA (11).

21. U. S. NATIONAL BUREAU OF STANDARDS/INSTITUTE FOR APPLIED TECHNOLOGY
    1967 U. S. Naval Weather Service World-Wide Airfield Summaries: Canada-Greenland-Iceland. 682 p.

22. ENCYCLOPAEDIA BRITANNICA (11).

23. ENCYCLOPAEDIA BRITANNICA (11); ALEKSANDROVA (6).

24. THOMAS, M. K.
    1961 A survey of temperatures in the Canadian Arctic. *In* G. O. Raasch, ed., Geology of the Arctic, 2:942-955. University of Toronto Press, Toronto.

25. U. S. ENVIRONMENTAL SCIENCE SERVICES ADMINISTRATION (16).
    RUSIN, N. P.
    1964 Meteorological and radiational regime of Antarctica. Translated by Israel Program for Scientific Translations, Jerusalem. 355 p.

26. GIOVINETTO, M. B.
    1964 Distribution of diagenetic snow facies in Antarctica and in Greenland. Arctic 17:32-40.

27. TEDROW, J. C. F.
    1966 Polar desert soils. Soil Science Society of America, Proceedings 30:381-387.

28. GUNN, B. M., and G. WARREN
    1962 Geology of Victoria Land between the Mawson and Mulock Glaciers. Antarctica. New Zealand Geological Survey, Bulletin n.s. 71. 157 p, 2 maps.

29. GUNN and WARREN (28).

30. GERASIMOV and ZIMINA (5).

31. HODGE and DUISBERG (2).

32. BERG, T. E., and R. F. BLACK
    1966 Preliminary measurements of growth of nonsorted polygons, Victoria Land, Antarctica. *In* J. C. F. Tedrow, ed., Antarctic soils and soil forming processes. Antarctic Research Series 8:61-108.

33. NICHOLS, R. L.
    1966 Geomorphology of Antarctica. *In* J. C. F. Tedrow, ed., Antarctic soils and soil forming processes. Antarctic Research Series 8:1-46.

34. FRISTRUP (6).

35. NICHOLS (33).

36. CALKIN, P. E.
    1964 Geomorphology and glacial geology of the Victoria Valley system, southern Victoria Land, Antarctica. Ohio State University, Institute of Polar Studies, Report 10. 66 p. 42 figs.

37. McCRAW, J. D.
    1967 Soils of Taylor Dry Valley, Victoria Land, Antarctica, with notes on soils from other localities in Victoria Land. New Zealand Journal of Geology and Geophysics 10:498-539.

38. DAVIES, W. E.
    Personal communication, August 23, 1968. U. S. Geological Survey, Washington, D. C.

39. FRISTRUP (6).

40. TEDROW and UGOLINI (7).

41. BUTZER, K. W., and C. R. TWIDALE
    1966 Deserts in the past. p. 127-144. *In* E. S. Hills, ed., Arid lands. Methuen and Company, London; Unesco, Paris.

42. FRISTRUP (6).

43. CALKIN (36).

44. FRISTRUP (6).

45. TEDROW, J. C. F., F. C. UGOLINI, and H. JANETSCHEK
    1963 An Antarctic saline lake. New Zealand Journal of Science 6:150-156.

46. NICHOLS (33).

47. LUSTIG, L. K.
    1968 Appraisal of research on geomorphology and surface hydrology of desert environments. p. 95-283. *In* W. G. McGinnies, B. J. Goldman, and P. Paylore (2).

48. GOLDMAN, C. R., D. T. MASON, and J. E. HOBBIE
    1967 Two Antarctic desert lakes. Limnology and Oceanography 12:295-310.

49. BULL (18); RUBIN (15).
    SOPOLOV, A. V.
    1967 Oases in Antarctica (translated title; English abstr.) Meteorology 6. 143 p. Nauka, Moscow.

50. BULL (18).

51. CAMERON, R. E., J. KING, and C. N. DAVID
    in press Microbiology, ecology, and microclima-
        tology of soil sites in dry valleys of south-
        ern Victoria Land, Antarctica. Antarctic
        Biology Symposium, 2nd, Cambridge, July
        30, 1968, Paper presented at SCAR.
52. WEYANT (12).
53. BULL (18).
54. RUBIN (15).
55. BULL (18).
56. TEDROW 1966 (27).
57. REITAN, C. H., and C. R. GREEN
    1967 Appraisal of research on weather and cli-
        mate of desert environments. p. 21-92. In
        W. G. McGinnies, B. J. Goldman, and P.
        Paylore (2).
58. TEDROW and BROWN (6).
59. TEDROW, J. C. F., P. F. BRUGGEMANN, and G. F.
    WALTON
    1968 Soils of Prince Patrick Island. Arctic In-
        stitute of North America, Research Paper
        44. 82 p.
60. TEDROW, UGOLINI, and JANETSCHEK (45).
    CAMPBELL, I. B., and G. G. C. CLARIDGE
    1969 A classification of frigic soils—the zonal
        soils of the Antarctic continent. Soil Sci-
        ence 107:75-85.
61. CAMERON, R. E.
    1966 Soil sampling parameters for extrater-
        restrial life detection. Arizona Academy of
        Science, Journal 4:3-27.
62. TEDROW and UGOLINI (7); McCRAW (37); TEDROW,
    UGOLINI, and JANETSCHEK (45).
    UGOLINI, F. C.
    1968 Environmental factors, soils, In V. C.
        Bushnell, ed., Terrestrial life of Antarctica.
        Antarctic Map Folio Series, Folio 5:3-5.
63. CAMERON, KING, and DAVID (51).
64. CAMERON, R. E., and G. B. BLANK
    1963 Soil organic matter (organic matter in
        Southwestern U. S. Desert soils). Jet
        Propulsion Laboratory, Pasadena, Cali-
        fornia, Technical Report 32-443. 14 p.
65. GUNN and WARREN (28).
66. MIKHAYLOV, op. cit.; TEDROW 1968a&b, (6); 1966 (27).
67. TEDROW 1966 (27).
68. TEDROW 1966 (27); 1968a (6).
69. TEDROW 1966 (27).
70. TEDROW and UGOLINI (7); U. S. ENVIRONMENTAL
    SCIENCE SERVICES ADMINISTRATION (16).
    LLANO, G. A.
    1962 The terrestrial life of the Antarctic. Scien-
        tific American 207:212-230.
    LLANO, G. A.
    1965 The flora of Antarctica. p. 331-350. In T.
        Hatherton, ed., Antarctica. Fredrick A.
        Praeger, New York.
    RUDOLPH, E. D.
    1965 Antarctic lichens and vascular plants:
        their significance. Bioscience 15:285-287.
    RUDOLPH, E. D.
    1966 Terrestrial vegetation of Antarctica: past
        and present studies. In J. C. F. Tedrow, ed.,
        Antarctic soils and soil forming processes.
        Antarctic Research Series 8:109-124.
71. DREGNE, H. E.
    1968 Appraisal of research on surface materials
        of desert environments. p. 287-377. In
        W. G. McGinnies, B. J. Goldman, and P.
        Paylore (2).

72. LLANO 1965 (70).
73. ALEKSANDROVA (6); PORSILD (6); LLANO 1962, 1965
    (70); RUDOLPH 1965 (70).
74. McGINNIES, W. G.
    1968 Appraisal of research on vegetation of
        desert environments. p. 381-566. In W. G.
        McGinnies, B. J. Goldman, and P. Paylore
        (2).
75. PORSILD (6).
76. DOWNES, J. A.
    1964 Arctic insects and their environment.
        Canadian Entomology 96:279-307.
77. GORODKOV (6).
78. ALEKSANDROVA (6).
79. ENCYCLOPAEDIA BRITANNICA (6).
80. LLANO 1962, 1965 (6).
    KOOB, D.
    1968 The flora, algae distribution. In V. C. Bush-
        nell, ed., Terrestrial life of Antarctica.
        Antarctic Map Folio Series, Folio 5:13-15.
81. GREENE, S. W.
    1964 Plants of the land. p. 240-253. In R.
        Priestley, R. J. Adie, and G. de Q. Robin,
        eds., Antarctic research. Butterworth,
        London.
82. DODGE, C. W.
    1965 Lichens, p. 194-200. In J. Van Mieghem,
        P. Van Oye and J. Schell, eds., Biogeogra-
        phy and ecology in Antarctica. W. Junk,
        The Hague.
83. HIRANO, M.
    1965 Freshwater algae in the Antarctic regions.
        p. 127-193. In J. Van Mieghem, P. Van Oye
        and J. Schell, eds., Biogeography and
        ecology in Antarctica. W. Junk, The
        Hague.
84. DROUET, F., and W. A. DAILY
    1956 Revision of the Coccoid Myxophyceae.
        Butler University, Botanical Studies
        12:1-218.
    DROUET, F.
    1968 Revision of the classification of the Oscil-
        latoriace. Academy of Natural Sciences,
        Philadelphia, Monograph 15. 370 p.
85. CAMERON, R. E., J. KING, and C. N. DAVID
    1968 Soil microbial and ecological studies in
        southern Victoria Land. Antarctic Journal
        of the U. S. 3:121-123.
86. CAMERON, R. E., and G. B. BLANK
    1966 Desert algae: soil crusts and diaphanous
        substrata as algal habitats. Jet Propulsion
        Laboratory, Pasadena, California, Techni-
        cal Report 32-971. 41 p.
    SHIELDS, L. M., and L. W. DURRELL
    1964 Algae in relation to soil fertility. Botanical
        Review 30:93-128.
87. CAMERON, KING, and DAVID (51, 85).
88. CAMERON, KING, and DAVID (51, 85).
    BENOIT, R. E., and C. L. HALL
    in press The microbiology of some dry valley soils of
        of Victorialand, Antarctica. Antarctic
        Biology Symposium, 2nd, Cambridge,
        July 30, 1968, Paper presented at SCAR.
    CAMERON, R. E., and H. P. CONROW
    1969a Soil moisture, relative humidity, and
        microbial abundance in dry valleys of
        Southern Victoria Land. Antarctic Jour-
        nal of the U. S. 4:23-28.
    CAMERON, R. E., and H. P. CONROW
    1969b Antarctic dry valley soil microbial incuba-

tion and gas composition. Antarctic Journal of the U. S. 4:28-33.

BOYD, W. L., J. T. STALEY, and J. W. BOYD
1966    Ecology of soil microorganisms of Antarctica. *In* J. C. F. Tedrow, ed., Antarctic soils and soil forming processes. Antarctic Research Series 8:125-177.

89. BOYD, W. L.
1967    Ecology and physiology of soil microorganisms in polar regions. JARE Scientific Reports, Special Issue No. 1. Proceedings of the Symposium on Pacific-Antarctic Sciences. Department of Polar Research, National Science Museum, Ueno Park, Tokyo, Japan. p. 265-275.

90. CAMERON (*61*).

91. CAMERON, KING, and DAVID (*51,85*).

92. IVARSON, K. C.
1965    The microbiology of some permafrost soils in the MacKenzie Valley, N. W. T. Arctic 18:256-260.

93. DI MENNA, M.
Personal communication, August 29, 1968. Department of Agriculture, Ruakura Agricultural Research Centre, Hamilton, New Zealand.

94. SWAN (*13*).

95. JENSEN, H. L.
1951    Notes on the microbiology of soil from northern Greenland. Meddelelser om Grønland 142:23-29.

96. BENOIT and HALL (*88*).

97. MISHUSTIN, E. N., and V. A. MIRZOEVA
1965    The microflora of Arctic soils. Problems of the North 8:181-211. Translated from Problemy Severa 8:170-199.

98. DOWNES (*76*).

99. GRESSITT, J. L.
1964    Ecology and biogeography of land arthropods in Antarctica. p. 211-22. *In* Biologie Antarctique. Hermann, Paris.

100. JANETSCHEK, H.
1967    Arthropod ecology of South Victoria Land. *In* J. L. Gressitt, ed., Entomology of Antarctica. Antarctic Research Series 10:205-293.

101. GRESSITT, J. L.
1968    The fauna. *In* V. C. Bushnell, ed., Terrestrial life of Antarctica. Antarctic Map Folio Series, Folio 5:17-21.

102. CAMERON, KING, and DAVID (*51,85*).

103. DOWNES (*76*).

104. SWAN (*13*).

105. LOWE, C. H.
1968    Appraisal of research on fauna of desert environments. p. 569-645. *In* W. G. McGinnies, B. J. Goldman, and P. Paylore (*2*).

106. ENCYCLOPAEDIA BRITANNICA (*6*).

107. BRIDGWATER, W., and S. KURTZ (eds.)
1967    Greenland. p. 870. *In* The Columbia encyclopedia, 3rd ed. Columbia University Press, New York.

108. BRIDGEWATER and KURTZ (*107*); ENCYCLOPAEDIA BRITANNICA (*6*).

109. NATIONAL SCIENCE FOUNDATION.
1968    U. S. Antarctic Research Program Personnel Manual. Washington, D. C. 79 p. Appendix.

110. National Science Foundation, Division of Environmental Sciences, Office of Antarctic Programs
n.d.    Survival in Antarctica. Washington, D. C. 66 p.

111. EDHOLM, O. G.
1964    Man and the environment. p. 39-60. *In* R. Priestley, R. J. Adie, and G. de Q. Robin, eds., Antarctic research. Butterworth, London.

112. NATIONAL SCIENCE FOUNDATION (*109*).

113. LEWIS, R. S.
1965    A continent for science. Chapter 10: The Antarcticans. Viking Press, New York. 300 p.

114. BILLING, G., and G. MANNERING
1965    South. Man and nature in Antarctica. A. H. and A. W. Reed Wellington. 207 p.

115. BILLING and MANNERING (*114*); LEWIS (*113*).

116. ibid.

117. LEWIS (*113*).

118. BISSELL, R., and G. HAMILTON
1968    Antarctic safari. Venture 5:52-63.

119. RUDOLPH 1966 (*70*).

120. JOHNSON, R. W., and P. M. SMITH
in press    Antarctic research and lunar exploration. Advances in Space Science and Technology, 10.

STUHLINGER, E., and W. VON BRAUN
Personal communication and Antarctic trip report, January 1-11, 1967.

STUHLINGER, E.
1969    Antarctic research, a prelude to space research. Antarctic Journal of the U. S. 4:1-7.

121. MURZAYEV, E. M.
1967    Nature of Sinkiang and formation of the deserts of Central Asia. U. S. Department of Commerce, Translation 67-30944. 617 p.

# ARID-LAND STUDIES IN AUSTRALIA

CLIFFORD S. CHRISTIAN
Member of the Executive, CSIRO,* Australia

and

RAYDEN A. PERRY
Rangelands Research Programme, CSIRO,* Australia

*Commonwealth Scientific and Industrial Research Organization.

# ABSTRACT

About 75 per cent of the continent of Australia can be classified as arid. Individual investigators have contributed to existing knowledge of the land resources of the region, and studies have been made pertaining to its climate, water resources, vegetation, native and domestic animals, insects, and economics.

A growing general knowledge of arid Australia and droughts in populous areas have been responsible for a substantial increase of interest in arid-zone studies. New projects are under way and new facilities are being constructed to aid in furthering arid-land research. A growing liaison between scientists and economists and a general interdisciplinary bent are becoming increasingly evident in these studies. Research on arid Australia is likely to receive the coordination and comprehension approach that the importance of the region justifies.

# ARID-LAND STUDIES IN AUSTRALIA

Clifford S. Christian and Rayden A. Perry

## INTRODUCTION

The importance of aridity in the Australian scene is indicated by the fact that, using criteria such as those of Meigs (1), about 75 per cent of the continent (Fig. 1) is classified as arid (2). This huge area has presented a challenge both to land utilization and management and to scientific study. This paper sets out to give a broad picture of the present status of arid-zone studies in Australia, to indicate some of the more active fields of research, and to describe some interesting plans for the future. It has been compiled from information received from a number of colleagues with the CSIRO.

Rainfall isohyets form a concentric pattern over the continent, with higher rainfall at the coasts and with one large continuous arid area occupying all but the southern, eastern, and northern coastal margins. In the south the arid area is bounded by the 10-inch isohyet which extends to the coast at the Great Australian Bight and in the west. In the North, the arid zone extends into areas with quite high (25-30 inches) but ineffective rainfall. Here the rainfall is ineffective because it is restricted to the summer period when evaporation is high and because the infrequent falls of rain do not provide a sufficiently long reliable period of available soil moisture for productive plant growth.

This large continuous tract extends for a distance of 2,000 miles from east to west and over 1,000 miles from north to south. Its latitudinal spread, from 16°S to 35°S stretches from southern temperate areas where rainfall is more or less equally divided between summer and winter to the northern dry tropical areas with summer rainfall predominant. In consequence, the range of habitats and environments is wide, and the plant and animal communities that occupy them are numerous and varied.

In the main, the land surface is formed on or from the products of an ancient weathered surface, and many of the soils are highly leached and very infertile. The range of relief is small, and only minor areas in the center of the continent exceed 2,000 feet in altitude. While these mountains result in a small local increase in rainfall, the Australian arid regions lack the humid highlands of many other arid parts of the world, and, with the exception of one major river system in the southeast, they are not traversed by streams from higher rainfall areas. Thus, the only substantial irrigation developments in the Australian arid zone occur within the major river system in the southeast of the zone. Apart from a few short coastal streams in the west, drainage is mostly discontinuous and terminates in internal distributary drainage basins.

Prior to settlement by Europeans, Australia was occupied by a sparse aboriginal population which did not practice any form of arable agriculture or animal husbandry. At the time of European pioneering settlement of these arid areas, the aboriginal population and the native fauna could be regarded as in a state of long-term equilibrium with the environment.

Since settlement of these areas began little more than one-hundred years ago, a large part has been occupied for grazing by domestic stock, with sheep predominating in the south and cattle in the center and north. Apart from sheep and cattle, camels, donkeys, and horses also have been introduced into the arid areas and still persist in small numbers in feral communities. Land is held in the form of large fixed holdings known as *stations,* on which the one herd or flock will graze for the year long. There is no true nomadic grazing such as is common elsewhere, excepting that a part of the flock or herd may be removed in drought time and that periodical movement of stock occurs from the stations, especially to market for fat cattle or to other areas for those requiring fattening.

Although scattered natural watering-points do occur, particularly near the northern and southern margins, water for grazing animals is largely provided from bores and partly from surface catchment tanks. Watering-points are usually many miles' apart. On the one hand, this provides a form of control of stock and reduces the necessity for fencing under these extensive grazing conditions; on the other, this wide spacing leads to uneven distribution of grazing with an uneven impact on the vegetation. There is usually over-utilization near watering-points and under-exploitation away from them.

At the margins of the arid areas, cultivation for crop production has been practiced, particularly for wheat but with some sorghum in the north-east. In the South and East, in particular where the environment of the marginal country is more favorable, a mixed system of land use is widely practiced, crops of wheat being rotated with sheep grazing. The pasture and forage phase consists mainly of self-regenerating species, including a number of

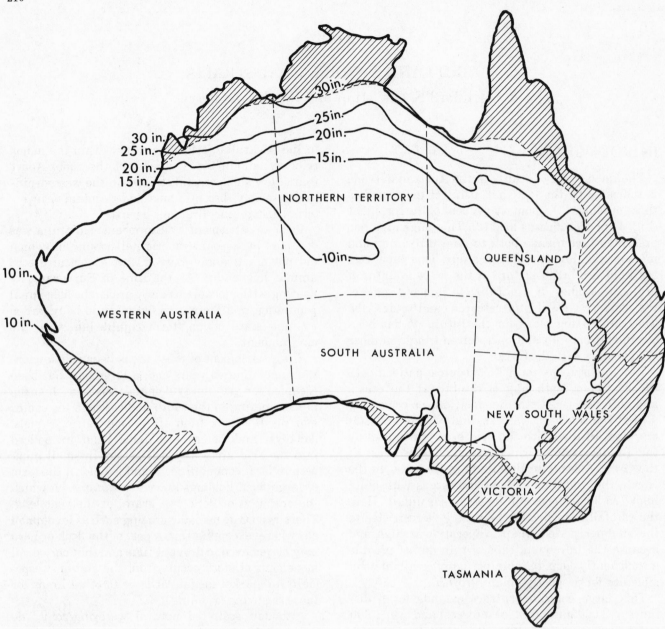

Fig. 1. Map of Australia showing arid area (not hatched) in relation to rainfall isohyets.

introduced species. In the main, however, land cultivation is not practiced in the arid regions, and there is no form of pasture improvement available for the greater part of them. The native vegetation thus remains the main resource on which the grazing animals depend, and apart from mining and tourism it is the main resource on which development and utilization must be based.

## LAND RESOURCES

Among the many investigations that have contributed to the existing knowledge of the land resources of the regions, the most extensive and systematic studies have been those of the CSIRO Division of Land Research. This work began in 1946 when CSIRO was given the task of making surveys and assessments of the land resources of the vast

areas of northern and central Australia. Because of the size of these areas, this task called for new approaches, and the outcome was what is now known as "the integrated survey" based on the recognition of land systems and land units (3). Although this type of resource survey has been shown to have wide appreciation in many different environments, it was largely developed and has been widely applied in the semiarid and arid areas of Australia.

These surveys aim to recognize in the landscape similar and dissimilar units of land, land being regarded as a whole complex of surface, near subsurface, and atmospheric features of a site, including both the physical and biological components. The land unit thus represents a distinctive habitat or a limited array of habitats and, similarly, is an area of land with a characteristic array of problems or possibilities for land use which differentiate it from other land units (4). Land units are associated in relatively small numbers to form recognisable patterns of landscape to which the term *land system* has been applied. Land systems can be identified on aerial photographs, which not only aid mapping but also enable the essential field traverses to be planned in the most appropriate and economic manner.

A map of the land systems of the region, combined with diagrammatic cross-sections illustrating the land units which comprise the pattern of each, together with descriptions of these units in terms of size, geology, soils, vegetation, climate, hydrology, and other pertinent land resource data, conveys a wealth of information about a region. This is quite adequate at the reconnaissance stage for decision making in relation to choice of areas or problems most deserving of immediate attention, or the kind of research or other effort which should follow. Land system mapping avoids the time-consuming effort of mapping detail and aggregating it into mapping units. It presents patterns but describes the detail. Detail can be mapped in selected areas subsequently in whatever from is most urgent and most appropriate.

The concept of land systems and land units has a strong genetic component in the sense that a land system is confined to an area with a reasonably homogeneous geological and morphogenetic history. Boundaries of land systems coincide with major geological or geomorphic boundaries; thus, different occurrences of the same land unit within the pattern of a land system may be expected to have similar qualities in many respects other than those which are directly observable, because they have been formed in the same way from similar parent geological material.

An area of over 700,000 square miles has now been mapped by this method in Australia (5). Apart from broad resource surveys the approach has been adapted for more specific purposes, for example, in the evaluation of land surface from the point of view of road engineering.

The land systems maps produced have been used for a variety of purposes, such as the selection of areas for particular kinds of land use development, the planning of new road systems, the location of watering points for stock grazing under extensive systems in low rainfall areas, the selection of priorities in research planning, and soil erosion and range vegetation surveys.

A map of Australian soils (6) covering the whole of the continent has recently been compiled, and geological surveys, accelerated considerably in the last decade because of the interest in mineral and oil exploration, now cover a large portion of the arid regions.

A climatic data collection network extends throughout the whole area, but recording points are of necessity widely spaced in the less inhabited regions. To some extent, however, the broad concentric zonation of climate in the region aids interpretation.

The paucity and therefore importance of rainfall over much of Australia has stimulated an interest in the analysis of rainfall occurrence in relation to plant and animal production (7). Some recent statistical work by Cornish of the CSIRO Division of Mathematical Statistics (private communication) is of a wider application than to arid-zone studies only. This work has aimed at overcoming the statistical problems of skewness or kurtosis of the rainfall distribution and of the occurrence of zero falls of rain, and of the consequent loss of information incurred by the use of inefficient statistical methods, especially where records are few, as is often the case in arid situations. New methods have been developed that utilize all the information available relevant to a specific point and which make possible precise statements of accuracy of the derived quantities. Such methods have their application in many determinations, such as the commencement of growth periods, likely length of growing season, and probabilities of occurrence of any specific phenomenon related to quantity of rain.

More intensive studies of a microclimatic nature, by Slatyer (8) at Alice Springs, examined in depth the fate of rainfall and solar energy received by the two main plant communities of the region. They measured the partitioning of the various components of the energy and water-balance budget, showed the pattern of utilization by the vegetation, and pointed out the nature of the losses involved. In the process some key adaptative features of the species to arid conditions were revealed. Such studies were the forerunners of further investigations in the field of plant-environment relationships.

## SURFACE WATER RESOURCES

Little quantitative information is available for stream flow in the Australian arid zone, over 90 per cent of which lies in three drainage divisions, the large Western Plateau system of uncoordinated drainage, the extensive system of streams draining to Lake Eyre, and the much smaller Bulloo-Bancannia system of internal drainage (9). The stream-gauging network reflects emphasis on potential for development of surface water resources. In 1965 only about 50 gauging stations, of a total of 1,500 for the continent, were in the arid zone (10). Most of these stations are located in the upper reaches of streams draining to Lake Eyre, and about 30 per cent of the basin is gauged. The estimated average annual runoff of the whole Lake Eyre basin is 3,600,000 acre-feet, which corresponds to a water depth of 0.15 inches over the catchment. Actual annual runoff is of course highly variable and ranges from zero in many years to possibly over one inch equivalent depth in the wettest years.

When flows occur in the braided channel systems of rivers draining into Lake Eyre from the north to northeast, there is a dramatic effect on the stock carrying capacity of the adjacent alluvial soils. It is only by the provision of improved road communications for stock entry at several points that this resource can be effectively exploited.

The concentration of surface runoff into stream flow in the Australian arid zone gives rise to three main resources: highly concentrated plant production in flood-out areas, recharge to usually shallow groundwater resources, and surface storages of several kinds (stock ponds, town supply reservoirs, natural waterholes, and terminal lakes with catchments ranging from a few square miles up to the 440,000 square mile catchment of Lake Eyre). The greatest scope for increased use of the surface water resource (whether through surface, soil, or groundwater storages) appears to be in small catchments, but little information is yet available on relations between surface runoff and the physical and biological characteristics of catchments.

One experiment started in the Alice Springs area in 1962 involves detailed recording of rainfall and stream level in 18 catchments, ranging in size from 1.2 to 68 square miles, each of which is entirely within one of the land systems defined for the area. Preliminary indications are that surface runoff may be at least as high in degraded catchments with little relief as in catchments with much greater relief.

## GROUNDWATER RESOURCES

The groundwater resource in the Australian arid zone can be broadly classified into two main types: (a) generally deep-seated water, often of poor quality, in crystalline rocks and consolidated sediments, which is in flow systems with renewal times of at least 1,000 years, and (b) generally shallow water, more frequently of good quality, in unconsolidated sediments, with renewal times usually of the order of 10 to 100 years. Both resources are extensively used for stock watering, and in very few places the good quality and yield of the shallow water has been exploited for irrigation of market gardens, citrus, and fodder crops.

Most of the deep-seated ground-water occurs in major sedimentary forms, of which the best known (and most exploited for stock water) is the Great Artesian Basin (9) recharged from streams outside or near the eastern boundary of the arid zone. Groundwater investigations have normally been related to recurring stock or town supplies in a particular area, but the hydrology of the Great Artesian Basin has been studied in greater detail and the safe yield estimated as 110 million gallons per day.

Investigations of shallow ground-water have also generally been on an ad hoc basis, but more detailed hydrologic studies have been undertaken in the Alice Springs area and near Wiluna in Western Australia.

## VEGETATION

Both the fauna and flora of Australia show many unique features, the significance of which to arid land utilization and management have not yet been adequately studied. There are no succulents and few spiny species in the vegetation. Large parts of the area are perennially evergreen, consisting of sclerophyllous hummock grassland covered by *Triodia* and *Plectrachne* species. Phyllodineous *Acacia* species are prominent in the shrubs and trees. *Myoporaceae, Proteacea,* and *Caesalpinaceae* are common in the shrub vegetation throughout, and *Chenopodiaceae* in the southern parts. *Eucalyptus* trees and mallee species are prominent in both the southern and northern margins but are sparse in the central lower rainfall areas, excepting along stream lines and in favored habitats (11). The main perennial species (grasses and browse plants) are relatively unpalatable to grazing stock and thus are protected to some degree after tains when stock concentrate more on palatable short-lived components of the vegetation. While botanical collections of the vegetation have been extensive and include many of the native species, there is still a good deal of taxonomic work waiting to be done.

Fire is not a frequent occurrence in the arid areas because fuel is generally limiting; however, after good seasons, fires are important and can cause widespread destruction of trees and shrubs.

The flora is specially adapted to the harsh arid environment. Most species are short lived, existing only as seeds during dry periods and germinating and growing rapidly after rains. Many arid species have some sort of inhibitory mechanism that prevents germination of seeds after small falls of rain (12). The leaves of most of the perennial grasses die with the onset of dry conditions, and only perennating organs such as rhizomes and stem bases survive through droughts. Some shrubs and trees are dry-period deciduous. Most of the *Acacia* species, the most common dominant shrubs and trees throughout the arid zone, have tough phyllodes instead of leaves. At least one of these,

*Acacia aneura* (mulga), intercepts rainfall and channels it down the stems into the ground immediately below the trunk (*13*).

The two grass genera, *Triodia* and *Plectrachne,* are of unusal interest in that, unlike most perennial grasses, they are evergreen. They have hard, lignified leaves with the stomates in sunken grooves in the lower surface. In addition the leaves are longitudinally folded about the midrib to give the appearance of being cylindrical. The lower surface is within a long narrow slit in the cylinder (*14*). At least some of the plants (for example, *Acacia aneura* and *Triodia basedowii*) have the ability to extract water from soils against tensions of up to 150 bars, which is far higher than what is normally considered as wilting point (*15*).

Considering the generally poor state of knowledge of Australia's arid zone plants, a surprisingly large amount has been written about plants poisonous to stock. While a great number of species are known or suspected of being poisonous, most are unpalatable and are not normally grazed in quantities large enough to be serious. However, two species are responsible for large losses in most years; these are a forage tree, *Acacia georginae* (gidgee), which grows on calcareous soils in an area astride the Queensland / Northern Territory border, and the herb *Indigofera dominii* (indigo), which is widespread in central Australia and western Queensland. Gidgee has been shown to contain sodium fluoroacetate (*16*) and causes severe losses in ruminants. Indigo causes a chronic disease (Birdville disease) in horses (*17*).

As might be expected in such a large varied area, there are a vast number of vegetation communities. Little is known of the precise details of their floristic composition and dynamics or of their environmental limitations. About half the area has been covered by reconnaissance surveys, and for this part general descriptions of the major communities and their distribution are available. The most important of these surveys are the series of natural resource inventories of the CSIRO Division of Land Research previously referred to. In addition the major plant communities of southwestern Queensland, western New South Wales, and a part of arid South Australia have been described by Blake (*18*), Beadle (*19*), and Jessup (*20*) respectively.

In general the major communities are of four main types.

*a. Tussock Grasslands.* These are mainly pure grasslands dominated by *Astrebla* species (Mitchell grasses) (Fig. 2). They occur on cracking clay soils in the northern part of the zone where rainfall is of predominantly summer incidence. The dominant perennial grasses are tussock grasses 18 to 36 inches high and 3 to 9 inches in diameter. Other perennial grasses include *Eragrostis* spp., *Aristida*

species, and *Dichanthium* species. In favorable seasons the interspaces between the perennial tussocks are occupied by annual grasses and forbs.

As far as animal industries are concerned these grasslands are the most productive of the arid communities, as they support stocking rates several times greater than any other communities. They are used for grazing cattle and wool sheep.

*b. Chenopod Shrublands.* These are restricted to the southern half of the zone where summer and winter rainfall are roughly equal, on the average. The most widespread dominants are *Atriplex vesicaria* (bladder saltbush) (Fig. 3), which occurs mainly on cracking clay or texture-contrast soils, and *Kochia sedifolia* and *K. pyramidata* (bluebushes), which are more common on calcareous soils. The bushes are low (2 to 4 feet) and spreading. As with the grasslands, the interspaces are occupied by short-lived grasses and forbs in favorable seasons.

The shrublands are used mostly for grazing wool sheep, but cattle are carried in northern South Australia.

*c. Low Woodlands.* These occur on medium-textured acid or calcareous soils throughout the whole zone. The dominants are tall shrubs or low trees (6 to 30 feet high but mostly 15 to 20 feet high) and over most of the area consist of almost monospecific stands of various *Acacia* species (Fig. 4).

The understory of the woodlands varies considerably, especially in its perennial components. In the south, perennial low shrubs similar to the shrubland communities occur under some stands. Long-lived perennial grasses are not common, and throughout all the woodland communities an understory of short grasses and forbs is characteristic in favorable periods. Grasses such as *Aristida contorta* (kerosene grass) and *Enneapogon* species, and forbs such as *Helipterum floribundum,* predominate after summer rains; in the south, *Bassia* species predominate after winter rains. The floristic composition varies greatly with the time of the year at which rains occur.

The woodlands are used for wool sheep grazing in the south and cattle grazing in the north.

*d. Spinifex.* Vast areas are characterized by spinifex (*Triodia* or *Plectrachne* species) (Fig. 5). The commonest habitat is sand plains and sand dunes on which *Triodia basedowii* and *Plectrachne schinzii* are the most wide spread, with *Triodia pungens* common in the higher rainfall tropics.

The majority of the spinifex areas are not used for grazing because the spinifex and the associated shrubs and low trees are not, or are only slightly, palatable. Towards the northern and northwestern margins, however, the *Triodia pungens* communities have a low grazing capacity, particularly where they occur in association with adjacent better types of country.

214

Fig. 2. Mitchell grass (*Astrebla pectinata*) grassland.

Fig. 3. Saltbush (*Atriplex vesicaria*) shrubland.

Fig. 4. Mulga (*Acacia aneura*) low woodland.

Fig. 5. Spinifex (*Triodia basedowii*) grassland.

Thus, broadly speaking, of the four major types, one (spinifex) is not used for grazing and the other three are used for grazing wool sheep in the south and cattle in the north. Sheep extend much further north on the Mitchell grass country in Queensland and in the far northwest of the zone. The three grazed types have in common the fact that the major part of the forage is provided by short-lived grasses and forbs which are present during, and for some time after, favorable rainy periods. In long dry periods the less palatable perennial components (Mitchell grass, chenopodiaceous shrubs, or shrubs and low trees of *Acacia* species) provide forage. Thus, because they are less palatable than the short-lived plants, the perennial components normally have a period of natural deferment following rains.

Sheep and cattle grazing, aggravated to a very important degree in southern and southeastern areas by grazing of the introduced rabbit, have resulted in substantial changes in the composition of the native vegetation communities, although few detailed quantitative studies have been made. One consequence of this grazing by domestic stock in some areas has been a change in favor of the native kangaroos, which have reached pest proportions over wide areas, in spite of the fact that the capacity of the land to carry domestic stock has concomitantly been reduced.

More research has been done on the regeneration of degraded country than on developing management methods for maintaining range condition. The New South Wales Soil Conservation Service has been particularly active (*21*). Results from their work and from the Koonamore Vegetation Reserve (*22*) indicate that range regeneration in arid Australia is an extremely slow and episodic process and that exclusion of stock alone has little or no effect on badly degraded sites over very long periods. In northwestern Australia depleted spinifex pastures have been regenerated and maintained by strategic burning followed by deferred grazing (*23*).

The results of a survey by the New South Wales Soil Conservation Service of range condition in central Australia indicate that, under the relatively uniform stocking rates applied to all land in the region, each land system has tended to degrade to a particular condition on all properties on which it occurs. The condition is related to the genesis of the landscape. Thus, land systems of the Tertiary weathered surface, which accounts for the greater part of the grazing lands of the region, now are in good condition, no doubt due to the low erosion energy on these landscapes.

On the other hand, land systems formed on erosional landscapes are generally in poor condition, as are those on depositional land surfaces where disturbance of the natural vegetation allows larger quantities of water to flow over the land surface at greater velocities, thus causing accelerated erosion. The survey shows clearly that the present condition of land in central Australia is related more to land types and their genesis than to stocking history, a result which illustrates the value of land-system surveys in natural grazing regions.

## NATIVE ANIMALS

Although a whole range of indigenous vertebrates of all classes have adapted successfully to the high evaporation and low, unreliable rainfall of arid and semiarid Australia, little work has been done on their ecology. Knowledge of the majority of species has not progressed beyond incidental descriptive natural history. There has been some interest in the ecology of breeding in desert birds (for example, *24*), many of which have no definite breeding season and breed only after rains at any season, but the only vertebrate species that have been studied ecologically to any extent are the large kangaroos. The studies were initiated because of a general belief that kangaroos were serious competitors of domestic stock.

From 1953-1960 Ealey (*25*) studied the euro (*Macropus robustus*) in the Pilbara region of northwestern Australia, an area that had been considered formerly as good sheep country. Under the influence of sheep-grazing, the softer grasses, which had been a minor component of the spinifex vegetation and had been dominant in small favored areas, had been largely supplanted by spinifex. This situation favored the euro, and it increased markedly in numbers. Population densities were measured by ground transect counts and were found to range as high as one euro per six acres in particularly favorable areas, although in general they numbered about one per 30 to 40 acres. In most areas euros considerably outnumbered sheep.

The euro is a relatively sedentary kangaroo, but about half the population disperses following summer rains, although most return to their old home ranges in autumn. Euros in the study utilized both natural and man-made water sources, but few drank regularly even when day temperatures were very high; many apparently never drank at all and presumably satisfied their water requirements from food plants. The animals avoided the worst of the high day temperatures by remaining immobile in refuges among piles of granite boulders.

Storr (*26*) studied the diets of the euro and sheep in the best part of the Pilbara and found that they differed considerably in their food preferences. The euro breeds all the year round in spite of high temperatures and changes in food availability, but in times of severe food shortage caused by drought many females enter anestrus and cease breeding

(27). It was concluded by Ealey and Main (28) that populations were controlled as a result of periodic crashes in some areas and regular die-offs in others caused by low protein availability.

Pastoral settlement in inland New South Wales has led to a change in ground vegetation and a decrease in tree density. The present short-grass community has favored the increase of the red kangaroo (*Megaleia rufa*) so that it is now considerably more abundant than in pre-settlement days. The red kangaroo was studied by Frith (29) in New South Wales and southwestern Queensland, from 1960, for several years. The animals were sampled widely to provide reproductive and age-structure data, and population estimates were made on study areas by means of aerial counting. A drought of steadily increasing severity occurred over the study period, and a further complication was commercial shooting.

The red kangaroo population declined markedly in all habitats in New South Wales over the study period; between April 1960 and September 1963 in the most favored habitat the density declined from 6.3 to 2.0 per mile of traverse. The chief mortality was among the young newly emerged from the maternal pouch. At the end of the study, in the most drought stricken area only 17 per cent of the young survived (30). It was thought that inadequate nutrition of the mothers was the cause of mortality of young. The animal breeds all the year round.

Among a number of marked animals the maximum distance moved was only 20 miles; however, further studies since Frith's have shown that under the stimulus of severe drought and consequent lack of food, some animals will move six times that distance. Other aspects of red kangaroo ecology, such as sociality and habitat utilization, have been studied by Caughley (31) at Cunnamulla, Queensland, which was also one of Frith's study areas. Griffiths and Barker (32) examined food preferences of red and grey kangaroos (*Macropus giganteus*) and sheep at Cunnamulla. The three species had different food preferences and in general were not in direct competition for food.

Newsome (33) investigated the red kangaroo in central Australia at approximately the same time as Frith and used essentially similar methods. Drought periods in central Australia were apparently more severe than those suffered on Frith's study areas. An important finding in the central Australian study was that the environment was ameliorated in the kangaroo's favor by cattle which cropped the long grasses and thereby exposed the short green shoots. Kangaroos seek green food even in droughts. Although the red kangaroo breeds continuously in favorable times, when nutritious food was scarce during droughts, a high proportion of females entered anestrus and ceased breeding. This is in contrast to Frith and Sharman's finding that few females were anestrous, although the proportion rose considerably in western New South Wales after Frith and Sharman's paper appeared, following an increase in the severity of the drought. During the course of the long drought in central Australia (1958-1966), numbers of red kangaroos decreased considerably; on about 2,500 square miles they decreased roughly from 4,900 to 2,800 between 1962 and 1966 (34).

Rather more species have been investigated in physiological studies than in ecology because of an interest in adaptations to desert survival, particularly in regard to the problems of poikilothermic vertebrates in coping with high temperatures and low humidities.

Many desert rodents in various parts of the world live on dry seeds and do not drink water. MacMillen and Lee (35) have shown that some Australian desert rodents have a similar ability to live on dry seeds and conserve water by producing concentrated urine and dry faeces. In fact two species, *Notomys alexis* and *Pseudomys hermannsburgensis,* have the highest urine concentrations yet recorded for mammals. Kinnear *et al.* (36) have studied the water relations of the tammar (*Macropus eugenii*), a small wallaby living on the semiarid Abrolhos Islands of Western Australia. It is able to drink seawater and maintain weight while eating dry food. It subsists on a low-protein diet and recycles urea. It has a low water requirement and a renal concentrating ability as good as that of most eutherian desert rodents.

Some work has been done on physiology of arid-zone lizards, especially *Amphibolurus* (family Agamidae). The lizards of this genus are diurnal and consequently have problems with high temperatures and low humidities. Body temperature is regulated by a combination of behavioral attitudes and movement into shade, or retreats under rocks, or burrows. Desert-living species are no more tolerant of grossly elevated body temperatures than species from less severe habitats.

*Diporophora bilineata,* an agamid which does not have thermal refuges and is forced to put up with extreme day temperatures, is able to withstand 46°C for six hours without apparent ill effect, a temperature that kills the most thermal-adapted *Amphibolurus* in an hour or so. *Diporophora* climbs as high as possible on vegetation and faces the sun, thus presenting the least body area to the source of heat (37).

Some desert lizards in other families (geckoes and skinks) also have a considerable ability to withstand relatively high temperatures and low humidities, and lose relatively little water at temperatures up to 35°C in dry air, although in general the performance of desert agamids is better (38). *Amphibolurus inermis* is able to survive 42°C for as long as 24 hours and 44°C for eight hours, conditions never experienced for such lengthy periods in its natural habitat (39).

Bradshaw and Shoemaker (40) studied *Amphibolurus ornatus,* a species that lives on granite outcrops in semiarid parts of Western Australia and subsists largely on sodium-rich ants. During the dry summer in the absence of free water it stores sodium ions in the extracellular fluids, and there is a movement of water from the intracellular fluid to maintain osmotic equilibrium. When free water is available during summer thunderstorms, the retained ions are excreted. The ability to retain sodium ions at elevated concentrations in the body fluids is thus a water-conserving adaptation to survival in semiarid environments.

Main (41) has recently reviewed physiological and ecological adaptations of Australian desert frogs. All desert species need free water for breeding, but a short larval life is advantageous. Most desert species escape high temperatures and cut down water loss by spending the greater part of their lives estivating in burrows. They have a rapid water uptake and respond quickly to rain and come to the surface for feeding (and breeding if sufficient rain falls and temperatures are high enough). *Notaden nichollsi,* which lives in *Triodia*-dune country, digs burrows as deep as five or six feet. Desert species can survive a relatively high degree of desiccation when compared with non-desert forms, and some frogs and their larvae can withstand high water temperatures—up to 40°C for several hours. *Cyclorana platycephalus,* a species living in desert claypans and which digs only short burrows because of the hardness of the soil, enters estivation with its bladder and coelom filled with water. Water loss is further reduced in this and other desert species by a cocoon with which the estivating frog invests itself (42).

## DOMESTIC MAMMALS

The lack of field data for domestic animals in arid environments is in marked contrast to the increasing knowledge for native mammals. For domestic animals, physiological, behavioral, and nutritional data must be procured from the field before an adequate understanding of the relationship of the animal with its total environment can be gained.

Few studies of the behavior of sheep in arid environments are in progress. Flocks tend to split into small subgroups during the day and re-form into a single flock at night. Sheep walk distances up to 9 or 10 miles per day between watering, feeding, and night camping sites. While provision of drinking water is an important feature of management, during cool weather Merinos drink infrequently (up to 10-day intervals between drinking).

The fleece of sheep affords them considerable protection from the heat load of solar radiation. Respiratory cooling is the major thermoregulatory mechanism during exposure to heat loads. Removal of the fleece during summer increases water turnover rates and respiratory rate (43). The rate of sweating in Merinos is very low and has only a minor role in temperature regulation.

Merino sheep are relatively well adapted to resist dehydration in a hot environment (44). Survival for up to 9 days without water at air temperatures of 40°C have been demonstrated under hot room conditions. The renal concentrating ability is about equal to that of the camel. Sheep can stand considerable losses of body fluids during dehydration, up to 26 per cent of body weight and 44 per cent of extracellular fluids (45).

Saline water supplies are common in much of arid Australia. Sheep can become accustomed to drinking saline water with little effect on the level of feed intake. There appears to be a physiological adaptation to drinking saline water, which is rapidly acquired and readily lost (46). Very high intakes of sodium chloride can be achieved when salt is included in the feed (47). Water intakes rise considerably to enable the excess salt intake to be excreted; this is particularly notable in sheep feeding on saltbush pastures (48).

Studies of walking behavior of cattle are just beginning; initial results indicate marked differences between individuals. Zebu cattle are able to graze in the sun for longer periods than European breeds and can conserve water by sweating less than European cattle, although they may sweat more vigorously in extreme conditions (49).

Little is known of the diets of either sheep or cattle in arid Australia. Initial studies using esophageal fistulas on sheep have revealed that plant species which form a minor component of the pasture available may form a major proportion of the feed consumed (50). Using rumen contents, Chippendale (51) has shown that even during a long drought when grass was very scarce, it formed the major proportion of cattle diets.

The problem of drought feeding is of considerable importance. Recent work on the feeding of dietary proteins specially protected from degradation by rumen microorganisms could be highly important (52).

Apart from the fact that marked differences in lamb marking or calf branding percentages occur in various regions and various seasons, little is known about sheep and cattle reproduction in arid areas. Donaldson (53) reports some evidence indicating that a rainfall deficiency 12 to 15 months prior to a branding adversely affects conception and calf drop, even when rainfall in the following summer is adequate. The Csiro Division of Animal Physiology is conducting a study of the reproductive performance, and the factors which influence it, of cattle herds in arid Queensland. In the Kimberley area in Western Australia a dramatic response to supplementation with whole cotton seed for four months has been demonstrated both in cow survival and calf drop (54).

Apart from plant poisonings, no major diseases of sheep and cattle are characteristic of the arid zone. In general, both ectoparasites and endoparasites constitute only minor problems; indeed, animals reared in the arid zone may not acquire resistance to endoparasites and thus may be highly susceptible to infection when moved to humid zones. Some infectious diseases such as bovine tuberculosis and contagious bovine pleuropneumonia can spread readily in arid zone herds if suitable control measures are not taken (55). Phosphorus deficiency is common in arid zone cattle and is, as usual, associated with botulism (56).

## INSECTS

Little is known of the insects of arid Australia. Harvester termites, particularly a species of *Drepanotermes* which harvests grass, are common, and presumptive evidence is strong that they are important in relation to the processes of soil formation and pasture productivity. Taxonomic and biological studies have been commenced on *Drepanotermes* to lay a foundation for ecological investigations. Preliminary conclusions indicate that termites are harmful in that their extensive hardened nest areas prevent water percolation and pasture establishment, they entirely denude areas surrounding their nests when pastures are dry, and their seed harvesting activities reduce the capacity of pastures to recover after rain. Tentative estimates indicate that the biomass per unit area and food turnover of termites is of the same order as that of domestic stock in arid Australia.

The native bushfly, *Musca vetustissima,* and the Old World *Musca sorbens* are both severe nuisances to man and domestic animals, and both can possibly transmit ophthalmic and enteric diseases. CSIRO has recently initiated studies of the reproductive biology of the bushfly and its capacity for overwintering in various parts of the continent.

A number of species of grasshoppers occur throughout arid Australia. Of these, the Australian plague locust, *Choitoicetes terminifera,* has been studied fairly extensively (57). During plague years it is thought to emigrate from arid areas as swarms and to invade agricultural areas. In order to reproduce, it needs fresh green feed, usually summer grasses, so the timing of its life cycle and its survival depend on rain. General rains lead to increased turnover of generations and high numbers. The high-density populations fly by day, adopting a rolling form of flight in which individuals

land and feed between successive flights. This enables them to maintain their water balance.

In the past considerable research on means of reducing the importance of sheep blowfly, *Lucilia cuprina,* has resulted in the perfection of surgical and management techniques that deal effectively with the most important section of the problem (58). For various reasons chlorinated hydrocarbons and organophosphorus compounds are no longer suitable for the prevention of blowfly strike on other areas of the body, and research is proceeding on other means of control, including the possibility of employing meiotic drive and of making the sterile male approach more effective.

## ECONOMICS

Although there has long been an understandable interest among Australian economists in the financial and management problems of agriculture under conditions of high weather variability, few empirical investigations were undertaken until comparatively recently. Much challenging work remains to be done. As the character of the pastoral industry in the arid areas has changed from virtual feral hunting to an integrated industry, so the involvement of economists in arid zone research has increased. Important stimuli have been the changing management patterns made possible by increased use of road transport and the development of inland slaughtering which have markedly affected management procedures and profitability in a period of optimistic price expectations and nationwide interest in the role to be played by the more remote regions in Australian economic development.

Much mundane but fundamental work in collecting basic data on the existing structure of the pastoral industry and its complementary secondary industries remains to be done, although a good deal has already been achieved.

## FUTURE RESEARCH

Up to the present time, research specifically for arid Australia has proceeded largely as separate studies undertaken by individually motivated scientists from many different organizations and departments and mostly operating from the capital cities in the coastal areas. Many of the studies have been made by scientists interested in arid areas merely as extreme environments.

Recent severe droughts in the populous areas to the south and east of the arid areas, coupled with the growing general knowledge of arid Australia, have motivated a change in the climate of public opinion which has led to a substantial increase in interest in arid-zone studies in Australia. As a result, in addition to the projects which have been referred to, new facilities for arid-zone studies, particularly pastoral studies, have been established at Charleville by the Queensland Department of Primary Industries; an existing research station at Fowler's Gap (near Broken Hill) is being developed by the University of New South Wales; the Northern Territory Administration has established an Arid Zone Research Institute at Alice Springs; and there have been notable developments in a number of activities in Australian universities which will have direct relevance to arid-zone studies, particularly at the Australian National University in Canberra, Monash University in Melbourne, and the University of New South Wales in Sydney.

Perhaps the most important new development is the increasing realization by many people that, as well as individual disciplinary studies, integrated studies are needed on the whole arid-land ecosystem. This has led to the establishment by CSIRO of a Rangelands Research Programme the objective of which is to establish the principles of, and the processes operating within, the climate/land/vegetation/animal ecosystem. This is an especially important development, because, although the main aspects of the arid-zone ecosystem are similar to those of any grazing ecosystem (viz.; rain and solar energy provide both plant growth and cause erosion; plant growth provides both forage and ameliorates erosive action; animals in consuming forage grow and reproduce but also reduce the landscape protection afforded by vegetation), opportunities for economic manipulation of the arid ecosystem are severely restricted. In fact, generally the only aspect of the ecosystem which can be managed by man is the animal factor; all other aspects are controlled by natural influences and by the impact of animal management.

Another exciting development is the possibility of being able to simulate numerically the pastoral ecosystem with modern fast computers. This will, of course, require knowledge in depth of the processes operating in the ecosystem.

The increasing involvement of biologists in arid-zone research tends to be mirrored by economists, and a close and growing liaison between scientist and economist in land research is probably more evident in the arid-zone work in Australia than in any other sphere. This liaison is most evident in ecological research, where a number of workers from both fields are cooperating in attempts to synthesize, from biological and economic principles, suggested stocking, management, and settlement patterns that may be stable and financially viable over the long run of seasons. It is hoped thus to pinpoint the critical physical and economic constraints in such systems and to provide guidance for further research.

Perhaps the most important development of all is the proposed "Australian Arid Lands Research Association"—an association of CSIRO Divisions, University Departments, Commonwealth and State Government Departments, and the Australian Academy of Science with the purpose of furthering arid land research in Australia and of making the best uses of existing facilities.

In the past, research on arid Australia has lacked the coordination and comprehensive approach that the importance of the region would justify in the Australian scene. Current and proposed developments point to more and better directed efforts in the near future, and the development of a number of stimulating areas of study is already under way (59).

## REFERENCES AND NOTES

1. MEIGS, P.
   1953 World distribution of arid and semi-arid homoclimates. In Reviews of research on arid zone hydrology. Unesco, Paris. Arid Zone Program 1:203-210.
2. PERRY, R. A.
   1967 The need for rangelands research in Australia. Ecological Society of Australia, Proceedings 2:1-14.
   PERRY, R. A.
   1968 Australia's arid rangelands. Annals of Arid Zone 7.
3. CHRISTIAN, C. S., and G. A. STEWART
   1964 Methodology of integrated surveys. Conference on principles and methods of integrated aerial survey studies of natural resources for potential development. Unesco, Paris.
   PERRY, R. A.
   1967 Integrated surveys of pastoral areas. Second International Seminar on Integrated Sur-

veys of Natural Grazing Areas. Publications of the ITC-Unesco Centre for Integrated Surveys, Delft, Netherlands.
4. CHRISTIAN, C. S.
   1958 The concept of land units and land systems. Pacific Science Congress, 9th, Bangkok, 1957, Proceedings 20:74-80.
5. CHRISTIAN, C. S., et al.
   1954 Survey of the Barkly Region, Northern Territory and Queensland, 1947-48. CSIRO. Australia Land Research Series 3:1-182.
   PERRY, R. A.
   1960 Pasture Lands of the Northern Territory, Australia. CSIRO. Australia Land Research Series 5:1-55.
   PERRY, R. A., et al.
   1962 General report on lands of the Alice Springs area, Northern Territory, 1956-57. CSIRO. Australia Land Research Series 6:1-280.
   MABBUTT, J. A., et al.
   1963 General report on lands of the Wiluna-

Meekatharra area, Western Australia, 1958. CSIRO. Australia Land Research Series 7: 1-215.

SPECK, N. H., et al.
1964 General report on lands of the West Kimberly Area, W.A. CSIRO. Australia Land Research Series 9:1-220.

PERRY, R. A., et al.
1964 General report on lands of the Leichhardt-Gilbert area, Queensland, 1953-54. CSIRO. Australia Land Research Series 11:1-224.

GUN, R. H., et al.
1967 Lands of the Nogoa-Belyando area, Queensland. CSIRO. Australia Land Research Series 18:1-190.

6. AUSTRALIA COMMONWEALTH SCIENTIFIC AND INDUSTRIAL RESEARCH ORGANIZATION, DIVISION OF SOILS
1960- Atlas of Australian soils, sheet 1-10. Mel-
1968 bourne University Press, Melbourne.

7. ANDREWS, J., and W. H. MAZE
1933 Some climatological aspects of aridity and their application to Australia. Linnean Society of New South Wales, Proceedings 58:105-20.

DAVIDSON, J.
1934 The monthly precipitation–evaporation ratio in Australia as determined by saturation deficit. Royal Society of South Australia, Proceedings 58:33-36.

DAVIDSON, J.
1934 Climate in relation to insect ecology in Australia. I: Mean monthly precipitation and atmospheric saturation deficit in Australia. Royal Society of South Australia, Proceedings 58:197-210.

DAVIDSON, J.
1935 Climate in relation to insect ecology in Australia. II: Mean monthly temperature and precipitation–evaporation ratio. Royal Society of South Australia, Proceedings 59:107-24.

DAVIDSON, J.
1936 Climate in relation to insect ecology in Australia. III: Bioclimatic zones in Australia. Royal Society of South Australia, Proceedings 60:88-92.

PRESCOTT, J. A.
1934 Single value climatic factors. Royal Society of South Australia, Proceedings 58:48-60.

PRESCOTT, J. A.
1936 The climatic control of Australian deserts. Royal Society of South Australia, Proceedings 60:93-95.

PRESCOTT, J. A.
1938 Indices in agricultural climatology. Australian Institute of Agricultural Science, Journal 4:33-40.

PRESCOTT, J. A.
1943 The value of harmonic analysis in climatic studies. Australian Journal of Science 5: 117-119.

PRESCOTT, J. A., J. DAVIDSON, and H. C. TRUMBLE
1937 Climatology in relation to biology and agriculture. Australian Institute of Agricultural Science, Journal 3:77-79.

PRESCOTT, J. A., and J. A. THOMAS
1948 The length of growing season as determined by the effectiveness of rainfall. Royal Geo-

graphical Society of Australasia, South Australian Branch, Proceedings 50:42-46.

TRUMBLE, H. C.
1937 Climate control of agriculture in South Australia. Royal Society of South Australia, Proceedings 61:41-42.

TRUMBLE, H. C.
1948 Rainfall, evaporation and drought frequency in South Australia. Journal of Agriculture, South Australia 52:55-64.

TRUMBLE, H. C., and E. A. CORNISH
1936 The influence of rainfall on the yield of natural pasture. Council for Scientific and Industrial Research, Australia, Journal 9: 12-28.

FARMER, JOAN N., S. L. EVERIST, and G. R. MOULE
1947 Studies in the environment of Queensland. I: The climatology of semi-arid pastoral areas. Queensland Journal of Agricultural Science 4:21-59.

8. SLATYER, R. O.
1962a Methodology of a water balance study conducted on a desert woodland (Acacia aneura F. Muell) community in central Australia. In Plantwater relationships in arid and semi-arid conditions, Proceedings of the Madrid Symposium. Unesco, Paris. Arid Zone Research 16:15-26.

SLATYER, R. O.
1962b Internal water balance of Acacia aneura F. Muell in relation to environmental conditions. In Plant-water relationships in arid and semi-arid conditions, Proceedings of the Madrid Symposium. Unesco, Paris. Arid Zone Research 16:137-146.

9. AUSTRALIA, DEPARTMENT OF NATIONAL DEVELOPMENT
1965 Review of Australia's water resources 1963. Published for the Australian Water Resources Council. Government Printer, Canberra.

10. AUSTRALIA, DEPARTMENT OF NATIONAL DEVELOPMENT
1967 Stream gauging information, Australia, June 1965. Published for the Australian Water Resources Council. Government Printer, Canberra.

11. PERRY et al. (1962), ibid. (5)

12. BEADLE, N. C. W.
1952 Studies in halophytes. I: The germination of seed and establishment of the seedlings of five species of Atriplex in Australia. Ecology 33:49-62.

BURBIDGE, N. T.
1945 Germination studies of Australian Chenopodiaceae with special reference to conditions necessary for regeneration. I: Atriplex vesicaria. Royal Society of South Australia, Proceedings 69:73-85.

BURBIDGE, N. T.
1946 Germination studies of Australian Chenopodiaceae with special reference to conditions necessary for regeneration. II: (a) Kochia sedifolia F.v.M., (b) K. pyramidata Benth, (c) K. georgei Diels. Royal Society of South Australia, Proceedings 70:110-120.

13. SLATYER (1962a), ibid. (8)

14. BURBIDGE, N. T.
1945 Morphology and anatomy of the Western Australian species of Triodia R.Br. I: Gen-

eral morphology. Royal Society of South Australia, Proceedings 69:303-308.

BURBIDGE, N. T.
1946 Morphology and anatomy of the Western Australian species of *Triodia* R.Br. II: Internal anatomy of leaves. Royal Society of South Australia, Proceedings 70:221-234.

BURBIDGE, N. T.
1946 Foliar anatomy and the delimitation of the genus *Triodia* R.Br. Blumea (suppl.) 3: 83-89.

15. SLATYER (1962b), *ibid.* (8)

16. WHITTEM, J. H., and L. R. MURRAY
1963 The chemistry and pathology of Georgina River poisoning. Australian Veterinary Journal 39:168-173.

17. BELL, A. T., and W. T. K. HALL
1952 "Birdsville disease" of horses; feeding trials with *Indigofera enneaphylla.* Australian Veterinary Journal 28:141-144.

18. BLAKE, S. T.
1938 The plant communities of western Queensland and their relationships, with special reference to the grazing industry. Royal Society of Queensland, Proceedings 49: 156-204.

19. BEADLE, N. C. W.
1948 The vegetation and pastures of western New South Wales with special reference to soil erosion. Government Printer, Sydney. 281 pp.

20. JESSUP, R. W.
1951 The soils, geology and vegetation of northwestern South Australia. Royal Society of South Australia, Proceedings 74:189-273.

21. BEADLE, N. C. W.
1948 Studies in wind erosion. III: Natural regeneration on scalded surfaces. Soil Conservation Service of New South Wales, Journal 4:123-34.

CONDON, R. W.
1960 Scald reclamation experiments at Trida, N. S. W. Soil Conservation Service of New South Wales, Journal 16:288-302.

JAMES, J. W.
1956 Scald reclamation experiments in the Bourke District. Soil Conservation Service of New South Wales, Journal 12:44-54.

JONES, R. M.
1966 Scald reclamation studies in the Hay District. I: Natural reclamation of scalds. Soil Conservation Service of New South Wales, Journal 22:147-160.

JONES, R. M.
1966 Scald reclamation studies in the Hay District. II: Reclamation by ploughing. Soil Conservation Service of New South Wales, Journal 22:213-230.

JONES, R. M.
1967 Scald reclamation studies in the Hay District. III: Reclamation by ponding banks. Soil Conservation Service of New South Wales, Journal 23:3-17.

KNOWLES, G. H.
1954 Scald reclamation in the Hay District. Soil Conservation Service of New South Wales, Journal 10:149-156.

STANNARD, M. E.
1956 Regeneration areas in western New South Wales. Soil Conservation Service of New South Wales, Journal 12:73-79.

STANNARD, M. E.
1961 Studies on Kiacatoo regeneration area. I: Soils, vegetation and erosion. Soil Conservation Service of New South Wales, Journal 17:253-263.

STANNARD, M. E.
1962 Studies on Kiacatoo regeneration area. II: The effect of stocking on recovery. Soil Conservation Service of New South Wales, Journal 18:10-20.

STANNARD, M. E., and R. W. CONDON
1958 Studies on Trida regeneration area. Soil Conservation Service of New South Wales, Journal 14:159-176.

22. HALL, A. A., R. L. SPECHT, and C. M. EARDLEY
1964 Regeneration of the vegetation on Koonamore vegetation reserve, 1926-1962. Australian Journal of Botany 12:205-264.

23. NUNN, W. M., and H. SUIJENDORP
1954 Station management—the value of deferred grazing. Department of Agriculture, Western Australia, Journal 3:585-587.

24. KEAST, J. A., and A. J. MARSHALL
1954 The influence of drought and rainfall on reproduction in Australian desert birds. Zoological Society London, Proceedings 124: 493-499.

25. EALEY, E. H. M.
1967 Ecology of the euro, *Macropus robustus* (Gould), in north-western Australia. I: The environment and changes in euro and sheep populations. CSIRO. Australia Wildlife Research 12:9-25.

EALEY, E. H. M.
1967 Ecology of the euro, *Macropus robustus* (Gould), in north-western Australia. II: Behavior, movement and drinking patterns. CSIRO. Australia Wildlife Research 12:27-51.

EALEY, E. H. M.
1967 Ecology of the euro, *Macropus robustus* (Gould), in north-western Australia. IV: Age and growth. CSIRO. Australia Wildlife Research 12:67-80.

26. STORR, G. M.
1968 Diet of kangaroos (*Megaleia rufa* and *Macropus robustus*) and merino sheep near Port Hedland, Western Australia. Royal Society of Western Australia, Journal 51:25-32.

27. EALEY, E. H. M.
1963 The ecological significance of delayed implantation in a population of the hill kangaroo (*Macropus robustus*). pp. 33-48. *In* A. C. Enders, ed., Delayed implantation. University of Chicago Press.

SADLEIR, R. M. F. S.
1965 Reproduction in two species of kangaroos *Macropus robustus* and *Megaleia rufa* in the arid Pilbara region of Western Australia. Zoological Society London, Proceedings 145: 239-261.

28. EALEY, E. H. M., and A. R. MAIN
1967 Ecology of the euro *Macropus robustus* (Gould), in north-western Australia. III: Seasonal changes in nutrition. CSIRO. Australia Wildlife Research 12:53-65.

29. Frith, H. J.
1964   Mobility of the red kangaroo, *Megaleia rufa*. Csiro. Australia Wildlife Research 9:1-19.
30. Frith, H. J., and G. B. Sharman
1964   Breeding in wild populations of the red kangaroo, *Megaleia rufa*. Csiro. Australia Wildlife Research 9:86-114.
31. Caughley, G.
1964   Social organization and daily activity of the red kangaroo and the grey kangaroo. Journal of Mammalogy 45:429-436.
Caughley, G.
1964   Density and dispersion of two species of kangaroo in relation to habitat. Australian Journal of Zoology 12:238-249.
32. Griffiths, M., and R. Barker
1966   The plants eaten by sheep and by kangaroos grazing together in a paddock in south-western Queensland. Csiro. Australia Wildlife Research 11:145-167.
33. Newsome, A. E.
1965   The abundance of red kangaroos, *Megaleia rufa* (Desmarest), in central Australia. Australian Journal of Zoology 13:269-287.
Newsome, A. E.
1965   The distribution of red kangaroos, *Megaleia rufa* (Desmarest), about sources of persistent food and water in central Australia. Australian Journal of Zoology 13:289-299.
Newsome, A. E.
1965   Reproduction in natural populations of the red kangaroo, *Megaleia rufa* (Desmarest), in central Australia. Australian Journal of Zoology 13:735-759.
34. Newsome, A. E., D. R. Stephens, and A. K. Shipway
1967   Effect of a long drought on the abundance of red kangaroos in central Australia. Csiro. Australia Wildlife Research 12:1-8.
35. MacMillen, R. E., and A. K. Lee
1967   Australian desert mice; independence of exogenous water. Science 158:383-385.
36. Kinnear, J. E., K. G. Purohit, and A. R. Main
1968   The ability of the tammar wallaby (*Macropus eugenii*, Marsupialia) to drink sea water. Comparative Biochemistry and Physiology 25:761-782.
37. Bradshaw, S. D., and A. R. Main
1968   Behavioral attitudes and regulation of temperature in *Amphibolurus* lizards. Journal of Zoology 154:193-218.
38. Warburg, M. R.
1966   On the water economy of several Australian geckos, agamids, and skinks. Copeia 2:230-235.
39. Warburg, M. R.
1965   Studies on the environmental physiology of some Australian lizards from arid and semi-arid habitats. Australian Journal of Zoology 13:563-575.
40. Bradshaw, S. D., and V. H. Shoemaker
1967   Aspects of water and electrolyte changes in a field population of *Amphibolurus* lizards. Comparative Biochemistry and Physiology 20:855-865.
41. Main, A. R.
1968   Ecology, systematics, and evolution of Australian frogs. *In* Advances in Ecological Research. Academic Press, London.

42. Lee, A. K., and E. H. Mercer
1967   Cocoon surrounding desert dwelling frogs. Science 157:87-88.
43. Macfarlane, W. V., R. J. H. Morris, and B. Howard
1958   Heat and water in tropical merino sheep. Australian Journal of Agricultural Research 9:217-228.
44. Macfarlane, W. V.
1962   Adaptation of merinos to the arid tropics. *In* A. Barnard, ed., The Simple Fleece. pp 106-121. Melbourne University Press. Melbourne.
Macfarlane, W. V.
1964   Merino sheep as desert animals. *In* Environmental physiology and psychology in arid conditions, Proceedings of the Tucknow Symposium. Unesco, Paris. Arid Zone Research 24:259-265.
45. Macfarlane, W. V., C. H. S. Dolling, and B. Howard
1966   Distribution and turnover of water in merino sheep selected for high wool production. Australian Journal of Agricultural Research 17:491-502.
Macfarlane, W. V., B. Howard, and R. J. H. Morris
1966   Water metabolism of merino sheep during summer. Australian Journal of Agricultural Research 17:219-225.
Macfarlane, W. V., R. J. H. Morris, and B. Howard
1956   Water economy of tropical merino sheep. Nature 178:304-305.
Macfarlane, W. V. *et al.*
1959   Extracellular fluid distribution in tropical merino sheep. Australian Journal of Agricultural Research 10:269-286.
Macfarlane, W. V. *et al.*
1961   Water and electrolyte changes in tropical merino sheep exposed to dehydration during summer. Australian Journal of Agricultural Research 12:889-912.
46. Pierce, A. W.
1957   Studies on salt tolerance of sheep. I: The tolerance of sheep for sodium chloride in the drinking water. Australian Journal of Agricultural Research 8:711-722.
Pierce, A. W.
1958   Studies on salt tolerance of sheep. II: The tolerance of sheep for mixtures of sodium chloride and magnesium chloride in drinking water. Australian Journal of Agricultural Research 10:725-735.
Pierce, A. W.
1960   Studies on salt tolerance of sheep. III: The tolerance of sheep for mixtures of sodium chloride and sodium sulphate in the drinking water. Austhan Journal of Agricultural Research 11:548-556.
Pierce, A. W.
1962   Studies on salt tolerance of sheep. IV: The tolerance of sheep for mixture of sodium chloride and calcium chloride in the drinkwater. Australian Journal of Agricultural Research 13:486-479.
Pierce, A. W.
1963   Studies on salt tolerance of sheep. V: The tolerance of sheep for mixtures of sodium

chloride, sodium carbonate and sodium bi-carbonate in drinking water. Australian Journal of Agricultural Research 14:815-823.

PIERCE, A. W.
1966   Studies on salt tolerance of sheep. VI: The tolerance of wethers in pens for drinking waters of the types obtained from under-ground sources in Australia. Australian Journal of Agricultural Research 17:209-218.

47. WILSON, A. D.
1966   The tolerance of sheep to sodium chloride in food or drinking water. Australian Jour-nal of Agricultural Research 17:503-514.

48. WILSON, A. D.
1966   The value of *Atriplex* (saltbush) and *Kochia* (bluebush) species as food for sheep. Aus-tralian Journal of Agricultural Research 17:147-153.

WILSON, A. D.
1966   The intake and excretion of sodium by sheep fed on species of *Atriplex* (saltbush) and *Kochia* (bluebush). Australian Journal of Agricultural Research 17:155-163.

49. TURNER, H. G.
1958   Factors affecting heat tolerance of cattle. *In* Climatology and microclimatology, Proceedings of the Canberra Symposium. Unesco, Paris. Arid Zone Research 11:243-246.

TURNER, H. G., and A. V. SCHLEGER
1960   The significance of coat type in cattle. Aus-tralian Journal of Agricultural Research 11:645-663.

SCHLEGER, A. V., and H. G. TURNER
1965   Sweating rates of cattle in the field and their reaction of diurnal and seasonal changes. Australian Journal of Agricultural Research 16:92-106.

DOWLING, D. F.
1958   The significance of sweating in heat toler-ance of cattle. Australian Journal of Agri-cultural Research 9:579-586.

ALLEN, T. E.
1962   Responses of Zebu, Jersey, and Zebu × Jer-sey crossbred heifers to rising temperature, with particular reference to sweating. Aus-tralian Journal of Agricultural Research 13:165-179.

ALLEN, T. E., Y. S. PAN, and R. H. HAYMAN
1963   The effect of feeding on evaporative heat loss and body temperature in Zebu and Jer-sey heifers. Australian Journal of Agri-cultural Research 14:580-593.

50. LEIGH, J. H., and W. E. MULHAM
1964   Dietary preferences of sheep in two semi-arid ecosystems. Australian Society of Animal Production, Proceedings 4:251-255.

LEIGH, J. H., and W. E. MULHAM
1966   Selection of diet by sheep grazing semi-arid pastures on the Riverine Plain. I: A bladder saltbush (*Atriplex vesicaria*)—cotton-bush (*Kochia aphylla*) community. Australian Journal of Experimental Agriculture and Animal Husbandry 6:460-467.

LEIGH, J. H., and W. E. MULHAM
1966   Selection of diet by sheep grazing semi-arid pastures on the Riverine Plain. II: A cotton-

bush (*Kochia aphylla*)—grassland (*Stipa variabilis*—*Danthonia caespitosa*) com-munity. Australian Journal of Experimental Agriculture and Animal Husbandry 6:468-474.

LEIGH, J. H., and W. E. MULHAM
1967   Selection of diet by sheep grazing semi-arid pastures on the Riverine Plain. III: A blad-der saltbush (*Atriplex vesicaria*—pigface) (*Disphyma australe*) community. Australian Journal of Experimental Agriculture and Animal Husbandry 7:421-425.

ROBARDS, G. E., J. H. LEIGH, and W. E. MULHAM
1967   Selection of diet by sheep grazing semi-arid pastures on the Riverine Plain. IV: A grass-land (*Danthonia caespitosa*) community. Australian Journal of Experimental Agri-culture and Animal Husbandry 7:426-433.

51. CHIPPENDALE, G. M.
1962   Botanical examination of kangaroo stomach contents and cattle rumen contents. Aus-tralian Journal of Science 25:21-22.

52. REIS, P. J., and P. G. SCHINCKEL
1964   The growth and composition of wool. II: The effect of casein, gelatin, and sulphur-con-taining amino-acids given per abomasum. Australian Journal of Biological Sciences 17:532-547.

HOGAN, J. P., and R. H. WESTON
1967   The digestion of two diets of differing pro-tein content but with similar capacities to sustain wool growth. Australian Journal of Agricultural Research 18:973-981.

FERGUSON, K. A., J. A. HEMSLEY, and P. J. REIS
1967   Nutrition and wool growth: the effect of protecting dietary protein from microbial degradation in the rumen. Australian Jour-nal of Science 30:215-217.

McDONALD, I. W.
1968   Nutritional aspects of protein metabolism in ruminants. Australian Veterinary Jour-nal 44:145-150.

53. DONALDSON, L. E.
1962   Some observations on the fertility of beef cattle in north Queensland. Australian Veterinary Journal 38:447-454.

54. ARMSTRONG, J. *et al.*
1968   Preliminary observations on the produc-tivity of female cattle in the Kimberley region of north-western Australia. Aus-tralian Veterinary Journal 44:357-363.

55. ROSE, A. C.
1966   Herd management and disease control in north Australia. Australian Veterinary Journal 42:91-95.

56. ROSE, A. C.
1954   Osteomalacia in the Northern Territory. Australian Veterinary Journal 30:172-177.

BARNES, J. E., and R. B. JEPHCOTT
1955   Phosphorus deficiency in cattle in the Northern Territory and its control. Aus-tralian Veterinary Journal 31:302-316.

57. CLARK, L. R.
1949   Behavior of swarm hoppers of the Aus-tralian plague locust *Choitoicetes termin-ifera* (Walker). Council for Scientific and Industrial Research, Australia, Bulletin 245.

CLARK, L. R.
1953   An analysis of the outbreaks of the Aus-tralian plague locust, *Choitoicetes termin-*

*ifera* (Walker) during the seasons 1940-41 to 1944-45. Australian Journal of Zoology 1:70-101.

58. GRAHAM, N. P. H., J. H. RICHES, and I. L. JOHNSTONE
1941 Fly strike investigations — experimental studies of the surgical removal of breech folds in lambs, the Mules' operation at "marking time." Council for Scientific and Industrial Research, Australia, Journal 14:233-241.

ANONYMOUS
1944 Recent advances in the prevention and treatment of blowfly strike in sheep. Council for Scientific and Industrial Research, Australia, Bulletin 174:1-20.

GRAHAM, N. P. H., and I. L. JOHNSTONE
1947 Studies on fly strike in merino sheep. VIII: A surgical operation for the control of tail strike. Australian Veterinary Journal 23: 59-65.

59. The authors wish to acknowledge the help of a number of their colleagues in providing information on which this paper is based. Particular thanks are due to T. G. Chapman (CSIRO Division of Land Research), J. H. Calaby (CSIRO Division of Wildlife Research), I. W. McDonald (CSIRO Division of Animal Physiology), K. R. Norris and D. P. Clark (CSIRO Division of Entomology), and E. J. Waring (Bureau of Agricultural Economics).

# VARIABILITY IN HUMAN RESPONSE TO ARID ENVIRONMENTS

DOUGLAS H. K. LEE
Associate Director, National Institute of Environmental Health Sciences*
National Institutes of Health
U. S. Department of Health, Education, and Welfare

*Research Triangle Park, North Carolina.

# ABSTRACT

Historically, deserts have often been used to great advantage, particularly by organized groups; but tragedy has frequently intervened, particularly for small or troubled bands. Individual variability is a critical factor.

The dominant environmental factors impinging on man are aridity, clear skies, high reflectivity, high air and ground temperatures, sparse vegetation, elusive fauna, infective agents, and isolation. Their effects on man include a high heat load, dehydration, increased incidence of malignant skin conditions, malnutrition, and psychological strains. The net effect of the total heat load can be estimated.

Those human characters which show continuous as opposed to discrete variation are often susceptible to environmental influence. At low levels of applied stress the variability is usually symmetrically distributed over a fairly wide range. As the stress increases, the population tends to split into two groups, the majority falling into a subgroup with decreasing variability and a minority constituting a long "tail" into the higher reactions.

Variation of genetic origin generally shows poor correlation with prevailing climatic conditions: dark skin pigmentation gives protection against the effects of ultraviolet radiation but increases the absorption of solar heat; sweat gland density varies little with environment or race; the significance of body form is doubtful.

Variation in cultural patterns may markedly affect reactions to environment. In many areas cultural mixing is reducing the importance of cultural patterns. Individual experience, on the other hand, produces considerable variability. Acclimatization and habituation signify improved adjustment, but aggravation may occur instead when the stress is too great or too frequent. Mental attitude toward the environment is an important factor in success or failure of adjustment.

Changing physiological states, such as the level of activity, disease, dehydration, age, and the phase of circadian rhythm can markedly affect adjustment to environmental stress.

Knowledge of the nature and causes of variability can be put to use in development of arid lands, particularly in such matters as selection of personnel, provision of motivation, house design, clothing, living patterns, consumption of alcohol and food, hydration, communications, and economic matters.

# VARIABILITY IN HUMAN RESPONSE TO ARID ENVIRONMENTS
## Douglas H. K. Lee

The ancient Greeks, Herodotus tells us (1), believed that the habitable earth (oikumene) was surrounded by deserts, peopled precariously, if at all, by strange and therefore inferior types, perhaps adapted like the burnt-faced Ethiopians to solar scorch, but certainly not to be compared with the favored ones of Aegean culture. It was a bitter blow when the unadapted barbarians from the northern and eastern deserts overwhelmed the apotheosis of civilization, underscoring the often forgotten and paradoxical superiority of some unfit.

That fitness and cultural superiority are but relative matters was, ironically enough, being demonstrated at the very time that the concept of oikumene prevailed. If Murdock's (2) analysis is correct, the semiarid, and perhaps the arid, zone of Africa has several times since the fifth millenium B.C. supported the diffusion of revolutionary agricultural developments. Many times since then the hordes of conquerors like Darius, Alexander, and Genghis Khan have proved that organized, experienced, and adequately motivated armies could not merely cross, but actually live if necessary, in all but the most inhospitable reaches of arid land. Conversely, those on the defensive have frequently put ruggedness of terrain to good use, as is well illustrated by the cyclic history of the Tibesti (3), surging forth when times are opportune to harass the plains folk, retreating to the Saharan fastnesses when pickings are thin. To use a trite phrase, the desert is not everyone's cup of tea, but some have made it so.

The desert does not yield to the timid, the vacillating, or the ignorant. Between abhorrence and conquest of the desert lies a middle zone fraught with tragedy. War is only one of the four horsemen; famine and pestilence are just as devastating and share in the destruction of many an erstwhile flourishing civilization. As one flies over parts of the Arabian peninsula and sees the remnants of repeated attempts at civilization, one wonders how much the cycle of habitation–conservation–malarialization–depopulation contributed to the downfall of Sumerian and Mesopotamian cities as well as that of desert outposts. Affluent carelessness or biological ignorance can humble the best of intentions – a lesson that we in our technologically euphoric era are coming to realize is still valid.

With our modern techniques we can literally conquer mountains and make the desert bloom – at a price, and with side-effects that are not entirely predictable. We still have areas of ignorance, but we are not above being complacent. The ubiquity today of environmental pollution, as of youthful intransigence, serve to remind us that surprises are still in store. Particularly cogent is the fact that in both cases individualism is a dominant feature, the habit in the one and the right in the other of the individual to behave as though he alone occupies the space around him. Absurd though these attitudes may appear, they emphasize the fact that people are individuals, and they do vary. The variability must be recognized, the causes ascertained, and the consequences assessed. Nowhere is human variability more important, as history plainly reveals, than in the effective occupation of arid lands. But before we consider the types of variability and their effects, let us summarize what it is that man faces in the arid regions.

## DOMINANT ENVIRONMENTAL FACTORS

Aridity is an inescapable fact, but it is relative rather than absolute. There is moisture in the soil, even in the dry seasons, but at depth. Similarly there is moisture in the air. In absolute quantities, the common vapor pressure of 10 mm Hg on a summer day is equivalent to saturation at 12°C (53°F), or twice as much as is present over a freezing stream in winter; but it represents only 10 per cent of what air at 49°C (120°F) could hold (4). If the air were cooled below 12°C, dew would form. It is not uncommon for vapor from nearby water surfaces, such as the Red Sea, Persian Gulf, and even the Gulf of California, to raise the vapor pressure to 20 or 25 mm Hg, permitting dew formation at 22 or 26°C (72 or 78°F). Free water occurs, however, only sporadically after rain, as seepage, or as outflow from wells or bores. Man, however, may need water – potable water – in amounts up to 12 liters (3 gallons) per day, or more under unusual circumstances, merely to replace sweat and maintain urinary flow. It should be noted incidentally, that the higher humidities mentioned above in no way reduce man's need to evaporate sweat, they simply make it harder and consequently make conditions appear hotter.

Clear skies permit maximum transmission of solar radiation, which may add a heat load of 500 kcal/hr (2,000 BTU/hr) to a standing man at midmorning or midafternoon. This is equivalent to the heat produced by a slow run, or to that added by a rise of 7-10°C (13-18°F) in air temperature (5). (The foreshortening of the profile presented to the high-angle sun reduces the direct solar load at noon.) The clear skies, however, also facilitate long infrared radiation to outer space, so that, by contrast, the same standing man may lose 250 kcal/hr (1000 BTU/hr) to the sky on a winter's night (4).

Clear skies also permit the transmission of a high proportion of the ultraviolet component of solar radiation, and the diffuse scattering of a still larger proportion. The radiation received by an exposed person is a combination of that coming directly from the sun and that scattered by the air particles. Calculations of the comparative incidence for different angles of the sun and altitudes, reported by Buettner (6), are given in Table 1.

Desert terrain has a fairly high *reflectivity* for solar radiation (7). Sandy loam reflects 18 to 28 per cent, and salt flats up to 42 per cent of the incident radiation. A gravel surface has a somewhat lower reflectivity because of the shadowing effect of its irregularities. There is little vegetation to screen an exposed person from the reflected solar radiation. In the Arizona desert we found (5) that reflection added 150 kcal/hr (600 BTU/hr) at midday. The reflectivity of terrain for ultraviolet is usually lower than that for the visible portion of the spectrum, as indicated in Table 2 giving data reported by Buettner (6).

*Air temperatures* in summer frequently rise above 45°, C (113°F), and occasionally above 55°C (132°F). At these temperatures, air movement increases the addition of heat to the body by conduction-convection and increases the need for compensatory cooling by evaporation of sweat.

*Ground temperatures* can go much higher, commonly to the 63°C (145°F) observed at midafternoon in Arizona and to 84°C (183°F) under exceptional conditions.

The paucity of *vegetation* reduces natural resources of food and timber to a minimum. Xerophytic forms are of little use beyond lending relief and authenticity to travel photographs. The outburst of flora following winter rains, and the vegetation of oases, are striking by their very contrast, but provide nutriment only in limited quantities. It should be noted, in passing, that water vapor evaporated from irrigation or seepage is barely detectable a hundred meters or so down wind.

*Animal life* is more plentiful than the daylight hours would suggest, since the inhabitants live under rather than on the desert and forage only at night. The following quotation from Lowe (8) puts this resource in perspective:

Reptiles, birds and mammals occur throughout the interiors of all of the deserts and constitute the principal vertebrate foods eaten by all aborigines and by others. Poisonous snakes and skunks are not neglected. . . .

In many of the references cited in this section for the Namib-Kalahari and Australian Deserts, it is stressed that a European would promptly starve to death in these desert areas where the aborigines keep themselves fed and animals appear to be nonexistent. This statement is also (and especially) true for the Sahara, and to some degree for every desert of the world. A nomadic Bedouin and his family . . . [were] well supplied daily with fresh meat (mainly reptiles and small mammals) *obtained by hand* from this essentially vegetationless sea of sand. I have had no greater respect for any man, for, in this habitat, this is an impossible feat for a modern American or European, and even when armed with gun and shovel it is essentially impossible.

*Infective agents,* by and large, find the desert as inhospitable as do higher forms, but special circumstances may provide a suitable environment or mode of dissemination. Irrigated areas, for example, may harbor pond snails suitable as intermediate hosts for schistosomes; wind may spread desiccation-resistant agents, as probably occurs with coccidia (8) in the San Joaquin Valley; or social factors, such as poor hygiene (9, 10), may conspire with flies to spread trachoma.

TABLE 1

### VARIATION OF INCIDENCE IN ULTRAVIOLET RADIATION WITH SUN ANGLE AND ALTITUDE (ARBITRARY UNITS)*

| Altitude | Sun angle above horizontal | | | | |
|---|---|---|---|---|---|
| | 20° | 30° | 40° | 50° | 60° |
| Scattered from sky | | | | | |
| Sea level | 1.4 | 3.0 | 4.8 | 6.0 | 7.0 |
| 3,300 meters | 1.4 | 3.1 | 4.8 | 6.2 | 8.0 |
| Direct from sun | | | | | |
| Sea level | 0.5 | 1.7 | 3.5 | 5.3 | 7.0 |
| 3,300 meters | 1.6 | 4.1 | 7.0 | 10.0 | 13.0 |
| Total | | | | | |
| Sea level | 1.9 | 4.7 | 8.3 | 11.3 | 14.0 |
| 3,300 meters | 3.0 | 7.2 | 11.8 | 16.2 | 21.0 |
| Total, horizontal surface | | | | | |
| Sea level | 1.6 | 3.7 | 7.0 | 10.0 | 13.0 |
| 3,300 meters | 1.8 | 5.0 | 9.3 | 14.0 | 18.0 |

*On a plane normal to the solar beam, except for last two lines [After Buettner (6)].

TABLE 2

### DIFFERENTIAL REFLECTIVITY OF DIFFERENT SURFACES TOWARD TOTAL SOLAR RADIATION AND ULTRAVIOLET COMPONENT*

| Surface | Per Cent Reflectivity | |
|---|---|---|
| | Total Solar | Ultraviolet |
| Fresh snow | 89 | 85 |
| Bright dry dune sand | 37 | 17 |
| Bright wet sand | 24 | 9 |
| Sandy grass area | 17 | 2.5 |
| Water above sand | 12 | 10 |
| Water | 9 | 0–5 |

*After Buettner (6).

It has been said that "the civilization of a people can be assessed by the number of its trachoma patients," and that "trachoma recedes before the advance of civilization." These statements are probably correct if civilization means sanitary education, a higher standard of living, decent housing conditions, and, above all, an abundance of clean water and facilities for treating all classes of the population. (11)

The role of malaria in the degradation of irrigated communities has already been mentioned, but differs only in its dramatic setting from the much more widespread ravages of the disease in the lowland tropics. In short, disease is much more attributable to the presence of adventitious circumstances than to the relatively sterile arid conditions themselves.

The term "desert," in its etymology, stresses the fact of *isolation*. Isolation, however, is relative and hard to measure. It could be argued that the increasing network of communications and technology of transportation are rapidly reducing true isolation to a sometime thing. On the other hand, through the difficulties they present, desert environments still impede these developments and keep the ease of communication *relative* to that in more amenable areas not very different from what it was.

To sum up, man in arid regions faces environmental burdens in the form of shortage of water, high solar radiation by day but radiative cooling at night, high reflectivity of terrain, high air temperatures on summer afternoons, still higher ground temperatures, paucity of vegetation, elusive animal resources, and isolation—a formidable array! We need now to review the principal effects that these environmental factors, singly and in conjunction, have upon man's physiology and well-being.

## EFFECTS ON MAN

Physical activity, solar radiation and high air temperatures combine to impose a marked *heat load* that can be met only by the evaporation of water from the skin and, to a minor extent, from the respiratory passages. A healthy, acclimatized man can meet the load imposed by all but the most unusual natural conditions, but only by bringing his full physiological resources to bear. His reserve is often reduced to a slim margin, which is easily exceeded if he is provoked to excessive activity, encounters unusual humidity levels, or experiences any loss of physiological efficiency.

To supply up to two liters of sweat per hour to meet peak loads, the blood circulation through the skin must be maintained at high levels, and the lost fluid, of course, must be replaced by an equivalent water intake. The amount of salt lost in the sweat by an unacclimatized individual may exceed the daily intake by seven grams or more, with a chain of effects running from further cardiovascular inadequacy to violent muscular cramps (4). Fortunately, part of the acclimatization process involves a reduction in the salt concentration of sweat, so that the risk of a deficiency is minimized.

As might be expected, the cardiovascular system does not always meet all of the high demands put upon it, so that local inadequacies of blood flow occur. Because of its anti-gravitational position and its sensitivity to oxygen lack, the brain provides the most dramatic and most frequent demonstrations of relative circulatory inadequacies. Heat exhaustion (4), in its various degrees from headache and dizziness to fainting, is simply the result of cerebral anoxia. On one occasion at least the author experienced something similar to an anoxic pilot's euphoria, while still having his feet planted on the floor of Death Valley. Other areas of the body such as the alimentary canal and renal tissue may experience deficiencies as blood is shunted to the skin or as the total blood volume drops.

Because no one of the various responses that man makes to hot conditions is a good and sufficient measure of his total state of strain, and because there is no logical way of quantitatively compounding the various measurable responses, several attempts have been made to arrive at an index of net strain from calculations of the imposed heat stress. The system advocated by the author and his colleagues (12) is developed out of the basic concepts formulated by Belding and Hatch (13). It uses a *Relative Strain Index*, which is defined as the ratio between: (a) the rate of evaporation needed to compensate for the environmental interference with heat balance and (b) the maximum rate at which evaporation could take place from a completely wet skin under these environmental conditions. The calculation takes into account the metabolic rate of the individual and the resistance of his clothing to the transfer of heat and water vapor, as well as the environmental factors of radiant heat, air temperature, air humidity, and air movement. Charts are provided for interpreting the significance of the ratio for people of different degrees of acclimatization, age, and health.

Figure 1 is a conventional psychrometric chart, with temperature and vapor pressure coordinates, on which are drawn lines indicating combinations of temperature and humidity resulting in equal degrees of relative strain. On this are superimposed some illustrative climagrams, indicating the degree of relative strain likely to be encountered by a person performing moderate work in the shade wearing light clothing, and exposed to gentle air movement at the location indicated. The significance of the degrees of relative strain for different types of people is indicated in Figure 2. In Figure 3 appears a map of Australia, prepared by Hounam (*14*), indicating the number of days in the year in which the index is likely to rise above 0.3 (a rather critical value for continued living) at 3 p.m. for men doing light work.

Any degree of *dehydration*, of course, will accentuate circulatory inadequacies. Urine volume is usually reduced, but renal excretory function will be little affected as long as increased reabsorption of water from the tubules can maintain blood volume. But some loss can be expected if the blood volume is reduced to the point that blood flow through the glomeruli is diminished; and a marked rise of body temperature, of course, may interfere with kidney as with many other cell activities. It is reasonable to expect that the risk of stone formation will be enhanced in a concentrated urine, and there is some evidence to support this (*15*). Any salt de-

ficiency that occurs will tend to let more water escape in the urine and add to the circulatory embarrassment. Severe dehydration (a loss equivalent to more than five per cent of the body weight) will result in marked reduction of blood flow and threaten a complex cumulative breakdown of cell function and death, unless relieved.

The *skin*, as the interface between man and his environment (*16*), receives many insults. Those particularly associated with arid conditions include desiccation from dry winds as well as any inadequacy of blood flow, and abrasion from blown sand and dust, splintered dry wood, and rough surfaces. Such infective agents as are present in spite of the aridity, heat, and ultraviolet radiation will of course have an opportunity to invade insulted tissues. Poor hygiene greatly increases the risk, as is well seen in the case of trachoma (*9-11*). The clearest

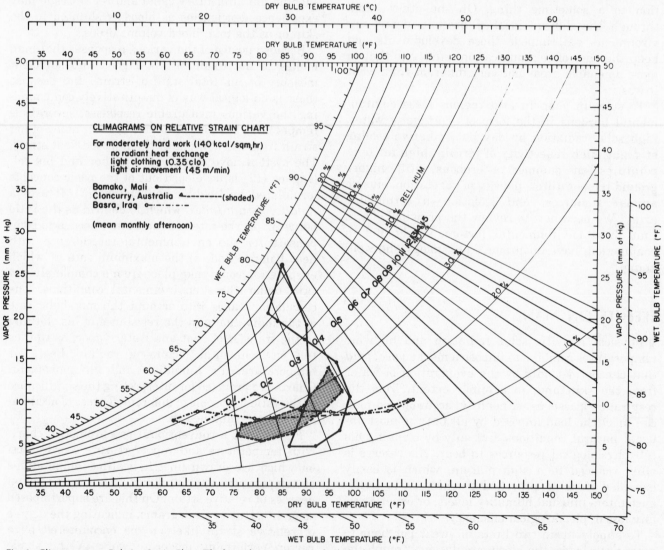

Fig. 1. Climagrams on Relative Strain Chart. The base diagram is a standard psychrometric chart relating the four variables of dry bulb temperature, wet bulb temperature, relative humidity, and vapor pressure, which are interdependent, so that if the values of any two are known the corresponding values of the other two can be read. On this is imposed a set of relative strain lines, (significance indicated in Figure 2), permitting an estimate of the strain imposed on the body by any combination of temperature and humidity. On this again are superimposed climagrams for three locations: *Bamako, Mali*—encompassing mean monthly values for 1:00 p.m.; *Cloncurry, Australia*—encompassing mean monthly values for 3:00 p.m.; *Basra, Iraq*—encompassing mean monthly maximum values (approximately equivalent to values for 3:00 p.m.).

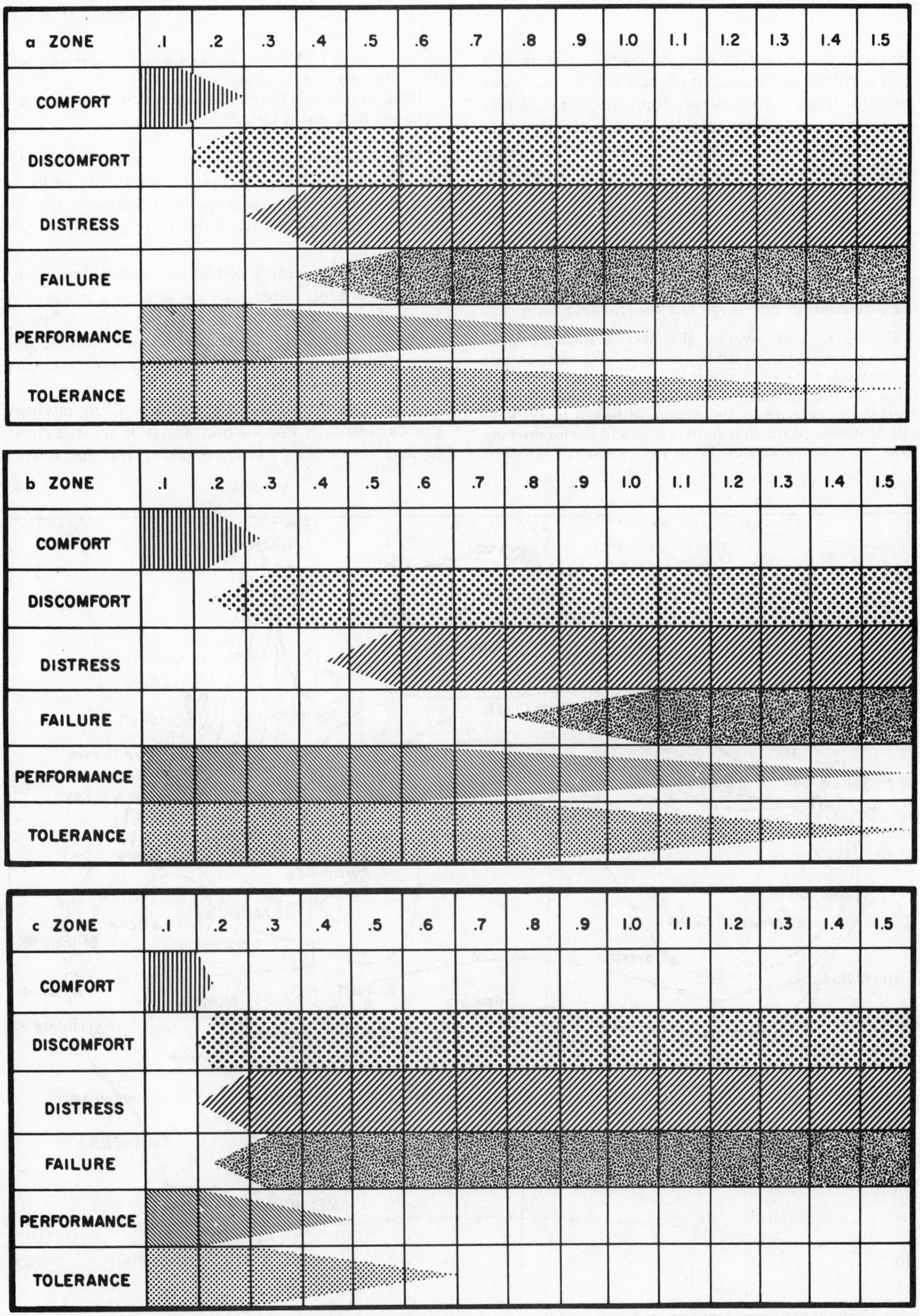

Fig. 2. Significance of relative strain values for (a) unacclimatized young man, (b) acclimatized young man, and (c) unacclimatized elderly man. The lower boundary of each zone corresponds to the strain value shown. [After Lee and Henschel (12)].

threat to the skin, however, comes from the ubiquitous ultraviolet component of solar radiation. Horny growths (hyperkeratoses) are common in chronically exposed persons, particularly those with poor tanning reactions. These frequently develop the malignant changes characteristic of rodent ulcer, and sometimes those of frank cancer (17, 18).

Poor socioeconomic conditions and isolation combine with minimal local resources to produce varying degrees of *malnutrition* in many arid indigenes. The following quotation from May (19) illustrates the situation in one large part of the arid zone.

The nutritional level of the Middle Eastern countries . . . , with the possible exception of Israel, cannot be considered satisfactory for the majority of the inhabitants. At the present time there is no marked general shortage of calories in the region, although inadequate caloric intake occurs in certain groups and the intake may show seasonal variations. Diets are, in general, deficient in proteins of high biological value and in certain essential vitamins and minerals. . . .

Clinical and subclinical disease due to malnutrition is common throughout the region.

Table 3, taken from the same source, illustrates the prevailing low per capita availability of foods, the high dependence on cereals and other vegetable sources, and the low availability of animal protein foods.

The primary causes of the bad nutritional situation have been listed (20) as poverty, pressure of population, land tenure systems, unsatisfactory hygiene, ignorance and bad dietary habits, and nomadic customs. Climate serves as a strong warp in this complex fabric.

It is evident that the range of prevailing environmental burdens must evoke fairly marked *psychological responses* in those exposed, but the nature

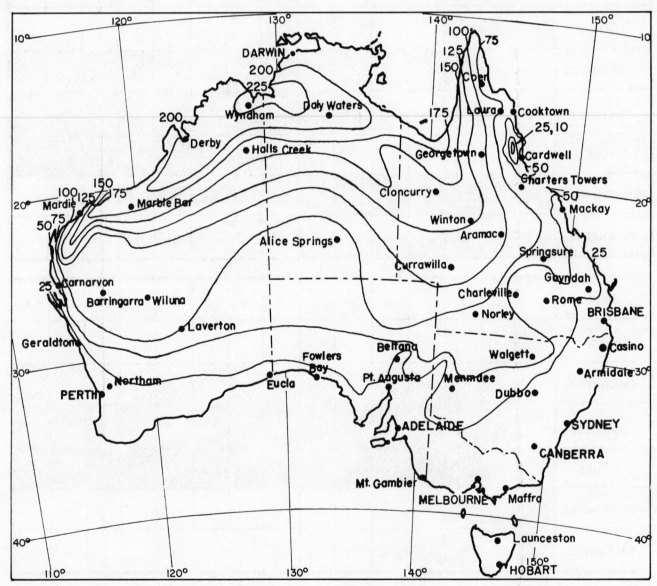

Fig. 3. Frequency of occurrence of Relative Strain value of 0.3 for a man performing light work, in Australia, at 3 p.m. [With permission of C. E. Hounam (*14*)].

of these reactions is more guessed at than studied. As with any set of stresses, there are individuals who adjust and those who do not (21). Individuality and persistence are valuable assets in adjustment, but the willingness to make the effort lies in the deep unfathomed caves of personal motivation. For those who succeed there is a sense of freedom and victory that binds them to the region. For those who do not adjust but are kept there by force of circumstance, the environment provides endless excuses for failure and outlets for complaint. There is nothing peculiar to arid regions in this; such differentiations can be found in many other stressful situations. There is no good reason to believe that any feature of arid environments induces psychological abnormalities in a previously healthy person; but there are many that can be blamed, and some that may evoke a previously latent condition. On the other hand, it is clear that the isolation of arid regions sometimes offers the illusion of escape to harried souls, and perhaps attracts some unadjusted individuals. Whatever the reason for immigration—predisposition, active adjustment, or flight—the resultant population will show certain differentiating characteristics relatable to the environment—an inevitable and on the whole a desirable outcome.

## NATURE OF VARIABILITY

Variability of human response is probably greater in respect of climatic than almost any other type of environmental stress. Variability may be seen at every level of the complex reacting system. In heat exchange there are differences in absorption of heat, particularly that of incident radiant energy, in efficiency of heat loss, and in rate of heat production. The intensity, efficiency, and persistence of compensatory reactions such as sweating vary over a wide range. Secondary effects of thermal compensation, such as cardiovascular insufficiency or even alkalo-

sis from increased breathing, range from the trivial to the incapacitating under environmental circumstances that are quite comparable. And beyond all of that, certain individuals seem to accept bodily conditions that others find intolerable.

The degree of variability in response to stress is well exemplified by Figure 4, developed by South African investigators (22). When the stress is mild, the responses are distributed fairly evenly around the mean, and the total range is moderate. As the stress increases, the responses tend to bunch together into a rather narrow range, except for a small proportion constituting a long tail to the high side. This is in accord with general experience that, within a healthy population, the variability in response is reduced as the stress increases, except for a small number of relatively unfit, who distinguish themselves by departing markedly from the rest: an experience popularly summarized as "sorting the men from the boys."

The causes of variability constitute an equally complex system of genetic, cultural, experiential, and phasic factors acting together in innumerable combinations which often defy analysis. For simplicity of description the more important factors will be considered separately, but the reader must remember that the causes of variation themselves act in variable combinations and variable fashion. A natural urge to simplify nature has produced much misunderstanding and not a few fallacies. That the separation of genetic from environmental causes of variation is a rhetorical device to be handled carefully is borne out by the following quotation from Boyce (23):

The characteristics of human beings which vary among populations can conveniently be divided into two types. First there are those characters which show smooth, continuous variation from one individual to another. They are the result of the interplay of many genes interacting with the environment during development. A large number of the morphological, physiological, and behavioral characteristics of human population show this kind of

TABLE 3

## SUPPLIES OF CERTAIN FOODS AVAILABLE FOR HUMAN CONSUMPTION IN SELECTED COUNTRIES
### (Kilograms per caput per year at the retail level)

| Country and Period | Cereals | Starchy Roots | Pulses | Sugar | Fats, oils | Fruit | Vegetables | Meat | Eggs | Fish | Milk* |
|---|---|---|---|---|---|---|---|---|---|---|---|
| Iran 1953-55† | 96 | 2 | .. | 14 | 8 | 40 | 55 | 6 | .. | 1 | 45 |
| Iraq 1953-55† | 119 | 3 | 13 | 23 | 5 | 58 | 55 | 10 | 2 | 1 | 83 |
| Israel 1954-55 | 141 | 43 | 10 | 24 | 18 | 100 | 114 | 18 | 15 | 13 | 104 |
| Lebanon 1953-55† | 126 | 17 | 18 | 20 | 7 | 136 | 22 | 15 | 4 | 1 | 80 |
| Turkey 1954/55† | 195 | 32 | 12 | 12 | 6 | 86 | 76 | 16 | 2 | 3 | 55 |
| United Arab Republic | | | | | | | | | | | |
| Egyptian Province 1955/56 | 186 | 6 | 12 | 18 | 4 | 66 | 65 | 13 | 1 | 6 | 59 |
| Syrian Province 1953-55† | 118 | 9 | 14 | 9 | 10 | 91 | 32 | 8 | 2 | 1 | 84 |

*Milk and milk products in terms of fresh milk.
†Provisional figures pending government approval.
Sources: Israeli data from: Statistical Abstract of Israel, 1958/59. Other data from: Joint F.A.O./W.H.O. Committee on Nutrition in the Near East. Report of First Meeting, 1958.
[From May, J. M., (19)]

variation. In contrast, there are those characters which show discrete rather than continuous variation and which are due to differences in a single pair of genes. Many biochemical and serological characters are of this kind. Unlike characters which show continuous variation, these characters are more or less independent of the environment. . . .

[M]any of the characters which vary among populations are continuously varying characters which . . . are often greatly affected during the lifetime of an individual by the environment. For example, differences in diet, or in the incidence of diseases in different populations, can bring about quite marked changes in morphology. Also, many physiological differences . . . are directly due to contrasts in environments. The interpretation of similarities and differences in continuously varying characters in genetic terms must therefore be made with caution.

In the following sections we will consider the extent to which each of the four types of causation contribute to the better known instances of variability in human reaction to arid conditions.

## VARIABILITY OF PREDOMINANTLY GENETIC ORIGIN

For a long time physiologists have been bothered by the paradox that, whereas black surfaces readily absorb heat, the dark-skinned races abound in the hottest parts of the globe. Several ingenious explanations have been offered to resolve the dilemma, but it is only recently that a fairly satisfactory hypothesis has been advanced—that the poleward spread of man from his equatorial origin put a premium on those whose less-pigmented skins permitted adequate amounts of vitamin D to be formed, in spite of reduced insolation in winter. A black skin is thermally disadvantageous in hot climates if the possessor is exposed to the sun (6, 24), but in his native state he seldom exposes himself and thus avoids the potential genetic burden.

Skin pigmentation plays an important role in protecting the deep germinative layer of the epidermis from carcinogenic stimulation by solar ultraviolet radiation. In Hawaii, for example, there is a great preponderance of skin cancer in the caucasian inhabitants (138/100,000) as compared with the more darkly pigmented ethnic groups (3.1/100,000) (25).

A search for racial variations in the density distribution of sweat glands (glands per unit area of skin) has been largely negative (26, 27). Differences in the rate at which sweat is produced by the glands are generally attributed to differences in the amount of sweat required for cooling, resulting from variations in heat production or in other aspects of thermolysis.

Aborigines of the central Australian plateau have long been renowned for their tolerance of cold (28-30). Once asleep with the aid of small fires, they permit their body temperatures to fall well below levels at which Caucasians wake up and shiver. This, however, could be an ingrained cultural rather than a genetic characteristic. So far no white families have volunteered their children as a control group.

Such studies as have been carried out on racial groups living under comparable conditions of diet, work, and clothing reveal only a low correlation between genetic background and tolerance of hot, dry conditions (31).

The effect of body form, whether genotypic or phenotypic, has been hotly disputed. Bergman's and Allen's rules have a sound theoretical base, but actual observations reveal numerous exceptions as well as confirmations. Baker (32), Schreider (33), and Newman (34) believe that the greater proportional surface area of the ectomorph permits easier heat loss, and that the distribution of this type is correlated with prevailing temperatures. The author, on the other hand, tends to join Wilber (35) and Scholander (36) in scepticism. Factors such as nutritional status, which influence body type, also influence other aspects of heat balance. Low nutrient levels, for example, tend to be more common in arid regions and thus to favor the appearance of ectomorphic trends. This argument, like so many others on human effects, will probably remain unresolved for lack of statistically unconfounded groups.

Congenital abnormalities, whether truly genetic or teratogenic, can, of course, have marked effect upon adaptation of an individual to hot, arid conditions. A congenital absence of sweat glands, for example, would necessitate artificial cooling whenever air temperatures rose much above 24°C (75°F). Ichthyosis would have a similar effect. Cardiovascular defects would interfere with heat exchange, and cystic fibrosis would increase the threat of salt depletion in hot weather as well as interfere with the bodily economy in general. But here, as in dealing with body form, we pass insensibly from genetic to intercurrent factors.

## VARIABILITY OF CULTURAL ORIGIN

All of us, to some extent, are prisoners of our culture; we hesitate to stretch our relations with the community too far. The few who do go to extremes often band together, setting up a subculture to which they then become slaves. Our mode of dress, housing, work patterns, and food habits largely conform to the customs of the group, even when they are palpably at odds with common sense. People from temperate and well-watered regions moving into hot and arid lands frequently continue practices that are quite unsuited to the new environment. An 8 a.m. to 5 p.m. workday in the summer sun is almost as absurd as the proverbial evening dress of the colonial. The high-pressure, time-conscious, up-and-at-it life to which many of us are accustomed originated in cool temperate climates, where activity saves on fuel, and accumulation of necessities against a long and adverse winter is a virtue. Imported to an area with hot summers and benign winters, this behavior pattern created physiological and psychological problems ameliorated only by the advent of air-conditioning. (The problems were admittedly greater in the humid tropics, but the mid-latitude arid areas felt them as well.)

A survey of arid areas today would disclose extensive variability of cultural origin. Indigenous cultures vie with imported, incomers vary all the way from conservative to liberal, cultural mixes change with economic and political events, and new conformal patterns are superimposed on old. Variations from place to place are giving way to variations within the place; as diversity within regions increases, that between regions is obscured.

Hygienic practices, so important in meeting any stressful conditions, are intimately related to prevailing cultures, and to the prevailing levels of education and socioeconomic status. Unfortunately, ignorance, poverty and conservatism form a triad that is easily reinforced by isolation. The result in some arid areas is a relatively uniform low level of hygiene, in contrast to the mixed but steadily improving practices of more densely settled regions. Paradoxically, the congregation of the poor and ignorant into local townships may serve to accentuate rather than ameliorate the isolation and misery. Fortunate indeed are those communities with a strong economic base and easy external communications which can combine the spaciousness of arid lands with the material benefits of modern technology.

## VARIABILITY FROM INDIVIDUAL EXPERIENCE

Within the prison of his culture each person is further restricted by the fetters of his individual experience. For each there is a resultant range of reactions to stress, shared to a certain extent with his fellows, but highly individualistic in at least the details. Statements of average reaction to a given stress, even for a narrowly defined group, are of limited value, often transcended by the actual response of any one member.

Probably the most marked and most frequent example of experience affecting reactions to arid conditions is the process of *acclimatization*. (The word is used here in the physiological sense of adjustment by the individual, a process to which ecologists often apply the term "acclimation.") Extensive and critical changes in reaction follow

repeated exposure to heat, quickly differentiating the older hand from the new arrival. Further, although more subtle, changes accrue with continued residence (7, 37).

Over the first five to ten days there may be a ten to twenty per cent increase in sweat rate, a fifteen per cent increase in blood volume to compensate for the dilated skin blood vessels, and a marked decrease in the concentration (and therefore the loss) of chlorides in the sweat. The heart rate, stimulated to excessive rise by almost any exertion, returns to more normal responses, and the body temperature shows progressively less indication of disturbed heat balance. It has been the author's experience in hot deserts that, somewhere about the tenth day, one experiences a feeling akin to the athlete's "second wind," and the conviction that from there on he "has it made." There is no doubt that acclimatization makes a new man. This rejuvenation is demonstrated not only by improved efficiency and reduced discomfort but also in the markedly diminished risk of clinical heat effects (Fig. 4). Heat stroke is seen in the acclimatized only

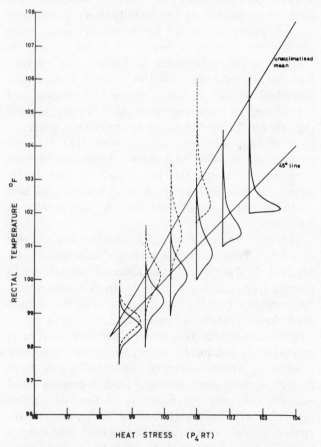

Fig. 4. Distribution of rectal temperature responses of men at various levels of heat stress. (Distribution curves should be read in plane vertical to the page) ($P_4RT$ is an index of heat stress commonly used by British investigators). Broken line indicates unacclimatized persons; solid line indicates acclimatized persons. (Reproduced with permission of Munro et al. (22) and Pergamon Press.)

rarely and under special circumstances of infection, dehydration, or emotional exertion; it is most common in the course of heat waves in normally more temperate climes (*38, 39*) and in persons suddenly introduced to and compelled to work under hot conditions (*40*). Heat cramps, associated with salt depletion, are likewise seldom seen in well acclimatized persons even when they are carrying out heavy work in the sun. Heat exhaustion, almost inevitable in some degree for newly exposed persons, is much less frequent after acclimatization, although it will still intervene as an escape mechanism when the body fluids or circulatory efficiency are reduced by indiscretions.

Acclimatization to cold also may be seen in arid regions, particularly in winter, but this relates more to comfort than to any real physiological threat. It should be noted that acclimatization to both heat and cold can co-exist to a large extent, provided that there is frequent exposure to both extremes. De-acclimatization is brought about more by lack of reinforcement than by intermittent subjection to the opposite stress.

*Physical fitness,* too often a vague and over-worked concept, assumes concrete and desirable form when defined as the ratio between the rate of oxygen uptake required for a person's customary job and the maximum rate of which he is capable. (This relative expression of fitness has certain features in common with the relative heat strain described earlier in this chapter.) Wyndham and his colleagues have confirmed previous suggestions that this relative fitness is an important determinant of heat tolerance during work (*41*). Persons who are overweight also show a reduction in heat tolerance at work, but how much of this interference is due to loss of physical fitness and how much to reduction of heat flow to the surface is not clear.

The later processes of acclimatization come closer to what Glaser (*42*) has termed "habituation"—a gradual diminution of responses or sensations to repeated or continued stimuli." In Glaser's words, "presumably habituation involves only the simple diversion of (nervous) pathways, so that a given stimulus reaching the central nervous system is increasingly conducted across synapses mediating inhibition, or increasingly conducted away from pathways mediating arousal." Such a process could account not only for damping of the high pulse rate of early heat exposure—a fairly useless response, since a mere increase in heart rate cannot increase the output of blood when the volume coming back from the body is deficient—but also for the progressive acceptance of the hot situation as "normal."

This brings us to *attitude* as a determinant of variation. It is defined (*43*) as "an organismic state

of readiness to act that is often accompanied by considerable affect and that may be activated by an appropriate stimulus into significant or meaningful behavior." Under stress approaching, but not exceeding, physical tolerance, attitude can have a marked effect on tolerance and performance, but in either direction. Attitude itself may change in response to environmental conditions that are long continued, or to variations in the physiological state of the individual. It may be consciously changed through reason or in response to social pressure. Whatever the end result of the complex forces operating on it, attitude largely determines the location of individual behavior in the spectrum of responses: surrender, avoidance, aggressive response, psychological adaptation, and constructive behavior. And, closing the loop, the behavior pattern once expressed becomes an influential determinant of still later attitudes. Recognition of this loop, and of the points at which it can be entered, provides an important means of influencing the psychological adaptation of the individual to conditions which, though stressful, are within his physiological tolerance.

Up to this point we have been assuming that continued or repeated exposure will result in better adaptation to conditions. Unfortunately, this is not always the case. *Aggravation* rather than acclimatization may prevail. Exposure to stress evokes two types of response: (1) displacement of function in the direction of the stress (for example, body temperature rises as a result of raised environmental temperature), and (2) compensatory reactions counteracting the stress (for example, sweating) or the effect of the stress (for example, increased blood volume restoring the circulation). The relative development of these two types of response at any moment determines how far from *normal* the individual will be. If, after an initial time lag, the compensatory responses start catching up on the displacements, adjustments will improve and we can speak of acclimatization. But if displacement goes faster and increases its lead over compensation, aggravation will ensue (*44*). Because the individual processes making up each type of response are so numerous and complex, and because each is itself subject to so many bodily conditions, the net result is highly variable. It may shift from a net acclimatization to a net aggravation if bodily conditions change (for example, with an intercurrent infection). It may at times show a mixed picture of acclimatization in one respect with failure in another. Certainly minor failures, such as *prickly heat,* can occur in otherwise well-adjusted persons. Response can quite easily pursue a biphasic course, an early worsening being followed by marked improvement. The whole process of adjustment to repeated climatic stress is a fascinating, dynamic, and highly individualistic phenomenon.

## VARIABLE STATES

The wide range of inter-individual variation of genetic, cultural, and experiential origins so far considered is further extended and complicated by an almost equal range of intra-individual variability. No one of us is the same from day to day, from morning to night, or even from moment to moment. We lead dynamic existences, we operate with extremely complex and delicate mechanisms, and we are subject to infinite vicissitudes. Some of the personal variability can be traced back to environmental stimuli, but much of it is bound up with our physiological constitution. In many instances it would be difficult to disentangle the two.

The most obvious and frequent personal variable, of course, is that of *activity*. The rate of heat production may vary from 70 kcal/hr at rest to 1,200 during short periods of intense effort. The overall daily rate may vary from 1,700 kcal or less in persons confined to bed to 5,000 in men performing heavy manual labor. In a bodily system operating close to its upper thermal limits, such changes in the rate of internal heat production through exercise have considerable significance. An important point under such circumstances is the efficiency with which objectives are obtained. Unnecessary movements, awkward postures, and superfluous operations cannot be freely indulged if the net stress upon the individual is to be kept in reasonable bounds. (The term *stress* is used by environmental physiologists in the physical sense of force applied to the body; the deformation of bodily function resulting from that application is properly termed *strain*.) Part of the ultimate acclimatization process is learning to accomplish tasks in the most efficient fashion, even if it means discarding cherished notions of proper behavior.

*Disease*, of course, can seriously disturb the thermal balance of the individual or his attempts at compensation. Febrile conditions, particularly those which are accompanied by inhibition of sweating, can have disastrous effects. Hyperthyroidism, by raising the basal metabolic rate, adds to the patient's difficulties. Cardiovascular insufficiency, whatever the cause, will interfere with transfer of heat to the skin and often with the sweating ability as well. Congestive cardiac failure may markedly reduce evaporation from the respiratory tract (45). Vomiting and diarrhea will increase fluid and chloride loss, making compensation for sweat loss even more difficult. Adrenal insufficiency will also impair water and electrolyte balance. Extensive skin affections may seriously interfere with sweating, as is often seen when repeated attacks of *prickly heat* result in blockage of sweat glands (anhidrosis) (46).

*Physiological disturbances* other than classical disease can intervene, with easily understood results. Inadequate water intake will aggravate the tendency to circulatory inadequacy and hasten the onset of heat exhaustion or worse. Diuretic substances may have similar but less dramatic effects. The effects of a beer-drinking session may be discernible for two or three days in men required to work hard in the sun. Inadequate salt replacement, most likely to occur in unacclimatized persons, working hard, in hot and dry environments, will accentuate any water deficiency and, if sufficiently severe, induce the characteristic heat cramps in limb and abdominal muscles. Heavy meals add to the demands on the blood circulation and increase the chances of inadequacy. Vasodilator drugs such as alcohol, as well as those taken for therapeutic purposes, can induce dramatic circulatory inadequacies under hot conditions.

The effect of *age* on heat tolerance is a matter of some dispute. In unacclimatized people, and probably in a cross section of the population at large, decreased heat tolerance is apt to develop after age forty (47). But in acclimatized persons, tolerance for heat may be much less affected than that for exercise (48). Wyndham (41) found no effect of age among his miners. Studies on elderly persons by Henschel et al (49) showed remarkably little loss of tolerance, but such persons may well owe their longevity in part to a better-than-average fitness. In periods of exceptionally hot weather, by contrast, the aged members of an exposed population show a disproportionately high incidence of and death from heat stroke (38, 39, 50, 51).

The importance of *circadian rhythm* is now well established (52). Numerous physiological processes show a regular fluctuation around the clock, and the susceptibility of the organism to environmental stresses varies likewise (53). A marked departure of environmental conditions from the normal rhythm is likely to find the exposed organism out of phase, and therefore unduly susceptible. A similar increase in susceptibility ensues if the organism is displaced in time, as readily happens in these days of high-speed flight. A person moving from the Arizona desert to a corresponding environment in Africa, for example, would be out of phase and unduly susceptible to stresses in general. During the days following such a shift, moreover, the various functions move to the new time pattern at different rates. As a result they are out of phase with one another, so that the organism may have a still greater increase in susceptibility to stress, or at least an increased difficulty in dealing with it.

## DEVELOPMENTAL IMPLICATIONS

The last half century has witnessed a radical change in attitude toward the potentiality of arid lands, largely based on the promise of developing technology. But there have been periods of euphoria before—does the present optimism signify a permanent transition, or is this just another oscillation in time? Must progress and decay follow in dreary succession, or can we really hope that man is now the master?

The major potential causes of failure are known—inadequate water, wide range of climatic conditions, disease, economic failure, and strife. The modes of control, or at least of compensation, are known. Man can stabilize the cycle of success and failure, but will he? How can plans for development of arid lands steer highly variable and individualistic man in the direction of continued success?

With increasing recognition of the importance of environmental stresses for human welfare there is a growing interest in urban and developmental planning in general. The developer, however, confronted with an infinitude of human variability, may feel that the only practical solution is a simplistic and somewhat arbitrary approach. He may be sorely tempted to set up a hypothetical median man, design for his needs, and hope that the well-advertised human capacity for adjustment will take care of the rest. The simplistic nature of stress assessment schemes (see "Effects on Man," p. 231) may encourage the trend. The idealists among the planners, on the other hand, may feel that not enough is known about the nature, causes, and control of human variability, and that more research is needed.

Antitheses, however, should be avoided. Greater understanding of man and his needs is certainly desirable, but enough is already known to furnish useful guides to action that cannot be postponed. From what has been said earlier, we can, right now, provide suggestions for:

(a) reducing the incidence of factors provoking adverse variations,

(b) capitalizing on those factors that favor desirable variations,

(c) maintaining flexibility in operating conditions to meet a fairly wide range of variations,

(d) detecting and counteracting adverse reactions at early stages.

some specific instances will illustrate the possibilities.

For new developments or special undertakings, *selection of personnel* is both desirable and practicable. Good health is, of course, important, and particularly freedom from dysfunctions of the cardiovascular, renal, or endocrine systems. A good familial history of health is important for those who are to become long-term residents. A history of adverse reactions to heat would certainly be an indication for non-selection, but a single test applied to an unacclimatized person would not provide good grounds for exclusion. For those who would be required to do heavy work in exposed situations, a high level of physical fitness is essential. Familiarity with the work to be performed is highly desirable. The practice still employed in some hot, dry areas of hiring outdoor labor on a daily basis is unrealistic as well as un-physiological.

Physical suitability, however, is not enough. For tasks that demand drive or perseverance the "gleam in the eye" is still a necessity. For long-term residence constructive attitudes, or at least willing acceptance of conditions, are good assets. Existing techniques of psychological evaluation could presumably be put to good use in selecting low-risk populations, although the author does not know of any instances in which they have been systematically employed. Trial and error still seems to be the prevalent mode.

For inhabitants once there, whether by birth or by recruitment, *motivation* is of considerable importance. With contemporary weakening of family ties and occupational traditions, and with increasing facilities for translocation, the dissatisfied are less and less likely to stay, or, if they stay, to work efficiently or adjust to local conditions. Since the whole range of human needs in involved, many aspects require attention—remuneration, opportunity, social amenities, housing, community spirit, and education facilities. While these apply to any community they have special implication in hot and dry regions, where physiological strains are apt to induce some reduction in initiative, increase in sensitivity, and relaxation in personal discipline.

In those communities where the dream of air-conditioning is becoming reality, attention to thermal protection in *house design* is of decreasing importance, but for the less fortunate it should still be given major consideration. Tables 4 and 5, published on more than one previous occasion, are repeated here; the physical and physiological bases are given in earlier publications (4). The principles can be observed and variations provided in the design of even low-cost housing (Fig. 5); money largely buys elegance and refinement in their application, and, of course, the possibility of installing air treatment.

Communities with long experience of arid conditions have developed appropriate *clothing*. Sophistication, however, is apt to disturb the adjustment with an overriding desire for change, and an urge to copy the admittedly more exciting designs suitable to warm humid conditions. Healthful principles, however, remain unchanged, as set out in Table 6, also taken from previous publications. "Clothing can provide a significant degree of protection against environmental heat gain under hot open-desert conditions. The problem of the designer is to maximize this protection and to avoid undoing protective effects by unnecessary interference with heat loss"

TABLE 4

## THERMAL PRINCIPLES OF HOUSING FOR HOT, DRY CONDITIONS*

| Objective | Principles | Application |
|---|---|---|
| Reduce heat production. | Minimize heat and vapor addition from cooking and other procedures. Functional convenience. | High operating efficiency in cooking and other heat-liberating devices. Isolation of "wild heat" by insulation or conductive-convective removal. |
| Reduce radiation gain and promote loss. | Minimize solar projection. Shade. High reflection of shorter wave lengths but emission of longer ones. Convection over surfaces heated by radiation. Insulation. | Labor saving layout and facilities. General design cubical. Trees, bushes suitably placed. Eaves, shades on sun-exposed walls. White paint on sun-exposed surfaces. Heat storage insulation in roof and sun-exposed walls. |
| Reduce conduction gain. | Insulation. Wind exclusion. Air cooling. | Permit air flow over sun-exposed surfaces, vents in hot-air traps. Wall openings closable on hot days, openable on cool nights. |
| Promote evaporation. | Wetting. Maintain sufficient convection over evaporating surface. | Evaporative cooling devices. |

*From Lee (4), with permission of the American Physiological Society.

TABLE 5

## APPLICABILITY OF DESIGN PRINCIPLES FOR HOT, DRY CONDITIONS TO ALTERNATING CONDITIONS*

| Item | Applicability to | |
|---|---|---|
| | Cool, Temperate Conditions | Warm, Humid Conditions |
| Reduced heat production by man | Not important | Still important |
| External shade | Not desirable | Still important |
| Reduced ground radiation | Not desirable | Still important |
| Attached shade | Minimal effect desired | Still important |
| Water cooling | Not desirable | Operative on roof |
| Minimal solar projection | Greater projection required | Still important |
| High reflectivity | Not desirable | Still important |
| High reemissivity | Not desirable | Immaterial |
| External convection | Not desirable | Still important |
| Insulation | Desirable against reverse heat flow | Still important in roof, immaterial in walls |
| Internal convection | Not desirable | Still desirable |
| Low internal emissivity | Desirable | Immaterial |
| Controlled ventilation | Closed | Wide open |
| Roof space ventilation | Closed | Open |
| Ground cooling | Now operates for heating | Nonoperative |
| Evaporative cooling | Not required | Relatively ineffective |
| Refrigerant cooling | Not required | For dehumidification |
| Reduction of heat liberation | Heat may be required | Still important |

*From Lee (4), with permission of the American Physiological Society.

TABLE 6

## THERMAL PRINCIPLES OF CLOTHING FOR HOT, DRY CONDITIONS*

| Objective | Principles | Application |
|---|---|---|
| Reduce heat production | Light weight Absence of restriction Functional convenience | Loose fit Functional design Extensive coverage Shade to sun areas |
| Reduce radiation gain | Shade Reflection Insulation | Critically distributed insulation Controllable ventilation Artificial wetting |
| Reduce conduction gain | Insulation Wind exclusion | |
| Promote evaporation | Maintain sufficient perflation Wetting | |

*From Lee (4), with permission of the American Physiological Society.

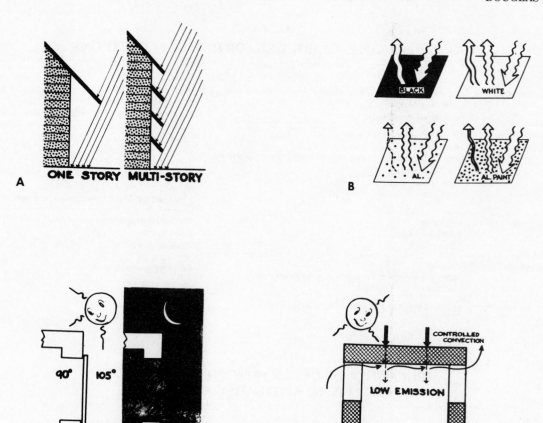

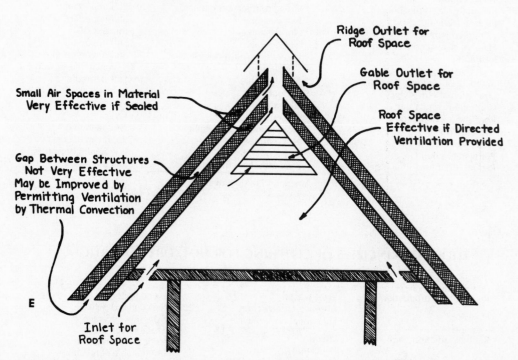

Fig. 5. Important principles of housing for hot dry climates. (A) Shade to wall from high-altitude sun by roof and wall projections. (B) Effect of surface on absorption of solar heat. Black reflects none, but has good emission of long infrared; white reflects visible and has equally good emission of long infrared; polished aluminum reflects visible and some short infrared but emits very little in long infrared; aluminum paint reflects less but emits rather better in long infrared. (C) Wall openings should be closed by day when outdoor air is hotter than indoor, but open by night when reverse gradient prevails. (D) Controlled ventilation over inner surface of heated structure very desirable. Controlled convection can be used to remove heat from inner surfaces of heated structure, without mixing into room space. Importance is increased and resultant protection is greater when inner surface has low emissivity for infrared. (E) Conventional roof space is effective in protecting room beneath when designed so that convection, free and forced, removes trapped heat. (From "Physiological Objectives in Hot Weather Housing," U. S. Housing and Home Finance Agency, 1953.)

(4)—or unnecessary increase of heat production through heavy or restrictive articles. Details of the thermal and physiological considerations involved are contained in the reference quoted; but greater emphasis could be placed on the need for protecting lightly pigmented skin from the adverse effects of ultraviolet radiation.

For every individual the guiding principle in *living patterns* is "adaptation without submission." The easiest way to balance the heat budget is to reduce activity, and this can very easily be made to justify minimum activity. But a positive attitude toward environmental problems is essential if the Greco-Roman disasters are not to be repeated. To build a house beneath a durian tree and wait for the fruit to fall, however helpful in acquiring edible durians, is hardly compatible with world pressures. Cultures almost everywhere are becoming more activist, and effective output cannot be greatly reduced with impunity. Other means of adaptation must be sought, such as increasing the efficiency of action patterns—an important mode of heat conservation. Most freely developed action patterns include movements not essential to the task and can be eliminated by rearrangement of components or by training of operators. The resistance to work analyses often encountered when the beneficiary is an employer should be much less when, as in hot environments, it is the operator who receives a direct and immediate benefit. Such major activity as is necessary, either for production or recreation, can often be planned for the cooler periods of the day, or under conditions where the environmental heat load is low. Advantage can be taken of radiative heat loss by exposing sleeping quarters as widely as possible to the night sky. Forward planning of daily activities can minimize stress and heat-producing frustration, while promoting morale and helping to maintain achievement.

A desirable ingredient of the living pattern that is usually and conveniently forgotten in these days of increasing environmental control is preservation of natural defense against the stresses of hot and arid conditions. Man's heritage of adaptability must not be dissipated. Alternation between an air-conditioned house and an air-conditioned office, in an air-conditioned car, is hardly conducive to continued acclimatization, or to facile handling of emergencies. The horse-and-buggy driver knew the limitations of his equipment; he carried water in the desert and told people where he was going. The modern motorist ignores the possibilities, does not know how to cope with desert heat, and runs the risk of dying if away from the beaten path. It is a pity that the story of SOS Sahara (54) is not better known.

Because *alcohol* puts an additional burden on the body by increasing heat production, promoting diuresis, and further dilating blood vessels, its con-

sumption during the hotter hours, or at times when cerebral function is critical, is contraindicated. *Heavy meals,* by increasing heat production and diverting blood from other areas, likewise are not to be advocated at times of peak heat.

But the adaptive measure that is most constantly necessary, and yet so often inadequately maintained, is *bodily hydration.* That immigrants to arid regions may be slow to learn the necessity of adequate water intake is indicated by the high incidence of urinary calculus reported among nonnative as compared with native Israelis (55). Some fluctuation in bodily hydration within the day is permissible, provided that the lack at no time exceeds two per cent of the body weight, and provided also that the deficit is made up within a few hours. Experienced workers often prefer to incur some dehydration during the first hour of work and to maintain a steady level thereafter until its completion, when sufficient fluid is taken to restore the balance. Such individuals may experience nausea if compelled to maintain body weight by drinking. A daily urine volume of one liter is a good index of adequate hydration.

*Communications* have greater than usual import in combatting any sense of isolation and in making the community a part of the general culture and economy. All media—radio, television, telephone, postal services, and other news media—share in this added importance and call for attention that may be out of proportion to the size or wealth of the group. Movement of people through the group, even as visitors but preferably as temporary participants, likewise fosters identification. The input by all of these channels needs to reflect the variability of the recipients, which we have been emphasizing.

The one parameter of arid zone development that inevitably rises is *cost.* The additional items or changes in design required for protective purposes, the generally higher unit cost of cooling as compared with heating, the provision of adequate water supply, the longer average transportation distance for supplies, the higher remuneration required to attract employees and their families, and the deterioration of many materials through heat and dryness, all contribute, although in proportions varying with circumstances, to a more than usually costly investment and recurrent expenditure. All of this means that there has to be some special economic incentive or strategic purpose to warrant the increased cost, which in turn means that, at least in this competitive world, high population densities are much less likely to occur in arid as compared with temperate and well-watered zones. The sentiments applied to viticulture by Guyot in 1865 (56) might hold for other highly desired commodities a century later:

The raw products of poor soils, which realize the highest prices, are best adapted for the development of any colony, because by the culture of such products the colonist is certain to ensure the cultivation of all the others. The capitalist, that powerful agent of progress, will not waste his means in furthering the cultivation of inferior products. He will soon discover that one acre of the Chateau-Lafitte or of the Clos-Vougeot, gives more wealth for the public than one hundred acres of poor wastes planted with a forest or turned into an ordinary farm. To speak more precisely, in poor soils the production of bread and meat will not create wealth, whereas wealth will always produce bread and meat.

And what greater environmental force in varying man's condition is there than wealth?

\* \* \*

*Acknowledgments.* The assistance of the following in providing new information is gratefully acknowledged: F. Urbach, Temple University; F. Daniels, Jr., New York Hospital / Cornell Medical Center; J. S. Weiner, Environmental Physiology Research Unit, M. R. C.

# REFERENCES AND NOTES

1. HEIDEL, W. A.
   1937 The frame of the ancient Greek maps. American Geographical Society, Research Series 20. 141 pp. New York.
2. MURDOCK, G. P.
   1959 Africa—its peoples and their cultural history. McGraw Hill, New York. 456 pp.
3. CHAPELLE, J.
   1957 Nomades noirs du Sahara. Librairie Plon, Paris. 449 pp.
4. LEE, D. H. K.
   1964 Terrestrial animals in dry heat: man in the desert. *In* D. B. Dill, E. F. Adolph, and C. G. Wilber, eds., Adaptation to the environment. American Physiological Society, Washington, D. C., Handbook of Physiology 4:551-582.
5. LEE, D. H. K., and J. A. VAUGHAN
   1964 Temperature equivalent of solar radiation on man. International Journal of Biometeorology 8:61-69.
6. BUETTNER, K. J. K.
   1968 The effects of natural sunlight on human skin. pp. 237-249. *In* F. Urbach, ed., The biological effects of ultraviolet radiation. Pergamon Press Oxford.
7. LEE, D. H. K.
   1968 Human adaptations to hot environments pp. 517-556. *In* G. W. Brown, ed., Desert biology. Academic Press, New York.
8. LOWE, C. H.
   1968 Fauna of desert environments. pp. 569-645. *In* W. G. McGinnies, B. J. Goldman, and P. Paylore, eds., Deserts of the world: an appraisal of research into their physical and biological environments. University of Arizona Press, Tucson.
9. SIDKY, M. M., and M. J. FREYCHE
   1949 World distribution and prevalence of trachoma in recent years. World Health Organization Epidemiological and Vital Statistics Report 2(11-12):229-277.
10. MARSHALL, C. L.
    1968 The relationship between trachoma and piped water in a developing area. Archives of Environmental Health 17:215-220.
11. FREYCHE, M.-J.
    1958 The ecology of trachoma. pp. 271-298. *In* J. M. May, ed., The ecology of human disease. MD Publications, New York.
12. LEE, D. H. K., and A. HENSCHEL
    1966 Effects of physiological and clinical factors on response to heat. *In* The biology of human variation. New York Academy of Sciences, Annals 134:743-749.
13. BELDING, H. S., and T. F. HATCH
    1955 Index for evaluating heat stress in terms of resulting physiological strain. Heating, Piping and Air Conditioning 27(8):129-136.
14. HOUNAM, C. E.
    1968 Climate and air conditioning requirements in sparsely occupied areas of Australia. Presentation to World Meteorological Organization Symposium on urban climates and building climatology. Brussels. (Advance copy)
15. FRANK, M., *et al.*
    1959 Epidemiological investigation of urolithiasis in Israel. Journal of Urology 81: 497-505.
16. ARNDT, K. A., D. H. K. LEE, and M. M. KEY
    *in press* Skin—the interface between man and tropical environments. International Review of Tropical Medicine.
17. BELISARIO, J. C.
    1959 Cancer of the skin. Butterworths, London.
18. ROBERTSON, D. F.
    1968 Correlation of observed ultraviolet exposure and skin cancer incidence in the population in Queensland and New Guinea. pp. 619-623. *In* F. Urbach, ed., The biologic effects of ultraviolet radiation. Pergamon Press, Oxford.
19. MAY, J. M.
    1961 The ecology of malnutrition in the far and near east. Hafner, New York.
20. FOOD AND AGRICULTURE ORGANIZATION
    1948 Report of Standing Advisory Committee—nutrition in the Middle East. Washington, D. C.
21. LEE, D. H. K.
    1962 Applications of human and animal physiology and ecology to the study of arid zone problems. *In* The problems of the arid zone, proceedings of the Paris symposium. Unesco, Paris. Arid Zone Research 18:213-233.
22. MUNRO, A. H., A. S. SICHEL, and C. H. WYNDHAM
    1967 The effect of heat stress and acclimatization the body temperature response of men at work. Life Sciences 6:749-754.
23. BOYCE, A. J.
    1968 Race, rhyme and reason. 1: The how and why of human diversity. New Scientist 39: 128-130.
24. HARDY, J. D., H. T. HAMMEL, and D. MURGATROYD
    1956 Spectral transmittance and reflectance of excised human skin. Journal of Applied Physiology 9:257-264.
25. QUISENBERRY, W. B.
    1963 Ethnic differences in skin cancer in Hawaii.

*In* conference on biology of cutaneous cancer, 1st, Philadelphia, 1962. National Cancer Institute Monograph 10:181-189.

26. THOMSON, M. L.
1954 A comparison between the number and distribution of functioning eccrine sweat glands in Europeans and Africans. Journal of Physiology 123:225-233.

27. COLLINS, K. J., and J. S. WEINER
1964 The effect of heat acclimatization on the activity and numbers of sweat-glands: a study on Indians and Europeans. Journal of Physiology 177:16-17P.

28. HAMMEL, H. T.
1961 Comparative study of temperature regulatory mechanisms in primitive man. American Journal of the Medical Sciences 241: 812-813.

29. HICKS, C. S., and W. J. O'CONNOR
1938 Skin temperature of Australian aboriginals under varying atmospheric conditions. Australian Journal of Experimental Biology and Medical Sciences 16:1-18.

30. SCHOLANDER, P. F., *et al.*
1958 Reactions of central Australian aborigines to cold. Journal of Applied Physiology 13: 211-218.

31. BAKER, P. T.
1958 American negro-white differences in heat tolerance. U. S. Army, Quartermaster Research and Engineering Center, Natick, Mass., Technical Report EP-75. 23 pp.

32. BAKER, P. T.
1955 Relationship of desert heat stress to gross morphology. U. S. Army, Quartermaster Research and Engineering Command, Natick, Mass., Technical Report EP-7. 23 pp.

33. SCHREIDER, E.
1963 Physiological anthropology and climate variations. *In* Environmental physiology and psychology in arid conditions. Unesco, Paris. Arid Zone Research 22:37-73.

34. NEWMAN, M. T.
1961 Biological adaptation of man to his environment: heat, cold, altitude and nutrition. *In* Genetic perspectives in disease resistance and susceptibility. New York Academy of Sciences, Annals 91:617-633.

35. WILBER, C. G.
1957 Physiological regulation and the origin of human types. Human Biology 29:329-336.

36. SCHOLANDER, P. F.
1956 Climatic rules. Evolution 10:339-340.

37. BASS, D. E., *et al.*
1955 Mechanisms of acclimatization to heat in man. Medicine 34:323-380.

38. HENSCHEL, A., *et al.*
1968 An analysis of the heat deaths in St. Louis during July 1966. Department of Health, Education, and Welfare, Occupational Health Program, TR-49. 23 pp.

39. BRIDGER, C. A., and L. A. HELFAND
1968 Mortality from heat during July 1966 in Illinois. International Journal of Biometeorology 12:51-70.

40. SCHICKELE, E.
1947 Environment and fatal heat stroke. Military Surgeon 100:235-256.

41. WYNDHAM, C. H.
1968 Factors affecting the heat tolerance of men at work in hot and humid mines. Abstracts of Symposium on Physiological and Behavioral Temperature Regulation, New Haven, 98-99. (To be published as Proceedings later.)

42. GLASER, E. M.
1963 Circulatory adjustments in the arid zone. *In* Environmental physiology and psychology in arid conditions. Unesco, Paris. Arid Zone Research 22:131-151.

43. LEE, D. H. K.
1967 The role of attitude in response to environmental stress. Journal of Social Issues 22: 83-91.

44. LEE, D. H. K.
1958 Proprioclimates of man and domestic animals. *In* Climatology. Unesco, Paris. Arid zone Research 10:102-125.

45. BURCH, G. E., and N. P. DE PASQUALE
1962 Hot climates, man and his heart. C. C. Thomas, Springfield, Illinois.

46. ROTHMAN, S.
1954 Physiology and biochemistry of the skin. University of Chicago Press, Chicago, Illinois. 741 pp.

47. LOFSTEDT, B.
1966 Human heat tolerance. University of Lund, Department of Hygiene, Sweden.

48. DILL, D. B., and C. F. CONSOLAZIO
1962 Responses to exercise as related to age and environmental temperature. Journal of Applied Physiology 17:645-648.

49. HENSCHEL, A., M. COLE, and O. LYCZKOWSKJI
1968 Heat tolerance of elderly persons living in a subtropical climate. Journal of Gerontology 23:17-22.

50. SHATTUCK, G. C., and M. M. HILFERTY
1933 Causes of death from heat in Massachusetts. New England Journal of Medicine 209: 319-329.

51. SCHUMAN, S. H., C. P. ANDERSON, and J. T. OLIVER
1964 Epidemiology of successive heat waves in Michigan 1962. Journal of the American Medical Association 189:733-738.

52. ASCHOFF, J.
1967 Human circadian rhythms in activity, body temperature and other functions. *In* A. H. Brown and F. G. Favorite, eds., Life sciences and space research, vol. 5, p. 159-173. North Holland Publishing Co., Amsterdam.

53. KLEIN, K. E., H. M. WEGMANN, and H. BRUNER
1968 Circadian rhythm in indices of human performance, physical fitness and stress resistance. Aerospace Medicine 39:512-518.

54. CROCE-SPINELLI, M., and G. LAMBERT
1961 S. O. S. Sahara. L'actuel Collection, Flammarion, Paris.

55. FRANK, M., and A. DE VRIES
1966 Prevention of urolithiasis. Archives of Environmental Health 13:625-630.

56. GUYOT, J.
1861 Cultare de la vigne et vinitication. 2^me Ed. Librairie Agricole de la Maison Rustique, Paris.

# BIBLIOGRAPHICAL SOURCES FOR ARID LANDS RESEARCH

PATRICIA PAYLORE
Office of Arid Lands Studies
The University of Arizona
Tucson, Arizona, U. S. A.

# BIBLIOGRAPHICAL SOURCES FOR ARID LANDS RESEARCH

## Patricia Paylore

Bibliography is a necessary nuisance and a horrible drudgery that no mere drudge could perform. It takes a sort of inspired idiot to be a good bibliographer and his inspiration is as dangerous a gift as the appetite of the gambler or dipso-maniac – it grows with what it feeds upon and finally possesses its victim like any other invincible vice.

Coues (1892)

The origin of the present compilation lies partly in this acknowledgment of how bibliographies grow out of other bibliographies. I confess to being no less committed to this vice than was Coues more than three quarters of a century ago. Bibliographers justify this proliferating refinement process by pointing out what a time-saver our work is, as though lurking in the depths of our subconscious is a longing to be equated with the computer. We still have a reprieve, however, from our eventual extinction, for my computer friends, sophisticated beyond ordinary understanding, admit with some reluctance that machine-readable information is still, alas, dependent on the input of some "inspired idiot." So here is an old-fashioned man-readable compilation of some sources of arid lands research information accumulated over several years at the Office of Arid Lands Studies, University of Arizona, based in part upon the bibliographic resource produced under U. S. Army Natick Laboratories contract DA49-092-ARO-71, *An Inventory of Geographical Research on World Desert Environments,* and published in 1968 by the University of Arizona Press as *Deserts of the World: An Appraisal of Research into their Physical and Biological Environments.*

*Deserts of the World* contains nearly six thousand prime references on the vegetation, fauna, geomorphology and surface hydrology, surface materials, weather and climate, coastal deserts, and ground-water hydrology of the desert areas that comprise such a great portion of the Earth's surface. Analysis of these references revealed a good many bibliographies among them, and the idea of segregating this category as a more effective tool for the use of arid lands scientists evolved therefrom.

This paper reports research undertaken in cooperation with the U. S. Army Natick (Mass.) Laboratories under contract DAAG17-67-C-0199 and has been assigned No. TP 573 in the series of papers approved for publication. The findings in this report are not to be construed as an official Department of the Army position. A first gathering was issued in October 1967 as "A Bibliography of Arid-Lands Bibliographies" by Natick Laboratories,

Earth Sciences Laboratory, as Technical Report 68-27-ES (also available as AD-663 843), but the information offered here is a quite different presentation, for in addition to some bibliographies retrieved from the Pertinent Publications sections accompanying the chapters of *Deserts of the World,* a great many more references were discovered, analyzed, and annotated for inclusion by the compiler. For those references retained from the above-cited book, I have eliminated or cut or revised the annotations to emphasize the bibliographical information pertinent to the present compilation.

In using this bibliography, it should be borne in mind that all references are related to arid lands studies and research, and that even though the pertinence may not be apparent from the title, its presence in this bibliography is itself an indication of relevance. In most cases of general material, an annotation is furnished that helps explain the contents in relation to aridity. Where no annotation is included, it may be assumed that the bibliographical importance of the reference is established.

Whereas citations in the bibliographies accompanying *Deserts of the World* were often to chapters or sections of larger works, those in this bibliography are, for the most part, rather to the larger work itself, with subject references in the annotation to assist the user in analyzing the contents for his particular interest (for example, several citations under Unesco authorship).

The format used calls for the following order of information: author, date, title (followed by translated title if required), source, annotation. If certain standard tools – *Biological Abstracts* (BA), or *Meteorological and Geoastrophysical Abstracts* (MGA), for instance – were used to verify, these commonly recognized symbols with the pertinent reference follow the source. The arrangement of the bibliography itself is a numbered alphabetical one, with subject index references to item number.

Users will be aware of the necessity, however tedious and unrewarding, of regular checking for current information in the above-cited tools, as well as in the several abstracting services of the Commonwealth Bureau, *Zoological Record, U. S. Government Research and Development Reports,*

*Abstracts of North American Geology, Index and Bibliography of Geology Exclusive of North America, Bibliography of Agriculture, FAO Documentation-Current Index, Geographical Abstracts (A: Geomorphology), Current Geographical Publications,* and *Bibliographie Géographique Internationale,* as well as others.

I could wish, and I do, that there were some less time-consuming and less wasteful way to recommend to arid-lands scientists around the world, but at the moment, vast computerized networks notwithstanding, I know of no such. Working at present under a National Science Foundation grant,

however, I hope to experiment with the possibility of creating such an interdisciplinary vocabulary that, applied to the worldwide literature of arid lands research, will indeed produce through the wonders of electronics a tool for this purpose. Hopefully, by the time the next issue of the present publication appears, I can report a breakthrough on this bibliographical front.

Meanwhile, here is this non-printout type which I hope will not be too strange to a computer-oriented user group to be useful. I add the usual request, with great sincerity, that omissions be called to my attention.

# REFERENCES AND NOTES

1. ABELL, L. F., and W. J. GELDERMAN
   1964  Annotated bibliography on reclamation and improvement of saline and alkali soils (1957-1964). International Institute for Land Reclamation and Improvement, Wageningen, Bibliography 4. 59 p.
   For updated version, *see also* van Alphen and Abell (1967).

2. ABERNETHY, G. L.
   1957  Pakistan, a selected, annotated bibliography. American Institute of Pacific Relations, New York. 29 p.
   A selected annotated bibliography of English-language books, Government publications, and periodical articles, arranged in that order.

3. AFRICAN STUDIES ASSOCIATION
   1961  American doctoral dissertations concerned with Africa. African Studies Bulletin 4(1).
   A valuable and complete listing to date of publication.

4. AGERTER, S. R., and W. S. GLOCK
   1965  An annotated bibliography of tree growth and growth rings, 1950-1962. University of Arizona Press, Tucson. 180 p.
   Covers tree growth, growth factors, taxonomy and distribution, and tree rings. Includes a list of bibliographies, a section on semi-popular literature, another on tree rings and archeology; and an author index.

5. AGUILAR Y SANTILLÁN, R.
   1898  Bibliografía geológica y minera de la República Mexicana [hasta el año 1896]. Instituto Geológico, Boletín 10. 158 p. (For English-language version, see 1902, below).

6. ———
   1908  Bibliografía geológica y minera de la República Mexicana [hasta el año 1904]. Instituto Geológico, Boletín 17. 330 p.

7. ———
   1918  Bibliografía geológica y minera de la República Mexicana 1905-1918. 97 p.
   1635 references. Author arrangement. Subject index includes geographical terms.

8. ———
   1936  Bibliografía geológica y minera de la República Mexicana 1919-1930. 83 p.

9. ———
   1902  Bibliography of Mexican geology and mining. *In* American Institute of Mining and Metallurgical Engineers, Transactions 32: 605-680.

10. AKADEMIIA NAUK S.S.S.R., BOTANICHESKII INSTITUT
    1956  Rastitel' nyi pokrov SSSR. E. M. Lavrenko, V. B. Sochava, eds. 2 vols.
    Bibliography: vol. 2, p. 869-917.

11. AKADEMIIA NAUK TURKMENSKOI S.S.R.
    1962  Mezhrespublikanskaia nauchniaia sessiia po osvoeniiu pustynnykh teritorii Srednei Azii i Kazakhstana. (Papers read at a 1962 conference on the deserts of the Soviet Union and their development) Ashkhabad. 3 vols.
    v. 1: Natural conditions, 486 p.; v. 2: Water and land resources of the arid zones, 180 p. Bibliographies following each chapter.

12. ALIMEN, H.
    1965  The Quaternary era in the northwest Sahara. *In* H. E. Wright, Jr. and D. G. Frey, eds., International studies on the Quaternary. Geological Society of America, Special Paper 84:273-291.
    A review of current Quaternary research in Algeria. Includes references to many pertinent works on desert sands, p. 288-291.

13. ALLEN, R. H., and E. L. SPOONER
    1968  Annotated bibliography of BEB and CERC publications. U. S. Army Coastal Engineering Research Center, Washington, D. C., Miscellaneous Paper 1-68. 141 p.
    Covers Beach Erosion Board publications from 1940-1963, and Coastal Engineering Research Center publications from 1963-1967. Includes a list of Beach Erosion Control Reports published as House Documents. Author, title, and subject indexes. While the geographic areas covered are limited as far as desert coastal zones are concerned, much scientific and technical information applicable to arid-lands problems is brought out under such headings as dunes, hydrographic surveys, littoral transport studies, sand, waves, and wind.

14. ALLOUSE, B. E.
    1954  A bibliography on the vertebrate fauna of Iraq and neighboring countries. I: Mam-

mals. Iraq Natural History Museum, Publication 4.

15. ———
1956 A bibliography on the invertebrate fauna of Iraq and neighboring countries. V: Mollusca. Iraq Natural History Museum, Publication 8.

16. ALMELA SAMPER, A.
n.d. Bibliografía de Guinea Española y del Sahara Español. Instituto Geológico y Minero de España, Madrid.

17. AMERICAN METEOROLOGICAL SOCIETY
1965 Agricultural meteorology. Meteorological Monographs 6(28). 188 p.
Several articles constitute excellent reviews of arid-zone climatology, especially those on Radiation, by Gates (references p. 24-26), Transport by wind, by Chepil (references p. 131-132), and Evapotranspiration, by Thornthwaite and Hare (references p. 179-180).

18. ANASTOS, G.
1957 The ticks or Ixodides of USSR, a review of the literature. U. S. Public Health Service, Washington, Publication 548.
Bibliography: p. 269-397.

19. ANONYMOUS
1955 A bibliography of African bibliographies, covering territories south of the Sahara. 3rd rev. ed. South African Public Library, Cape Town, Grey Bibliographies 6. 169 p.
An excellent list to 1955, topically arranged.

20. ———
1964 Bibliografia para el Peru. Biota 5(43): 307-314. BA(47) 65159.
Bibliography on Peruvian botany and zoology, 1962-1965.

21. ———
1966 Afrika — Schriftum. — Literature on Africa. — Etudes sur l'Afrique; bibliographie deutschsprachiger wissenschaftlicher veröffentlichungen über Afrika südlich der Sahara. v. 1, text. Franz Steiner Verlag, Wiesbaden. 688 p.
A bibliography of scientific publications on Africa south of the Sahara, in the German language, covering: geography, ethnology, linguistics, tropical medicine, zoology, and botany.

22. APARICIO, F. DE and H. A. DIFRIERI, (eds.)
1958 La Argentina, suma de geografía. Edi-
1963 ciones Peuser, Buenos Aires. 9 vols.
This handsome work, with its profusion of photographs, drawings, and maps, is nevertheless difficult to use for lack of volume-by-volume contents and indexes. There are bibliographies which appear to be comprehensive that follow many sections of the work throughout, and there is, mercifully, a tremendous analytical index at the end of vol. 9. A sampling under Chubut, for instance, revealed an entire column of references to this arid area in all nine volumes.

23. ARGENTINA. CONSEJO FEDERAL DE INVERSIONES
1962- Evaluación de los recursos naturales de la
1963 Argentina. 9 vols. Kraft, Ltd., Buenos Aires.
Extensive bibliographies (e.g., v. 3, Suelo y flora, 1,337 references).

24. AUDRY, P. and C. ROSSETTI
1962 Observations sur les sols et la végétation

en Mauritanie du sud-est et sur la bordure adjacent du Mali (1959 et 1961). FAO, Rome, 267 p. (UN Special Fund Project: UNSF/DL/ES/3)
Bibliography: p. 261-264.

25. AVNIMELECH, M.
1965 Bibliography of Levant geology, including Cyprus, Hatay, Israel, Jordania, Lebanon, Sinai and Syria. Daniel Davey and Co., Inc., New York. 208 p.
Approx. 4500 entries arranged by alphabetical order of authors, chronological index, and analytical subject index. The latter lists all publications under headings of geographical and stratigraphical units, paleontological categories, various problems of applied geology, especially those of oil and water geology, and mineral resources. A note indicates that it is intended "to continue the publication in yearly or two-yearly issues in a similar arrangement of entries, but with the addition of short annotations."

26. BANTA, B. H.
1965 An annotated chronological bibliography of the herpetology of the State of Nevada. Wasmann Journal of Biology 23(1/2): 1-224.
300 titles, annotated, arranged chronologically from 1852, with an alphabetical author index that gives dates of publications of works included in the bibliography proper. For systematic references, a list of authors and dates by generic category is included. There is also a geographical index giving names of authors who reported on material from specific Nevada counties, as well as a list of published localities giving the institution of deposition. It includes most works published through the spring of 1965.

27. BARRETT, E. C.
1957 Baja California, 1935-1956; a bibliography of historical, geographical, and scientific literature relating to the peninsula of Baja California and to the adjacent islands in the Gulf of California and the Pacific Ocean. Bennett and Marshall, Los Angeles. 284 p.

28. BARTON, L. V.
1967 Bibliography of seeds. Columbia University Press, New York. 858 p.
Covers world literature on seed research and technology and on seed agronomy through June 1964. Over 20,000 references in alphabetical author arrangement, with plant name index (cross-indexed by common and scientific names; entries with many references are sub-classified) and subject index. In the latter, 19 references were found cited under "desert seeds," but the bibliography's usefulness to an arid lands investigator is certainly not limited to these few. He will want, rather, to work from the plant name index, assuming he is familiar with desert plants.

29. BESPALOV, N. D.
1964 Soils of Outer Mongolia. Israel Program for Scientific Translations, Jerusalem. 320 p. Originally published by the Akademiia Nauk S.S.R., Moscow, 1951. Translation also cited as: OTS 64-11073.
Bibliography: p. 305-314.

30. BEZANGER-BEAUQUESNE, L.
1955 Contribution des plantes à la défense

BEZANGER-BEAUQUESNE (*cont*)

de leurs semblables. (Contribution of plants to their self defense) Société Botanique de France, Bulletin 102:548-575. BA(10)30277.

Includes a review of literature pertaining to the means by which plants protect themselves against plant and animal enemies.

31. BLAKE, S. F., and A. C. ATWOOD

1942 Geographical guide to floras of the world. I: Africa, Australia, North America, South America, and Islands of the Atlantic, Pacific, and Indian Oceans. U. S. Department of Agriculture, Miscellaneous Publication 401. 336 p.

An annotated selected list of floras and floristic works relating to vascular plants, including bibliographies and publications dealing with useful plants and vernacular names. Indexed by authors and by geographic areas.

32. BLÁSQUEZ LÓPEZ, L.

1959 Hidrogeología de las regiones desérticas de México. Universidad Nacional Autónoma de México, Instituto de Geología, Anales 15. 172 p.

Bibliografía: p. 169-172.

33. BLAUDIN DE THÉ, B. M. S.

1960 Essai de bibliographie du Sahara français et des régions avoisinantes. Arts et Métiers Graphiques, Paris. 2nd ed. 258 p.

34. BOGOMOLOV, G. V. (ed.)

1963 Problemy kompleksnogo izuchiniia zasushlivykh zon SSSR. Akademiia Nauk S.S.S.R., Moscow, 242 p.

Bibliographies covering natural resources, including underground water, of Soviet Central Asia.

35. BOLTON, A. R. C.

1959 Soviet Middle East studies, an analysis and bibliography. Chatham House Memoranda. Distributed for the Royal Institute of International Affairs by the Oxford University Press. 8 parts.

Emphasis is on current affairs, social anthropology and social studies, and economics.

36. BONNET, P.

1945- Bibliographia araneorum; analyse mé-
1961 thodique de toute la littérature araneologique jusqu'en 1939. Les Frères Douladoure, Toulouse. 3 vols. in 7 pts.

In addition to author lists, there is a geographical section with cross-references to authors of appropriate citations.

37. BOOCOCK, C., and O. J. VAN STRATEN

1962 Notes on the geology and hydrology of the central Kalahari region, Bechuanaland Protectorate. Geological Society of South Africa, Transactions 65:125-171.

This paper is the best modern source of information on the general geology and hydrology of the central Kalahari region. The most pertinent references to previous work on the Kalahari are provided. *See also* Geological Survey of South West Africa (n.d.).

38. BORCHARDT, D. H.

1963 Australian bibliography; a guide to printed sources of information. F. W. Cheshire, Melbourne. 72 p.

Bibliographical essays on library catalogues and general retrospective bibliographies, current national bibliographies and indices to newspapers, bibliographies of subject areas, regional bibliographies, government publications, and bibliography in Australia. Includes a "list of works referred to," p. 55-72.

39. BOYKO, H. (ed.)

1966 Salinity and aridity, new approaches to old problems. W. Junk, The Hague. 408 p. (Monographiae Biologicae 16)

Extensive bibliographical contributions by chapter authors for a total of 646 references dealing with deserts or aridity and salinity problems.

40. BRASSEUR, P.

1964 Bibliographie générale du Mali (anciens Soudan français et Haut-Senegal-Niger). Institut Française d'Afrique Noire, Catalogues et Documents 16. 461 p.

A briefly annotated bibliography of Mali with approx. 5000 entries. Topics covered include geology, geomorphology, hydrology, and others. Author-subject index, including geographic terms, in a chronological arrangement within subject classifications.

41. BRIDGMAN, J., and D. E. CLARKE

1965 German Africa; a select annotated bibliography. Hoover Institution, Stanford University, Bibliographical Series 19. 120 p.

German Southwest Africa, p. 82-95 (103 references).

42. BRITISH WEST AFRICA. METEOROLOGICAL SERVICES

1954 Bibliography of publications relating to the meteorology of British West Africa. Lagos. 14 p. mimeo.

Complete lists in alphabetical order, with titles but no abstracts, of the published and manuscript material available in the Headquarters library at Lagos.

43. BURGESS, R. L.

1965 Utilization of desert plants by native peoples; an overview of southwestern North America. *In* J. L. Gardner, ed., Native plants and animals as resources, in arid lands of the southwestern U. S. AAAS Committee on Desert and Arid Zones Research, Contribution 8:6-21.

A summary of the ethnobotany of the southwestern U. S. Discusses plant uses under the headings of food uses, drink plants, drugs and medicinal uses, fiber uses, building materials, etc. An excellent bibliography, p. 17-20, on a specialized subject.

44. BURGESS, R. L., A. MOKHTARZADEH, and L. CORNWALLIS

1966 A preliminary bibliography of the natural history of Iran. Pahlavi University, Science Bulletin 1. 220, 143 p.

1719 references, repeated in Farsi. Arrangement is by categories: bibliographies and dictionaries; geography, travel, and exploration; environment; floristics; fauna; miscellaneous. It includes author, geographical, and chronological indexes.

45. BYKOV, B. A.

1959 Geobotanicheskie i floristicheskie issledovaniya v Kazakhstane. p. 7-18. *In* Botanika v Kazakhstane. Akademiia Nauk Kazakhskoi S.S.R., Alma-Ata, Institut Botaniki. Referativnyi Zhurnal, Biologiya, 1961, No. 1V4. BA41(6)24245. (*continues*)

Valuable contribution on the history of flora and geobotanical studies of present-day Kazakhstan; includes a bibliography of 114 references.

46. CAIRO. DAR AL-KUTUB AL-MISRIYAH. QISM AL IRSHAD.
1960 Qaimah bi-al-kutub wa-al-maraji an al-Iraq. (A bibliography of works about Iraq) 86 p.
Includes publications in Arabic and Western languages.

47. CAIRO. SCIENTIFIC AND TECHNICAL DOCUMENTATION CENTRE
1955- Abstracts of scientific and technical papers published in Egypt, and papers received from Afghanistan, Cyprus, Iran, Iraq, Jordan, Lebanon, Pakistan, Saudi Arabia, Sudan and Syria. v. 1, 1955- to date (?). Issued as a Bulletin of the Centre.
Supersedes Unesco's "List of Scientific Papers published in the Middle East."

48. CALCUTTA. NATIONAL LIBRARY.
1960 Indian scientific and technical publications, exhibition 1960; a bibliography. Council of Scientific and Industrial Research, New Delhi. 2 parts. 198, 195 p.
Pt. 1, over 2900 publications in Indian languages; pt. 2, publications in English. Arrangement by Dewey Decimal classification, indexed, like the English Catalogue, to the Dewey numbers. The most useful references to arid lands are in the botanical and agricultural sciences classifications. The compilers lament that publishers failed to cooperate in sending more than "sample" publications, perhaps a universal complaint. Nevertheless, the bibliography has usefulness as a precursor of the more sophisticated Indian Science Abstracts (q.v.).

49. CARRAWAY, D. M.
1956 Bibliography on the climate of Angola. U. S. Weather Bureau, WB/BC-8. 26 p. Also cited as AD-669 410.

50. ———
1961 Annotated bibliography of climatic maps of Jordan. U. S. Weather Bureau, Washington, D. C., WB/BM-35. 13 p. Also cited as PB-176 027; AD-664 718.

51. CENTRAL ASIAN RESEARCH CENTRE, LONDON
1956 Bibliography of Russian works on Afghanistan. 12 p.
Includes articles and monographs in an author arrangement, with no index. Titles are translated. "All subsequent Russian books and articles on Afghanistan will be entered in the Bibliographical Note of Central Asian Review. An analysis of Soviet publications on Afghanistan was contained in a 1956 issue of Central Asian Review, 4(2):161-200."

52. ———
1957- Bibliography of recent Soviet source material on Soviet Central Asia and the
1960 borderlands. 8 vols.
Issued as supplements to Central Asian Review. Earlier bibliographies appeared in v. 1-4 of the Review.

53. CENTRE D'ANALYSE DOCUMENTAIRE POUR L'AFRIQUE NOIRE (CADAN)
[This agency analyzes periodicals and other publications on social and human sciences on Africa south of the Sahara, prepares about 4000 abstract cards per year, usable either on index cards or processed by automatic data methods (SYNTOL) worked out by the Centre National de la Recherche Scientifique in its automatic data section. The address is 293 Ave. Daumesnil, Paris-12e.]

54. CHAPMAN, V. J.
1960 Salt marshes and salt deserts of the world. Leonard Hill, London; Interscience Publishers, New York. 392 p.
Bibliography: p. 353-371.

55. CHENERY LIBRARY, BOSTON
1960 Catalog of African Government documents and African area index.
A good resume of publications by the various African governments.

56. CHEPIL, W. S., and N. P. WOODRUFF
1963 The physics of wind erosion and its control. Advances in Agronomy 15:211-302.
Covers such topics as surface wind, equilibrium forces on soil grains, cycle of wind erosion, soil properties that influence wind erosion, wind erosion control, and the wind erosion equation. References: p. 299-302. Author/source information only, making selection of appropriate titles difficult.

57. CHI, Y. S.
1947 Bibliography of geology and geography of Sinkiang. Hsiao-Fung Li, Nanking. 213 p.

58. COOPER, C. F.
1963 An annotated bibliography of the effects of fire on Australian vegetation. Soil Conservation Authority, Victoria. 21 p.

59. CREASI, V. J.
1960a Bibliography of climatic maps for Australia. U. S. Weather Bureau, Washington, D. C., WB/BM-16. 44 p. Also cited as PB-176 007; AD-665 178.
Lists 86 reports, with descriptive comments.

60. ———
1960b A selected bibliography on the climate of Australia. U. S. Weather Bureau, Report WB/BC-44. 112 p. Cited as AD-664 746.
219 reports listed by title and author, with descriptive comments.

61. CURSON, H. H.
1947 Notes on the eastern Caprivi Strip. South African Journal of Science 43:124-157.
An excellent geographical study, with a good bibliography.

62. DAHLBERG, R. E. and B. E. THOMAS
1962 An analysis and bibliography of recent African atlases. African Studies Bulletin 5(11):22-23.
This and the following supplement constitute excellent listings and reviews of recent atlases.

63. ———
1963 Supplement #1. African Studies Bulletin 6(May):6-9.

64. DAVEAU, S.
1965 Bibliographie pratique pour l'étude du relief en Afrique Occidentale. Revue de Géographie de l'Afrique Occidentale 1/2: 229-233.

65. DAYTON, W. A.
1951 Historical sketch of barilla (Halogeton glomeratus). Journal of Range Management 4(6):375-381.
Review literature covering the toxic properties, habitat, distribution, and life history of this introduced livestock-poisoning plant. Includes brief but pertinent bibliography of 30 titles.

66. DEKEYSER, P. L., and J. DERIVOT
1959 La vie animal au Sahara. A. Colin, Paris. 220 p. Bibliography: p. 205-218.

67. DICKSON, B. T. (ed.)
1957 Guide book to research data on arid zone development. Unesco, Paris. Arid Zone Research 9.
A somewhat dated but still useful compilation of research pertaining to arid regions. Bibliographies for most chapters in the two main sections: physical and biological factors; human factors.

68. DIXEY, F.
1963 Geological bibliographies of African territories. Unesco, Paris. Natural Resources Research 1:84-86.
A brief list of sources responsible for geological information, arranged by countries, including Algeria, Angola, French Somaliland, Libya, Morocco, Somalia, South Africa, Spanish Sahara, Sudan, Tunisia, and UAR. The references are bibliographically imprecise, but the leads are very valuable. A note indicates that Dixey, "with the help of a number of African Geological Surveys, has also established bibliographical lists which are filed at the offices of the Economic Commission for Africa in Addis Ababa."

69. DOLAN, R., and J. M. McCLOY
1964 Selected bibliography on beach features and nearshore processes. Louisiana State University Press, Coastal Studies Series 11. 59 p.

70. DOST, H.
1953 Bibliography on land and water utilization in the Middle East. Wageningen Agricultural University College, Wageningen, Netherlands. 115 p.

71. DUIGNAN, P. (ed.)
1963- U. S. and Canadian publications on Africa in 1961- to date. Hoover Institution on War, Revolution, and Peace, Stanford University, Bibliographic series 14(1961), 15(1962), 20(1963), and 25(1964)- to date.
This annual topical bibliography, compiled since the issue covering the year 1963 by Liselotte Hofmann, is thorough and excellent. An earlier number for the year 1960 was issued by the Library of Congress, Washington, D. C.

72. DZERDZEEVSKII, B. L. (ed.)
1957 Sukhovei ikh proiskhozhdenie i bor'ba s nimi. Izdatel'stvo Akademii Nauk SSSR, Moskva. Translated, 1963, as "Sukhovei and drought control," by Israel Program for Scientific Translations. Also cited as OTS-63-11140. 366 p.
Bibliography, p. 344-366 of the translated version, includes 415 papers on sukhovei and droughts, and measures for combating them, for the period 1917-1955. Translated titles are provided for all Russian-language works cited.

73. EDWARDS, E.
1961 Lost oases along the Carrizo. Westernlore Press, Los Angeles. 126 p.
A descriptive bibliography of the Colorado Desert, p. [53]-104.

74. EGYPT. NATIONAL RESEARCH COUNCIL
1952- Classified list of Egyptian scientific papers
1954 published in 1951-1953. 3 vols.

75. ELLENBERG, H.
1958 Uber den wasserhaushalt tropischer nebeloasen in der küstenwüste Perus. Geobotanische Forschungsinstitut Rübel, Berichte, p. 47-74.
Bibliography on mist deserts.

76. ENGEL, C. G., and R. P. SHARP
1958 Chemical data on desert varnish. Geological Society of America, Bulletin 69:487-518.
41 references, p. 517-518.

77. ENGLISH, P. W.
1966 City and village in Iran: settlement and economy in the Kirman Basin. University of Wisconsin Press, Madison. 204 p.
Considerable data on climate, soils, crops, land and water ownership in this scholarly work. Bibliography: p. 181-191.

78. EWAN, J.
1936 Bibliography of the botany of Arizona. American Midland Naturalist 17:430-454.
208 references, with emphasis on taxonomic and floristic works, in a chronological arrangement. About one-half of the references cover the period up to 1907. Annotated, with geographical and author indexes.

79. FÀNTOLI, A.
1952 Le pioggie della Libia, con particolare riguardo alle zone di avvaloramento. Ministero dell'Africa Italiana, Ispettorato Meteorologico. Rome. 528 p.
Meteorological bibliography on Libya and adjacent regions (arranged by country), p. 513-521.

80. FERREIRA, H. A.
1957 Bibliografia meteorológica et geofísica de Angola. Angola, Serviço Meteorológico, Luanda. 14 p.

81. FIELD, H.
1953- Bibliography on Southwestern Asia. Uni-
1962 versity of Miami Press, Coral Gables, Florida.
I: 3,016 anthropogeographical and natural history titles, and author index; II, 3,292 titles; III, 6,661 titles; IV, 12, 149 titles; V, 6,739 titles; VI, 8,364 titles; VII, 7,492 titles. Cumulative subject index to Bibliographies I-V, and another covering VI-VII. These indices are in 3 parts for each cumulation: anthropogeography, zoology, botany. Coverage: Turkey, Syria, Lebanon, Israel, Egypt, Jordan, Yemen, Saudi Arabia, Aden, Dhufar, Oman, Trucial Oman Coast, Qatar Peninsula, Bahrain Island, Kuwait, Iraq, Iran, Caucasus, Afghanistan, and West Pakistan.

82. FIESE, M. J.
1958 Coccidioidomycosis. Charles C Thomas, Springfield, Illinois, 253 p.
Includes a comprehensive chronological bibliography of 968 articles covering the period 1892-1957. See also Symposium on Coccidioidomycosis, Phoenix, 1965 (1967) for later information.

83. FISHER, W. B.
1963 The Middle East, a physical, social, and regional geography. 5th ed. Methuen, London. 568 p.
Bibliography, p. 553-339, follows the organization of the book: land forms, climate, soils and natural vegetation for Persia, Asia Minor, the Tigris-Euphrates lowlands, coastlands of the Levant,

the Arabian Peninsula, the lower Nile Valley, and Libya. Included are references to books, journal articles, and reports.

84. FRENCH SOMALILAND. SERVICE DES TRAVAUX PUBLICS
    1960 Bibliographie géologique et minière (1930-59). (Geologic and mining bibliography) Djibouti.

85. FURON, R.
    1960 Géologie de l'Afrique. 2 ed. Payot, Paris. 350 p., maps.

86. _____
    1963 Geology of Africa. Translated from the 2nd French edition. Oliver & Boyd, Edinburgh. 377 p.
    A basic guide to the geological literature on Africa, with regional divisions making it convenient to locate specific arid areas.

87. GANSSEN, R.
    1963 Südwest-Afrika, böden und bodenkultur. Verlag von Dietrich Reimer, Berlin. 160 p. Bibliography: p. 125-127 (46 refs.).

88. GEOLOGICAL SURVEY OF INDIA
    1938 Geographical index to the Memoirs, v. 1-54; and Records, v. 1-65, of the Geological Survey of India. 576 p.

89. _____
    1943 Catalogue of publications and index of geological maps of the Geological Survey of India up to June 1941. Memoir 77. 114 p.

90. GEOLOGICAL SURVEY OF PAKISTAN
    1965 List of reports issued (as of 31st January 1965). Quetta. 8 p.
    A basic source of information on all reports published by the Geological Survey. See Offield (1964) for a general bibliography of geology of Pakistan and the PANSDOC reference lists (Pakistan National Scientific and Technical Documentation Center, 1957, 1958, 1959, 1960, 1961, 1965) on nongeologic subjects.

91. GEOLOGICAL SURVEY OF SOUTH AFRICA
    1962- Bibliography and subject index of South African geology, 1959-  to date. Pretoria.
    An annual, issued at least since the coverage for 1959 (publication lags about 3 years behind). Annotated, author arrangement, separate section of theses. Subject index includes geographic terms. South West Africa and Bechuanaland are covered.

92. GEOLOGICAL SURVEY OF SOUTH WEST AFRICA
    n.d. Bibliography of geology of South West Africa. Windhoek.
    An unpublished bibliography which contains references to nearly every work on the surficial features of South West Africa. Periodic additions make this bibliography a most useful source of information.

93. GERMAN GEOLOGICAL MISSION TO AFGHANISTAN
    1964 Bibliography of the geology of Afghanistan up to 1964. Kabul. 18 p.

94. GHANI, A. R.
    1951 Pakistan: a select bibliography. Pakistan Association for the Advancement of Science, University Institute of Chemistry, Lahore. 339 p.

95. GLEESON, T. A.
    1952 Bibliography of the meteorology of the Mediterranean, Middle East and South Asia areas. Florida State University, Department of Meteorology, Scientific Report 1 (Appendix). 37 p.

96. GOLD, H. K.
    1962 An annotated bibliography on the climate of Israel. Air Weather Service Climatic Center, Washington, D. C. 50 p. Report No. WB/BC57. Also cited as AD-660 872.

97. GOOD, R.
    1964 The geography of the flowering plants. 3rd ed. John Wiley, New York. 518 p. BA46(2)7961.
    A revised two-part edition with updated information. The first part contains 15 chapters on the floristic regions of the world; the second contains 8 chapters covering factors of distribution and the theory of tolerance. Includes a bibliography of 838 references and an index of plant names.

98. GOULD, S. H. (ed.)
    1961 Sciences in Communist China; a symposium presented at the New York meeting of the American Association for the Advancement of Science, December 26-27, 1960. American Association for the Advancement of Science, Publication 68. 872 p.
    While neither the chapters nor their bibliographies are arranged by geographical areas, so that references pertaining to deserts cannot easily be retrieved, there are substantial bibliographies included for sections covering "Science and Society," "Biological and Medical Sciences," "Atmospheric and Earth Sciences," "Mathematics and Physical Sciences," and "Engineering Sciences and Electronics," and the book does include a geographical index.

99. GRIMES, A. E.
    1959 Bibliography of climatic maps for Ethiopia, Eritrea, and the Somalilands. U. S. Weather Bureau, Washington, D. C., WB/BM-17. 13 p. Also cited as PB-176 030; AD-665 179.

100. _____
    1960a An annotated bibliography on the climate of the Arabian Peninsula. U. S. Weather Bureau, Washington, D. C., WB/BC-42. Also cited as PB-176 017; AD-664 696.
    The bibliography consists of sources containing meteorological and climatological data for the Arabian Peninsula, including Saudi Arabia, Kuwait, Qatar, Yemen, Oman, Trucial Oman, Aden Colony and Aden Protectorate. Lists 124 reports by title and author, and includes descriptive comments concerning the content of these reports.

101. _____
    1960b Bibliography of climatic maps of the Arabian Peninsula. U. S. Weather Bureau, Washington, D. C., WB/BM-15. 13 p. Also cited as PB-176 041; AD-664 712.

102. _____
    1960c An annotated bibliography on climatic maps of Lebanon. U. S. Weather Bureau, Washington, D. C., WB/BM-14. 14 p. Also cited as PB-176 005; AD-664 708.
    Consists of sources containing climate maps of Lebanon and the following areas: Syria, the Near East, southwest Asia, northeast Africa, the Middle East, and the Levant. This bibliography and

Grimes (1961a) below were compiled from available sources in the Weather Bureau Library, the Library of Congress, the Geological Survey Library, the National Library of Medicine, the Department of Agriculture Library, and the Middle East Institute Library.

103. _____

1960d Bibliography of climatic maps of Morocco. U. S. Weather Bureau, Washington, D. C., WB/BM-19. 11 p. Also cited as PB-176 032; AD-664 709.

Includes Spanish Morocco.

104. _____

1960e Bibliography of climatic maps of Tunisia. U. S. Weather Bureau, Washington, D. C., WB/BM-21. 11 p. Also cited as PB-176 033; AD-664 710.

105. _____

1961a An annotated bibliography on the climate of Lebanon. U. S. Weather Bureau, Washington, D. C., WB/BC-50. 45 p. Also cited as PB-176 004; AD-664 698.

Lists 125 reports by title and author, and includes descriptive comments concerning the content of these reports.

106. _____

1961b An annotated bibliography on climatic maps of Libya. U. S. Weather Bureau, Washington, D. C., WB/BM-28. 9 p. Also cited as PB-176 011; AD-664 715.

107. _____

1962 An annotated bibliography of climatic maps of the United Arab Republic (Egypt). Air Weather Service Climatic Center, Washington, D. C. 23 p. Also cited as AD-660 866.

The bibliography of 65 references has been compiled from available sources in the U. S. Weather Bureau Library, Library of Congress, U. S. Geological Survey Library and the U. S. Navy Hydrographic Office Library. Areas covered include Egypt, sections of Egypt, or areas which include Egypt, such as northeastern Africa and southwestern Asia or the Mediterranean Area. Abstracts from the 1959 Bibliography of Climatic Maps of Egypt, have been incorporated.

108. _____

1964 An annotated bibliography of climatic maps of India. Air Weather Service Climatic Center, Washington, D. C. Report no. WB/BM-65. 90 p. Also cited as AD-660 832.

The 266 references listed were compiled from available sources in various libraries of the Washington Metropolitan Area with maps of India, India and adjacent areas, and South Asia. Map scales are given in the abstracts.

109. _____

1965 An annotated bibliography of climatic maps of Pakistan. Air Weather Service Climatic Center, Washington, D. C. Report WB/BM-71. 77 p. Also cited as AD-660 827.

222 references, including all sources in the 1962 Annotated Bibliography of Climatic Maps of Pakistan, together with abstracts from additional available sources in various libraries of the Washington Metropolitan area, on Pakistan, adjacent areas, and South Asia, through April 1964. Map scales are given.

110. GROOT, J. J., and C. R. GROOT

1948 An annotated bibliography on the environment of desert and semi-desert areas of western Africa. U. S. Department of the Army, Quartermaster General, Research and Development Branch, Environment Protection Series Report 128. 96 p.

592 references. Regional studies include Algeria, Mauritania, Niger, Sudan. Other topics: bibliographies, geology, landforms, hydrography, itineraries and transportation problems, climate, flora and fauna, agriculture, ethnography, and health.

111. GUEST, E. R., and R. A. BLAYLOCK

1956 Botanical bibliography of Iraq. Kew. 10 p.

112. GUPTA, R. K.

1966 Bibliography on the ecology (synecology and phytosociology) of the arid and semi-arid regions of India. Excerpta Botanica, B, 7(3):178-190.

113. HADAČ, E.

1966 Bibliographia phytosociologica: Iraq. Excerpta Botanica, B, 7(2):102-104.

114. HANELT, P.

1964 Bibliographia phytosociologicae: China. Excerpta Botanica, B, 6(2):106-134.

115. HANSTRÖM, B., P. BRINCK, and G. RUDEBECK

1955- South African animal life. Results of the
1967 Lund University Expedition in 1950-1051. Distributed by Swedish Natural Science Research Council, Stockholm. 13 vols. (through 1967)

Vol. 1 includes general information about the Expedition, its origins, purpose, itinerary, with localities investigated keyed to maps and briefly described. South West Africa is included, with extensive coverage of species found in both the Namib and Kalahari. Each volume covers a different order. In English, with voluminous bibliographical information for each section in all volumes.

116. HARMON, R. W., and C. B. POLLARD

1948 Bibliography of animal venoms. University of Florida Press, Gainesville. 340 p.

4157 references in a chronological arrangement, 1875-1946, with author index. See also Russell and Scharffenberg (1964).

117. HARRIS, C. D.

1962 Geography, resources and natural conditions in the Soviet Union, an annotated bibliography of selected basic books in Russian. University of Chicago, Department of Geography. 45 p. mimeo.

118. HARRIS, G. L.

1957 Egypt. Human Relations Area File, New Haven. 370 p.

Bibliography: p. 354-365.

119. _____

1958a Iraq: its people, its society, its culture. Human Relations Area File Press, New Haven. 350 p.

Bibliography: p. 331-338.

120. _____

1958b Jordan, its people, its society, its culture. Grove Press, Inc., New York. 246 p., maps.

This interdisciplinary appraisal constitutes a comprehensive summary of all phases of Jordanian life, with the emphasis on cultural rather than physical geography. Bibliography: p. 235-237.

121. HARSHBERGER, J. W.
1911 Phytogeographic survey of North America. G. E. Stechert and Company, New York. 790 p. (Die Vegetation der Erde, 13)
A classic report on the vegetation of North America based on the best knowledge available at the time. Bibliographies are arranged by regions, with that for the deserts appearing on p. 75-78 under the heading "Southwest Arid States and Great Basin," although some additional appropriate references are scattered through the section labelled "Mexico and Central America," p. 82-87. Chief value of these references now is historical. There is a useful index of plant names.

122. HAUG, P. T., and G. M. VAN DYNE
1968 Secondary succession in abandoned cultivated fields: an annotated bibliography. Oak Ridge National Laboratory, Oak Ridge, Tennessee, ORNL-TM-2104. 68 p.
This bibliography of 120 references was undertaken preliminary to a comprehensive review and analysis of old-field succession. The objectives of the present report are to give abstracts of important papers and to provide a computer-generated index of key words in the titles of those articles. The annotations are lengthy and detailed. In their introduction, the compilers make this chilling observation: "Apart from purely scientific interest in revegetation of old fields, there is a more sobering reason for intensive exploration of ecological processes in secondary succession. In the event of thermonuclear attack, there is a possibility that vast areas might be involved in wholesale degradation of the landscape. Much of the acreage now under cultivation would be abandoned. Widespread conflagration could alter the environment extensively, and fallout could differentially affect plant species attempting to recolonize abandoned lands. The numerous possible micro- and macro-climatic modifications following such attack make it important to understand thoroughly the processes concerning secondary succession in important ecosystems."

123. HAYWARD, H. E., and C. H. WADLEIGH
1949 Plant growth on saline and alkali soils. Advances in Agronomy 1:1-38.
A review. References: p. 35-38. Author/source information only, making selection of appropriate titles difficult.

124. HEATH, J. O.
1965 Bibliography of reports resulting from U. S. Geological Survey participation in the United States Technical Assistance Program, 1940-1965. U. S. Geological Survey, Bulletin 1193. 51 p.
670 reports listed chronologically under each country, including several desert areas in Africa, Latin America, the Middle East, Central Asia, etc. Author and subject indexes are included.

125. HILL, R. L.
1939 A bibliography of the Anglo-Egyptian Sudan from the earliest times to 1937. Oxford University Press. 213 p.
For continuation, see Nasri (1962), and Ibrahim and Nasri (1965).

126. HILL, R. W.
1959 A bibliography of Libya. University of Durham, Department of Geography, Research Paper 1. 100 p.
A useful bibliography of Libyan studies with emphasis on geographic reports.

127. HOLDSWORTH, M.
1961 Soviet African studies, 1918-1959; an annotated bibliography. Distributed for the Royal Institute of International affairs by Oxford University Press. 2 pts.
Part 1, General functional studies. 282 references. 80 p. Part 2, Regional studies. 216 references. 70 p. Author index.

127a. HOLMES, R. F. and J. J. FOOTEN
1968 Selected bibliography of the terrain sciences. 2nd rev. ed. Raytheon Company, Space and Information Systems Division, Autometric Operation, Alexandria, Virginia. 39 p.
A compilation of approximately 500 references to articles, papers, reports, and texts, arranged alphabetically within sections covering geology, geomorphology, soils, vegetation and land use, engineering photogrammetry, color, multi-sensor interpretation, and general subjects and bibliographies. "Oriented to those professional terrain scientists who use photo-interpretative techniques in the exploitation of the information content of the growing family of remote imaging systems." While the arid lands user will have to hunt for those of particular interest to him, the search will be rewarding.

128. HUDSON, A. E.
1938 Kazakh social structure. Yale University Publications in Anthropology 20. 109 p.
Bibliography: p. 106-109.

129. HUMLUM, J.
1959 La géographie de l'Afghanistan, étude d'un pays aride. Scandinavian University Book, Oslo. 421 p.
A standard geography of Afghanistan that provides a summary of the physiography, climate, and allied topics. Bibliography: p. 385-396.

130. HURST, E.
1942 The poison plants of New South Wales. Compiled under direction of the Poison Plants Committee of New South Wales. The Snelling Printing Works, Sydney. 498 p.
Bibliography: p. 433-468.

131. HUTCHINSON, J.
1946 A botanist in southern Africa. P. R. Gawthorn, Ltd., London. 686 p. BA25(12)37424.
The final part of this 5-part publication reviews South African botanical literature, p. 554-608.

132. IBRAHIM, A, and A. R. EL NASRI
1965 Sudan bibliography, 1959-1963. Sudan Notes and Records 46:130-166.
Supplements Nasri's Bibliography of the Sudan, 1938-1958 (q.v.).

133. INDACOCHEA G., A. J.
1946 Bibliografía climatológica del Perú. Instituto Geológico del Perú, Boletin 4. 81 p. MGA 2-10:156.
Approx. 750 unannotated items arranged alphabetically by author. In addition to meteorology, includes hydrology, economic geography, agricultural meteorology, and oceanography. Majority of the references are to Spanish language works, with a few in English and German.

134. INDIA. MINISTRY OF EDUCATION
1964 Bibliography of arid zone research.
An annotated list, arranged in 72 sections ac-

India (*cont*)
cording to Dewey classification numbers, largely in 551 and 631.

135. INDIAN NATIONAL SCIENTIFIC DOCUMENTATION CENTER, DELHI

1965- Indian Science abstracts, v. 1, 1965-         to date.

Issued monthly. Dewey Decimal classification arrangement, with the usual difficulty in narrowing the search for pertinent arid lands references. There are now, however, computer-produced author and keyword indexes in each issue, and presumably there will be an annual cumulated index. Reference is to a serial number assigned each citation in sequence as it appears in the classified list. INSDOC services, by the way, include microfilm, paper photocopies, translations, and bibliographies. Prices available upon request.

136. INSTITUT EQUATORIAL DE RECHERCHES ET D'ETUDES GÉOLOGIQUES ET MINIÈRES, BRAZZAVILLE

1961    Bibliographie géologique et minière de l'Afrique Equatoriale Française. Direction des Mines et de la Géologie de l'Afrique Equatoriale Française, Brazzaville.

137. INSTITUT NATIONAL DE LA RECHERCHE AGRONOMIQUE, RABAT

1965    Bibliography of studies and investigations on arid zones, particularly in Morocco (translated title). Cahiers de la Recherche Agronomique 19:71-130.

571 references grouped by subjects: meteorology and climatology; agricultural botany, phytogeography, and ecology; pastures, lucerne, and other forage plants; various cultivated plants.

138. INSTITUTO DE INVESTIGACIÓN DE RECURSOS NATURALES

1967    Informacion bibliográfica de recursos naturales, 1945-1965. En colaboración con Centro Nacional de Información y Documentación (CENID) del Consejo de Rectores de las Universidades Chilenas. Santiago de Chile. 294 p.

139. INSTITUTO MEXICANO DE RECURSOS NATURALES RENOVABLES

1955-    Los recursos naturales de México. 3 vols.
1961

Extensive bibliographies in each volume. Volume 2, for instance, on Suelo y Agua, has a bibliography extending from p. 89-145, plus a list of studies on file in the archives of the Secretaría de Recursos Hidráulicos, arranged by states (Sonora, p. 179; Baja California, 148-149; Chihuahua, 153-156, etc.).

140. ISRAEL, RESEARCH COUNCIL

1953    Desert research; proceedings of the International Symposium held in Jerusalem, May 7-14, 1952, sponsored by the Research Council of Israel and Unesco. Research Council of Israel, Special Publication 2. 641 p.

A somewhat dated but still useful compilation of papers on many aspects of desert research, largely pertinent to the Middle East. Bibliographies for each paper.

141. ISTITUTO AGRONOMICO PER L'AFRICA ITALIANA

1953    Contributo ad una bibliografia italiana su Eritrea e Somalia con particolare riferimento all' agricoltura ed argomenti affini (fino al dicembre 1952). Firenze. 239 p.

A comprehensive bibliography pertaining to Italian studies in Eritrea and Somalia. Useful to ecology and more so to agriculture.

142. JONES, G. N.

1966    An annotated bibliography of Mexican ferns. University of Illinois Press, Urbana. 297 p.

The principal basis of selection of authors/titles was taxonomic and phytogeographic, but many articles on morphology, ecology, explorations, economic botany, bibliography, biography, herbaria, and botanical gardens are included. There is a plant-name and author index. The geographical index cites, as examples of arid areas, 59 references to Baja California, 39 to Chihuahua, and 18 to Sonora.

143. JONES, N. E.

1966    Bibliography of remote sensing of resources. U. S. Army Corps of Engineers, Ft. Belvoir, Virginia, for NASA, Earth Resources Program Office. 40 p. N68-1 870.

371 entries covering the open literature between 1960 and June 1966, excluding geophysics, meteorology, and electromagnetic sensing. Not annotated. Index terms of interest to arid lands users include agriculture, geography, geology, hydrology, soils, and vegetation.

144. KASAPLIGIL, B.

1955    A bibliography on the botany and forestry of the Hashemite Kingdom of Jordan. Expanded Technical Assistance Program of FAO. Amman. 17 p.

145. KEAST, A., R. L. CROCKER, and C. S. CHRISTIAN (eds.)

1959    Biogeography and ecology in Australia. W. Junk, The Hague. 640 p. (Monographiae Biologicae, 8)

Compendium-type work, with extensive chapter reference lists, covering such topics as birds, reptiles, insects, dipterofauna, termites; vegetation; soil ecology, soil erosion; conservation, etc.

146. KELDANI, E. H.

1941    A bibliography of geology and related sciences concerning Egypt up to the end of 1939. Government Press, Cairo. 428 p.

147. KING, L. J.

1966    Weeds of the world: biology and control. Interscience Publishers, New York. 526 p.

The extensive bibliographies following each chapter, with the total references numbering over 5000, reflect the range, variety, and special approaches throughout the world to the problems associated with these plants considered troublesome, harmful, or otherwise annoying to man and to his agriculture. Topics covered include parasitic weeds, uses of weeds, harmful aspects of weeds, injurious interactions of weeds and crop plants, phytosociology and world distribution of weeds, herbicides, non-chemical methods for the control of weeds, etc. The subject index includes scientific plant names; the author index lists some 2500 personal names.

148. KIRMIZ, J. P.

1962    Adaptation to desert environment; a study on the jerboa, rat and man. Butterworths, London. 168 p.

A bibliography of over 250 references, p. 149-157.

149. KÖHLER, J.

1962    Deutsche dissertationen über Afrika; ein verzeichnis für die jahres 1918-1959. K. Schroeder, Bonn.

A topical arrangement.

150. KOROVIN, E. P.
    1961- Rastitel'nost Srednei Azii. 2nd ed. Aka-
    1962  demiia Nauk Uzbek S.S.R., Tashkent. 2
          vols.
          Originally published in 1934. This important con-
          tribution to Russian botanical literature on the
          vegetation of Central Asia and southern Kazakh-
          stan has an extensive bibliography in this greatly
          enlarged edition.

151. KRADER, L.
    1963  Peoples of Central Asia. University of
          Indiana Press, Bloomington. 319 p.
          Bibliography: p. 279-303.

152. KRAMER, H. P.
    1952  A selective annotated bibliography on
          the climatology of Northwest Africa. Me-
          teorological Abstracts and Bibliography
          3(1):37-79.
          This extensive bibliography itself designates fur-
          ther bibliographic sources by coding pertinent
          entries. 302 references.

153. KRAMER, P. J.
    1944  Soil moisture in relation to plant growth.
          Botanical Review 10(9):525-559.
          This general review of literature on soil moisture
          covers conditions of desert environment and in-
          cludes very useful information on the relation of
          soil moisture to plant growth. Includes a bibliog-
          raphy of 108 titles.

154. KREEB, K.
    1961  Bibliographia phytosociologica: Iraq. Ex-
          cerpta Botanica, B, 3(1):78.

155. KÜCHLER, A. W. (ed.)
    1965- International bibliography of vegetation
          maps, v. 1, 1965-  University of Kansas,
          Lawrence, Library Series 21.
          v. 1, Vegetation maps of North America, 453 p.,
          with J. McCormick; v. 2 (1966), Vegetation maps
          of Europe, 584 p.

156. ———
    1967  Vegetation mapping. The Ronald Press
          Company, New York. 472 p.
          Has excellent bibliography.

157. LANGMAN, I. K.
    1964  A selected guide to the literature on the
          flowering plants of Mexico. University
          of Pennsylvania Press, Philadelphia.
          1015 p.
          It is a pity to carp about such an immensely useful
          and comprehensive work as this, and I do not have
          any quarrel with the information but only with
          the arrangement. For instance, in examining the
          voluminous index for references to Sonora, a
          typically arid portion of Mexico, I found a sub-
          stantial list of plant names as cross-references;
          picking *Nitrophila occidentalis* at random, I was
          referred from this species, in its place in the index,
          to *Nitrofila,* from which, without further informa-
          tion, I was referred to Chenopodiaceae, which
          led in turn to a sub-heading under this family
          name, "*Nitrophila occidentalis,* range extension
          to Sonora, FOSBERG, F. R., 1942." Perhaps there
          is a no less cumbersome way to unlock the many
          treasures of this bibliography, for patience will
          turn the key. The main body of the work is an
          author arrangement; the index, referred to above,
          includes plant names, geographic terms, and
          subject terms, many of which are cross-references.
          There is a bibliography of 359 references.

158. LEITHEAD, C. S., and A. R. LIND
    1964  Heat stress and heat disorders. Cassell &
          Co., London. 304 p.
          Bibliography: p. 257-271.

159. LEOPOLD, L. B., M. G. WOLMAN, and J. P. MILLER
    1964  Fluvial processes in geomorphology. W. H.
          Freeman, San Francisco. 522 p.
          Comprehensive bibliographies follow each section
          of this text: evolving landscape, process and form
          (weathering, water and sediments in channels,
          channel form and process, hillslope character-
          istics and processes), effects of time (geochron-
          ology, drainage pattern evolution, channel
          changes with time, evolution of hillslopes); and
          to the extent such geomorphological concepts are
          important to arid lands research, these references
          are an excellent guide.

160. LITTLE, E. C. S., and G. W. IVENS
    1965  The control of brush by herbicides in tropi-
          cal grassland. Herbage Abstracts 35(1):
          1-12.
          A very good review of various methods used for
          control of brush. Although oriented toward tropi-
          cal and subtropical grassland, the materials and
          methods discussed are also applicable to desert
          conditions. Includes a bibliography of 179 ref-
          erences.

161. LIU, J. C.
    1930  Important bibliography on the taxonomy
          of Chinese plants. Peking Society of Nat-
          ural History, Bulletin 4(3):17-32. BA6(1)
          2109.
          Cites the principal papers containing descriptions
          of Chinese plants and, in most cases, gives the ap-
          proximate number of new Chinese plants de-
          scribed, so that some idea of the value of the work
          can be obtained.

162. LOBOVA, E. V.
    1960  Pochvy pustynnoi zony SSSR. Akademiia
          Nauk SSSR, Pochvennyi Institut im. V. V.
          Dokuchaeva. 362 p. (Translated 1967 by
          Israel Program for Scientific Translations,
          Jerusalem, as "Soils of the desert zone of
          the USSR," 405 p. Also cited as TT 67-
          51279.)
          Bibliography (original), p. 351-363; (translation),
          p. 373-403.

163. LOWE, C. H. (ed.)
    1964  The vertebrates of Arizona; annotated
          check lists of the vertebrates of the State:
          the species and where they live. University
          of Arizona Press, Tucson. 259 p.
          Part 1, "Landscapes and habitats," by C. H. Lowe,
          includes an extensive bibliography, p. 110-132.

164. LUST, J.
    1964  Index Sinicus, a catalogue of articles re-
          lating to China in periodicals and other
          collective publications, 1920-1955. W.
          Heffer and Sons, Ltd., Cambridge, England.
          663 p.
          Section on Sinkiang, p. 546-572.

165. LYDOLPH, P. E.
    1964  The Russian sukhovey. Association of
          American Geographers, Annals 54:291-
          309.
          Includes a brief but very specialized bibliography,
          p. 309, on this atmospheric phenomenon known
          elsewhere variously as khamsin, sirocco, etc. *See
          also* the earlier publication by Dzerdzeevskii
          (1957).

166. MACRO, E.
 1958 Bibliography of the Arabian peninsula. University of Miami Press, Coral Gables, Florida. 14, 80 p.

167. ——
 1960 Bibliography of Yemen, and notes on Mocha. University of Miami Press, Coral Gables, Florida. 63 p.

168. MAGIN, G. B., JR., and L. E. RANDALL
 1960 Review of literature on evaporation suppression. U. S. Geological Survey, Professional Paper 272-C.

169. MAICHEL, K.
 1962- Guide to Russian reference books. Hoover
 1964 Institution on War, Revolution, and Peace, Stanford University Press. 2 vols.
 Volume 1 lists general bibliographies and reference books; the second volume contains bibliographies published in the fields of Soviet history, world history, auxiliary historical sciences, enthnography, and geography. This is one of the most exhaustive, up-to-date bibliographies of its kind. Annotated.

170. MARGAT, J.
 1958 Bibliographie hydrogéologique du Maroc, 1912-1958. Ed. du Service Géologique du Maroc, Rabat, Notes et Mémoires 142. 591 references.

171. ——
 1964 Guide bibliographique d'hydrogéologie; ouvrages et articles en langue française. Bureau de Recherches Géologiques et Minières, Paris, suite Hydrogéologie. 113.
 Only the more important and recent (1950 and later) are included. The section listed as "Hydrogéologie des regions arides et semi arides," p. 47-52, is arranged geographically and cites approx. 100 references.

172. MARTÍNEZ, M.
 1936 Plantas útiles de México. 2nd ed. México. 400 p.
 This excellent reference replaces an earlier edition published in 1928. Lists the most useful plants of Mexico alphabetically by common name, with a bibliography for each. Includes information on where they grow, their cultivation, chemical composition, industrial application, food value, and medicinal value. Includes many drawings and photographs.

173. ——
 1959 Plantas útiles de la flora mexicana. [3rd ed.?] Ediciones Botas, México, D. F. 621 p.
 Descriptions and discussion on useful plants other than those commonly cultivated. Bibliography.

174. MASSON, H.
 1954 La rosée et les possibilitiés de son utilisation. Institut des Hautes Etudes, L'Ecole Supérieure des Sciences, Dakar, Senegal, Annales 1. 44 p. MGA 6B-196.
 A general review of the literature on dew and the instrumentation for measuring it. A complete and comprehensive though poorly organized bibliography of 320 references is included.

175. MAXIMOV, N. A.
 1929 The plant in relation to water. Authorized English translation, edited, with notes, by R. H. Yapp. George Allen and Unwin, Ltd., London. 451 p.

One of the earlier comprehensive studies of the physiological basis of drought resistance. It provides a great deal of fundamental information on plant-water relations, such as absorption of water by the plant, loss of water by the plant, and water balance and drought resistance of plants, particularly the chapter on xerophytes (p. 249-283). Bibliography: p. 403-434.

176. MCGILL, J. T.
 1960 Selected bibliography of coastal geomorphology of the world. Los Angeles. 50 p.
 This bibliography of 933 titles consists principally of references used in the preparation of McGill's recent (1958) map of coastal land forms of the world; references are regional and topical; many of the regional references cover more than the coastal portion of a given country or region.

177. MCGINNIES, W. G.
 1968 Publications related to the work of the Desert Botanical Laboratory of the Carnegie Institution of Washington, 1903-1940. University of Arizona, Office of Arid Lands Studies, Tucson. 50 p.
 Various sources were used to compile this list of publications representative of the work of this famous Laboratory: mainly they were the Year Books of the Carnegie Institution and material supplied by the U. S. Forest Service which took over the Laboratory grounds, buildings, and library in 1940. 530 references in author arrangement. No index since material is specialized. Mimeo.

178. MCGINNIES, W. G., B. J. GOLDMAN, and P. PAYLORE, eds.
 1968 Deserts of the world, an appraisal of research into their physical and biological environments. University of Arizona Press, Tucson. 780 p.
 Sections by individual scientists on weather and climate, geomorphology and surface hydrology, surface materials, vegetation, fauna (with desert disease information), desert coastal zones, and groundwater hydrology, each accompanied by extensive annotated bibliographies (e.g. geomorphology, 2455 references; vegetation, 1062 references, etc.), for a total of nearly 6000 prime citations.

179. MEIGS, P.
 1966 Geography of coastal deserts. Unesco, Paris. Arid Zone Research 28. 140 p., 15 maps.
 Identifies the world's coastal deserts in the text and on 15 maps. Discusses in detail climate and terrain. No data for the offshore areas. Bibliography, p. 134-140.

180. MERRILL, E. D., and E. H. WALKER
 1938 A bibliography of eastern Asiatic botany. Harvard University, Arnold Arboretum, Jamaica Plain, Massachusetts. 719 p.
 A valuable reference. See Walker (1960) for continuation.

181. MIAGKOV, N. I.
 1957 Bibliograficheskii ukazatel' literatury po limatu Turkmenii. (Bibliography on the climate of the Turkmen Republic) Upravlenie Gidrometsluzhby Turkmenskoi SSR. Ashkhabadskaia Gidrometeorologicheskaia Observatoriia, Ashkhabad. 104 p. MGA 11:8-295.                      (continues)

1429 entries, divided into 15 sections, including radiation and heat balance, air and soil temperature and humidity, precipitation and runoff, moisture exchange, agricultural meteorology, synoptic processes, air pressure, winds, etc.

182. MIDDLE EAST INSTITUTE
1956- Current research on the Middle East,
1958 1955-1957. Washington, D. C.
Research in progress, presented by subject fields, with geographical and researcher indexes. Sections on geography and regional surveys in each issue.

183. MIDDLETON, C.
1965 Bechuanaland; a bibliography. University of Cape Town, School of Librarianship. 37 p.

184. MILLER, A. H., and R. C. STEBBINS
1964 Lives of desert animals in Joshua Tree National Monument. University of California Press, Berkeley. 452 p.
Covers birds, mammals, amphibians, reptiles, with a chapter on faunal analysis. Bibliography: p. 435-441.

185. MISRA, R.
1955 Progress of plant ecological studies in India. University of Saugar, Botanical Society, Bulletin 7(2):90-101. BA30(12) 33660.
The chief value of this paper is its bibliography of 177 entries.

186. MOGADISCIO. Camera di Commercio, Industria et Agricoltura della Somalia
1958 Bibliografia somala. Scuola Tipografica Missione Cattolica. 135 p.

187. MONOD, T.
1957 Les grandes divisions chorologiques de l'Afrique. (The large mapping divisions of Africa) Commission for Technical Cooperation in Africa/Scientific Council for Africa, Publication 24. 147 p.
Contains an extensive bibliography (p. 127-137).

188. ──────
1963 Notice sur les titres et travaux scientifiques de Théodore Monod. Imprimerie Protat Frères, Macon. 126 p.
This is the only such entry in this compilation for a bibliography of a single person, but the significance of Monod's scientific investigations in arid Africa over a period of 45 years cannot be overestimated. In addition to a list of his "grades, functions, titres divers," there is an interesting list of his principal voyages and missions, 1921-1963. The "exposé général" is a narrative account of his work in northern Africa, for the most part, arranged by disciplines: geography, geology, botany, zoology, conservation, etc., with numbered references to appropriate citations in the bibliography which follows. The publications list, numbering 475, is a chronological one, 1921-1963, with notice of additional works in press and in preparation. An appendix cites 286 other "marginal" publications, including more popular works (listed as Sér. S). And finally there is a list of publications of others (Sér. R., over 300 citations) whose work utilizes the findings of Monod's lifetime investigations in arid lands.

189. ──────
1967 Orientation bibliographique sur le Sud-Est du Désert Lybique (triangle ᶜUweinat-Erdis-Merga). In T. Monod, Rapport sur

une mission exécutée dans le nord-est du Tchad du 19 décembre 1966 au 20 janvier 1967. Institut National Tchadien pour les Sciences Humaines, Fort-Lamy, Etudes et Documents Tchadiens, sér. A, annexe 3. 50 p. mimeo.
Annotated (in French). A valuable selection of often obscure references to this specific area, many from Monod's personal library of Saharan materials.

190. MORELLO, J.
1955- Estudios botánicos en las regiones áridas
1956 de la Argentina. Revista Agronómica del Noroeste Argentino 1:301-370, 385-524; 2:79-152.
A total of 160 references support the three parts of this work on the ecology and phytogeography of the arid zones of Argentina, principally the Monte, with emphasis on water balance of the vegetation. English summaries.

191. MORIN, P.
1965 Bibliographie analytique des sciences de le terre: Maroc et régions limitrophes (depuis le début des recherches géologiques à 1964). Morocco, Service Géologique, Notes et Mémoires 182. 2 vols.
Anonymes, documents complémentaires, cartes.

192. MULLER, D. J.
1953 The Orange River from the confluence of the Vaal and Orange Rivers to the mouth of the Orange in the Atlantic Ocean; a bibliography. University of Cape Town, School of Librarianship. 21 p.
An excellent bibliography.

193. MUMA, M. H.
1951 The arachnid order Solpugida in the United States. American Museum of Natural History, Bulletin 97:35-141.
Includes a brief but especially useful bibliography of 30 references, largely to N. A. fauna.

194. MUÑOZ CRISTI, J.
1955 Bibliografía geológica de Chile (1927-53). Universidad de Chile, Santiago de Chile, Instituto de Geológica, Publicación 5. 121 p.
A classified, partly annotated list of over 600 titles on the geology of Chile.

195. EL NASRI, A. R.
1962 A bibliography of the Sudan, 1938-1958. Oxford University Press, London. 171 p.
Over 2800 references arranged by broad categories such as agriculture (broken down by provinces), anthropology (broken down by tribes), botany, forestry, geography, geology, medicine, meteorology, zoology, and others. It has a subject and author index, and a bibliography. Based on R. L. Hill's 1939 work (q.v.), and is itself continued by Ibrahim and Nasri (1965), q.v.

196. NATIONAL ACADEMY OF SCIENCES
1966 Weather and climate modification, problems and prospects. II: Research and development. Publication 1350. 198 p.
A comprehensive analysis of the current status of this controversial subject. Includes bibliography, p. 147-159.

197. NATIONAL BOOK CENTRE OF PAKISTAN
1965 Books on Pakistan, a bibliography. 2nd ed. Karachi. 71 p.
Includes books on Pakistan published in Pakistan

NATIONAL BOOK CENTRE OF PAKISTAN (*cont*)
and elsewhere. Emphasis is on cultural aspects.
The arrangement is by subject, with indexes of
titles and publishers.

198. NATIONAL IRANIAN OIL COMPANY
1959   Geological map of Iran, scale 1:2,500,000,
with explanatory notes. Tehran.
The explanatory notes that accompany the map
include an extensive bibliography of Iranian
geology.

199. NAZAREVSKIY, A. (ed.)
1957   Kazakhskaya SSR; ekonomiko-geogra-
ficheskaya kharakteristika. Gosudarst-
vennoye Izdatel'stvo Geograficheskoy Literatury, Moskva. 733 p.
A good detailed and comprehensive geographical
analysis of the Kazakh Republic, including discussions by region and oblast. Maps, and an extensive bibliography (p. 701-732).

200. NEAL, J. T. (ed.)
1965   Geology, mineralogy, and hydrology of
U. S. Air Force Cambridge Research Laboratories, Environmental Research Papers
96. 176 p. Also cited as AD-616 243.
A series of 6 papers, each accompanied by a bibliography, for a total of over 100 references, with
only slight overlapping. Since the papers relate to
U. S. areas, the references reflect this circum-
scription.

201. NUTTONSON, M. Y.
1958   The physical environment and agriculture
of Australia with special reference to its
winter rainfall regions and to climatic and
latitudinal areas analogous to Israel.
American Institute of Crop Ecology, Washington, D. C. 1124 p. MGA 9.10-96.
One of several studies compiled by Nuttonson
(see his 1961*a, b,* and *c*) featuring excellent general discussions comparing characteristics of
various arid regions to Israel. All include substantial bibliographies.

202. ———
1961*a* Introduction to North Africa and a survey
of the agriculture of Morocco, Algeria, and
Tunisia with special reference to their
regions containing areas climatically and
latitudinally analogous to Israel. (Vol. 1 of
a Survey of North African Agro-Climatic
Counterparts of Israel). American Institute
of Crop Ecology, Washington, D. C. 608 p.

203. ———
1961*b* Physical environment and agriculture of
Libya and Egypt with special reference to
their regions containing areas climatically
and latitudinally analogous to Israel.
(Vol. 2 of a Survey of North African Agro-
climatic counterparts of Israel). American
Institute of Crop Ecology, Washington,
D. C. 452 p.
Bibliography: p. 439-452.

204. ———
1961*c* The physical environment and agriculture
of the Union of South Africa with special
reference to its winter-rainfall regions containing areas climatically and latitudinally
analogous to Israel. A study based on official records, material, and reports of the
Department of Agriculture and of other
government and provincial agencies of the

Union of South Africa. American Institute
of Crop Ecology. Washington, D. C. 459 p.
Bibliography: p. 452-459.

205. OBERDORFER, E.
1961   Bibliographia phytosociologica: Chile. Excerpta Botanica, B, 3(1):79-80.

206. OFFICE DE LA RECHERCHE SCIENTIFIQUE ET TECH-
NIQUE OUTRE-MER
1962(?)-Bulletin bibliographique de pédologie. 1,
1962(?)-   to date. ORSTOM, Paris.
References on a variety of aspects of soil science
are abstracted or annotated (in French). Many
relate to ORSTOM missions in arid lands.

207. OFFIELD, T. W.
1964   Preliminary bibliography and index of
the geology of Pakistan. Geological Survey
of Pakistan, Records 12(1).

208. OPPENHEIMER, H. R.
1961   Essai d'une révision des trèfles de la Palestine. Société Botanique de France, Bulletin
108(1/2):47-71. BA38(5)19860.
Abundant references to the literature on trifolium
and a bibliography of 53 citations.

209. ORNI, E., and E. EFRAT
1966   Geography of Israel. 2nd ed., rev. Israel
Program for Scientific Translations, Jerusalem. 363 p.
The first geography of Israel in English, translated
from Hebrew. Comprehensive and particularly
well done are the sections on physical geography.
Valuable though brief information on the Negev
is dispersed throughout the text. Includes a useful
bibliography (p. 333-344).

210. OZENDA, P.
1958   Flore du Sahara septentrional et central.
Centre National de la Recherche Scientifique, Paris. 486 p. BA 33(8)30868.
One of the best descriptions of vegetation and
conditions of the Sahara. Part III includes definitions of botanical terms used, a bibliography, and
an alphabetical index of families, genera, and
synonyms.

211. PAKISTAN NATIONAL SCIENTIFIC AND TECHNICAL
DOCUMENTATION CENTRE
1957   Sediment control in rivers, canals, and
reservoirs. Karachi. Bibliography 18.

212. ———
1958   Dams and reservoirs. Karachi. Bibliography 75.

213. ———
1959   Flood control, a select bibliography (1900-
1958). Karachi. Bibliography 101. 116 p.

214. ———
1960   Engineering geology. Karachi. Bibliography 212.

215. ———
1961   Scientific and technical periodicals of
Pakistan. Bibliography 303. 12 p.

216. ———
1965   List of PANSDOC bibliographies. Karachi. 27 p.

217. PALMER, E.
1966   The Plains of Camdeboo. Collins, London.
320 p.
A popular discussion of the Karroo, with an excellent bibliography.

218. PATAI, R.
1957   Jordan, Lebanon and Syria, an annotated
bibliography. Human Relations Area Files,
New Haven. 289 p.

219. PAYLORE, P.
1966 Seventy-five years of arid-lands research at The University of Arizona: a selective bibliography, 1891-1965. University of Arizona, Office of Arid Lands Research. 95 p.
1609 references, arranged chronologically within the 18 subdivisions. These include geochronology and geology, ancient and modern peoples and land use, tree-rings, atmospheric environment, water: resources, uses; soils and fertilizers, agronomy, horticulture, plant breeding and genetics, entomology, plant diseases, plant chemistry, native vegetation, range management, animal science and industry, wildlife and wilderness studies, environmental studies, and a separate section of theses and dissertations. Author index.

220. PÉREZ MOREAU, R. A.
1965 Bibliografía geobotánica patagónica; contribución as la bibliografía botánica argentina. Instituto Nacional del Hielo Continental Patagónico, Buenos Aires, Publicación 8. 110 p.

221. PETERSON, A. D.
1957 Bibliography on the climate of Iran. U. S. Weather Bureau, Washington, D. C. 26 p.

222. PETROV, M. P.
1939 Ecology of desert plant culture (translated title). Voprosy Ekologii i Biotsenologii 5/6:3-39. BA15(8)16414.
Reviews work at the Repetek Sandy Desert Station in the southeast Kara-Kum Desert. Includes a bibliography of 65 titles.

223. ——
1966 A review of literature on the reclamation of desert- and semi-desert sands in the USSR for 1965 (translated title). Akademiia Nauk Turkmenskoi SSR, Izvestiya, ser. Biologicheskikh Nauk 6:79-87. HA(37)1338.
This is a list of 106 annotated references to the natural conditions and vegetation of desert- and semi-desert sands and to their reclamation by the planting of windbreaks or sowing herbage plants.

224. PHILLIPS, J.
1956 Aspects of the ecology and productivity of some of the more arid regions of southern and eastern Africa. Vegetatio 7(1):38-68. BA 32(7)22191.
Describes the drier regions of southern and eastern Africa, including the Namib, Karroo, and Kalahari. Includes a bibliography of about 150 references.

225. PICKWELL, G.
1939 Deserts. Whittlesey House, London; McGraw-Hill, New York. 174 p.
A popular book, with brief bibliography on deserts, desert plants and animals.

226. PIERCE, R. A.
1966 Soviet Central Asia, a bibliography. University of California, Center for Slavic and East European Studies. 3 vols.
Pt. 1, 1558-1866; pt. 2, 1867-1917; pt. 3, 1917-1966.

227. PITHAWALLA, M. B.
1953 The problem of Baluchistan; development and conservation of water resources, soils, and natural vegetation. Pakistan Ministry of Economic Affairs, Karachi. 166 p.

A good review paper on the soils and water resources of this region with data on drainage features and soil erosion. Bibliography: p. 153-162.

228. ——
1959 A physical and economic geography of Sind (the Lower Indus Basin). Aage Kadam Printery, Karachi. 389 p.
Bibliography: p. 375-389.

229. PLAAT, A. F.
1951 List of books and pamphlets in German on South Africa and South West Africa published after 1914, as found in the South African Public Library, Cape Town. University of Cape Town, School of Librarianship. 61 p.
A good bibliography though coverage incomplete.

230. POLISH ACADEMY OF SCIENCES
1963 Problems of geomorphological mapping. Geographical Studies 46.
This is the best summary paper available on the methods and goals of geomorphological mapping. The arid areas that are included among the regions treated in individual articles are parts of Sudan and Morocco. This work contains many references and an appendix which presents the symbols used to depict various types of landforms.

231. POLLER, R. M.
1964 Swakopmund and Walvis Bay: a bibliography. University of Cape Town. 29 p.
An excellent compilation, with helpful cross-references.

231a. PSUTY, N. P., W. BECKWITH, and A. K. CRAIG
1968 1000 selected references to the geography, oceanography, geology, ecology, and archaeology of coastal Peru and adjacent areas. 3rd rev. Paracas Papers 1(1). 52 p. ONR Geography Branch contract N00014-67-A-0320. Also cited as AD-671 870.
Compiled in support of a long-range program to study marine desert ecology, particularly as it occurs along the central coast of Peru.

232. PURI, R. K.
1961 Bibliography relating to geology, mineral resources, paleontology, etc., of the Somali Republic. Somali Republic, Survey Report RKP/1, Hargeisa.
A good bibliography on Somalia through the date of publication. For current information on mapping, geological studies, and allied topics, the "Annual Reports" of the Geological Survey Department of the Ministry of Industry and Commerce, at Hargeisa, should be consulted. These reports list all current projects and their completion status, and all publications issued during the year involved.

233. RAGONESE, A. E.
1955 Plantas tóxicas para el ganado en la región central Argentina. Universidad Nacional de La Plata, Facultad de Agronomía, Revista, 3a. ép., 31:133-336.
Systematic study of poisonous plants, native and introduced, in central Argentina, with details about their toxic properties. An 18-page bibliography.

234. RAGONESE, A. E., and R. MARTINEZ CROVETTO
1947 Plantas de la Argentina con frutos o semillas comestibles. (Indigenous plants of Argentina with edible fruits or seeds)

RAGONESE AND CROVETTO (*cont*)
Revista de Investigaciones Agrícolas
1:147-216.
A systematic account of 233 species of Argentine flowering plants with edible seeds or fruits including many of the arid zones. A brief description, many illustrations of the edible parts, special references, and bibliography are given. Particularly useful for a subsistence manual.

235. RAHEJA, P. C.
1966   Aridity and salinity, a survey of soils and land use. p. 43-127. *In* H. Boyko, ed., Salinity and aridity, new approaches to old problems. W. Junk, The Hague. 408 p. (Monographiae Biologicae 16)
A worldwide survey of salt-affected soils in arid lands, with particularly detailed description of Indian soils, characteristic vegetation, and land use. Includes an extensive bibliography of 261 references.

236. RAYNAL, R.
1961   Plaines et piedmonts du bassin de la Moulouya (Maroc oriental); étude géomorphologique. Imframar, Rabat. 617 p.
Bibliography: p. 577-586.

237. RAYSS, T., and S. BORUT
1958   Contribution to the knowledge of soil fungi in Israel. Mycopathologia 10:142-174.
A list with descriptive notes of 107 species of soil fungi isolated from arid soils of the Judaean Desert and the northern Negev. 40 references.

238. RAZUMOVA, L. A.
1963   Izmenenie agrometeorologicheskikh uslovii pod vliianiem polezashchitnogo lesonasazhdeniia. (Changes in agrometeorological conditions due to shelter belts) Tsentral'nyi Institut Prognozov, Trudy 131:64-100. MGA 15:8-72.

239. REPARAZ, G. DE
1958   La zone aride du Pérou. Geografiska Annaler 40(1):1-62.
An extensive study of the physical and biological geography of the arid coast, with a good bibliography, p. 60-62.

240. REYNOLDS, H. G.
1962   Selected bibliography on range research in Arizona. American Society of Range Management, Arizona Section, Proceedings, p. 50-68. (Reprinted as a separate for the U. S. Forest Service.)
237 references. Includes a 5-page narrative summary by topics, with sources from the bibliography indicated by number. Historical development sketched briefly on such subjects as revegetation, poisonous plants, soil and water conservation, climatic influences, and woody plant control.

241. RICE, M. L.
1961   Bibliography on meteorological articles (1949-1960) [to accompany M. Rigby's article, "Meteorology, hydrology, and oceanography."] *In* Sciences in Communist China. American Association for the Advancement of Science, Publication 68: 558-601.

242. RICEMAN, D. S.
1953   Minor element deficiencies and their correction. International Grassland Congress, 6th, 1952, Proceedings 1:7101717.
A review of 49 references to trace-element deficiencies in Australia.

243. RICHARDS, H. G., and R. W. FAIRBRIDGE
1965   Annotated bibliography of Quaternary shorelines (1945-1964). Academy of Natural Sciences, Philadelphia, Special Publication 6. 280 p.
This bibliography covers the world by countries and includes the desert coastal areas. Extremely useful for the coverage of foreign literature.

244. RIKLI, M.
1942-  Das pflanzenkleid der Mittelmeerländer.
1948   Hans Huber, Bern. 1418 p. 3 vols.
This important work deals not only with vegetation of the countries bordering the Mediterranean but also of those into which Mediterranean flora is penetrating. An excellent source of information for many vegetational (particularly plant-geographical) problems, as well as for the desertic regions of Iran, Iraq, Arabia, and the Sahara. Hauptliteraturverzeichnis: v. 3, p. 1114-1309.

245. RODIER, J.
1963   Bibliography of African hydrology. Unesco, Paris. Natural Resources Research 2. 166 p.
Though suffering from lack of an author index, this is otherwise an extremely useful bibliography providing references to most of the important hydrological investigations undertaken in Africa. Arranged by countries under topics such as precipitation, evaporation, run-off, hydrological balance, infiltration, soil moisture, solid transports, groundwater.

246. RODIN, L. E.
1961   Dinamika rastitel'nosti pustyn. (Dynamics of desert vegetation) Akademiia Nauk S. S. S. R., Botanicheskii Institut. 227 p.
Bibliography: p. 208-221.

247. ——
1963   Rastitel'nost pustyn' Zapadnoi Turkmenii. (Desert vegetation of West Turkmenistan) Akademiia Nauk S. S. S. R., Botanicheskii Institut. 309 p.
Bibliography: p. 273-294. English contents.

248. RODINOFF, N. R. (ed.)
1951   Soviet geography, a bibliography. Library of Congress, Washington, D. C. 2 vols.

249. ROMAN, S. J.
1958   Bibliography of climatic maps for Iraq. U. S. Weather Bureau, Report WB/BM-5. 19 p. Cited as PB-176 038; AD-665 176.
The majority of the sources reviewed contain maps, not for Iraq alone, but for the Near or Middle East. Maps for some countries, such as India, Iran, Israel, and Turkey, overlap and include all or parts of Iraq.

250. ROUGERIE, G.
1962   Systèmes morphogéniques et familles de modelés dans les zones arides. Centre de Documentation, Paris. 167 p.
Bibliography: p. 74-77 (57 references); p. 157-163 (167 references).

251. ROUKENS DE LANGE, E. J.
1961   South West Africa, 1946-1960, a selective bibliography. University of Cape Town, School of Librarianship. 51 p.
322 references. *See* Welch (1946) for 1919-1946.

252. ROUSSINE, N., and C. SAUVAGE
1961   Bibliographia phytosociologica: Afrique du Nord. Excerpta Botanica, B, 3(1):34-51.

253. ROZANOV, A. N.
1961   Serozems of Central Asia. Israel Program

for Scientific Translations, Jerusalem. 550 p. (Translated from the Russian "Serozemy Srednei Azii," published in 1951 by Izdatel'stvo Akademii Nauk S. S. S. R.). Bibliography: p. 497-541.

254. RUSSELL, F. E., and R. S. SCHARFFENBERG
1964 Bibliography of snake venoms and venomous snakes. Bibliographic Associates, Inc., West Covina, California.
5829 citations arranged by broad subject categories including biological effects, physiopharmacology and toxicology (cardiovascular, blood, and nervous system), immunology and antivenin, snakebite, pathology and bacteriology, experimental therapeutics, etc. The references in sections on "general biology of venomous snakes" and "snakebite" are arranged geographically; others are arranged by families: Elapidae and Colubridae, Crotalidae, Viperidae, and Hydrophiidae. There is a separate section of 163 references to works published prior to 1850. There is also a complete author list to the whole work.

255. RZHANITSYN, N. A.
1960 Morphological and hydrological regularities of the structure of the river net. U. S. Agricultural Research Service. 380 p.
This report is illustrative of the application of quantitative drainage analysis to hydrologic problems in the U. S. S. R. It contains an extensive bibliography of allied work that has been reported in the Russian literature.

256. SAID, R.
1962 The geology of Egypt. Elsevier Publishing Company, Amsterdam. 377 p.
A modern and comprehensive survey of the geology of Egypt which provides more current data than the original compendium by Hume (1925-1939). Bibliography, p. 334-348.

257. SAMPSON, A. W.
1939 Plant indicators: concept and status. Botanical Review 5(3):155-206.
An excellent discussion of plants in relation to climate, soils, grazing use, fire, and other factors. Also a good review of literature and an extensive bibliography.

258. SASSON, A.
1967 Recherches écophysiologiques sur la flore bactérienne de sols de régions arides du Maroc. Institut Scientifique Chérifien, Rabat, Travaux, sér. Botanique et Biologique Végétale 30. 231 p.
Bibliographie: 203-213 (349 refs.).

259. SCHMIDT-NIELSEN, K.
1964 Desert animals; physiological problems of heat and water. Oxford University Press, London. 277 p.
Bibliography: p. 253-270.

260. SCHOELLER, H.
1959 Arid zone hydrology: recent developments. Unesco, Paris. Arid Zone Research 12. 125 p.
Bibliography: p. 116-125.

261. SCHOFF, S. L.
1962 Hydrologic investigations for northeastern Brazil. U. S. Geological Survey, Administrative Report. Recife.
A valuable summary of available information on past and current hydrological work, including description of the "drought polygon." A good bibliography appended.

262. SCHWEIGGER, E.
1959 Die westküste Südamerikas im bereich des Peru-stroms. (The west coast of South America in the vicinity of the Peru Current) Keysersche Verlagsbuchhandlung, Heidelberg. 513 p. MGA 12. 9-6.
240 references.

263. SELTZER, P.
1946 Le climat de l'Algérie. Institut de Météorologie et de Physique du Globe de l'Algérie, Travaux, Hors série. 219 p.
A comprehensive and detailed study of the Algerian climate. An extensive bibliography is included.

264. SHREVE, F.
1942 The desert vegetation of North America. Botanical Review 8(4):195-246.
Thorough literature review.

265. SIBBONS, J. L. H.
1962 A contribution to the study of potential evapotranspiration. Geografiska Annaler 44:279-292.
An excellent review article, with bibliographical footnotes throughout.

266. SINCLAIR, P. C.
1966 A quantitative analysis of the dust devil. University of Arizona, Department of Meteorology. Ph.D. dissertation. 292 p.
Includes bibliography.

267. SINGHVI, M. L., and D. S. SHRIMALI
1962 Reference sources in agriculture; an annotated bibliography. Rajasthan College of Agriculture, Udaipure. 428 1.

268. SLATYER, R. O., and J. A. MABBUTT
1964 Hydrology of arid and semiarid regions. Sec. 24:1-46. In V. T. Chow, Handbook of applied hydrology. McGraw-Hill, New York.
117 references, p. 42-46.

269. SNYDER, T. E.
1954 Annotated subject-heading bibliography of termites, 1350 B.C. to A.D. 1954. Smithsonian Miscellaneous Collections 130. 305 p.

270. ——
1961 Supplement to the annotated, subject-heading bibliography of termites, 1955-1960. Smithsonian Miscellaneous Collections 143:1-137.

271. SOMMER, J. W.
1965 Bibliography of African geography, 1940-1964. Dartmouth College, Hanover, N. H., Geography Publications at Dartmouth 3. 139 p.

272. SOUTH AFRICA. WEATHER BUREAU
1965 Climate of South Africa, Part 8: General Survey. Pretoria. 330 p. MGA 17.1-394.
This volume is a complete discussion of the surface climate of South Africa. "The exhaustive text [is] duly documented."

273. SPARN, E.
1964 Cuarta contribución al conocimiento de la bibliografía meteorológica y climatológica de la Republica Argentina. Años 1949-1955. Academia Nacional de Ciencias, Córdoba, Miscelánea 44:1-39.
Fourth contribution to the knowledge of meteorological and climatological bibliography of Argentina, 1949-1955.

274. SPATE, O. H. K., and A. T. A. LEARMONTH
     1967  India and Pakistan, a general and regional
           geography. 3rd ed., rev. Methuen, London.
           877 p.
     A full-scale revision of earlier editions, necessi-
     tated by changes during the last decade. Text is
     arranged by sections on the land, the people,
     the economy, the face of the land. There are
     voluminous annotated evaluative bibliographical
     footnotes throughout. There is a Russian trans-
     lation of the 2nd (1957) edition, "Indiya i Pakis-
     tan; obshchaya i regionalnaya geografiya," avail-
     able from Foreign Literature Publishing House,
     Moscow.

275. SPOHR, O. H.
     1950  Catalogue of books, pamphlets and peri-
           odicals published in German relating to
           South Africa and South West Africa, Uni-
           versity of Cape Town. 71 p.
     A rather selective work, but well worth con-
     sulting.

276. STAMP, L. D. (ed.)
     1961  A history of land use in arid regions.
           Unesco, Paris. Arid Zone Research 17.
           388 p.
     Chapter bibliographies.

277. STANDING COMMITTEE ON LIBRARY MATERIALS ON
     AFRICA
     1964  Theses on Africa accepted by universities
           in the United Kingdom and Ireland. W.
           Heffer, Cambridge. 74 p.
     Comprehensive for its narrow scope.

278. STEPANOVA, N. A.
     1951  Climatological aspects of the Virgin Soil
           Project in the U. S. S. R. U. S. Weather
           Bureau, Washington, D. C. 35 p. Also
           cited as PB-176 045.
     Marginal weather conditions in newly-developed
     areas of Kazakhstan. This survey is designed to
     give an elementary introduction to pertinent
     climatological literature published by Soviet
     scientists.

279. STEPHENS, C. G., and C. M. DONALD
     1958  Australian soils and their responses to
           fertilizers. Advances in Agronomy 10:167-
           256.
     130 references, p. 253-256, on the Australian
     soil landscape and its deficiencies, relationship
     of genetic factors to soil fertility, and soil classi-
     fication. Author/source information only, making
     selection of appropriate titles difficult.

280. STEVENS, P.
     1947  Bechuanaland bibliography. University of
           Cape Town, 27 p.
     An excellent bibliography of 305 references.

281. STUNTZ, S. C., and E. E. FREE
     1911  Bibliography of eolian geology. In E. E.
           Free, The movement of soil material by
           the wind. U. S. Department of Agriculture,
           Bureau of Soils, Bulletin 68:174-263.
     An alphabetical arrangement by author used as
     an author index to Free's text. The subject index
     to the text will lead the user to author citations
     on various subjects.

281a. SYMPOSIUM ON COCCIDIOIDOMYCOSIS, PHOENIX,
      1965
      1967  Coccidioidomycosis, Papers from the 2nd
            Symposium on Coccodioidomycosis, Phoe-
            nix, Arizona, December 8-10, 1965. Edited

     by Libero Ajello. University of Arizona
     Press, Tucson. 434 p.
     Literature cited: p. 415-430. While it is difficult
     to compare this with Fiese (1958), q.v., because of
     the differences in arrangement, it appears that
     the earliest reference cited in this list is to a 1921
     publication (57 of Fiese's references occur before
     this date). Of the total of 543 papers listed in the
     Symposium, the majority were published during
     the late 1950's and up to 1965, so that we may
     consider this bibliography in effect an up-dating
     of the Fiese work.

282. SYMPOSIUM ON REMOTE SENSING OF ENVIRONMENT
     1962-  Proceedings, 1st, 1962-  to date. Uni-
            versity of Michigan, Institute of Science
            and Technology, for the Office of Naval
            Research. (1st-2nd, February, October
            1962, also cited as AD-274 155 and AD-
            299 841)
     Many papers delivered at these symposia and
     published as their Proceedings are accompanied
     by bibliographies of varying length and impor-
     tance. For information on this relatively new field
     of research, this source can be helpful.

283. TABORSKY, O., and G. THURONYI
     1960  Annotated bibliography on weather modi-
           fication. Meteorological Abstracts and
           Bibliography 11(12):2181-2415.

284. TENRI CENTRAL LIBRARY
     1960-  Africana; catalogue of books relating to
            Africa in the Tenri Central Library. Tenri
            Central Library, Tenri, Japan, Library
            Series 24-  and continuation.
     An excellent thorough regional bibliography.

285. TERRY, R. D.
     1955  Bibliography of marine geology and ocean-
           ography, California coast. California Divi-
           sion of Mines, Special Report 44. 131 p.
     Covers sedimentation, submarine topography,
     beach erosion and its control, marine engineering
     problems, coastal sand dunes, marine geophysics,
     salt-water intrusion, and physical and chemical
     oceanography.

286. THEAL, G. M.
     1912  Catalogue of books and pamphlets relating
           to Africa south of the Zambezi, in the
           English, Dutch, French, and Portuguese
           languages. Cape Times Limited, Cape
           Town. 408 p.
     An author arrangement, with poor subject and
     geographical indexes.

287. TÖTEMEYER, G.
     1964  Südafrika, Südwestafrika. — South Africa,
           South West Africa; eine bibliographie. — a
           bibliography, 1945-1953. Freiburg im
           Breisgau. 284 p. (Materialien des Arnold-
           Bergstraesser Instituts f. Kulturwissen-
           schaftliche Forschung)

288. TOUPET, C.
     1962  Orientation bibliographique sur la Mauri-
           tanie. Institut Français d'Afrique Noire,
           Bulletin, sér. B, 24:594-613.
     An extremely useful bibliography of work in
     Mauritania.

289. TRAVIS, B. V., and R. M. LABADAN
     1967a  Arthropods of medical importance in Asia
            and the European USSR. Pt. 1: Introductory
            and explanatory material; data on mos-
            quitoes. Pt. 2: Data on Arthropods other

than mosquitoes. U. S. Army Natick Laboratories, Technical Report 67-65-ES. 694 p.

Like Travis, Mendoza, and Labadan (1967), below, for Africa, this Report covers large desert areas of Asia and the European U. S. S. R., including the Middle East and Soviet Central Asia. Literature cited: mosquitoes, p. 273-322; black flies, p. 340-343; sand flies, p. 370-379; midges, p. 404-406; horse flies, p. 488-494; non-biting flies, p. 508-509; fleas, p. 576-586; Hemiptera, p. 590-591; urticating and vesicating arthropods, p. 599; ticks, p. 643-655; mites, p. 672-675; miscellaneous arthropods, p. 683-684.

290. ———

1967b Arthropods of medical importance in Latin America. Pt. 1: mosquitoes. Pt. 2: arthropods other than mosquitoes. U. S. Army Natick Laboratories, Technical Report 68-30-ES. 491 p.

Summarizes literature for mosquitoes (p. 193-215), black flies (p. 241-244), sand flies (p. 264-267), midges (p. 288-291), horse flies (p. 366-370), biting flies (p. 373), non-biting flies (p. 381-383), fleas (p. 417-421), bugs (p. 433-436), urticating and vesicating arthropods (p. 442), ticks (p. 468-474), mites (p. 482-483), and miscellaneous arthropods (p. 490-491). Because of extensive humid tropic regions in Latin America, the scattered references to arthropods in the desert areas of Argentina, Chile, Mexico, and Peru are fewer, but patient searching will reveal some significant citations.

291. TRAVIS, B. V., C. E. MENDOZA, and R. M. LABADAN
1967 Arthropods of medical importance in Africa. Pt. 1: Introductory and explanatory material; data on mosquitoes. Pt. 2: Data on Arthropods other than mosquitoes. U. S. Army Natick Laboratories, Technical Report 67-55-ES. 804 p.

Summarizes all available literature for Africa, including large areas of desert.

292. TREWARTHA, G. T.
1961 The earth's problem climates. University of Wisconsin Press, Madison, Wisconsin. 334 p.

Bibliographical information in "Notes," p. 309-329.

293. UNDERHILL, H. W.
1965 Report to the Government of Jordan on the establishment of the National Hydrologic Service. FAO, Rome, EPTA Report 1998. 31 p.

One of the most recent reviews of water resources in Jordan, containing basic hydrologic data, gaging station locations, and water-balance estimates for the country. Many pertinent references to earlier work are included.

294. UNESCO
1948- Index translationum. Répertoire internation to    tional des traductions. International bibdate liography of translations. n.s. v. 1- to date. Unesco, Paris.

Bibliographies of translations, published annually in English/French (no. 19, 1966), arranged by countries, and within the country presented under the 10 major headings of the UDC. Author index. Books and monographs only. For the year 1966, e.g., there were 39,630 entries for 70 different countries. Not a particularly useful tool for arid lands works, considering the searching required for few relevant references.

295. ———

1953 Reviews of research on arid zone hydrology. Unesco, Paris. Arid Zone Programme 1. 212 p.

Bibliographies for each geographical area covered.

296. ———

1955 Plant ecology, reviews of research. Unesco, Paris. Arid Zone Research 6. 377 p.

Extensive bibliographies for each geographical area covered.

297. ———

1956 Utilization of saline water, reviews of research. Unesco, Paris. Arid Zone Research 4. 102 p.

Bibliographies: p. 29-35 (245 refs.), 64-70 (256 refs.), 92-100 (307 refs.), on plant growth under saline conditions, utilization of sea water, plant tolerance to salt water, etc.

298. ———

1957 Human and animal ecology, reviews of research. Unesco, Paris. Arid Zone Research 8. 244 p.

Bibliographies: Effects of environment on human communities in the arid regions, p. 38-42 (196 refs.); Influence of environment in arid regions on the biology of man, p. 88-99 (696 refs.)—for journal articles in this list there is author/source information only, no titles; Ecology of mammals in arid zones, p. 134-137 (154 refs.); Ecological data on the avifauna of the Arabo-Saharian desert zone, p. 152-163 (611 refs.); Aridity factor in the ecology of locust and grasshoppers of the Old World, p. 190-198 (308 refs.); Ecology of arid zone insects (excluding locusts and grasshoppers), p. 224-240 (649 refs.).

299. ———

1958a Climatology. Reviews of research. Unesco, Paris. Arid Zone Research 10. 190 p.

Papers written as review reports for the Canberra symposium (see Unesco, 1958b). Of particular interest to the arid zone are those on Evaporation and the water balance, by Deacon et al (207 references p. 29-34), Radiation and the thermal balance, by Drummond (77 references p. 72-74), Climates and vegetation, by Vernet (89 references, p. 99-101), and Climatological Observational requirements in arid zones, by Gilead and Rosenan (15 references, p. 188).

300. ———

1958b Climatology and microclimatology, Proceedings of the Canberra symposium. Unesco, Paris. Arid Zone Research 11. 355 p.

One of several titles in Unesco's Arid Zone Research series devoted to weather and climate. Evaporation and the water balance, radiation, and requirements for climatological observations are three pertinent sections to the arid zone. Proceedings such as this are also valuable for lists of participants and their affiliations. Bibliographies for most papers.

301. ———

1960 Plant-water relationships in arid and semi-arid conditions, reviews of research. Unesco, Paris. Arid Zone Research 15. 225 p.

Bibliographies: Precipitation and infiltration, dew,

301. UNESCO (*cont*)
      evaporation, soil water balance, etc., p. 30-36 (265 references); Soil water relations, p. 55-61 (309 references); Physiological and morphological changes in plants due to water deficiency, p. 95-104 (435 references); Xerophytism, p. 133-138 (264 references); Eco-physiological measuring techniques, p. 165-171 (280 references); Management of native vegetation, p. 183-190 (297 references).

302. ——
      1961 Salinity problems in the arid zones, Proceedings of the Teheran symposium. Unesco, Paris. Arid Zone Research 14. 395 p.
      Contents: Hydrology with reference to salinity; physiology of plants and animals in relation to consumption of saline water; use of brackish water in irrigation and saline soils; demineralization of saline water. Chapter bibliographies.

303. ——
      1962*a* Plant-water relationships in arid and semi-arid conditions, Proceedings of the Madrid symposium. Unesco, Paris. Arid Zone Research 16. 352 p.
      Contents: Water relation studies of plants; water sources for plants; water balance of plants; drought and heat resistance of plants; practical applications to agronomy. Chapter bibliographies.

304. ——
      1962*b* The problems of the arid zone, proceedings of the Paris symposium. Unesco, Paris. Arid Zone Research 18. 481 p.
      Bibliographies: Surface water, including sedimentation, p. 21-22; Geology and geomorphology, and groundwater hydrology, p. 46-52; Climatology and hydrometeorology, p. 78-81; Microclimate, p. 107-113; Soils, p. 134-137; Salt-affected soils and plants, p. 167-174; Plant physiology, p. 193-195; Plant ecology, p. 208-211; Human and animal physiology and ecology, p. 229-233; Insect fauna, p. 245-248; Energy sources, p. 257-258 and 269-270; Saline water conversion, p. 290-297; as well as briefer reference lists on nomadism, and alternative uses of limited water supplies.

305. ——
      1963*a* Geological bibliography of Africa. N. Y. 169, 114 p. UN Document E/CN.15/INR/48.
      Supplements chapter 2 (Geology) of the book "Review of the Natural Resources of the African Continent."

306. ——
      1963*b* Nomades et nomadisme au Sahara. Unesco, Paris. Arid Zone Research 19. 195 p.
      Bibliographical footnotes throughout; additional chapter bibliographies.

307. ——
      1963*c* A review of the natural resources of the African continent. Unesco, Paris. Natural Resources Research 1. 437 p.
      Subject coverage for arid zones of Africa includes topographic mapping, geology, climate and meteorology, hydrology, soils, flora, and fauna, each with extensive bibliographies. Both the list of international, governmental, and private agencies and experts, and the list of abbreviations (including many acronyms) of organizations are useful appendices.

308. UNESCO/FAO
      1963 Ecological study of the Mediterranean zone. Explanatory notes to accompany Bioclimatic map of the Mediterranean zone. Unesco, Paris. Arid Zone Research 21. 58 p.
      Extensive bibliography: p. 47-58.

309. UNESCO/WMO
      1963 Changes of climate, Proceedings of the Rome symposium. Unesco, Paris. Arid Zone Research 20. 488 p.
      Most sections have lists of references, many of them substantial.

310. U. S. DEPARTMENT OF AGRICULTURE
      1962 A bibliography of publications in the field of saline and sodic soils through 1961. Agricultural Research Service 41-80, 47 p.
      Papers are grouped under the following main headings: the occurrence, development, and properties of salt-affected soils; the effects of salts and sodium on plants; tolerance of crops to salts and to sodic soils; evaluating soils for crop production with reference to salinity and sodium; management practices for saline and sodic soils; reclamation of saline and sodic soils; and water quality.

311. U. S. DEPARTMENT OF STATE. BUREAU OF INTELLIGENCE AND RESEARCH
      1964 The tribes of Yemen. External Research Paper 146.
      A list of western language books and periodical articles containing information on the tribes of Yemen.

312. U. S. DEPARTMENT OF STATE. DIVISION OF LIBRARY AND REFERENCES SERVICES
      1952 Pakistan; a basic list of annotated references to evaluate programs for economic development. Bibliography 65. 80 p.

313. U. S. GEOLOGICAL SURVEY
      1962 Annotated bibliography on hydrology and sedimentation: U. S. and Canada, 1955-1958. Water-Supply Paper 1546. 236 p.
      Author arrangement, annotated, with subject index. Preceded by U. S. Interagency Committee on Water Resources (1955), *q.v.*

314. U. S. GEOLOGICAL SURVEY, OFFICE OF INTERNATIONAL ACTIVITIES
      1967 Interim bibliography of reports related to overseas activities of the Water Resources Division, 1940-67. Open file report. [26 p.].
      Contains references to ground-water investigations and appraisals made in most of the major deserts (except Australia) by personnel of the Geological Survey, including references to unpublished, open-file reports. With few exceptions, however, the investigations deal with local or small-scale problems.

315. U. S. INTERAGENCY COMMITTEE ON WATER RESOURCES
      1955 Annotated bibliography on hydrology 1951-54 and sedimentation 1950-54 (U. S. and Canada), compiled and edited under auspices of Subcommittee on Hydrology and Sedimentation by American Geophysical Union, National Research Council of National Academy of Sciences. Joint Hydrology Sedimentation Bulletin 7. 207 p.

316. U. S. LIBRARY OF CONGRESS. GENERAL REPORT AND BIBLIOGRAPHY DIVISION
      1951 Iran; a selected and annotated bibliography. 100 p.

317. U. S. Weather Bureau
     1955 Bibliography on the climate of Peru. 13 p. WB/BC-5. Also cited as PB-176 726; AD-670 031.
     45 references.

318. ———
     1956 Bibliography on the climate of Chile. Supplement 1. 25 p. WB/BC-23. Also cited as PB-176 739; AD-670 043.

319. ———
     1958a Bibliography on the climate of Iraq. Rept. no. WB/BC-32. 14 p. Cited as PB-176 035; AD-665 182.
     Lists 50 reports by author and title and includes descriptive comments concerning the contents. Also includes comments concerning recommendations for best sources of information within the bibliography.

320. ———
     1958b Secondary bibliography on the climate of Iraq. Report WB/BC-35. 14 p. Also cited as PB-176 023; AD-665 185.
     Sources by author/subject.

321. ———
     1967 Bibliography on the climate of Chile. 22 p. WB/BC-12. Also cited as PB-176 732; AD-670 036.
     58 references.

322. Universidad de Chile. Instituto de Geografía
     1951 Informaciones geográficas. v. 1, 1951- to date. Santiago de Chile.
     Each issue contains a section called "Bibliografía Geográfica Chilena" (Año 11-14, in 1 no., for instance, covered the years 1961-1963), listing references treating the geography of Chile. The Bibliografía is annotated, includes maps, and is arranged by subject.

323. University of Southern California
     1958 An annotated bibliography for the Mexican desert. Prepared under contract for the U. S. Army Engineer Waterways Experiment Station, Vicksburg, Mississippi.

324. ———
     1959 An annotated bibliography for the desert areas of the United States. Prepared under contract for U. S. Army Engineer Waterways Experiment Station, Vicksburg, Mississippi. 387 p.

325. Uteshev, A. S.
     1959 Klimat Kazakhstana. (Climate of Kazakhstan) Gidrometeoizdat, Leningrad. 366 p. MGA 12.5-11.
     A complete discussion of the characteristics of climate in this semiarid and arid part of the Soviet Union. 246 references.

326. Valentin, H.
     1952 Die küsten der erde; beiträge zur allgemeinen und regionalen küstenmorphologie. Petermanns Geographische Mitteilungen, Ergänzungsheft 246. 118 p., 9 figs., 2 map pls.
     Includes extensive bibliography, p. 102-118.

327. van Alphen, J. G., and L. F. Abell
     1967 Annotated bibliography on reclamation and improvement of saline and sodic soils (1966-1960). International Institute for Land Reclamation and Improvement, Wageningen, Bibliography 6. 43 p.
     An updated version of Bibliography 4 (Abell and Gelderman, 1964). References arranged by topics: leaching and drainage, application of chemical and organic amendments, tillage, influence of crops and crop rotation, costs, bibliographies.

328. Varley, D. H.
     1936 A bibliography of Italian colonisation in Africa with a section on Abissinia. London.

329. Venter, R. J.
     1950 Bibliography of regional meteorological literature. Vol. 1: Southern Africa 1486-1948. Union of South Africa Weather Bureau. 412 p.
     2900 references. Classified by regions, with author index, including tabulation of observational data available for each region by type and chronological period.

330. Viney, N. M.
     1947 A bibliography of British Somaliland. War Office, London.

331. Vitale, C. S.
     1963 Bibliography on the climate of Sinkiang, China. Air Weather Service Climatic Center, Washington, D. C. Report on WB/BC-71. 46 p. Also cited as AD-660 813.
     Sources for over a period of time were abstracted for this bibliography of climatic data. Since Sinkiang does not have a very good meteorological coverage, sources with data for very short or older periods (such as those obtained from Expeditionary trips or from narrative books on the whole of China) were also used in order to obtain as much material as possible.

332. ———
     1967 Bibliography of the climate of the Somali Republic. U. S. Weather Bureau, Report WB/BC 94. 60 p. Also cited as AD-670 048.

333. Walker, E. H.
     1960 Bibliography of eastern Asiatic botany, Supplement I. American Institute of Biological Sciences, Washington, D. C. 552 p.
     A fine list of publications, although does not include much material relating to deserts. Gives journal titles. Includes a general subject index, geographical index, and a systematic botany index. See Merrill and Walker (1938) for original compilation.

334. Wallace, J. A., Jr.
     1961a An annotated bibliography on the climate of Jordan. Air Weather Service Climatic Center, Washington, D. C. 33 p. Also cited as AD-662 589.
     95 references. Climatology, atmospheric precipitation, humidity, wind, thunderstorms, fog, hail, evapotranspiration, clouds, dew, atmospheric temperature. Indexed by authors and by subjects.

335. ———
     1961b An annotated bibliography of climatic maps of People's Republic of China. U. S. Weather Bureau, Washington, D. C. 21 p. Also cited as PB-176 047.
     50 references for period 1901-1961.

336. ———
     1962a An annotated bibliography of climatic maps of Angola. Air Weather Service Climatic Center, Washington, D. C. Report no. WB/BC-8. 14 p. Also cited as AD-660 851.
     Contents: Map of political divisions of Africa; Map of Angola; Sources with abstracts listed alphabetically by author; Alphabetical author index; Subject heading index with period of record; Subject heading index with map scales.

337. ———
1962b An annotated bibliography of climatic maps of Republic of South Africa. Air Weather Service Climatic Center, Washington, D. C. 24 p. Also cited as AD-660 863.

Only climatic maps, containing data summarized over a period were abstracted and included in this bibliography.

338. ———
1962c An annotated bibliography on climatic maps of Sudan. Air Weather Service Climatic Center, Washington, D. C. 9 p. Also cited as AD-660 868.

In the bibliography only climatic maps containing data summarized over a period of more than one year were abstracted. The U. S. Weather Bureau Library, Washington, D. C. and Library of Congress, Washington, D. C., were searched, and all sources located and examined and found to contain climatic maps are included. The search for sources terminated April 1962.

339. ———
1964 An annotated bibliography on the climate of Sudan. U. S. Weather Bureau, WB/BC-80. 41 p. Also cited as AD-660 804.

103 references to meteorological and/or climatological data for Sudan. All available sources up to Feb. 1964 in the Weather Bureau Library and the Library of Congress have been abstracted.

340. ———
1965 An annotated bibliography on the climate of Mongolia. U. S. Weather Bureau, WB/BC-89. 17 p. Also cited as PB-176 001; AD-664 702.

All available sources in the U. S. Department of Commerce, Environmental Science Services Administration, Weather Bureau Library, and the Library of Congress were examined and abstracted. Coverage: all aspects of Mongolian climate for period 1895 to the present.

341. ———
1966 An annotated bibliography on cloud information and data for mainland China. U. S. Weather Bureau, Washington, D. C., WB/BS-2. 45 p. Also cited as PB-176 043; AD-664 720.

108 references. Abstracted. Geographical location indices.

342. WANG, J. Y., and G. L. BARGER
1962 Bibliography of agricultural meteorology. University of Wisconsin Press, Madison. 673 p.

10,762 references arranged according to subjects: radiation, temperature, moisture, microclimate, instrumentation, environmental control, agrometeorological forecasting, etc. There is an author and subject index.

343. WARD, D. C.
1967 Geologic reference sources. University of Colorado Studies, Series in Earth Sciences 5. 114 p.

A compilation of selected reference sources of value in geologic research, summing up what is currently available in the form of abstracting services and bibliographies leading to specific information on a subject or region. Worldwide in scope. Current references sources stressed. Arranged by sections: general, subject, regional. Annotations minimal.

344. WARREN, A.
1969 A bibliography of desert dunes and associated phenomena. p. 75-99. In W. G. McGinnies and B. J. Goldman, eds., Arid lands in perspective. University of Arizona Press, Tucson.

Separate listings under 3 topics: sand dunes (excluding coastal dunes), sand movement by wind, and studies of aeolian sand. Approximately 700 entries in all, with coverage to 1967.

345. WEBBER, J. M.
1953 Yuccas of the Southwest. U. S. Department of Agriculture, Agriculture Monograph 17: 1-97. BA 28(7)16775.

Includes an index to species and a bibliography.

346. WEIGHT, M. L.
1963 An annotated bibliography of climatic maps of Syria. Air Weather Service Climatic Center, Washington, D. C. Report no. WB/BM-57. 26 p. Also cited as AD-660 826.

The bibliography of climatic maps of Syria consists mostly of sources in the Weather Bureau Library and the Library of Congress. Since the sources of Syrian climatic maps are limited the geographic area of the Middle East is included. Sixty-six references are listed.

347. WEIGHT, M. L., and H. K. GOLD
1962 An annotated bibliography of climatic maps of Iran. Air Weather Service Climatic Center, Washington, D. C. 17 p. Also cited as AD-660 857.

The bibliography of climatic maps consists of sources in the Weather Bureau Library, the Library of Congress and U. S. Hydrographic Office Library. Since few climatic maps for Iran are available, a larger geographic area has also been included: Middle East, Persian Gulf, Indian Ocean, etc.

348. WEIGHT, M. L., and E. J. SALTZMAN
1963 An annotated bibliography of climatic maps for Mexico. Air Weather Service Climatic Center, Washington, D. C. Report no. WB/WM-59. 30 p. Also cited as AD-660 829.

The bibliography of climatic maps consists of sources in the Weather Bureau Library, the Library of Congress and the U. S. Oceanographic Office Library. The area includes not only Mexico, but Central America as well. Eighty-two references are listed.

349. WELCH, F. J.
1946 South-West Africa, a bibliography (1919-1946). University of Cape Town, School of Librarianship. 33 p.

343 references. See Roukens de Lange (1961) for 1946-1960.

350. WELLISCH, H. (ed.)
1961 A selected bibliography on fluid mechanics, hydrology and hydraulic engineering, 1950-1960. Tahal Water Planning for Israel, Ltd., Tel Aviv. 70 p.

"Sources of information for the planning and construction of country-wide hydro-engineering schemes in the young state of Israel where water resources are scarce and have to be exploited to the utmost." Books and monographs only. Much information on pertinent translations. See Wellisch (1967) for revised and enlarged compilation.

351. WELLISCH, H.
1967 Water resources development: an international bibliography, 1950-1965. Israel Program for Scientific Translations, Jerusalem. 144 p.
A completely revised and much enlarged edition of the compiler's 1961 work. Lists 2000 books and monographs published throughout the world on the scientific and engineering aspects of water-resources development. Titles are translated into English. Special care has been taken to include international and national conferences, congresses, and symposia difficult to trace bibliographically. The arrangement is by UDC classification. Additional information is given on 142 periodicals with full data on publication, etc.; and finally there is a similar listing of 38 abstracting services. There are separate subject indexes to each of these 3 parts, and a combined name index.

351a. WEST, N. E.
1968 Ecology and management of salt desert shrub ranges: a bibliography. Utah Agricultural Experiment Station, Mimeograph Series 505. 30 p.
First printed in Wyoming Range Management 228:147-167, 1966; also printed in Salt Desert Shrub Symposium, Cedar City, Utah, Proceedings p. 269-292. The present version is corrected, with additions, to August 1968. Approx. 400 citations, author arrangement.

352. WHITE, A., R. A. DYER, and B. L. SLOANE
1941 The succulent Euphorbiaceae (Southern Africa). Abbey Garden Press, Pasadena, California. 2 vols. BA 19(5)9136.
A study of the South African species of *Euphorbia, Monadenium,* and *Synadenium,* with keys, descriptions, and many photographic illustrations. Includes a glossary, bibliography, and notes on Euphorbia culture.

353. WHITING, A. F.
1966 The present status of ethnobotany in the Southwest. Economic Botany 20:136-325.
Reviews the status of knowledge (123 citations) in an ethnobotanical review of a largely desert and semiarid region, not neglecting the related ethnozoology of the area. Citations of ten papers are included on ethnozoology of Indians in the American Southwest (1896 to 1965) that contain much detailed information on the use of native animal species as human food.

354. WHITING, A. G.
1943 A summary of the literature on milkweeds (*Asclepias* species) and their utilization. U. S. Department of Agriculture, Bibliographical Bulletin 2. 41 p.
Notes that those from arid areas have most promise for rubber production, but of low quality, and that milkweeds are valuable also for fibers and oils. Includes an excellent bibliography.

355. WILBER, D. N.
1962 Annotated bibliography of Afghanistan. 2nd ed. Human Relations Area Files, New Haven. 259 p.
A comprehensive bibliography of 1230 annotated references, arranged by categories such as art and archaeology, economic structure, geography (items #57-247), history, languages and literature, political structure, social evolution and institutions, social organization. Includes journal articles, many historical, and there is an author index and a bibliography of sources.

356. WILLIAMS, M. A. J., and D. N. HALL
1965 Recent expeditions to Libya from the Royal Military Academy, Sandhurst. Geographical Journal 131:482-501.
Cites 70 useful references on the geology, geomorphology, climate, and archaeology (with emphasis on rock-paintings) of the Gebel Archenu in southeastern Libya which the expedition investigated during the years 1962 and 1963. Map. *See also* Monod (1967).

357. WILLIMOTT, S. G., and J. I. CLARKE (eds.)
1960 Field studies in Libya. University of Durham, Department of Geography, Research Paper 4. 138 p.
Includes bibliographies.

358. WOLLE, M. V. S.
1953 The Bonanza trail, ghost towns and mining camps of the West. Indiana University Press, Bloomington. 510 p.
Most scholarly and reliable study of abandoned or near abandoned urban settlements. Bibliography: p. 483-489.

359. WORLD HEALTH ORGANIZATION
1958 Publications of the World Health Organization 1947-1957, a bibliography. 128 p.

360. ———
1963 Bibliography on climatic fluctuations.
Compiled from information received from 28 countries about published and unpublished meteorological studies on climatic fluctuations carried out in their countries during the past 10 years. Reproduced by the WMO Secretariat and distributed to participants in the Rome Symposium on Changes of Climate (*see* Unesco/WMO, 1963).

361. ———
1964 Publications of the World Health Organization, 1958-1962, a bibliography. 125 p.

362. ———
1965a Catalogue of meteorological data for research, Part 1. WMO 174, T. P. 86 (looseleaf).
Lists the current publications of meteorological data by countries, usually with a short description of the material contained in each publication.

363. ———
1965b Meteorology and the desert locust; proceedings of the WMO/FAO seminar on meteorology and the desert locust. Tehran, 25 November-11 December 1963. Technical Note 69. 310 p.
A current review of the synoptic climatology of the desert locust area of Africa through the Middle East to the Thar. References at the end of each of the 35 lectures given at this seminar.

364. WOZAB, D. H., and K. A. KAWAR
1960 Bibliography of geologic and groundwater reports. Jordanian Development Board, Amman. Circular 1.

365. WRIGHT, H. E., JR., and D. G. FREY (eds.)
1965 The Quaternary of the United States. Princeton University Press, Princeton, New Jersey. 922 p.
This volume must be considered a basic reference work on all aspects of the Pleistocene in the U. S. The reference lists that accompany each individual report are also useful as sources of further information.

366. YUAN TUNG-LI
      1961  Russian works on China, 1918-1960, in
            American libraries. Far Eastern Publica-
            tions, Yale University, New Haven, Conn.
            162 p.
            Sinkiang, p. 117-138.
367. ZIKEEV, N. T.
      1952  Annotated bibliography on dew. Meteoro-
            logical Abstracts and Bibliography 3:360-
            391.
            208 references.

368. ZOHARY, M.
      1959  Bibliographia phytosociologica: Palaestina,
            pt. 1. Excerpta Botanica, B., 1(3):202-212.
369. ———
      1961  Bibliographia phytosociologica: Palaestina,
            pt. 2. Excerpta Botanica, B, 5(2):157-160.
370. ———
      1962  Plant life of Palestine: Israel and Jordan.
            Ronald Press, New York. 262 p.
            Bibliography: p. 231-240.

# INDEX TO BIBLIOGRAPHICAL SOURCES

(Numbers refer to items; they are not page numbers.)

# ARID-LANDS KNOWLEDGE GAPS
# AND RESEARCH NEEDS

WILLIAM G. McGINNIES
Office of Arid Lands Studies
University of Arizona
Tucson, Arizona, U. S. A.

# ABSTRACT

The arid portion of the world as a problem area has been given full recognition only in recent years. Earlier the tendency was to write off arid lands as desert *inhospitable to man*. Because people moved ever deeper into arid regions, sometimes with disastrous results, they became better acquainted with desert regions; hence, a greater interest in scientific studies of deserts developed. This interest was nurtured by Unesco and blossomed into a worldwide program aimed toward the solution of arid-lands problems.

The research program has made a great deal of progress in spite of the difficulties imposed by less favorable physical conditions than in humid areas. The first major problem encountered was that of defining terms and developing communications that would enable a scientist in one part of the world to understand and to apply results obtained elsewhere. The land manager dependent on research was even more at a loss to apply results to his particular problems. Much work still remains to solve this problem.

Much information was needed about the arid and semiarid portions of the world. The problem of geographical and temporal variations in climate led to difficulties in relating plant and animal production to fluctuating environments. Little was known about soils of arid regions, their classification was sketchy, their chemical, physical and biological makeup not well understood, and the widespread salt-affected soils posed problems uncommon in soils of humid regions. Much progress has been made, but there are yet many gaps to be closed.

Water, the "lifeblood" of arid lands, nearly always limited in supply, has been a major subject for investigation and research. Surface and subsurface supplies, usually over-obligated and often mismanaged, have been limiting factors in economic production. There is a great deal of room for improvement in the management of water supplies for beneficial use and the avoidance of salinity increases and waterlogging. In this field research is important, but the great improvements could be made if advanced practices already known were put to use.

The vegetation, though it may be sparse, is the basis for all life in the desert, and an understanding of plant ecology and especially the impacts of past and present use is fundamental to occupancy, whether on a nomadic or sedentary basis and with or without the addition of exogeneous sources of water and energy.

There are broad gaps in our knowledge of desert animals and their beneficial or detrimental influences. This statement applies to both native and introduced species, including man, who has the power to make or destroy the desert biome, an environment offering many amenities making arid lands desirable places for future human population expansion.

# ARID-LANDS KNOWLEDGE GAPS AND RESEARCH NEEDS
## William G. McGinnies

## INTRODUCTION

### History of Arid-Lands Programs

In 1934 black clouds of dust rolled out of the Plains States, from Montana and New Mexico eastward. These clouds carried a message to the United States Congress meeting in Washington, D. C., a message of drought and destruction resulting from man's attempt to wrest a living from a land with little rainfall—a country, in the words of Zebulon M. Pike and Stephen H. Long who explored the western United States during the early years of the 19th century, that offered no opportunities for development, which would remain forever the domain of nomadic tribesmen.

The explorers, on the basis of their experiences in more humid climates, found the plains area to be inhospitable; but settlers moving westward in search of places of their own learned that in wet years the land could be very productive, and later, to their sorrow, that it could be cruelly harsh when winds whipped up the loose soil no longer protected by the short grass sod that through the ages had stabilized the land.

A new era was ushered in by the oral and written words of prophets (scientists) whose incantations reiterated the theme that the laws of nature were absolute and man's desires could not cause them to be set aside. It was recognized that we could not conquer nature in the sense of upsetting natural laws, but rather that man's welfare was dependent on knowledge of the laws and adherence to practices that were within the rather rigid structure established by nature. It was on that basis that the arid-lands program moved forward during the middle third of the 20th century in the United States.

A better understanding of deserts was one of the results of World War II, which introduced United States soldiers to the deserts of the world. Many were trained in the hottest, driest portion of the North American Desert and learned that it was possible to survive and retain mobility under what had been considered impossible conditions for the softened citizen raised in less arid and burning climates. These military men became acquainted with some of the deserts of the world, and many gained an appreciation of, and some a love of, the desert.

While occupancy of arid lands by "Europeans" was relatively new in the United States, in the Old World existence under arid conditions had long been an accepted way of life, and not too much thought had been given to modernizing life on the desert. Cultures had developed, civilizations had risen and fallen, and life went on as usual—but in the second quarter of the 20th century, changes stimulated by far-reaching political and military events arose. Peoples of arid lands exchanged ideas to an increased extent, and, with the development of such international organizations as FAO and Unesco, means of communication were improved, facilitating mutual aid and international cooperation. Out of this new situation the arid lands program of Unesco was born, an accelerated program that over two decades brought a greater understanding of arid-lands problems and progress in their solution than had occurred in the two thousand years preceding.

In 1948 the Indian delegation proposed at the Unesco General Conference in Beirut that an international institute of arid-zone research be established. In December, 1949, a committee of experts was convened in Paris to make detailed proposals for the Unesco Arid Zone Programme. The Committee recommended that an Arid Zone Research Council be organized. The recommendations of the Committee were sent to member states for comments, and in accordance with these recommendations and comments an International Arid Zone Research Council was convened in Paris in November 1950. The seven members of the Council were appointed from nominations received from the governments of India, Israel, Egypt, France, United Kingdom, United States of America, and Mexico. An Arid Zone Programme was drawn up by this International Council which recommended that Unesco establish an Advisory Committee on Arid Zone Research. This committee held its first meeting in Algiers in April 1951 (1). Under the program set up, activities were concentrated upon one particular subject of major importance in arid-zone studies each year. This plan was generally adhered to from the beginning until 1964, when the Arid Zone Programme was discontinued as a major project. By this time a wide variety of subjects had been covered.

### Appraisal of Progress

As a result of the impetus received from the Unesco program, interest and activity in arid lands increased greatly in the 1950s and 1960s. At the time this program was started there was not any state-supported research group with an arid-lands designation in the United States. Now there are four such, including the Desert Institute, University of Nevada; the Dry Lands Institute, University of California, Riverside; the International Center for Arid and Semi-Arid Lands Studies, Texas Technological College; and the Office of Arid Lands Studies at the University of Arizona. In addition many departments at other universities are involved in substantial arid-lands programs.

Internationally the situation is much the same. In 1949 there were scarcely any institutions set up specifically for arid-lands studies, and in 1953 Unesco's the *Directory of Arid Lands Institutions* (2) listed 90 institutions, whereas in 1967 the *Arid Lands Research Institutions: A World Directory* (3) listed 243 institutions that had returned forms indicating their activities and provided a supplementary list of 130 believed to be engaged in arid-lands activities.

### Status of Knowledge

A 1960 Unesco symposium in Paris, "Problems of the Arid Zone," (4) was specifically designed to review worldwide progress up to that date and to consider future needs. It included:

(a) critical appraisal of the state and perspectives of knowledge in the various scientific disciplines involved; (b) appraisal of Unesco's activities in promoting research and training related to arid zones, supplemented by reports on related activities of international governmental and non-governmental organizations and by reports from Member States in the major project area; (c) studies on selected problems illustrating difficulties in translating knowledge into action in arid zone development; (d) study of the perspectives for future international and national action in arid zone research, training, and development.

The papers presented at the Paris conference provide an excellent review and appraisal of arid-lands studies up to 1960 and serve as a base for an evaluation of the developments up to that time. This evaluation has been brought up to date in more recent publications, especially in *Deserts of the World: An Appraisal of Research into their Physical and Biological Environments* in 1968 (5).

In an inaugural address at the Paris conference Vittorino Veronese, Director-General of Unesco, outlined the theme of the conference in the following words:

What are the prospects which are beginning to open up before us? What part will the arid and semi-arid wastes of our planet play in the future? No one can yet tell. American, Russian, Australian and French achievements, to mention only a few, afford glimpses of the greatest promise. How fabulous has been the history of these arid lands! In Mesopotamia and Palestine, on the Nile and the Indus, they have seen the birth of the great civilizations which were to shape the destiny of mankind, only to be effaced for a while by man's conquest of lands

with different climates. Today, by a quirk of history, science is once more moving toward the countries where it was born, to give them back a new prosperity. As the Americas once opened up to meet the needs of Europe, will the arid lands, like a second New World, supply man with the space and wealth he needs today? It is for you, ladies and gentlemen, to tell us where we have arrived in this vast and noble undertaking.

## PROBLEMS OF COMMUNICATION

Perhaps the greatest problem we are faced with is that of communications. This is a two-dimensional problem, on the one hand a question of definition and on the other of conveying information from one individual to another across the barriers of distance and languages.

### The Arid Climate

As an example of the magnitude of the problem, so far there is not even a single, universally accepted definition of such terms as *extremely arid, arid, semiarid, desert, semidesert,* or *steppe* and *semisteppe*. One principal reason for this lack is that no indisputable standard exists by which these terms can be established; another involves the communication barriers between scientists scattered about the world.

To some people the discussion of the boundaries of arid lands and their subdivisions appears to be of academic interest only, but to the experienced it becomes of vital concern, with the same questions raised by scientists all over the world. What should the categories "arid-lands," "steppe," and "desert" include, and what should they exclude? Is there a need for a greater number of subdivisions, and, if so, on what considerations should they be based?

At the 1960 Paris meeting, Emberger and Lemée (6) stated:

Enormous progress would be made if international agreement could be reached on the definition of the major bioclimates into which the world is divided. The problem is fundamental, since these great climatic territories constitute, *ipso facto*, the major ecological divisions. There should also be some agreement about the meaning to be given to "zone," "region," "climate" and "arid" or "semi-arid." At present, wide liberties are taken in the matter, and we are prevented from getting views of the whole and from establishing general homologations.

These problems are basic to all reports on research, communication among scientists, development of programs geared to local conditions, improvements in native vegetation to benefit man, and occupation of arid areas.

As example, a common definition of "desert" in northern Africa is an area with less than two inches average annual rainfall and with a history of periods exceeding 12 months without rain.

We in the United States who desire to help solve problems of the Sahara on the basis of our research in the United States find that very little of our so-called desert will qualify, as most of it has a much higher rainfall. Only a small area around the upper end of the Gulf of California and portions of Death Valley will qualify climatically, and even these areas lack representative physiographic features prominent in the Sahara. On the basis of differences between the so-called desert areas in the United States and Africa, extrapolations may be useless or even dangerous.

Monod (7), in discussing the arid ecosystems of the Sahara, recognizes the various attempts that have been made to classify the area. He points out the differences and the terms proposed by various observers. These include those proposed by Mecke-lein: Trockensteppenzone or Dry Steppe (P = approx 200 millimeters), Halbwüste or Semidesert (P = approx 50 millimeters), Vollwüste or Desert Proper (P = approx 20 millimeters), and Extremwüste or Extremely Dry Desert (P = approx 5 millimeters). Contrasted with this is the classification of Meigs (8) that places the upper limit of *extremely arid* at approximately 50 millimeters.

Monod recognized a semiarid domain, or steppe, and an arid area, or desert; the latter could be subdivided into *plioaride* and *hyperaride* (extremely arid).

On the moist side, the boundary of the semiarid is established by various kinds of computations and is usually considered as the lower limit of dependable rainfall for crop production. What is considered "dependable" varies with different authorities; for example, Russell (9), who mapped the frequency of *dry* and *desert* years in western United States, showed that there was a great deal of diversity year by year and that dry years could extend far beyond the average line established as the boundary between semiarid and subhumid. Similarly, Thornthwaite (10) showed that at least 1 year in 25 can be expected to be semiarid as far east as the Mississippi River.

The boundary between semiarid and arid has been arbitrarily set by Köppen, as reported by Tre-wartha (11) as one-half the amount separating steppe from humid climates. According to Tre-wartha, two subdivisions of dry climates are commonly recognized: (a) the arid or desert type, and (b) semiarid or steppe type. He used Köppen formulas for calculating *BS* (steppe) and *BW* (desert) climates, taking into consideration summer, winter, and evenly distributed precipitation.

## Flora and Fauna as Indicators

Lowe (12) believes that vegetation can be used to mark the boundaries of deserts. He believes that the edge of the desert is a broad statistical edge locatable and mappable on the basis of vegetative statistics. His opinion is probably strongly condi-tioned by his experience with the creosotebush desert area of the Southwest. In that area the plant species may be a reasonably good indicator. On the other hand, the indicator value of big sagebrush, one of the prime species in the Great Basin Desert, is limited because its range extends from true desert well into the grassland (similar situations occur in other deserts). The natural range is from communities of big sagebrush with a little grass to communities of grasses with a little sagebrush, and under heavy grazing the grass disappears and "statistical" measurements might indicate that the latter communities were desert. In one sense they may be, as this elimination of grass may be con-sidered as a kind of desertification. Stated in broader terms, the difference between the big sagebrush and the creosotebush situation exemplifies the dif-ferences between desert areas world-wide; it is not always possible to find "creosotebushes" to serve as indicators.

For practical reasons it is important to know the kind and intensity of each important parameter. A desert vegetation may occur on poor soils under rainfall that might support a less arid type of vege-tation on better soils, or the desert vegetation may occur where grazing or other disturbance has re-moved other vegetation. If range improvement programs are to be initiated under such conditions, it is very important to know whether the particular site has a desertic appearance because of poor soils, overgrazing, or lack of precipitation.

It has been suggested by some scientists in the United States, notably Clements (13), that in general the absence of perennial grasses in the climax vegetation marks true desert — but this definition is unsatisfactory even in the United States: in the vicinity of Yuma, with a precipitation of 3.38 inches (85 millimeters), one of the codominants is *Hilaria rigida*, a woody, perennial grass. Clements' defini-tion would also not apply in Australia, where grasses called "spinifex" (*Triodia basedowii* and *Plectrachne schinzii*) dominate in arid areas.

## Delineation of Soils

According to Dregne (14) our information on soils is not sufficient to provide a basis for an arid-zone classification. This assessment is probably made because information on arid-zone soil properties is inadequate to permit a comprehensive understand-ing of the processes involved in soil development and the reaction of soils to natural and man-made forces. It has not yet been determined just what kinds of soils develop under arid and extremely arid conditions. In some cases it is thought that arid-zone soils showing more than moderate to weak development must be the product of an earlier, more moist environment. These soils may be redder and have a finer texture than typical arid soils. In some cases the degree of redness may be a result of parent materials, but in other cases this explanation does not seem to suffice in accounting for color differences.

Beyond the characteristics of the soils themselves, looms the controversy over classification and the lack of universal agreement of soil scientists. As Dregne has stated (15):

It appears that everyone knows what a desert soil is but no one can define it. Other points of nomenclature confusion include (1) the use of the terms "grey desert soil," "grey soil," and "sierozem" (which means grey soil) to refer to soils that may or may not be similar, (2) calling soils "semidesert," "subdesert," or "semiarid," without indicating what properties distinguish those soils, (3) not knowing what is meant by the term "brown" in Soviet, French, Australian, and American usage, and (4) uncertainty as to whether exchangeable sodium is or is not an essential constituent of solonetz soils.

### The Problems of Terminology

Stone (16) has made a contribution toward the development of a universal terminology in the publication of "A Desert Glossary." In the introduction he states:

The purpose of this paper is to provide a single source where much of the proliferation of arid region terminology is presented, to simplify the task of the scientist in his perusal of the literature and to enable other individuals who are interested in deserts and desert phenomena to more readily understand published information.

The Glossary entries are largely physiographic and geologic, but botanical, pedological, climatological, and other terms are included. The major section covers the deserts of the United States and Mexico. It is followed by a supplementary terminology list for other deserts.

While in general Stone's selections of definitions are intelligent, he suffers from the problem of the pioneer who has to make the best selection he can as many of the terms are not as yet stabilized as to definition. Some of his definitions may not be widely accepted, for example:

Arid Region: Area of scant rainfall, commonly 6 inches or less.
　Rainfall is seasonal and there are wide variations from the norm. [Geographers commonly use 10 inches].
Semiarid: Partially arid: on the basis of rainfall, a region in which the mean annual rainfall is 12 to 16 inches, and by some observers, between 10 and 20 inches. [Maybe there should be a semi-semiarid category to fill the gap between 6 inches of rainfall — moist end of Stone's "arid" — and 12 inches — dry end of his semiarid.]
Desert: A barren and uninhabited region: a region of low rainfall and high evaporation as the result of which plant growth is scanty and specialized, and erosion carves a distinctive topography.
Steppe: Vast areas in Southwestern Europe and Asia of generally level and unforested areas of semiarid climate and where scattered bushes and short-lived grasses furnish a scanty pasturage. [Allan (17) lists over fifty different uses of this term varying from grass steppe to forest steppe].

Veld: Grassland in which there may be scattered shrubs or trees: a desert plain. [According to Küchler (18), this formation is primarily grassland].

## CLIMATE GAPS IN KNOWLEDGE

It is obvious that more needs to be known about local climatic and microclimatic conditions. For most of the larger deserts specific knowledge is almost entirely restricted to the periphery, with a few scattered records in the desert itself, and those usually short-time or spotty. The data for the less arid portions are more numerous, more widely distributed, and of longer duration, but except for the more heavily populated areas these data are still far from adequate.

The Sahara could be mentioned as a typical example of a region where sparse population and a severe climate have limited the possibilities for the collection of climatic data. The Sahara includes an area as large as all Europe, where the number of climatological stations is at least a hundred times greater. According to Dubief (19), the number of climatological stations is only 125 for an area including the dry regions of Morocco, Algeria, Tunisia, Libya, Egypt, and the areas formerly known as French West Africa, French Equatorial Africa, and Anglo-Egyptian Sudan.

Even though in certain areas, available climatic observations may allow for a study of the recent development, in most arid regions of the world it is impossible to obtain climatic data of ancient historical periods and of geological times. To obtain an idea of the climate of several hundreds or thousands of years back, it is necessary to use other means than climatological observations to attack the problem. These involve geochronology, archaeology, dendrochronology, and historical records.

The problem of classification of environments is complicated by the seasonality of precipitation. The difference between mediterranean climate and tropical climate, such as shown in the Sahara, is a relatively simple case. More difficult problems are those such as posed by the Caatinga in Brazil, the Chaqueña and Chaco country in Argentina, and adjacent countries of South America, where there may be abundant rainfall during part of the year and very hot, dry conditions for a prolonged period. On the basis of total annual rainfall these areas possibly might not even be considered arid, but the conditions on the ground during six to nine months of the year would qualify the area as arid.

The role of dew as a water source for plants in arid regions is still a subject of dispute. Does it provide a direct source of moisture, and what effect does it have on evapotranspiration? Many problems remain to be solved, including: (a) the development of a suitable quantitative measuring method, (b) the amount and nature of dew in arid environments, and (c) the amount of water which may be available to plants through aerial and underground intake. Dew as a water source may constitute only an insignificant part of the daily transpiration of a plant well supplied by ground humidity, but on the other hand it may be effective in the survival of plants which have reached the wilting stage (20).

## Hydrology

From the viewpoint of hydrology, Dixey (21) pointed out that water, particularly ground-water, is the prime necessity of life in the arid zones, and knowledge of its occurrence, replenishment, and recovery is basic to the development of these regions. Surface waters, except when brought in by rivers from more humid terrains elsewhere, are normally scarce or absent, and ground-waters accordingly assume special importance.

Dixey (22) believed that water-resource development in the arid-zone countries is likely to remain their most important problem for many years, and the development of simple and widely applicable methods of artificial recharge of ground waters possibly represents the most important present-day challenge to hydrogeology of arid zones.

Surface water is a fluctuating and variable resource, and since its variability increases with aridity, this characteristic is most marked in the arid zones. Streamflow and sediment gauging in the arid zones are particularly difficult. Arid-zone hydrology involves largely the study of small flows, and the smaller the area, the greater is the effect of its character upon the run-off and sediment yield. For this reason more studies of the relations of terrain and climate to run-off and sediment yield are needed. Dixey (22) also noted the need for further research into and development of artesian waters.

Evaporation is the hydrologic process which characterizes the arid zone, and evaporation research has been one of the most popular forms of research in recent years. Basic research has provided universally applicable information on the mechanics of the process, but there is a need for more and more research in this field to meet the growing needs to make more efficient use of available water resources.

The salvage of water in arid zones usually involves methods of abating evaporation losses. Important among these are water lost from reservoirs, from stream channels and from phreatophytes, and evapotranspiration losses from the watershed. The prospects of reducing evaporation from small reservoirs appear promising, but large reservoirs still present a major problem (23).

A great amount of research has been done on evapotranspiration, and a great deal of progress has been made, but there is still a need for evapo-

transpiration measurements over much of the arid lands, especially for highly variable facets of complex landscapes.

Much has been learned about phreatophytes, a problem in semiarid and arid areas. Measurements of water losses have provided very good information as to the magnitude of the problem, and methods of control have been developed; nevertheless, the phreatophyte problem is still with us, and one of the principal obstacles to removal is public opinion which often favors maintenance of the favorable wildlife habitats provided by dense growth of phreatophytes.

The management of upland areas for optimum water production has improved greatly, but there is still a great deal of uncertainty as to the possible gains and losses due to removal of vegetation. Each vegetation type and each watershed area poses specific questions, and there is no common denominator that can be applied to all watersheds. Two things are clear: vegetation uses water and vegetation aids in the prevention of erosion and runoff. The desire to increase water yield needs to be governed by scientific and practical guidelines. In areas with very sparse vegetation resulting from scanty precipitation, little if any increase in water yield will result from removal of vegetation – and at the same time erosion and sedimentation may be greatly increased. At the other end of the range, where the vegetation cover is dense and precipitation is great, especially where snowfall makes up a major portion, the results of vegetation removal have a measurable effect on water yield. In between these two extremes there are many variations in precipitation-plant relations that need further study before optimal management can be applied.

Although water supplies increase with increasing humidity, sediment yield increases with aridity. The returns to be gained from more precise knowledge of sedimentation and fluctuations in water supply in the form of lower construction and operation costs, freedom from maintenance troubles, and longer project usefulness will more than offset the investment in the collection and study of hydrologic records (23).

## Soils

Dregne (14) has summarized the present status of knowledge on desert soil in a very comprehensive study. He believes that overall information on arid-zone soil properties is inadequate to permit a comprehensive understanding of the processes involved in soil development and in the reaction of soils to natural and manmade forces. The major reason for this spottiness undoubtedly lies in the inhospitable environment of the extremely arid and arid zones as contrasted with the semiarid and humid zones. Relatively little basic research has been done or is underway on arid-zone soils. The greatest amount of research, both basic and applied, has been done in the United States and the Soviet Union.

There are many deficiencies in the determination of physical properties for soil texture and soil aggregation, and there are many chemical methods needing improvement such as those for determination of exchangeable calcium and magnesium in calcareous soils and for exchangeable sodium in highly saline soils.

One of the problems in making soil information most usable is that soil surveys are usually pointed toward one particular soil use. Most soil surveys have been made on the basis of collecting information for agricultural purposes, and as a result the data do not give information that is needed by soil engineers and others. There is very little information in the field of soil mechanics indicating suitability of soils for brickmaking, oil-well mud, ceramics, road construction, and building construction. The problems of trafficability which have been considered by military organizations, have largely been ignored in other soil surveys.

Because of the many unknown factors involved in the plant-soil-water-climate relation, precise interpretation of the soil factor is difficult. On the basis of the present state of knowledge about soil mechanics, it does not appear there is sufficient information to permit more than a fairly reliable estimate of the engineering characteristics of soils.

Lustig (24) feels that there are a great many gaps in our knowledge as related to desert sands. Very little is known about the sand sources of the desert today and the relative rates of sand production through weathering of different lithologies under present climatic conditions. The general problem of the origin of desert sands and the rate of production from source areas is merely one part of the more general question of rock weathering in the desert areas or elsewhere and of the rates of weathering.

Dregne (25) points out that research on soil microbiology has been rather neglected in the United States as well as in most other countries outside Europe. The amount of information on microbial numbers and species in arid-zone soils is very small, and the research, as in many other cases, has raised more questions than it has answered. Aside from the need for survey data on just what is present in the soil under varying conditions of salinity, carbonate content, organic matter, pH, texture, nutrient conditions, maximum and minimum temperatures, and moisture content, the reasons for whatever differences may occur should be determined. The question of whether extremely arid soils actually are abiotic or only so outside the rhizosphere needs to be answered. Whether algal and lichen crusts

are significant factors in soil stabilization, particularly of sand dunes, in upland soils is unresolved, as is the effect of such crusts on water penetration and runoff, to say nothing of their effect on soil development. Another question relates to the function of termites, which are distributed worldwide in the arid zones, in affecting the soil environment. Just why ammonifying organisms are prevalent and active in arid-zone soils, whereas nitrite- and nitrate-producing organisms frequently are inactive, is a question with many practical implications.

Bernstein (26), reviewing the status of knowledge on salt-affected soils, pointed up the problems of salt-affected soils and plants and noted the very great magnitude of problems in arid countries because of the vastness of saline and water-logged soils. The situation is further alarming because large areas are still going out of cultivation very rapidly in most of the countries.

Although a good deal of work has been done on methods for studying salt-affected plants and soils, there is need for improving the methodology on the management and preparation of salt-affected soils in order to reclaim them on a permanent basis and at economical costs. Since water is involved in all movement of salt into, through, and out of the soil, the importance of soil-water dynamics is self-evident.

Very little is known about fertilizer requirements of and fertilizer effects on saline soils; systematic studies are needed to be carried out both in the greenhouse and in the field.

Salt tolerance of vegetable, field, and forage crops, and specific toxicity of ions should be tested in salt nurseries in different salt-affected countries for the development of adequate methods and techniques and for the selection of plant material such as fruit plants, forest trees, and forage plants that may be especially salt tolerant. Effect of salinity on the occurrence of plant diseases should also be studied.

## Vegetation

Evenari (27) pointed out that in studies of vegetation in arid zones, all the different disciplines of the plant sciences must play a role in the collection of basic floristic and phytosociological data concerning the vegetation. Moreover, the different branches of botany in conjunction with climatology, meteorology, and soil science are needed to elucidate the ecological relationship between vegetation and environment and between the individual plant and environment. This does not concern the natural vegetation only, but includes research on the environmental factors governing crop and pasture production under arid conditions.

Arid regions are characterized by a fragmentation

of physical environments resulting in a complex mosaic of micro-environments. The slightest topographical accident—for instance, even a small, isolated stone or a tiny depression—can represent an environment which is topographically very restricted but no less sharply individualized than wider surfaces, and characterized by special vegetation (28).

Because of the severity of arid environments, the problems of relationships to the environment tend to be mainly autecological and synecological aspects become secondary. Bare patches appear in the plant cover. As aridity increases, these characteristics become sharper. The covering of plant life becomes completely dissociated, and, finally, there are no competitive relationships among species. They live as isolated individuals; each plant is autonomous, and its relations with the environment are purely autecological (28).

More knowledge is needed about the exact mechanism by which the plant covering is dissociated under the influence of increasing aridity and how it might be increased by means of artificially formed depressions or small impermeable catchment basins.

The standardization of terminology, particularly in the domain of bioclimatology, and also of methods and units of measure, is essential for real progress in less time in ecological and phytosociological research, inasmuch as it will facilitate needed comparisons and reasoned generalizations.

Observations in different regions have shown wide variations in changes of the floristic composition of the plant production, according to the kind and number of livestock and duration of grazing. When pasture or range lands are exploited intensively, there is a regression of species consumed, extension of other species (especially shrubs), denudation of the soil fostering the extension of annuals, appearance or intensification of erosion phenomena, changes in organic matter, degradation of the structure and diminution of permeability, and modification of the microclimate at the level of the ground (28).

Little, or at least not enough, is as yet known about arid-zone species of plants. For the successful introduction of a desirable species and its extension or the prevention of the establishment of a noxious species, we must have information on production of viable seeds, germination requirements, requirements for the survival of young plants, and the amount of tolerance to unfavorable outside factors.

## Fauna

The kinds of venomous snakes that occur in most of the deserts of the world are known, although venomous snakes are nowhere treated as a single unit or a problem of the world deserts (29). Medical treatment of poisonous spiderbites, particularly for some of the most dangerous species, has been studied in some detail, but taxonomy of the responsible spiders that occur in deserts has received little attention. Antivenins for treatment of scorpion sting have received much greater attention than for spiderbite. Lowe (29) believes much more information is needed on the distribution and ecology of venomous animals of the worlds deserts. As for fauna as human food, information is incomplete, widely scattered, and often obscurely treated in the writings of travelers, ethnographers, ecologists, and others. It is known that insects are rich in animal proteins, animal fats, and calories, but little is known specifically as to their present or potential use as food.

In our knowledge of desert animals in particular, there are broad gaps in the inventory of animals, especially for invertebrates. There is no known listing of all of the invertebrate animals species—of insects, spiders, scorpions, centipedes, snails, and such—that occur in any desert of the world (29). The wild fauna is an important source of pests in developing agriculture, and studies are needed, prior to large-scale development in a desert, of the potential interest of the native animal species in new crop plants. Especially where widespread irrigation is involved, the resulting insect populations may increase suddenly and involve seasonally massive buildups of both desert and nondesert species. Van Wijk and de Wilde (30) have pointed out, "In the reclamation of waste lands when irrigation is applied microclimatic changes occur which may create conditions favorable for crop infestation. The response of certain fungi, viruses, insects and nematodes to change in microclimate is a research item of vital and direct practical importance."

Most, if not all, human and animal diseases in deserts are extensions of their diseases elsewhere, occasionally magnified by the harshness of desert environments and occasionally abetted by the meager living conditions of some native desert human populations (29).

It should be emphasized that much research is needed on all aspects of the scientific study of desert faunas. On the basis of the record, which reveals serious inadequacies and gaps, with the general exception of that portion of the North American Desert within the United States, efforts should be directed toward initiating and extending investigations in all deserts and in all species of environmental biology: taxonomy, morphology, physiology, genetics, ecology, and behavior (31).

Although considerable experience has been gained in the moving of animals from one desert to another, sometimes with good results and many times with bad or disastrous results, more information is needed. Care should be exercised in any program involving the moving of animals from one place to another because of the dangers involved.

## Occupancy and Utilization of Arid Lands

Gilbert White (32) expressed the opinion that the arid lands of the world encompass too many variations to be considered as the unit for problem solution, but that they should be divided into six subtypes including: (1) extremely arid, precipitation 50 to 100 millimeters, (2) sparse localized supplies, precipitation 100 to 400 millimeters, (3) semiarid supplies, sufficient rain in some years (but not most) for crop production, (4), exogenous, surface supplies—Indus, Syr Darya, Tigris-Euphrates, Nile, and Colorado systems, (5) exogenous ground-water supplies which may have a source outside the arid area, (6) localized exogenous supplies where sharp differences in relief and rainfall exist over short distances. As water is a limiting factor in development, the optimal use of sparse water can only be obtained by greater emphasis on modification and management patterns.

Within this framework there is a need to determine capacities of arid and semiarid lands under a variety of climatic conditions involving site and soil conditions, possibilities of improvement through reseeding, brush control, grazing management, water development, economic and social problems associated with agriculture in arid regions. Studies are also needed to establish the limits of nomadic and stable human populations whose livelihood must depend on the use of the soil and plant resources available in arid regions.

As has been pointed out earlier, great expanses of the arid zone have been insufficiently surveyed to provide adequate knowledge of geological conditions, hydrology, landforms, soil, and vegetation necessary for practicable schemes for using sparse localized water supplies and local exogenous supplies. More basic data are required, more mapping is urgent, and more basic research is essential to extend and interpret the physical knowledge already in hand (32).

According to White:

Water is a limiting factor in development but there are many areas where this is not yet so. What is apparently much more limiting is the hard frame of custom, thought and organization in which water use takes place.

To the extent that this is true, one of the promising lines of action for increasing the productivity of the arid zones lies in the direction of research and education designed to widen the range of choice for water use. Research is required in at least three sectors.

First, more refined methods are needed to assess the physical efficiency of water use and to estimate the economic efficiency of its application. . . .

The economic analysis deserves more concrete testing and appraisal in working situations to demonstrate its use and limitations. Quite aside from the problem of comparing alternative investment possibilities, this analysis must be extended if it is to serve as a guide to determining the net returns from different combinations of water use. . . .

Third, the processes by which social change can be achieved in resource management are still understood only imperfectly . . . unless means can be found to translate that technical knowledge of salt movement into the daily action of a farmer tending his irrigation ditch in a distant valley the expert's symposium will avail little. . . .

As is so well summed up by White (32):

In the extremely arid areas the water supplies are negligible, but in the other major types of supply areas it may be misleading to think of water supply as the sole, immediately limiting factor to improvement of living conditions. The quality of human ingenuity in applying available technology and social organization to the satisfaction of mounting human needs may be as severely limiting. This social dimension of arid zone research is one offering a great challenge and rich returns in human betterment.

## SUMMARY

It has been the aim of this paper to point out some of the problems facing the arid-lands researcher. He has been working in a relatively new scientific field and in geographic regions formerly believed to be of little promise. In the United States, the real interest in arid lands resulted from a disastrous drought, and the appreciation of the arid environment was a by-product of war.

The organization of a research program came out of man's efforts to unite in programs for peaceful development. The foundations that were laid with the support of the United Nations provided the base for the development of a research program, but that development was beset with problems because it embarked into new strata of problems relating to geology, climate, hydrology, vegetation, fauna, and the diverse activities of man.

A great deal of progress has been made, and it is in the spirit of providing a means of communication to promote further progress that this publication was undertaken.

## REFERENCES AND NOTES

1. WHITE, G. F.
   1960  Science and the future of arid lands. Unesco, Paris. 95 pp.
2. UNESCO
   1953  Directory of institutions engaged in arid zone research. Unesco, Paris. Arid Zone Programme 3. 110 pp.

3. PAYLORE, P.
   1967  Arid-lands research institutions, a world directory. University of Arizona Press, Tucson. 268 pp.
4. UNESCO
   1962  The problems of the arid zone, proceedings of the Paris Symposium. Unesco, Paris. Arid Zone Research 18. 481 pp.

5. McGinnies, W. G., B. J. Goldman, and P. Paylore, eds.
   1968 Deserts of the world: an appraisal of research into their physical and biological environments. University of Arizona Press, Tucson. 788 pp.
6. Emberger, L., and G. Lemée
   1962 Plant ecology. pp. 197-211. *In* Unesco, (*4*).
7. Monod, T.
   1964 Desérts. Union Internationale pour la Conservation de la Nature, Réunion Technique, 9e, Nairobi, 1963, Proces-Verbaux et Rapport 2:116-132. (Publications UICN n. sér. 4)
8. Meigs, P.
   1953 World distribution of arid and semi-arid homoclimates. *In* Reviews of research on arid zone hydrology. Unesco, Paris. Arid Zone Programme 1:203-209.
9. Russell, R. J.
   1932 Dry climates of the United States. II: Frequency of dry and desert years, 1901-1920. University of California, Publications in Geography 5(5):245-274.
10. Thornthwaite, C. W.
   1956 Climatology in arid zone research. *In* G. F. White, ed., The future of arid lands. American Association for the Advancement of Science, Washington, Publication 43:67-83.
11. Trewartha, G. T.
   1954 An introduction to climate. 3rd ed. McGraw-Hill Book Company, Inc., New York. 402 pp.
12. Lowe, C. H.
   1964 Arizona landscapes and habitats. pp. 1-132. *In* C. H. Lowe, ed., The vertebrates of Arizona. University of Arizona Press, Tucson.
   1968 Inventory of research on fauna of desert environments. pp. 569-645. *In* W. G. McGinnies, B. J. Goldman, P. Paylore (*5*).
13. Clements, F. E.
   1936 The origin of the desert climax and climate. p. 87-140. *In* T. H. Goodspeed, ed., Essays in geobotany in honour of William Albert Setchell. University of California Press, Berkeley.
14. Dregne, H. E.
   1968 Inventory of research on surface materials of desert environments. p. 287-377. *In* W. G. McGinnies, B. J. Goldman, P. Paylore (*5*).
15. Dregne (*14*), p. 343.

16. Stone, R. O.
   1967 A desert glossary. Earth-Science Reviews 3: 211-268.
17. Allan, H. H. B.
   1946 Tussock grassland or steppe? New Zealand Geographer 2(1):223-234.
18. Küchler, A. W.
   1957 Localizing vegetation terms. Association of American Geographers, Annals 37:197-208.
19. Dubief, J.
   1959 Le climat du Sahara, vol. 1. Université d'Alger, Institut de Recherches Sahariennes. 312 p.
20. Stone, E. C.
   1957 Dew as an ecological factor. II: The effect of artificial dew on the survival of *Pinus ponderosa* and associated species. Ecology 38:414-422.
21. Dixey, F.
   1964 The availability of water in semi-arid lands: possibilities and limitations. *In* Land use in semi-arid mediterranean climates. Unesco, Paris. Arid Zone Research 26:37-45.
22. Dixey, F.
   1962 Geology and geomorphology, and ground-water hydrology. p. 23-52. *In* Unesco (*4*).
23. Langbein, W. B.
   1962 Surface water (including sedimentation). p. 3-22. *In* Unesco (*4*).
24. Lustig, L. K.
   1968 Inventory of research on geomorphology and surface hydrology of desert environments. p. 95-283. *In* W. G. McGinnies, B. J. Goldman, P. Paylore (*5*).
25. Dregne (*14*), p. 320.
26. Bernstein, L.
   1962 Salt-affected soils and plants. p. 139-174. *In* Unesco (*4*).
27. Evenari, M.
   1962 Plant physiology and arid zone research. p. 175-195. *In* Unesco (*4*).
28. Emberger and Lemée, (*6*).
29. Lowe, 1968 (*12*).
30. van Wijk, W. R., and J. de Wilde
   1962 Microclimate. p. 83-113. *In* Unesco (*4*).
31. Lowe, 1968 (*12*), p. 613.
32. White, G. F.
   1962 Alternative uses of limited water supplies. p. 411-421. *In* Unesco (*4*).

# A WATER-QUALITY HIERARCHY FOR ARID LANDS

DANIEL A. OKUN
Department of Environmental Sciences and Engineering
School of Public Health
University of North Carolina at Chapel Hill, U. S. A.

# ABSTRACT

With population and urbanization growth, the pressure on existing water resources suggests that wastewaters be reused for public water supply. Aesthetic considerations and the presence of viruses and organic chemicals not removed in conventional water and wastewater treatment processes are deterrents to reuse where water for human consumption is involved.

A dual system of water quality is proposed for municipalities, with one system for drinking and cooking, about 10 per cent of the total, being provided from naturally pure sources or being subjected to intensive treatment, and with the second system utilizing treated wastewaters or polluted surface waters for such purposes as flushing, laundering, and lawn sprinkling. The second supply would be bacteriologically safe, so that accidental ingestion would pose little problem, but the long-term ingestion of organic chemicals would be avoided. The costs for a dual system are not much greater than for a conventional system, especially when the system is installed in new cities.

# A WATER-QUALITY HIERARCHY FOR ARID LANDS
## Daniel A. Okun

"I sell water . . . it isn't easy. When water is scarce I have long distances to go in search of it, and when it is plentiful, I have no income." These words of the water carrier, which open the play by Bertolt Brecht, *The Good Woman of Setzuan,* express our predicament with water. The same dilemma is observed in arid areas where water is scarce and valuable as compared with the profligate attitude toward water in humid areas.

With rapid population growth, increasing urbanization and industrialization, and increasing per capita use of water, even cities in humid areas are having "long distances to go in search of it." Table 1 indicates the increasing proportion of persons receiving water from public water supply sources, which generally means from metropolitan and community supply systems. The increased per capita use and the increased demand for water from public water supplies for domestic and commercial use are also illustrated in Table 1. In the arid West in the United States, the rate of population growth and urban growth may well exceed the rate for the country as a whole (1).

TABLE 1

## GROWTH IN POPULATION AND DOMESTIC WATER USE IN THE UNITED STATES

| Year | Total U.S. Population (millions) | Public Water Supply Users (millions) | (%) | Per capita use (gal./day) | Total Public Water Supply Demand (million gal./day) |
|------|------|------|------|------|------|
| 1930 | 123 | 70 | 57 | 110 | 7,700 |
| 1965 | 200 | 150 | 75 | 150 | 22,500 |
| 2000 (est) | 330 | 300 | 91 | 180 | 54,000 |

Just as ground-waters are being mined in arid areas, making it necessary to go longer and longer distances for surface waters, so too in the humid areas, as the populations have grown, local water resources have become polluted, and it has become necessary to go longer and longer distances to obtain adequate quantities of reasonably good water.

## WATER RECLAMATION AND REUSE

Whenever water is short, either because of limited quantity or limited quality, it is proposed that wastewaters be reused for public water supplies. Actually, almost 50 per cent of the people in the United States now use water taken from sources containing sewage. As populations increase and additional waters become polluted, these polluted sources will need to be used more extensively. Therefore, direct reclamation and reuse of wastewaters appear to be an attractive solution to the water supply problem.

The Research Report of the Fourth Arizona Town Hall on Arizona's Water Supply (2) states that the public is generally reluctant to accept the concept of planned water reuse. "If maximum beneficial use of our water resource economy is to be realized, this must be changed."

And elsewhere in the report: "Failure to consider greater reuse of a locally available water resource that represents more than 50 per cent of municipal water demand has been considered equivalent to an invitation to disaster. While the idea of purifying sewage is downgraded from an aesthetic viewpoint it provides too much promise as a future water resource to justify passing it up without a thorough examination."

Whether there is, in fact, any aversion to drinking or cooking with water of which a significant part is reclaimed wastewater is hard to establish. In cities where the antecedent history of its water is disreputable, people either are unaware of the source of their water supply, or they are resigned to it, as only city dwellers can be resigned to the many personal insults brought about by their proximity one to the other.

Studies of attitudes toward water in two Georgia towns taking water from the same source revealed that the people in one town found it satisfactory while those in the other found it unsatisfactory. The key to the difference was that the latter town received water occasionally from an alternative source with less pronounced taste. Thus, New Yorkers object to using the polluted Hudson River, because they've never had to, while Philadelphians accept polluted Delaware River water as a fact of life.

It is claimed (3) that "The challenge of public acceptance of a sewage-polluted water source is not a formidable one," but "The challenge of acceptance is greater with water utility managers." The waterworks industry is accused of being reactionary because it is slow to accept sewage-polluted sources for its supplies!

Another advocate of utilizing, in Professor Gordon M. Fair's idiom, "repentant" rather than "innocent" waters states (4): "Municipal water treatment processes are often incapable of destroying tastes, odors, and colors introduced by industrial wastes; hence the public must learn to tolerate conditions that are certainly unpleasant though not necessarily harmful."

In planning water-management systems for space vehicles, the psychological revulsion of astronauts to drinking water reclaimed from their own wastes has led to consideration of other alternatives for their water supply.

That people are becoming dissatisfied with the quality of public water supplies is evidenced by the growth of the bottled-water industry, which now claims 1 in 600 American homes as a customer, and 1 in 7 homes in Southern California (5). The industry reports a tenfold increase in retail distributors of bottled water—to 10,000 in the last ten years; a 50 per cent increase in business is expected by 1970. The growing demand is attributed to a demand for taste and purity of water and "because good reservoir water becomes scarce in periods of drought."

The public attitude is important. In a country as rich as the United States, with efforts being made on every side to improve the quality of living, should people be required to accept second-hand water for drinking?

## HEALTH CONSIDERATIONS IN WATER REUSE

The aesthetic consideration would be less important if there were no serious health implications in reusing wastewaters, either directly or after they have been discharged to a stream, for public water supplies. In particular, two types of health hazards— viruses and chemicals—present themselves when wastewaters become part of the water supply.

### Viruses in Water

In a comprehensive paper prepared for the World Health Organization, Chang (6) concludes that water may be a significant route for the transmission of enteric viral infections ". . . if polluted water is used as a source of municipal supply, and sporadic cases of infectious hepatitis (or other enteroviral disease) occur in the population using such supply."

Viral infections may be much more widespread than is indicated by clinical viral disease. Ninety to ninety-five per cent of infections due to infectious hepatitis and polio viruses go undetected. Therefore sporadic cases of these diseases may not be isolated incidences but are more likely the clinical manifestations from among a much greater number of subclinical cases resulting from exposure to viral infection. Consequently, water as a mode of transmission for viral infections must not be arbitrarily

limited to identifiable disease outbreaks. In his introduction to the volume *Transmission of Viruses by the Water Route*, Berg (7) points out that the primary sources of viruses of human origin are the secretions and excretions that enter water with domestic wastes. Because they normally multiply only in living cells, viruses occur in water only in low concentrations. This situation is likely to produce asymptomatic carriers, making it difficult to indict the water route; however subsequent contact transmission, with high concentrations of virus, from those infected by the water route would likely result in clinical illness.

What is of concern is that the occasional clinical cases of infectious hepatitis may, in fact, be only that part of the iceberg that shows, representing a much greater mass of infection that is not medically evident but that may have been triggered by inadequately treated water. For example, the infectious hepatitis virus, in the presence of organic matter, appears to be unusually resistant to water-treatment procedures (8). In a major European city, viruses were reported in 18 per cent of more than 200 drinking-water samples taken from the distribution system (7). A thorough assessment of the effectiveness of water treatment methods for the removal of the viruses of infectious hepatitis must wait until methods for their propagation in the laboratory can be established.

Because municipal wastewaters always contain enteric viruses, and because wastewater treatment processes are known to be ineffective in eliminating them, we must examine critically the possibility of virus transmission with current methods of water treatment where polluted waters or reclaimed wastewaters are used as the source of municipal supply and where occasional cases of infectious hepatitis or other intestinal virus diseases occur. The fact that the water meets the Public Health Service Drinking Water Standards (9) for coliform organisms is no assurance that the virus of infectious hepatitis is absent.

Although Mosley in the opening chapter in the Berg volume (7) finds limited epidemiological evidence to date that unrecognized viral disease is being transmitted by the water route, he concludes that: "We must consider . . . the possibility that present standards of water treatment are not adequate to prevent low levels of virus from producing what appear to be sporadic cases of infectious hepatitis and other viral diseases."

### Chemicals in Water

Literally hundreds of new chemical compounds are being introduced into the environment daily. Few of these are assessed for their potential impact on the health of man, particularly on the synergistic effect they may have when acting together or in concert with other kinds of environmental insults. Some of these chemicals may cause cancer, genetic damage, and/or birth malformations.

In a comprehensive paper on cancer hazards from water pollutants, Hueper (10) states: "It is obvious that with the rapidly increasing urbanization and industrialization of the country and the greatly increased demand placed on the present resources of water from lakes, rivers and underground reservoirs, the danger of cancer hazards from the consumption of contaminated drinking water will grow considerably within the foreseeable future."

The impact of the long-term ingestion of low levels of pollutants of water is difficult to ascertain. Studies in Holland showed that cancer death rates in municipalities with a water system tended to be lower than in those without a water system, and that municipalities receiving their drinking water from polluted rivers had a higher cancer death rate than those taking their water from purer underground sources. Studies in England reveal that wastewaters from municipalities contained two well-established carcinogens—2,4-benzpyrene and 1,2-benzanthracene—which might have originated in effluents from gas works, from washings from macadam roads, or from atmospheric soot washed from the air by rain.

In addition, Hueper lists radioisotopes, particularly in association with other chemicals, as possible contributory sources of cancer hazards to the general population. Among other probable cancer hazards, he lists petroleum products and aromatic amino and nitro components likely to be found in water, as well as numerous substances that may potentially be significant. He states (10):

Available knowledge on intentional and unintentional, actual, suspected and potential carcinogenic pollutants of water is highly defective not only concerning the chemical nature of substances but also with regard to the carcinogenic effectiveness on the consumer.

The worst that can at present be stated regarding a cancer hazard from specifically polluted water is that the prolonged or lifelong consumption of such water will contribute to the total carcinogenic burden from all sources for the exposed individuals.

Hueper concludes that sanitary measures are a small price to pay for the prevention and control of health hazards created by industry, industry which has brought many benefits and advantages to the people of all countries.

Radiation is known to cause genetic damage, and the principle in controlling radiation is that exposure to it should be kept at the lowest practicable level and that no additional exposure should be permitted unless the benefits from the exposure outweigh the anticipated damage.

The Genetics Study Section of the National Institutes of Health fears that some chemicals may constitute possibly a more serious risk than radiation, and exposure to them must be weighed against benefits in their use (11). A number of chemicals, including some that are widely used, are known to induce genetic damage in some organisms, and a chemical that is mutagenic to one species is likely to be mutagenic to others. The Study Section believes that special attention should be given to the low concentrations of highly mutagenic compounds that are brought into contact with large populations. They point out that even though the compounds may not be demonstrably mutagenic to man at the experienced concentrations, the total number of deleterious mutations induced in the whole population over an extended period of time could be significant.

According to Stokinger (12): "Chemicals with mutagenic potential reaching water sources may be expected to increase in number and amount as a result of development of certain newer industrial plastics (epoxy, diepoxy, possibly episulfide compounds), chemical sterilants (aziridines), missile fuel and other intermediates (nitrosoamines, certain sulfonates)."

The thalidomide episode vividly demonstrated the teratogenic potential of certain chemicals, and the Food and Drug Administration has found that certain pesticides are teratogenic (12). Combinations of teratogens have been found to produce greatly potentiated effects at the very low concentrations that could be experienced in water supplies.

Some organic chemicals have undergone rigorous studies to establish their toxicity or safety; however, if all organics as they are developed are required to undergo intensive and comprehensive investigation prior to their acceptance for general use, so that their discharges can be permitted into drinking water supplies, the costs both of making the tests and of the delays in authorizing the use of the chemicals would be prohibitive. Dubos (13) observed that new techniques and new substances are introduced so rapidly that a requirement for testing of all of them would paralyze progress.

That there are insufficient epidemiological data is clear, and it is also clear that we are not likely to come up with adequate data in the future. Studies of this kind are not attractive to epidemiologists because the impacts that are to be assessed are subtle both in the environment and in man. While we can urge that such studies are necessary, we cannot afford to wait for them in selecting from alternatives in water management. Again, Dubos (13) states: "In the case of environmental pollution, the situation may well become unmanageable if the accumulation of convincing epidemiological evidence is made a prerequisite of social action."

The earliest water sanitation measures, from the Biblical era to the sealing of the Broad Street well, were adopted before scientific evidence in their behalf was established.

Perhaps the most expedient course of action would be not to restrict the use of the many new chemicals created by industry but rather to avoid the necessity for ingesting these chemicals by avoiding the use of wastewaters for drinking-water supplies.

## A HIERARCHY OF WATER QUALITY

Those who advocate the reuse of wastewaters for public water supplies contend correctly that the technology exists for creating a virtually pure water from water of any quality; however, conventional water and wastewater treatment processes are not suitable for converting wastewaters to drinking-water quality. Organic chemicals and most salts and heavy metals are not removed, not is there any assurance that viruses are eliminated. Other processes similar to those used in desalination are required, such as distillation, membrane filtration, and carbon adsorption. Often a combination of these treatment processes would be necessary. The cost of renovating wastewaters for municipal domestic supply would be exorbitant, especially as most domestic uses do not need drinking-water quality.

It is appropriate, therefore, to consider the recommendation made by the U. N. Economic and Social Council (14): "No higher quality water, unless there is a surplus of it, should be used for a purpose that can tolerate a lower grade." Industry illustrates this principle quite well by using raw surface waters for cooling, a higher quality water for process, a demineralized water for boiler feed, and a bacteriologically safe water for drinking. Industry and agriculture, particularly in arid areas, are already using treated wastewaters on a large scale for cooling and process in the first instance and irrigation in the second. At Santee and Golden Gate Park in California, treated wastewaters are used for recreational purposes.

It is here proposed that municipalities in water-short areas, where less than 10 per cent of the water used is required to be of drinking-water quality, have a dual system. One system could consist of a high-quality potable supply for drinking and culinary purposes; a second system, with a lesser-quality supply, possibly from reclaimed wastewaters, would serve for flushing, washing, lawn irrigation, and other nonpotable uses.

When the prospect of dual water supply systems for urban communities is presented, inevitably two important questions are raised: one, the danger of cross-connections between the two systems, and two, the high cost of introducing a second distribution system.

### Cross-connections with Dual Supplies

We are all familiar with the dangers of dual water supply systems where one is potable and the other,

generally only used for emergency, is not, and the serious waterborne disease outbreaks that have resulted. In the dual system proposed here, the drinking water would be of highest quality, taken for naturally pure sources or produced from brackish or sea water by such processes as distillation. The second supply, providing the bulk of the water, would be of questionable chemical quality, might contain some viral contamination, and might be aesthetically less desirable because of taste or odor, but it would be bacteriologically safe through conventional treatment including disinfection. In fact, the nonpotable supply would be what many cities are now providing. Occasional ingestion through misadventure would create no problem, and even if this were not discovered for weeks or months, the health hazard would not be comparable to that from continuous ingestion of low levels of potentially toxic substances over a period of many years. Of course, adequate plumbing codes, the utilization of two different kinds of piping material, and the education of the public would generally lead to an appreciation of the two types of service.

Another approach would be through a discriminating pricing policy for the two waters with a much higher unit price for excessive use of high-quality drinking waters and a lower price for the second supply. In fact, something along this line has been suggested in the Soviet Union, where it was pointed out that 43 per cent of the pure water supply to Russian cities was used by industries which did not really require chemically pure water and that it would be useful to price drinking water beyond the reach of those who now use such water for industrial purposes (15).

### Costs of Dual Supplies

Haney and Hamann (16) show in an excellent study that the order of magnitude of costs for dual systems is about the same as for conventional systems. They assume that the potable supply would be used for drinking, cooking, dishwashing, cleaning, bathing, and laundering, requiring about 40 gallons per capita per day of a total requirement of about 150 gallons per capita per day. If this were modified to provide the high-quality water only for drinking and cooking, with all other purposes being served from the second supply, the requirements for the potable supply would be less than 10 gallons per capita per day, and the costs would be correspondingly less.

Another, even more important, factor in assessing the cost of dual systems is that Haney and Hamann assume in their estimates that a second drinking water supply system would be added to existing conventional systems. The cost of the dual system would be significantly less if the two systems were to be planned and installed at the same time, and this is feasible for new cities.

We will be building as much new urban housing in the next thirty years as exists already today in the United States. The number of people to be served by public water supply systems will more than double by the end of the twentieth century (Table 1). By 1980 new investments of more than 21 billion dollars will be required for distribution systems alone, to correct present deficiencies (4 billion), for obsolescence and depreciation (6 billion), and for growth (11 billion) (17). Shall these investments be made to bring a water of uncertain quality to consumers of the future, or shall we use this opportunity for new construction to anticipate other alternatives?

Private organizations and government both have shown considerable interest in the creation of new cities. According to Spilhaus (18), more than 200 *new cities* are either in the design stage or under construction in the United States alone. In many of the arid areas of the world—the southwestern United States, the west coast of Peru, West Pakistan, and the Near East, for example—rapid population growth is expected, and this growth will of necessity be concentrated in cities. In the design of these new population centers, innovation in managing water is mandatory. One feature of new cities is likely to be the utility tunnel, or utilidor. This would avoid some of the anarchy that now exists under streets in large cities where each of some ten or more utilities has its own trench or tunnel. Placing utilities in a single service tunnel under the street when the street is built would help reduce the cost of dual water systems.

In providing two different kinds of water, the materials for each distribution system can be better adapted to the specific service. The drinking water lines can be made of materials that are virtually corrosion-proof, not only assuring longer life for the pipe but eliminating the effects of corrosion on water quality. Each line could be operated at the most desirable pressure for the particular service.

The utilization of a hierarchy of water supplies would permit development of the optimal sources of water. Because the bulk of the water supply, 90 to 95 per cent of the toal, would not need to meet the drinking-water chemical standards, more accessible sources could be developed for the nonpotable supply, or reclaimed wastewaters could be utilized with profit. On the other hand, because relatively small quantities would be required for drinking, to produce this high quality water it would be feasible to go long distances or to invest in sophisticated treatment processes that would otherwise be excessive in cost.

The greatest advantages of the dual systems would be, first, the assurance that introduction of new chemicals or exposure of a community to enteric viral disease would no longer pose a threat to the drinking-water quality and that plans therefore could be made virtually in perpetuity for people to utilize water for drinking and cooking with complete confidence; and secondly, the removal of water availability as a restriction to metropolitan growth.

## PLANNING FOR DUAL SUPPLIES

In water-supply planning for arid areas, groundwaters would be reserved for drinking water while treated wastewaters or surface streams carrying wastewaters would be used for the second system. Depending on the consumptive use in the metropolitan areas, the drinking-water supply, about 10 per cent of the total, could be used for *makeup*. In an experimental wastewater conservation facility planned by the University of Arizona and the Tucson Airport Authority, the water input required to satisfy the consumptive use is estimated to be 7 per cent of the total use (19). Rather than adding high-quality makeup water to reclaimed wastewaters, a hierarchy of water supply would permit the makeup water first to be used for the drinking water supply, after which it would be added to total supply in the nonpotable system.

Where resources of fresh water are limited, the drinking water supply could be obtained from brackish or saline waters by demineralization. With dual supplies, economies of scale could be achieved because regional plants could be built with extensive distribution networks for the relatively small quantities of water required for drinking.

A triple system might be feasible, in some instances, in arid coastal areas. Desalinated water would be used for drinking and perhaps other household uses, reclaimed wastewaters for toilet-flushing and lawn irrigation or for industrial uses, and salt waters for fire protection.

## CONCLUSION

Hodge (1) states that the lack of water has not been a barrier to the expansion of population, employment, and income in the arid West of the United States, nor is it likely to be within the near future. He adds, however, that the lack of water "ultimately may become a serious deterrent, unless new sources of water are found or wiser use is made of existing supplies."

The discovery of *new* sources of water in arid regions is at best uncertain, but the wiser use of such water as is available is well within our technical capabilities. The development of a hierarchy of water supplies based on water quality offers a reasonable approach to conserving this precious resource.

# REFERENCES

1. HODGE, C. (ed.)
   1963  Aridity and man. American Association for the Advancement of Science, Publication 74. 584 pp.; pp. 17, 394.
2. ARIZONA ACADEMY
   1964  Research report, Fourth Arizona Town Hall, Phoenix, Arizona, 167 pp.; pp. 130, 137.
3. METZLER, D. F., and H. B. RUSSELMAN
   1968  Wastewater reclamation as a water resource. Journal of the American Water Works Association 60:95-102.
4. GURNHAM, C. F.
   1955  Principles of industrial waste treatment. John Wiley and Sons, N. Y. p. 3.
5. WASHINGTON POST
   1968  Increasing pollution dooms bottled water trade. April 14, 1968, p. F7.
6. CHANG, S. L.
   1968  Water-borne viral infections and their prevention, World Health Organization Bulletin 38:401-414.
7. BERG, G. (ed.)
   1966  Transmission of viruses by the water route. John Wiley and Sons, N. Y. 474 pp; p. 3.
8. PRESIDENT'S SCIENTIFIC ADVISORY COMMITTEE
   1965  Restoring the quality of the environment. The White House, Washington, D. C. 317 pp; p. 21.
9. U. S. DEPARTMENT OF HEALTH, EDUCATION AND WELFARE
   1962  Public Health Service drinking water standards. 61 pp.
10. HUEPER, W. C.
    1960  Cancer hazards from natural and artificial water pollutants. Conference on Physiological Aspects of Water Quality, Public Health Service, Washington, D. C., Proceedings, pp. 181-193.
11. NATIONAL INSTITUTES OF HEALTH
    n.d.  Report of chemical mutagens as a possible health hazard. Genetics Study Section, Bethesda, Maryland.
12. STOKINGER, H. E.
    1965  Chemical pollutants. Man's environment in the twenty-first century. University of North Carolina, Chapel Hill, Department of Environmental Sciences and Engineering, Publication 105:60-95.
13. DUBOS, R.
    1966  Promises and hazards of man's adaptability. Environmental quality in a growing economy. Johns Hopkins Press, p. 35-36.
14. U. N. ECONOMIC AND SOCIAL COUNCIL
    1958  Water for industrial use. Document E/3058 ST/ECA/50.
15. WASHINGTON POST
    1968  February 18, 1968, p. 632.
16. HANEY, P. D., and C. L. HAMANN
    1965  Dual water systems. American Water Works Association, Journal 57:1073-1098.
17. KOLLAR, K. L., and A. F. VOLONTE
    1968  Water and wastewater facility requirements, 1955-1980. Water and Wastes Engineering, February 1968, p. 48-51.
18. SPILHAUS, A.
    1968  The experimental city. Science 159:710-715.
19. UNIVERSITY OF ARIZONA, Tucson Airport Authority
    1967  Program plan. Proposed wastewater conservation and pollution control demonstration facility. Tucson, Arizona.

# INTERNATIONAL PROGRAM FOR IMPROVING ARID AND SEMIARID RANGELANDS

ROALD A. PETERSON
Pasture and Fodder Crops Branch
Food and Agriculture Organization of the United Nations
Rome, Italy

# ABSTRACT

An international program for the improvement of arid and semiarid range-lands must be aimed at increasing the contribution these lands can make to the economic stability and development of the countries concerned. That this contribution can be substantial is evident from the size of the area and the number of animals carried. The need for action is urgent, because of the continuing deterioration of these rangelands.

Any program, to be effective, must be focused on relevant and clearly defined problems, with action priorities taking realistically into account stage of development plus cultural, economic, administrative, and other factors that influence what can and should be done. The international significance of any given activity will depend on the possibility of transferring the results to other places and on the geographic spread of the occurrence of the problem.

To improve the basis for transfer (as well as to increase understanding of the problems), increased activity in the field of resource evaluation is needed. At this stage it is already possible to define certain major widespread problems in the improvement of arid and semiarid grazing. Each of these problems should be given a priority appropriate to the local context.

# INTERNATIONAL PROGRAM FOR IMPROVING ARID AND SEMIARID RANGELANDS

Roald A. Peterson

## IMPORTANCE OF RANGELANDS

Any program for the improvement of arid and semiarid rangelands must face the fact that governments tend to give these lands relatively low priority in their development plans. This is quite understandable inasmuch as the actual and potential production per unit area of arid land is low, and improvements are likely to be slow in comparison with more favored lands. In the long run, the resultant neglect is likely to be catastrophic. The production which is derived from arid rangelands is of very considerable importance, and their deterioration under current mismanagement is steady and sometimes alarming.

According to estimates derived from maps produced by Meigs (1), about 36 per cent of the earth's surface is classified (2) as *extremely arid, arid,* and *semiarid.* The *arid* and *semiarid* portions with which we are concerned here occupy about 30 per cent. This amounts to some 2,147 million hectares classified as arid and 2,071 million as semiarid, the total of which is almost equal to the area of the North and South American continents. This area receives about 40 per cent of the total solar energy and about 20 per cent of the rainfall that reaches the land surface of the earth.

All but a small part of the area is grazed, including the cultivated fields after harvest. If one assumes a stocking rate of 15 hectares per livestock unit for the semiarid area and 30 for the arid area, the total number of animals carried may be equivalent to more than 200 million livestock units. Assuming a 12 per cent annual offtake, some 2,400,000 livestock units would be supplied to the market each year, as well as substantial amounts of wool and milk (Fig. 1) and large quantities of game. Even though assumed stocking rates are rough estimates, the order of magnitude of the animals carried and annual offtake make clear the impressive contribution of arid lands.

In addition to being a source of animal products, the arid zones are important sources of water and they increasingly contribute to recreation, both factors which are strongly influenced by the grazing management practiced.

Many countries such as those immediately north and south of the Sahara, and most of the countries in the Near East, are virtually completely dependent on arid or semiarid lands. In these countries, a high percentage of the people are directly engaged in grazing. For example, in Saudi Arabia possibly 40 per cent of the people are dependent in one way or another upon grazing animals for a livelihood (3); in Algeria about 25 per cent (4). In Iran a growing lamb-finishing industry based on range animals provides additional employment and income, as well as more meat for the population. Similar trends are developing in other countries. In many countries rangeland production has an important influence on foreign-exchange earnings or expenditure.

It is evident, therefore, that the widely held assumption that range improvements merit relatively low priority in development is questionable and obviously erroneous in many cases.

The second major consideration justifying greater concentration on the improvement of the range resource is its steady deterioration.

Clearly, especially from a national point of view, investments need to be made not only to obtain increased yields and returns but also to prevent further losses. Such investments are in some sense analogous to maintenance costs.

During the past 20 years, many observers (5-8) have noted the destruction or alteration of the vegetation of the arid zones and the resultant progressive decrease in productivity. Desertification is becoming increasingly wide spread. In other places, the vegetation may remain intact but livestock production goes down and becomes less stable. The rate of recuperation, given good management, depends on the degree of deterioration and the environment; therefore, the more advanced the destructive process the longer and more costly will be the remedial measures required, and postponement of the needed action must inevitably have unfavorable effects upon the economy.

The practical alternatives that policy makers must face are to assure investments in range management and improvement or to saddle their countries with the resultant losses in production and income in the future, plus increasing erosion, floods, siltation of reservoirs, and other undesirable effects. The relevance of this choice is evident because only a small proportion of arid lands are suitable for other types of production.

## THE BASIS FOR AN INTERNATIONAL PROGRAM

International programs have the obvious advantages of providing external financial and technical assistance. They may also provide a more coordinated and concerted attack on the problem. Their value will largely depend upon their relevance to

the economic welfare of the countries concerned and the degree to which the results may be shared among countries. To be meaningful today, such programs should have an impact in developing countries.

It is a regrettable fact that numerous programs in range management which have been undertaken bilaterally or multilaterally have been marked by failure. The reasons include: inapplicability of the "solutions" proposed, lack of wide support by the graziers or the government, failure to maintain continuity of programs, and lack of incentives for the producers. Various factors limiting the impact

Fig. 1. Milking mares in Mongolia. Among the less well-known of the variety of animal products provided by rangelands is mare's milk fermented to make *kumis,* a much-appreciated drink in some countries. (Photo by R. A. Peterson)

of range management programs have been discussed in some detail by Peterson (9) and Pearse (10).

Despite continuing obstacles, I believe that we have learned enough from the past to improve the percentage of successes. In an international program, all concerned can concentrate on activities that will in the long run systematically contribute to range improvement in the widest sense. In this paper an attempt is made to indicate some aspects of a line of attack that would meet many national needs and at the same time contribute to international requirements.

## UNDERSTANDING THE PROBLEM

The first requirement for success is proper identification of the problems *at the relevant level,* taking into account such factors as the socioeconomic framework, the traditional system of grazing, and current productivity. Problem identification requires an understanding of the major components of the production system and their relation to one another, and identifying those elements that constitute major bottlenecks to production (11). For this purpose rangeland and arable lands must usually be considered jointly, because the solution of the problems of rangelands may depend on effective complementary use of arable lands, even though arable lands are only a small portion of the total land area.

After identification of the problems, priorities must be set that will make best use of the available staff and finances. Although often the primary requirement is to gather the data needed for defining the relevant problems, this task is generally neglected because countries are in a hurry to "get on with the job" without delay. Lacking clear orientation, it is not surprising that what gets done is often what the scientist or technician happens to know best, whether it deserves priority or not.

## EXAMPLES OF COUNTRY PROGRAMS

Identifying the relevant problems and deciding on what is to be done is by no means a simple matter. The process can be illustrated by a few examples, drawn from FAO experience.

### Kenya

Livestock production is one of the major agricultural activities in Kenya. Over four-fifths of the land surface is classified as rangeland (12). This supports, apart from countless wildlife, about half of the national herd. Only 2 per cent of the area has been developed in the past by expatriate ranchers. Apart from revenue gained from tourism, the contribution of the range area to the economy of Kenya is far below its potential. Due to mismanagement and overpopulation, famine has become a recurrent problem, involving the expenditure of nearly 5 million pounds for relief in 1961/62. Bush encroachment and range deterioration threaten the future productivity of these lands and would inevitably increase famine and affect Kenya's important tourist industry.

On the basis of these considerations, the government of Kenya created a Range Management Division within its Ministry of Agriculture in 1963 and requested the help of the United Nations Development Program to strengthen the new Division. A Special Fund project was activated in 1966. The support and priority given by the government were such that the attack could be launched on a number of important interrelated problems.

Range survey (including game, water development, and range economics) is being carried out in order to facilitate the rational organization of several types of ranches, some of which will be occupied and run by groups of Masai families who are now semi-nomadic. Research is being done on range management, bush control, game population movements and competition with livestock, livestock improvement, and plant physiology.

No real change in range use, however, can be effected unless the graziers themselves are involved. In order to reach them a sociologically oriented approach is employed which includes the training of pastoralists at specialized Farmers' Training Centres, and the use of mobile education units. The preparation of films and other audiovisual media of local relevance is one of the important activities of the project.

### Senegal

In 1961, at the request of the government of Senegal, FAO sent an ecologist, A. Naegelé, to study the grazing area of Northern Senegal. He concluded that the major problem was the deficiency of forage during the dry season, accentuated by a predominance of annual grass species which upon maturation lost quality and tended to disintegrate. In the meantime rapport had been established with the graziers, so the ecologist could discuss their problems freely with them. He found they were well aware of the feed deficiency during the dry season, but did not know how to solve the problem.

With Naegelé's assistance, the pastoralists decided to make hay from surplus feed available toward the end of the rainy season. The effect on animal responses was such that an ever-widening group of graziers became interested (Fig. 2). To simplify the work and increase hay production, a proposal was put forward, through the Freedom from Hunger Campaign, the secure mowing equipment which could be drawn by oxen. *Entraide et Fraternité* (Belgium) and *Misereor* (Germany) agreed to support the project, some 25 mowers and carts together with scythes and forks were provided, and production of hay was begun in significant quantities.

Obviously haymaking is only a step. In the meantime, however, a vital point of attack has been found which has aroused the interest and the involvement of the graziers. They are now becoming aware of the full significance of better feeding of their livestock. With continued and consistent effort, there seems to be every prospect of an increasingly rational use of these grazing lands.

## Syria

As in other Near Eastern countries, uncontrolled grazing is the rule in Syria. Flock numbers are decimated by drought every few years (Fig. 3). For example, the sheep population as reported in the 1962 FAO Production Yearbook (13) dropped from almost 6 million to 3.6 million between the years 1958/59 and 1960/61. Then it built up again to about 6.5 million head (14).

Over the years, the government has tried to develop a range management organization. Emergency feed deposits, strategically located in various parts of the country, help correct for variations in range forage growth. These, along with better veterinary practices, tend to maintain higher average grazing pressures. Development of watering-points (Fig. 4) has received considerable emphasis, but there has not been a corresponding control of grazing. As in other arid countries, the effect has too often been to extend the grazing damage.

FAO range specialists in Syria have been concerned with securing a knowledge of the range resources and ways of supplementing those resources. Currently, less than one-half of the grazing is derived from the range. Most of the rest comes from the grazing of crop residues (14).

It has become clear that vastly increased forage

Fig. 2. Senegalese semi-nomadic graziers learning to make hay from the native range to reduce losses during the long dry season. Traditional methods of grazing and animal husbandry can best be changed by introducing improvements aimed at solving problems of which the pastoralists are fully aware. (Photo by A. Naegelé)

production in the cultivated areas is an essential step in relieving grazing pressure on the range, reducing the impact of droughts, and meeting the growing demands for more animal products. Lamb gains are negligible on the range after the forage dries in the spring until the following spring (14). The resultant low output and the use of substantial amounts of range forage could be avoided by early transfer of lambs to feed lots or to irrigated pastures. Stratification of production between crop areas and rangeland is increasingly feasible with a wider use of high-yielding wheat varieties along with heavy fertilization and more effective use of irrigation water, as a means of making more land available for forage production. Furthermore, a great deal of land left in fallows could be used for forage production.

The beneficial effects of controlled grazing and spring resting of the range have been demonstrated on some 400,000 hectares, which with Bedouin cooperation were organized into a grazing and summer resting district.

Protection of parts of the range from grazing for use during times of feed shortage could be the most economic method of meeting most feed shortages. To this end, an attempt is now under way to reintroduce the *hema* system. This ancient system was once closely allied to Koranic teachings (15, 16). Some remnants of the system have been reported in Saudi Arabia. Essentially, it is based on setting aside certain lands for use under prescribed conditions agreed to and enforced by the nomadic people themselves. It is hoped that success on a limited scale could eventually lead to wide application with the support of the pastoralists, who are not noted for their willingness to accept government controls.

## Iran

As much as 88 per cent of animal feed in Iran is derived from rangelands (17). Pabot (5) points out that this basic supply is threatened by alarming range deterioration. Pearse (7), in an analysis of the problems, confirms Pabot's observations (5) and identifies technical, administrative, legal, sociological, and economic obstacles to improvement of the rangelands. The destructive trends have been highly accentuated by wide-scale plowing of rangelands after World War II when tractors came into wide usage. For example, the 1966 FAO Production Yearbook (2) reports 3,158,000 hectares of land in cereals in 1952, whereas 5,567,000 hectares were under these crops in 1961. One to two or more hectares (depending on rainfall) for every hectare of currently cropped land is a common pattern. Much of the plowed land is submarginal to permanent agriculture and commonly is cropped only in years of above average rainfall, or it may be abandoned for crop production over a long period of time. Pabot (5) estimated that in 1963 there were some 10 million hectares of formerly cultivated lands in Iran. Pearse (7) stated that since Pabot's estimation the area of abandoned cropland has probably doubled. Crop residues may partly compensate for lost range grazing while the land is being farmed; however, this is a highly seasonal form of compensation and necessarily increases the grazing pressure on the range during the spring when the crops are growing.

In 1964 a United Nations Special Fund Project was initiated with the Iranian Government. Following Pabot's analysis, the project was aimed primarily at trying to find ways of revegetating depleted and plowed rangelands as the most practical means for

Fig. 3. Landscape east of Damascus showing ephemerals and a few stunted *Artemisia herba-alba* (a relatively unpalatable subshrub) remaining after years of heavy grazing. Overgrazing of vast tracts in arid and semiarid regions has led to the disappearance, or greatly reduced density and production, of palatable sub-shrubs and desirable perennial grasses. (Photo by J. P. H. van der Veen)

Fig. 4. Motorized transport of water on the Syrian steppes with tractor-trucks. Transport can be achieved over a radius up to 60 kilometers and can be an effective way of improving production from the range, if adequate control is maintained over animal numbers and degree of utilization. (Photo by J. P. H. van der Veen)

reducing the stocking load on the ranges. Numerous species have been tested, as have methods for establishment. A key technical limitation has been found to lie in the difficulty of species establishment due to variable weather and crusting of the soil surface.

Only if the problem of establishment is solved can the reseeding of vast areas become economically viable. Such reseeded areas would also have to be accompanied by a vigorous management program, and further strengthening of the administrative organization for these lands. The first grazing association has been organized and a range analysis is under way. For longer-term planning, it is essential to identify those areas remaining in Iran where perennial forage species still predominate. This could provide the factual basis for setting priorities in grazing control to prevent further depletion of these areas. Much more intensive forage production will also need to be developed in the more favored rainfall zones and under irrigation, in order to draw more animals from the range for fattening, reduce weight losses of young stock due to migrations, and to provide hay for winter. Success will mainly depend on the continued and increased dedication of the Iranian government.

### Panama

The west coast of Panama, as well as that of other Central American countries, is subject to a dry season, varying mostly from 4 to 6 months. Although the area is not classified as semiarid, its problems are illustrative. At the request of the Panamanian government, an FAO consultant made an analysis of these problems.

Most of the grassland vegetation consists of the introduced but naturalized *Hyparrhenia rufa*. This grass loses quality rapidly as maturity occurs. Because grazing is heavy, severe malnutrition is the rule during the dry season.

A multiple-pronged attack was recommended to find alternative solutions to the dry-season feed shortage. As a follow-up, a United Nations development program project was initiated in 1967 in cooperation with the Panamanian government. This project seeks to find solutions to the dry season shortage through the use of fertilizers and tropical legumes to extend the green feed period further into the dry season, control of stocking to provide more standing hay at the end of the rainy season, conservation of forage, supplementation of dried forages by urea, and better animal management to secure a closer harmony between animal requirements and a seasonal forage supply. It is likely that the best solution will be an integrated use of various methods. These modern approaches to the solution of the dry season problem appear feasible in Panama due to relatively favorable livestock prices and the

level of existing management, which includes use of fences, watering, and farm machinery.

With these examples, I hope to have illustrated something of the diversity of the problems encountered, and the need to adopt an approach appropriate to each situation and the means at hand. My excuse for belaboring the point somewhat is that I consider lack of adequate problem definition and priorities of action the main weaknesses in virtually all programs in developing countries.

It is now appropriate to mention that the need for setting priorities, taking into account global, regional and country considerations, is becoming increasingly recognized. FAO is currently engaged in a global study known as the Indicative World Plan to determine on a broad basis steps which developing countries may need to take if the demand requirements for agricultural products are to be met in 1985. This should permit the gradual development of an overall strategy. Ultimately, however, the success of such a strategy will depend upon a thorough understanding of the needs and possibilities of each resource, such as the range, and the translation of this understanding into specific action programs.

## INTERCHANGE OF RESULTS AND EXPERIENCE

The assumption that results and experience can be transferred from one place to another is implicit in much of the philosophy of international technical assistance. Transfer is not a simple matter, because not only do the components of biological problems vary quantitatively and in kind from place to place, but so do such factors as human institutions, traditional modes of behavior, technical manpower availability. It is for this reason that major efforts must still continue to be placed on understanding the problems in the local context and designing programs accordingly. Only by identifying areas in which similar problems occur, however, can maximum use be made of results, secured in each place through transfer to other places.

Perhaps the most systematic work done to evolve a basis for the generalization of agricultural research and experience has been by M. Y. Nuttonson and his colleagues in The American Institute of Crop Ecology in their study of agroclimatic analogues (18). These studies take into consideration crop requirements as well as vegetation. Various climatic classifications have also attempted to take these relationships into account. All these studies have been handicapped by difficulties of securing, interpreting, and managing the vast amount of data involved. One may hope that with computers and vastly improved international communications, the process of identifying similar environments and fitting plants and management systems to those environments may be less onerous and gain precision.

In restricting this discussion to arid and semiarid environments and to range problems, a basis for generalization has already been assumed, at least in terms of broad problems, such as low and irregular precipitation, seasonal variations in forage supply, frequent droughts. Differences in rainfall patterns and quantity, temperatures, the nature of the range vegetation, soils, and other factors still make it extremely difficult to generalize results, however. A further handicap is the lack of adequate data in many countries on which to establish similarities.

It may be concluded, therefore, that an international program on rangelands must include an inventory of the resource in terms that will permit orderly comparisons. This needs to be done at various levels so that comparisons can be made according to the specific purpose in mind. Various activities under way in FAO bear strongly on this problem.

FAO, Unesco, and WMO have jointly carried out agroclimatological studies in part of the Near East (19) and the Sahelian and Sudanian zones of West Africa (20). A similar study of the highlands of eastern Africa is nearing completion. These studies relate to climate and crop growth and are relevant to a better understanding of the vast rangelands in those regions and in making comparisons to other regions.

A soil map of the world by FAO and Unesco is nearing completion. It will help to establish similarities in the soil environment.

In a number of countries surveys of the range have been, or are being, carried out by FAO and other organizations. More recently in FAO we are attempting to evolve a systematic approach to determine the extent, distribution, and role of grazing lands in developing countries. Numerous studies within individual countries contribute to the general goal of providing data that may be compared to other situations. The usefulness of these studies in this sense is still somewhat limited, due to the different methods and standards followed.

A committee was recently established by the American Society of Range Management for the study of rangeland biometeorology.

These efforts and others will gradually give us a more comprehensive view of the grazing lands of the arid and semiarid parts of the world, as well as other regions, and a better base for transferring results and experience. There is an urgency to speed up accumulation of this kind of knowledge, due to the continuing rapid rate of destruction of arid and semiarid ranges. The importance is such that an international conference specifically devoted to this problem appears justified.

Many other problems which are already known to be widespread and of great importance offer good possibilities for effective work of international consequence.

## LOW PRODUCTIVITY

How excessive grazing depresses yields is well known, although many details need to be better understood, especially the interactions between grazing, run-off and efficiency of utilization of soil moisture.

Although run-off is a function of the intensity of storms, intensity levels are very often not known. The efforts of the World Meteorological Organization and of the developing countries to correct this deficiency are of the greatest importance. Measures of intensity may be as important or more so than amounts and distribution of rainfall. More strictly on the range management side is the question of how the amount of runoff may be reduced by practical management of the vegetation and how production is influenced.

Efficiency of soil moisture use can also be radically influenced by the effect of grazing on the rate and time of growth of the vegetation. Relatively rapid growth at the beginning of the growing season, when temperatures are usually lower, will be most efficient for water use. This rate of growth depends on such factors as the root reserves in perennials, the leaf area per plant during the early part of the season, and the plant species. All of these may be strongly influenced by grazing. Low soil fertility will also limit the rate of growth, particularly at lower temperatures. Where low temperatures are not limiting, fullest efficiency of water use may depend mainly on the leaf surface available for photosynthesis and fertility levels.

Completeness of soil moisture utilization by plants is also of vital importance. Heterogeneity of the vegetation in terms of growing period and response to showers, root distribution, and capacity to extract moisture from the soil influence the fullness of moisture utilization. Heterogeneity, especially the presence of some types of shrubs, will also, through shading and slowing of air movements, influence evaporation from the soil surface. Continuous uprooting of small woody shrubs for fuel is in many places a major factor in desertification (5). Litter cover will also be of substantial importance in increasing moisture infiltration and reducing runoff and evaporation.

While all of these factors are more or less understood, not enough studies have been made in diverse environments, and not enough attention has been given to how these factors interact and how the results relate to practical management problems. This is a field of study in which developed countries could well take a strong lead, wherever possible in conjunction with developing countries.

On a less fundamental level, there is need for increased knowledge of the potential productivity of range sites. Lacking such knowledge, gathered under a great variety of conditions, setting production goals demands considerable research in individual countries. Minimum norms of the methodology to be followed, and complementary observations are needed. This is definitely a program which must depend for its success upon international cooperation. FAO and the International Biological Program have discussed collaboration on measurements of grassland production in general. It would be helpful if organizations particularly interested in arid lands could take a major responsibility for production studies on those lands.

## SEASONAL AND ANNUAL VARIATION OF FORAGE SUPPLY

Variation of forage supply is not, of course, discrete from production *per se*. Its importance lies in its influence on animal responses. In combination with year-to-year fluctuations, it is the major factor contributing to the instability of livestock production in arid zones and ultimately to the level of production. Decreasing this instability should be the first step to increasing production progressively in countries primarily dependent upon arid and semiarid lands. Dry farming practices, too, tend to remain less advanced than desirable, or necessary, because stubble fields are traditionally grazed at certain times by outside flocks. Farming then must be geared accordingly. Furthermore, due to current risks, graziers are reluctant to undertake improvements. Governments are also reluctant to take necessary action because of the risks, as any change which the government may carry out implies a responsibility for possible resultant failure. These circumstances constitute a formidable combination of factors to be overcome if the present stagnation in production is to be broken.

The normal seasonal variation in range forage supply is greatly accentuated by too heavy grazing during the growing season and almost total lack of reserves for the dry season. Where the vegetation is heterogeneous, with plants of different palatability and different season of growth, or where plants serve well as standing hay through the dormant season, the problem can be solved pretty well by control of stock numbers. But even in advanced countries the temptation for short-term gains makes it difficult to get a rational control of stocking rate. In developing countries, especially where nomadic grazing is practiced and where marketing and other services are inadequate, the difficulties are compounded.

Emergency feed storage (as in Syria) and better distribution of watering-points may be helpful in ameliorating the effects of variation in range forage supply. These methods have not usually involved the active participation of the graziers. Nor have they included means for gradually securing control of numbers. In the long run, they are likely to increase the deterioration of the rangelands without lasting benefits to the graziers.

To a large extent the solution of the problem will have to be sought outside of the rangelands by the development of supplementary forage supplies which will allow drawing a maximum number of animals from the range. In many cases, large-scale fattening operations on irrigated lands may help to assure a high demand for range animals for fattening (Fig. 5). Large-scale reseeding of marginal farmlands for early grazing or other special uses can help to provide the additional feed. There must, however, be a comparative advantage for the producer to use these lands for these purposes. Economically feasible ways of establishing sown species under relatively difficult environments and high efficiency in the utilization of these sown areas are key requirements. Government policies which will help to initiate this type of change in land use are needed. Such policies must take into account that so long as range forage is vastly cheaper than cultivated forages, change toward rational stratification will be difficult to achieve. Appropriate grazing fees may be an essential tool to achieve a better balance between costs of range and cultivated forages.

Additionally, we need to know more exactly to what extent ranges can carry livestock throughout the year and the characteristics which determine the responses. In a part of Syria, it has been shown

Fig. 5. Sheep grazing irrigated alfalfa, which is also cut for hay, near Teheran, Iran. Improvement of rangelands depends to a considerable extent on the development of additional forage resources. (Photo by C. K. Pearse)

that with proper stocking the range can in fact carry adult sheep throughout the year without great weight loss, except during extreme drought (14). This range contains a number of shrubs as well as perennial and annual herbaceous species. The problem is how to predict more efficiently the suitability of a range for a given season or for year-long use, based on composition and distribution of range types. Deficiency of information about what animals eat under diverse conditions is a major obstacle in answering this question.

The introduction and testing of new species will be a necessary part of many range improvement programs aimed at a better distribution of the forage supply. International standards for testing are essential to facilitate meaningful comparison of data. Some plants, such as certain saltbush species, are of such interest that regional projects for testing them would seem highly appropriate. Selection and breeding of types for specific purposes merits more attention. These types of studies lend themselves to collaborative efforts between developed and underdeveloped countries. FAO is prepared to assist in projects of this nature.

## NOMADISM AND TRANSHUMANCE

The nomadic and transhumance systems of grazing constitute a special and widespread problem in many of the arid and semiarid regions. There can be no question that these systems have demonstrated survival capacity and have been an effective means of harvesting a highly variable forage supply. While much variation exists within these general systems, they are characterized by a close interdependent relationship between man and beast. Sweet (21), for example, gives a good account of the relationship between the Bedouin of northern Saudi Arabia and camel pastoralism, and how the social organization itself has been formed accordingly. Generally, the internal organization maintains the integrity (and independence) of the system. Formalized moral codes may be strictly enforced. Traditionally these systems are essentially subsistence and marginal to the market economy.

These features, while admirable and adaptive, present inbuilt resistances to development and orderly administration of the rangelands. For example, if the animals serve mainly for prestige, purchase of brides, or meeting immediate food needs, there is little incentive for securing quick growth of the animals (as for early marketing) and for increasing the offtake. The main quality of an animal then becomes its capacity to survive, and there is no question that natural selection aided by the herdsman has been effective in producing some remarkably well adapted animals. Organization of grazing into districts will conflict with the free movement now practiced, especially during times of stress. Lack of roads, inadequate prices, poor communications, lack of schools, and other socioeconomic factors have reinforced the cultural and biological restrictions which limit the contribution made by nomadic people to the economy. At this point in time, the nomadic systems themselves are threatened by the encroachment of the plough and the decreasing carrying capacity of the rangelands.

The populations of the nomadic peoples are growing, while one of their most important sources of income, the camel, has been to a large extent replaced by the truck. Lack of educational opportunity has also tended to place the nomad in a relatively lower status than in the past (22). Governments are faced with the need to increase the contribution that nomadic people make to the economy and to take measures to assure that the range resources are not further deteriorated. Settlement of the nomads is often thought to be the most feasible solution. This is costly, and, considering the shortage of arable land and the long-established customs of the people, can hardly be recommended as a generalized solution. The problem rather calls for an open-minded experimental approach, guided by a number of considerations.

Perhaps the most important consideration is to recognize that, due to the integrated nature of nomadic systems, changes in one respect are likely to affect the rest of the system. This in turn may be expected to increase the conservatism and resistance to change of the system. Marketing of animals on a regular basis, for example, may restrict the availability of animals for other purposes, such as the purchase of brides. Control of watering-points and grazing interferes with the sharing of these resources, which is so important for survival during emergencies. The expression of individual rights may also conflict with religious belief. To transform nomadic grazing into more productive systems, then, it is necessary to discover those modifications that can be introduced with the cooperation of the people concerned.

Achieving cooperation will depend upon a better understanding of the problems encountered by the nomadic people and a knowledge of their felt needs. Integrated surveys, as suggested by Jager Gerlings (22), may offer the best means for arriving at this understanding. Once these problems and possibilities are understood, steps can be taken to transform gradually their system to make it more productive and more suitable to the modern state. Further, when the problems are known, the necessary research may be undertaken to find applicable solutions. The examples cited of the active participation by Senegalese graziers in haymaking, the interest of the Syrian Bedouins in grazing reserves, and the group ranches of the Masai in Kenya help to substantiate that this approach can give useful results, even when done on a modest scale.

The observation of Mesthene (23) applies to the problem of development in general, and is particularly apt in the context of the question of how to bring about change in nomadic systems:

Traditional ways (beliefs, institutions, procedures, attitudes) may be adequate for dealing with the existent and known. But new technology can be generated and assimilated only if there is technical knowledge about its operation and capabilities, and economic, sociological and political knowledge about the society into which it will be introduced.

## INTEGRATING AGRICULTURAL AND LIVESTOCK PRODUCTION

Reference has already been made to the need to integrate the use of arable lands with rangeland. This constitutes a major problem because of the traditional separation of livestock production and crop farming. The core of the problem is to find ways to make such integration or stratification of production mutually beneficial.

## INSTITUTIONAL OBSTACLES

### Range Organization

The scope and nature of the job to be done on managing the rangeland resource is, in its broad context, reasonably well understood. Only a few countries such as Kenya, Iran, Syria, and Algeria have formed organizations for this purpose. Many other countries carry on a certain amount of range research and administration within other agricultural or forestry services. In the latter case, the range activities are too diluted to have an effective impact and tend to perpetuate the low priority given rangelands within government programs.

### Education and Training

Most developing countries have no regular training in range management. Several countries, including Kenya, Mexico, Iran, Pakistan, and Syria, are receiving assistance for the introduction or strengthening of range management in their University curricula. Effective in-service training under FAO/UNDP projects is under way in short courses in some 15 countries involving some 25 to 30 trainees. Additionally, special training through short courses is given emphasis in Iran and Kenya. A review of current training in range management and pastures is to be undertaken by FAO in several African countries south of the Sahara.

Up until now the range management specialists available in developing countries have mostly been trained abroad, primarily in the United States. Their number is still too small to cope with the tasks

on hand. Training at the non-University (technician) level is even more deficient, thus further reducing the effectivity of the range specialist. Some countries are making an effort to correct this deficiency. This is an essential development which should be extended to other countries as well.

In meeting training needs, consideration should be given to: (a) training many more people within their own or a similar environment, and (b) better orientation of training in developed countries to meet the needs of the students from developing countries (11). The first point could perhaps be best met on a regional basis through special courses, and by building up the range management curriculum in certain selected universities.

I hardly venture to make recommendations on the second point because a great many professors of range management have had foreign experience and in general provide a fairly broad orientation. Still the penchant for narrow specialization in research too often wins the day. Methodology may be stressed too much as an end in itself. More consideration could be given to the identification and analysis of problems and priorities, so as to help the student learn how to evaluate what is relevant to his ambient.

### Administration

Developing countries for the most part are beset by a jungle of administrative procedure, which is enough to defeat all but the most patient and stouthearted. Authority may be diffused, or the most minor decision may have to be referred to high authority. The checks and controls are often such as to call forth more energy and creativity for finding ways around them than for doing the job itself.

Promotions generally follow strict seniority rules. Regard for achievement is likely to be secondary to other considerations. One consequence of this is that there is a strong trend to stay in the capital city where contacts can be maintained. The commendable practice used in Argentina, and some other countries, of differential compensations taking into consideration the favorableness of post location, deserves wide emulation.

Many other deficiencies could be cited, and there is no denying their relevance to any effective program. The practice, followed in some countries, of setting up semi-autonomous institutions has proved a rather effective but perhaps temporary solution. This problem is one in which scientists need the earnest assistance of administrative innovators, who can fit the administration to the job to be done. In the long run there is no substitute for a strong administrative structure at the local, provincial, and country level for lasting success in the improvement of the arid and semiarid range lands.

# REFERENCES

1. MEIGS, P.
   1953 World distribution of arid and semi-arid homoclimates. *In* Reviews of Research on Arid Zone Hydrology. Unesco, Paris. Arid Zone Programme 1:203-210.

2. FOOD AND AGRICULTURE ORGANIZATION OF THE UNITED NATIONS
   1966 Production Yearbook Vol. 20. FAO, Rome. 763 pp.

3. HEADY, H. F.
   1963 Report to the Government of Saudi Arabia on grazing resources and problems. FAO, Rome. TA Report No. 1614. 30 pp.

4. BAUMER, M.
   1963 Rapport au Gouvernement de l'Algérie sur les pâturages et l'elevage sur les Hauts Plateaux algériens. FAO, Rome. TA Report No. 1784. 94 pp.

5. PABOT, H.
   1967 Report to the Government of Iran on pasture development and range improvement through botanical and ecological studies. FAO, Rome. TA Report No. 2311. 129 pp.

6. LE HOUÉROU, H. N.
   1968 La Désertisation du Sahara septentrional et des steppes limitrophes (Libye, Tunisie, Algérie). Réunion technique sur l'Ecologie et la Conservation de la Nature (Programme Biologique International), Hammamet, Tunisie, 24-30 March 1968. 36 pp.

7. PEARSE, C. K.
   1968 A range, pasture and fodder crop research program for Iran. A problem analysis and working plan. UNDP/SF Pasture and Fodder Crop Investigations Project (Iran/10), Teheran. 73 pp.

8. SPRINGFIELD, H. W.
   1948 Forage problems and resources in Iraq. International Cooperation Administration, Washington. 37 pp.

9. PETERSON, R. A.
   1964 Improving technical assistance in range management in developing countries. Journal of Range Management 17(6):305-309.

10. PEARSE, C. K.
    1966 Expanding horizons in worldwide range management. Journal of Range Management 19(6):336-340.

11. PETERSON, R. A.
    1966 The training needs of grassland workers in developing countries. International Grassland Congress, 10th, Helsinki, 1966, Proceedings, p. 50-55.

12. PRATT, D. J.
    *in press* Rangeland development in Kenya. [To be published in Annals of Arid Zone, 1968].

13. FOOD AND AGRICULTURE ORGANIZATION OF THE UNITED NATIONS
    1962 Production yearbook, vol. 16. 493 pp.

14. VAN DER VEEN, J. P. H.
    1967 Report to the Government of Syria on range management and fodder development. FAO, Rome. TA Report No. 2351. 76 pp.

15. KLEMME, M.
    1965 Report to the Government of Saudi Arabia on pasture development and range management. FAO, Rome. TA Report No. 1993. 27 pp.

16. DRAZ, O.
    n.d. The ancient "hema" system of range reserves in the Arabian Peninsula. Its possibilities in range improvement and conservation projects in the Near East. (Unpublished manuscript).

17. HAYNES, J. L.
    1965 Crop zones of Iran. US AID, Division of Agriculture, Washington. 14 pp.

18. NUTTONSON, M. Y.
    1947 International cooperation in crop improvement through the utilization of the concept of agroclimatic analogues. American Institute of Crop Ecology, Washington, Reprint A [from Institute for International Collaboration in Agriculture and Forestry, Interagra 1 (3/4)].

19. PERRIN DE BRICHAMBAUT, G., and C. C. WALLÉN.
    1963 A study of agroclimatology in semi-arid and arid zones of the Near East. World Meteorological Organization, Geneva. Technical Note No. 56. 64 pp.

20. COCHEMÉ, J., and P. FRANQUIN.
    1967 An agroclimatology survey of a semiarid area in Africa South of the Sahara. World Meteorological Organization, Geneva. Technical Note No. 86. 136 pp.

21. SWEET, L. E.
    1965 Camel pastoralism in North Arabia and the minimal camping unit. *In* Man, culture and animals. American Association for the Advancement of Science, Washington. Publication 78: 129-152.

22. JAGER GERLINGS, J. H.
    1967 Problems of nomadism. ITC-Unesco Centre for Integrated Surveys, Delft. Publication No. S. 18. 16 pp.

23. MESTHENE, E. G.
    1968 How technology will shape the future. Science 161(3837):135-143.

# THE PASTORAL ETHIC
## A comparative study of pastoral resource appraisals
in
## Australia and America

RONALD L. HEATHCOTE
School of Social Sciences
Flinders University of South Australia
Bedford Park, South Australia

# ABSTRACT

From approximately the middle of the nineteenth century, American and Australian stockmen and their herds have moved into the occupation of the semiarid and arid grazing ranges of western United States and interior Australia. By the time of World War I, the initial occupation of most of the ranges had been completed and the first problems associated with that occupation had arisen. While the sequence and character of the occupation of the two ranges show striking parallels, the responses to the challenges of the problems show some parallels but also show important contrasts not only between the two countries but between the responses of interested parties within each country. A comparison of the sequence of the evaluation of the two ranges, and the legal background to their use by stockmen, illustrates similarities and contrasts. Despite contrasts in detail, there has been evidence of the existence of what might be called a *pastoral ethic* in the appraisal and use of the range in both countries.

# THE PASTORAL ETHIC

## a comparative study of pastoral resource appraisals in Australia and America*

### Ronald L. Heathcote

The latter half of the nineteenth century saw the occupation of the semiarid and arid areas of western United States of America and interior Australia by a system of commercial livestock grazing using both sheep and cattle. Despite the more intensive agricultural and other uses of isolated localities, the grazing of livestock on the natural range (1) persisted over most of the areas virtually unchanged until at least the middle of the twentieth century. Within the generally similar historical and geographical settings, however, significant differences evolved, particularly in the appraisal of the pastoral resources and the systems of land tenure set up to control their use. Certain of these differences were between the two national policies toward the range, but equally important were differences in each country in the attitudes to the resources of, on the one hand, the central governments, and on the other, the pastoralists themselves. In the case of the latter, however, I will try to show that sufficient similarities existed either side of the Pacific to suggest the existence of some kind of common *pastoral ethic* among the resource users—the stockmen.

## DEFINITIONS

The arid and semiarid ranges to be considered are shown in Figure 1. To allow a comparable area to be studied and to consider that area about which most is known, only the eastern half of the Australian range has been included.

*Pastoral resource appraisals* are the attitudes and policies toward those features of the natural environment that have value for the raising of domesticated livestock. Such attitudes and policies can be derived from private papers, journals, books, newspapers, official policy statements, reports, and legislation; they can be traced from these sources over the century of occupation considered here. The features of the natural environment that have value for the raising of livestock are limited here to the three basic requirements: *land, water,* and *forage.* These three resources as evaluated by the officials and the men on the land will be our main concern; the evaluation by the pastoralists themselves, their conscious or unconscious system of moral science or moral framework for resource use, is what I shall call a *pastoral ethic.*

*This paper is based on an address to the Australian and New Zealand American Studies Conference at Sydney, Australia, in August, 1968.

## HISTORICAL CONTEXT OF PASTORAL SETTLEMENT

The spread of Anglo-American Settlement into the arid western United States and Anglo-Australian settlement into interior Australia was but the extension of the occupation of the more humid eastern and coastal areas. The filling up of supposedly empty lands, and the bringing into use of supposedly untouched resources, were seen as merely spatial and temporal extensions of activities, the success of which was already proven (2).

As part of the general process, and accepted as one of the first possible extensive uses of the virgin lands, the grazing of livestock on the *waste* lands had developed and thrived in the humid eastern part of the United States and of Australia. The cowboy and sheepherder had pioneered from the Carolinas to Ohio, and from the Cow Pastures of the Nepean River to the plains of Bathurst. The techniques of overlanding cattle to market for slaughter, the roundups, the annual shearing and lamb-marking, were all part of a technology that was at first merely extended into the arid areas. The highly mobile capital—that is, the livestock—solved some of the transport problems and enabled the use of grazing resources as and when they occurred; little or no investment in land was needed, since most operations were beyond the surveyed and legally controlled districts; the natural increase of births over deaths was usually rapid, and surpluses soon accumulated.

From the older centers of settlement the herds were pushed further into the remote frontier districts. Improvements in land transport (principally the railway systems) and land communications (particularly the telegraph and later the telephone) enabled the pioneer producers to keep in touch with traditional markets and later to take advantage of the expansion of the international markets and to attract foreign (mainly British) capital and local funds from gold-mining successes, to support their occupation of the new lands. The telegraphed news of good rains or droughts brought solace or despair to distant Scottish and English shareholders alike from the Panhandle of Texas or the plains of the Warrego (3).

Beginning in the early 1860s and gathering momentum rapidly in both areas, the spread of settlement had occupied, at least initially if not permanently, all of the arid country of New South Wales and Queensland by approximately mid-1880s, and the western ranges of the United States by approximately mid-1890s. Already, by these dates, problems of resource appraisal had occurred, but to understand the difficulties we need to examine briefly the geographical context of the settlement.

## THE RANGES

Since we are concerned with the sequence of appraisal of pastoral resources, we must establish what is known now of the two ranges. With this as basis we may then compare the various contemporary assessments and try to date the recognition of individual characteristics by officials and settlers.

### Similarities in the Ranges

Both ranges include semiarid and arid climates, characterized by the occurrence of seasonal and often inter-seasonal (i.e. extended) droughts, which are the result of low absolute rainfalls, high evaporation especially of the summer falls, and extremely variable occurrence of significant falls. Because of

the generally concentric zoning of semiarid areas (approximately averaging 25-inch to 10-inch rainfalls) around the arid core (of less than 10 inches), settlers advancing from the east in both continents would experience the change of climate only gradually; because of the variability of seasons these settlers would be justifiably confused as to the true character of the climate. Thus the Great Plains east of the Rockies were labeled the *Great American Desert* for almost 50 years prior to 1860 on the basis of explorers' accounts from a few months' experiences. Similar misconceptions existed about Australia's interior for similar reasons immediately prior to the extension of settlement (4). In both cases contrary experience of good rainfalls about mid-century reversed the initial opinions—from *deserts* both became *gardens*.

Such arid and variable climates had made their mark on the vegetation. Alongside true xerophytic cactuses and succulents were ranged species needing more humid conditions. The combination of variable climates and variable local moisture conditions meant that the ranges had a great mixture

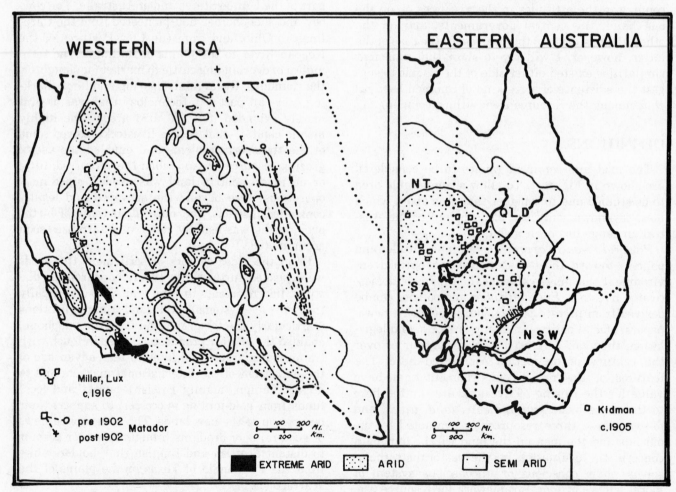

Fig. 1. Rangeland in the United States and in Australia. Locations controlled by the Miller and Lux Company are shown at about 1916. The headquarters and subsidiary stations of the Matador Land and Cattle Company are shown for the period pre-1902 and post-1902. *Sources:* C. Hodge and P. C. Duisberg, editors, *Aridity and Man,* American Association for the Advancement of Science, Washington, 1963; R. L. Heathcote (3); I. Idriess (21); W. D. Lawrence (20).

of vegetation species. Some plants were permanent inhabitants of the relatively moist floodplains, where streams brought runoff from tropical storms on the Queensland coast or snow-melt water from the central and southern Rockies; many of the others were ephemerals or annuals whose appearance closely reflected seasonal chance rainfalls (Fig. 2). After good rains a wealth of low herbs and annual grasses mingled with the edible bushes and low trees; in the drought, the bare ground cracked around the withered stumps of dying bushes and trees (Fig. 3).

The quality of the vegetation as forage for the livestock therefore varied considerably from place to place and from time to time. At no time, however, was it able to support a dense livestock population. At best, on the shortgrass plains of the United States and Mitchell grass plains of Australia, modern opinion suggests that one sheep needs about 5 to 10 acres to support it for a year; at the worst, in the Mohave Desert shrubs (for example; creosote-bush, *Larrea divaricata*) and acacia scrubs (for example; mulga, *Acacia aneura*) of Australia, about 40 to 50 acres are needed. Given that a flock of sheep, to be economical, would need to consist of several hundred (5), it is obvious that small-sized intensive pastoral operations would be impossible on such country.

### Contrasts Between the Ranges

Not all the qualities of the two ranges were similar, however. The geologically older and less-disturbed plains and low hills of interior Australia, rising to 600 feet above sea level (Fig. 4), contrast with the young, more contorted, rugged terrain of southwestern United States, where the high plains rise to 5,000 feet and the basins and ranges of the southern Rocky Mountains and canyon-plateau country vary from over 7,000 feet above sea level to over 200 feet below it (Fig. 5). Here a rapid vertical change in climate and associated vegetation is experienced, whereas in Australia the same change would need a journey of hundreds of miles. One of the most important contrasts climatically stems from the terrain contrasts, in that snowfall and blizzard conditions are possible in the United States arid areas and have at times provided as grave a threat to the pastoral industry there as drought in Australia (6).

For the pastoralist in western United States, a greater choice of seasonally varied forage was nearer at hand than in Australia, and his surface water supplies were more frequent, more reliable, and more useful for irrigation than was the case in Australia. In fact, from the superior quality of the soils and greater chances for irrigation, the possibilities of land uses other than for grazing were greater in the United States than in Australia.

The only advantage possessed by the Australian range seems to have been the quantity of its underground water supplies which (once the Artesian Basin began to be tapped in the 1880s) effectively extended the scope of grazing on the dry plains. By comparison the role played by underground water supplies was not so vital in the United States, although locally important.

In indigenous occupation there were significant contrasts. While the hunting-gathering economy of the Australian aborigines offered slight material resistance to the invading pastoralists, it offered little of material benefit. The varied indigenous cultures of the American arid lands, however, could be locally a serious impediment to the spread of pastoralism, as with the mounted tribes of the Great Plains, particularly the Sioux in the North and the Apache in the South. In some cases the tribes were already demonstrating the practicability of grazing, as were the Navajo with sheep and goats obtained from the Spanish, or the practicality of more intensive agricultural use, as were the Hopi and Pima. The Rio Grande ranges had had a century of Spanish and Mexican ranching experience prior to American *(Anglo)* occupation. As has been shown by many previous studies, this latter source was of importance in the refining of the basic techniques brought from the eastern humid United States (7).

A final contrast needs to be made between the irregular pattern of aridity in the western United States and the solid core of aridity in eastern Australia. In the United States the *oases* of the Rocky Mountain fringes and California have attracted and supported agriculture, industry, and urban land uses which have brought an associated investment of *social* capital, in communications and services, which is available to the local grazier at relatively little cost. In contrast, the Australian grazier has no equivalent, and his land use reflects this.

## THE APPRAISAL OF THE PASTORAL RESOURCES

To examine the sequence of appraisals of the three basic pastoral resources — land, water and forage — we shall examine the systems of land tenure proposed for the arid lands and see to what extent they were successful. In effect we shall be comparing the officially designated, or *Gesellschaft* systems, with the spontaneously developed, independent, or *Gemeinschaft* systems (8). The pastoral ethic — if such exists — will lie in the latter.

During the 1860s governments in both the United States and Australia introduced similar policies for the occupation of the rangelands. The initial intent was the disposal of all federal or Crown land to the citizens, after allowing for such reserves as might be needed by the public in general. This dis-

Fig. 2. Valley of the Virgin River, Hurricane, Utah, U.S.A. Timber and grasses along permanent water course and sagebrush and bare ground on valley slopes are dependent on the local average rainfall (about 12 inches).

Fig. 3. Valley of the Paroo River, Queensland, Australia. Mulga scrub (*Acacia aneura*) on lateritic ridge shows effects of drought, 1960. No grass or herbage cover exists in foreground; mulga is seen in valley in background. Annual rainfall is about 12 inches.

Fig. 4. Plains between Broken Hill and Menindie, west of the Darling River, New South Wales, Australia. Annual grasses and herbs, with salt-bushes, show good season growth. Annual rainfall averages approximately 8 to 10 inches.

Fig. 5. Arches National Monument, Utah, U.S.A., with snow-covered La Sal Mountains in background. Annual rainfall averages approximately 8 to 10 inches in foreground and about 30 inches in mountains.

posal was to be by sale at a reserve price in amounts at least sufficient to support a settler and his family in average circumstances. The Queensland Land Code of 1860, the New South Wales Land Act of 1861, and the American Homestead Act of 1862, were all intended to place on the soil settlers variously described by the legislators as a *reliable, self-dependent people*, or a solid *yeomanry*. Small, intensively farmed, family production units were the common aim. Pastoral land use was seen in mid-nineteenth century as a temporary initial occupation—currently on common open range in the United States and on cheaply rented grazing leases in Australia—which must in time give way before the beneficial influence of the plow.

Yet the facts of occupation were to disappoint the initial hopes. A hundred years after the legislation, 9 of every 10 acres on the Australian ranges are still in government ownership, approximately 1 acre as reserves and 8 leased to pastoralists, while in the western United States, 4 of every 10 acres are still in federal or state ownership: about 2½ as reserves and 1½ leased to pastoralists. In certain states the situation approximates the Australian conditions; for example, Nevada has 9 of every 10 acres and Utah 7 of every 10 acres still in federal hands. Most of the occupation is by large production units, often amalgamations of smaller original properties, often company owned, each supporting directly several families as labor force and indirectly several more families by dividend payments (table 1). Over a hundred years, the practical experiences of the two ranges has forced a reappraisal of the official land tenure system in the light of knowledge of the nature of the pastoral resources.

TABLE 1

## COMPARATIVE GRAZING-PROPERTY SIZES IN AUSTRALIA AND THE UNITED STATES

| | Australia* | | Southwestern United States† | |
|---|---|---|---|---|
| | New South Wales | Queensland | | |
| | Sheep Properties | | Cattle Ranch | Sheep Ranch |
| Average size (acres) | 29,524 | 35,465 | 10,000 | 12,000 |
| Natural unimproved forage area (% total) | 99.5 | 97.8 | 96.0(?) | |
| Invested capital ($) in land, buildings, livestock‡ | 168,788 | 150,884 | 150,000 | 200,000 |

*Source: *Australian Sheep Industry Survey 1960-1 to 1962-3*, Bureau of Agricultural Economics, Canberra, 1965.

†Source: M. Clawson, R. B. Held, and C. H. Stoddard, *Land for the Future*, Johns Hopkins Press, Baltimore, 1960, pp. 399-401.

‡Australian dollars for Australia; U. S. dollars for the United States.

What were the assumptions of the original land policies and why did they not work?

*a*. The sale of land at a reserve price implied that the land had an even and constant quality (or resource potential) for which the citizens were prepared to pay.

*b*. The sale of such land in fixed maximum amounts or blocks implied that the fixed size was sufficient for the purpose intended, that is, to support an average family.

*c*. The sale, by its intent, was implied to be a permanent transaction by which *bona fide* settlers were permanently established on the land.

These assumptions were rapidly proven invalid by events.

### The Problem of Land Quality

The reaction of pastoralists in both areas to attempts to confine them as farmers to small blocks of 160 or 320 acres was either to ignore the system and occupy land illegally, by squatting on the *waste lands* and grazing the open range without any payment (defending themselves if necessary by force against competitive intruders), or to circumvent the law by acquiring through agents several small freehold blocks, which strategically located would control the open range between them. Here they were capitalizing on their knowledge of the *uneven* spatial qualities of the range—the fact that control of the few waterholes or river banks meant effective control of the dry range around it. (Fig. 6).

The separate problem of the uneven quality of the range forage supplies—not only between different localities but at different times at the same locality—was met either by nomadic movement of livestock on the open range, or later, when the bulk of the range was legally occupied, by shifting livestock between stations under common ownership but far enough away to experience different conditions, or by illegally driving them slowly over the traveling stock reserves until conditions improved, or by the acquisition of holdings sufficiently large to allow variations at one part to be compensated by conditions at another (Fig. 1).

The lack of a constant pastoral land quality was illustrated in both ranges by the problem of deterioration of the range as a result of its use. By implication and in theory, settlement developed the resources latent in the land, and therefore the value of the land could only increase through time. In the arid lands, however, the facts have shown the reverse; continued occupation and grazing has led to the deterioration of the forage resource. Evidence in Australia has suggested a deterioration to about one half the original range livestock carrying capacities, and figures for the western United States are similar (9). This deterioration of the general range carrying capacity has taken place in spite of improvements

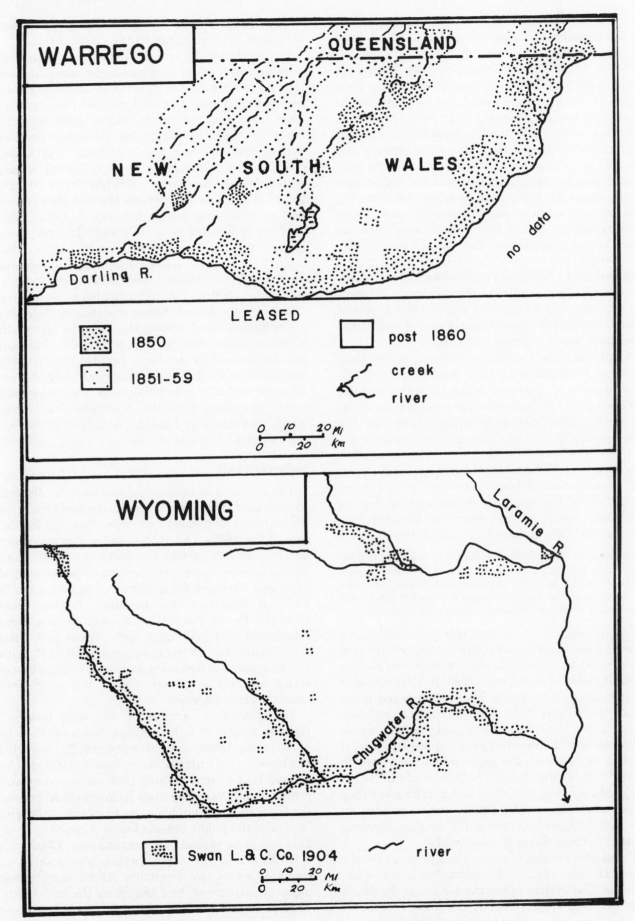

Fig. 6. Strategic freeholds and leases. *Sources:* R. L. Heathcote (3) and E. S. Osgood (6).

in the water resources, such as the sinking of wells to tap underground supplies, and tanks and dams to store elusive surface run-off.

While officials and pastoralists both have generally admitted the losses, evident by 1900 in New South Wales and Queensland and by 1930s in the United States, the reasons for the losses have been disputed. In Australia, the deterioration was stressed by the pastoralists at the turn of the century and claimed to be the result of a combination of accidental overgrazing, droughts, and the infestation of the range by rabbits. In fact, the deterioration was used by them as justification for their claims for a reduction of their rents, which were based on stock carrying capacities. The deterioration of the freely used open range of the western United States, documented in 1936, was questioned by the American pastoralists, and the official charge of overgrazing was met by protests that the report was of conditions during a particularly *bad* drought and the range would improve later (*10*).

Apart from the interesting contrast of reactions between Australians as legal occupiers, using the deterioration of the range to ask for relief, and the Americans as "illegal" occupiers, denying its existence, the deterioration question illustrated the major difficulties of the timing of the resource appraisals. It depended on when you looked as to what you saw. It also illustrated the difficulties of applying conservation policies developed in the more humid areas and implying restrictions on use, in situations where as one American observer (*11*) commented of the Mohave Desert in 1880:

I have seen . . . large bodies [of land where grass covered] the whole ground, and I have seen it the next season without a single spear of anything that looked like vegetation. It would support immense flocks of sheep in one season and would not support a rabbit the next.

The question in the arid ranges was whether the conservation doctrine of restricted but continuous yield could be applied to such widely contrasting natural yields of water and forage. In 1901 a grazier from Bourke, New South Wales, was asked if his losses would have been less if he had stocked less heavily. "No," he replied, "In a good year you cannot stock too heavily. In a bad year you are nowhere. If we had not stocked, the grass would have gone to waste" (*12*). Pastoralists on both sides of the Pacific very soon claimed that they had to take *what* they could, *while* they could, if they were to survive on the range. *Opportune use* of the fleeting resource seemed to them the only solution (*13*).

Faced by a refusal of pastoralists to purchase the whole of their range, officialdom has had to compromise. The system of leasing the range, by which only the forage is acquired by the pastoralist for an annual rent, was accepted as paralleling land sales in Australia from the 1830s; the system was stabilized as a preliminary means of land occupation by mid-century, and recognized as a permanent feature of the arid range by the 1880s (*14*).

In the United States, the dominance of the sale concept, the fears of land monopoly and of the setting up of a feudal tenantry, plus conflicting ideas between cattlemen and sheepmen (the one favoring state control of the range, which might be influenced by them as taxpayers, the other favoring federal control, since many paid no local taxes), combined to hold back a general federal leasing system of the open range until the Taylor Grazing Act of 1934. This was despite the fact that private leasing of grazing had been common in the 1880s, that the Indian tribes of Oklahoma Territory were leasing grazing to cattlemen in mid-1880s, that some of the western states (Wyoming, Colorado and Texas) had been leasing their educational lands at least since 1879, and that grazing leases were introduced into the federal forest reserves in 1897 (*15*).

Conflicting ideas among the resource users, and a federal reluctance to change policies that supposedly worked in the East, enabled the virtually nomadic free-for-all on the American range to continue for virtually a hundred years after the system which now dominates the Australian range and which operates over half the American range had been established in Australia.

## Sufficient Land?

The amounts of land made available for disposal to each settler were recognized to be insufficient for an economic livelihood very soon after the legislation of the 1860s. Officialdom reacted in two ways: the first was to enlarge the areas available. Thus, in the United States, the federal 1862 homestead area was enlarged from 160 acres to 640 acres in 1877 (if irrigation was possible). Despite pleas from the Public Lands Commission for a *pastoral* homestead of 2,560 acres, only Texas took heed (*16*). On the rest of the range, not until 1916, with the acceptance of 640 acres as a stock-grazing homestead, was any legal agricultural tenure of more than 160 acres allowed.

The American ranches of the day, however, claimed ranges of half a million acres or more as early as the 1880s, and were composed of a varied combination of illegal *homesteads* obtained by agents, land bought cheaply from railway company grants, land inherited free from Spanish grants, land leased from neighbors or from the state, and land officially public domain but so enclosed by their holdings as to be useless to anyone else (*17*).

In Australia the same picture was painted in terms of leases, not freeholds. Attempts at closer pastoral settlement had begun in the mid-1880s. The minimum size of the *living* or *home-maintenance* areas, supposedly sufficient for each family, was set in western New South Wales initially at 5,760 acres but was increased to 10,000 acres by 1897; it has been increased locally to between 20,000 and 60,000 acres in the twentieth century (18). In fact, however, the majority of the successful stations were much larger. For example, a combination of illegally obtained small freeholds and legally acquired long-term leases, together with original pastoral *homesteads* of 5,760 to 20,000 acres bought from bankrupted (?) or merely shrewd homesteaders, and leases acquired for special developmental projects such as scrub clearance or well drilling, had brought the total acreages of several properties in the Warrego country over the million acres mark by 1880 (19).

The second reaction of the officials was to attempt to classify the range according to its potential use and to mold land systems accordingly. In the United States this was proposed by the Public Lands Commission in 1879 and begun under John Wesley Powell's Irrigation Survey in 1888, when pastoral land was differentiated from irrigable, mining, and timberland, but legislation to follow up the survey did not affect the pastoral resource use. In New South Wales the classification and recognition of the special needs of the arid Western Division was made in 1883 and confirmed in the Western Lands Board (now Commission) of 1901. No equivalent classification appeared for Queensland, however.

Yet, with or without official help, the pastoralists, by astute legal and illegal amalgamations of properties, had built up their holdings and were ensuring their long-term survival in the face of the short-term stresses of the environment.

### Bona Fide Settlement?

Official policies in both countries were aimed at the *bona fide* permanent settlement of the range by families of the citizens who had purchased the land. However, partly because of an inevitable speculative element to be found in any real estate transactions, partly because of the impossibility of survival on the inadequately sized blocks, and because of the *natural* process of amalgamations, the settlement was often neither lasting nor legal.

What was forgotten in the official policies was that many of the citizens saw land as merely a commercial commodity and its use as a commercial proposition rather than a social or national responsibility. For the cattleman or sheepman anxious to take advantage of the inflow of capital (especially in 1880-90s) to extend his operations and achieve the necessary economies of scale, the end (profits) justified the means (devious).

For Henry Miller, on the road to his first million in Southern California at the turn of the century, the range was only *part* of *his* system. Cheap range land was bought, developed, and then subdivided for sale to small settlers who then traded at shops and got loans from banks that Miller had set up from the ranch profits. Profits from the shops and banks

went back into more cheap range land and so the system went on (20). In Australia, Kidman's system was to buy up bankrupted stations in the drought and to stock them as heavily as possible, ready for the drought to break (21). Like Miller's, it was a profitable short-term system with unforeseen and therefore ignored long-term consequences, particularly on the forage.

Even established and apparently successful small settlers were often prepared to sell up everything for the right price. A New South Wales official complained of farming settlers close by the Darling River in 1883 that "although these selectors appear to take up the land in good faith, yet they sell as soon as they can get a price . . . they have sold very nice homes with orchards and so on"—back to the neighboring pastoralist for grazing land (22). The attraction of ready cash and the chance to give up the difficulties of a remote and hazardous life for a fresh start in the city was already becoming attractive, and with the increase in the possible and accepted standards of living, the attractions of the pioneer life for the small operator were rapidly being reduced.

## A PASTORAL ETHIC?

The nature and demands of resource use on the semi-arid and arid ranges of western U. S. A. and interior Australia over the last century seem to have produced an attitude of the users to the areas' pastoral resources which might be loosely titled a Pastoral Ethic. This comprises an assemblage of at least three separate characteristics.

### Independent Self-made Men

The attitudes of the resource users are very obviously those of independently minded, self-made men. The necessity to control large areas of low-yielding range to support an economically viable herd; the need to employ and direct a wage labor force, if not continually, at least seasonally; the need to have either sufficient capital or credit reserves to cover seasons of no returns; the remote and inconvenient location of most of the operations requiring a large capital investment in basic facilities (communication systems and basic supporting amenities); and finally the highly mobile and vulnerable nature of the main capital asset, that is, the livestock, has forced over the years the survival of pastoralists who are men of strong character with considerable abilities of judgment and decision and what is politely called *business acumen* (23).

It is not surprising therefore to find in the semi-popular and even scholarly literature such epithets as I. Idriess's *The Cattle King*, L. Atherton's *The Cattle Kings*, and M. Durack's *Kings in Grass Castles*. It is also not surprising that such regal and independently minded decision makers should be unable to cooperate with their fellows on resource uses, or submit to external controls. The Bishop of Texas made an astute pastoral comment (24) on his particular flock at the failure of the 1890 Interstate Cattlemen's Convention, at Fort Worth, to agree on controls for the open range:

A cattleman within his own vast inclosure is a king in a small way, and he is accustomed to enforcing his will like law upon all men whom he meets on his land, and when he comes into contact with other men made up like himself and likewise accustomed to rule, he is not inclined to give up his will and his way to anyone; he would rather kick over the whole bucket of milk and go home.

At the convention, true to form, the cattlemen kicked hard and kept their independence.

For such men to be admonished by a remote and apparently unsympathetic government as to their use of the range, or to be refused support for rights claimed by usage as their own, was a grave insult to their pride. Unable to agree among themselves or get acceptable help from the authorities, such men were accustomed to take the law into their own hands, either to defend their considered interests (as the various *range wars,* illegal flouting of the land laws, and *sheep-shooting associations* bear witness), or even worse to prey on their neighbors, for cattle and sheep rustling was an early and continuing feature of both the American and the Australian ranges.

### The Commercial Viewpoint

The use of pastoral resources has always been in commercial terms—a feature not always recognized by successive governments. The timing of the opening up of these range lands in the latter half of the

nineteenth century, paralleling the development of international trade and capital exchange between Europe and the New World, meant that the ranges were viewed not as a resource for local market but as a competitor in an international exchange where official settlement policy had to be equated to the mundane essentials of an annual balance sheet. Since much of the capital to develop the range came from external, often international, sources and was in the form of investments expected to yield certain dividends, the importance of a profitable operation was uppermost in resource use.

In such a context the range was seen as a commodity by which an income could be provided. That income might result from the use of the land, or even possibly from its sale as a speculation in real estate (Fig. 7). Certainly in such circumstances there was a strong inclination to the short-term rather than long-term view. With the threat of eventual eviction by government-sponsored agricultural settlements hanging over their heads, it is not surprising that several of the resource users made as *much* money as possible, as *quickly* as possible. As an agent for a remote foreign company, the station manager was unmoved by pleas of national well-being, and as an agent, often paid by commission, unwilling to limit this year's profit for an unknown but quite likely negative, quantity next year.

The speculative attitude to resource use, shown in the exploitive attitude of some resource users, was encouraged in part by the variable quality of the range itself. Uncertain whether next year would bring drought and ruin, or floods and wealth, operators made opportune use of what they had. In some cases they legally had nothing and moved at

Fig. 7. Real estate development at Canyon Vista, Arizona, approximately 25 miles south of Grand Canyon National Park, April, 1968.

will—as *grass pirates*—over the open range of the western United States and the stock routes and camping and water reserves of Australia, grazing as and where conditions permitted. Even operators with legal claims to parts of the range competed for the common grazing with the nomad herders, and on their own ranges tended to graze all that grew, believing, often justifiably, that not to use it was to lose it.

Range lands were for a time an attractive financial investment; no one could ignore dividends of 20 to 30 per cent; but the rangelands were only one of several contemporary possibilities and often were only a part of a financial empire.

Not all the profits from pastoral properties have come from pastoral operations, and the commercial instinct of shrewd owners has seen many alternative opportunities. Oppenheimer (*25*) commented in 1966 that a survey of the wealthy ranchers in the United States would show that every one of them had got where he was by one or a combination of the following methods:

1) Had a great-grandfather who bought the land for 10c acre before the Civil War.
2) Discovered oil or uranium on the land.
3) Owned land adjacent to the suburban growth of Dallas or Omaha ten years before the expansion occurred.
4) Bought the property when it was 100 miles from the nearest railroad or highway and had one of the new super-highways go right through the middle of it, not only increasing its value, but giving a big condemnation [compensation] award at the same time.
5) Bought large herds on a 10% down payment on a distressed market and then had cattle prices double over the year. . . .
6) Had a family of ten children, all of whom were sons and all of whom worked for room and board until such time as they married. If your wife can time her pregnancies properly this procedure guarantees a free labor supply for about 30 years and gives you a tremendous head start on your neighbors.

Such methods have not been unknown in Australia.

The profitability of the range as a pastoral resource, however, has always depended upon the sharp control of investment per unit-area. The bulk of operations in both areas are still using the unimproved natural grasses and herbages for the bulk of the stock feed and are consciously limiting their unit-area capital investment. The feed is cheap, and the stock search for it and feed themselves with little assistance from the stockowner. Such a resource use policy, however, is clearly opposed to

official hopes for intensive land use and subdivision of the extensive operations; as a result it has been under continuous official *siege* over most of the century of settlement considered here. Settlers' attempts to make a living by reducing the competition, by amalgamation of smaller into larger and more economic production units, has been looked upon with disfavor by officialdom. Darwinian theory, apparently, was not thought to be relevant to the arid lands.

### The Pioneer Syndrome

The independent self-made men, "making a go" of it on the range, inherited a proud place in the national histories on both sides of the Pacific. They were and in many cases still are pioneers in the sense of people living on the edge of the currently inhabitable world; beyond them is the desert. The practical difficulties of such pioneering, however, brought some compensations.

Central governments, while impressed by the environmental hazards, have been torn between their functions as supporters of national development, viewing sympathetically any economic use of the waste lands, and their role as landlords, with responsibilities to future generations and concerned at the apparent misuse of the national range resources by an apparently feckless citizenry. Popular sentiments, however, have often been sympathetic to the pastoralists who have, as Hazard (*26*) suggested, acquired some of the aura of the picturesque hero, and who, confronted by catastrophic nature in the form of droughts have usually had a sympathetic hearing in their pleas for relief. Elsewhere I have suggested that this sympathy, what I called the *drought relief syndrome*, may on occasions have been exploited by unscrupulous operators to obtain very liberal drought relief measures from the governments (*27*).

The century of occupation of the arid ranges forced reappraisal of official settlement policies as well as reappraisal of the resources available. The success of settlement in fact was often in spite of, rather than because of, official policies, and owes much to the independent attitudes and abilities of the men on the land. In their attitudes and abilities can be traced a common pastoral ethic which enabled a difficult environment to be exploited with at least a short-term measure of success.

## REFERENCES AND NOTES

1. *Range* here is defined as that land area providing forage and water for domestic livestock. See:
   STRICKON, A.
   1965   The Euro-American ranching complex. *In*
   A. Leeds and A. P. Vayda, eds., Man, culture and animals. American Association for the Advancement of Science, Publication 78:229-258.

2. See:
PERRY, T. M.
    1963   Australia's first frontier. Melbourne University Press, Melbourne.
THOMPSON, J. W.
    1942   A history of livestock raising in the United States, 1607-1860. U. S. Bureau of Agricultural Economics, Agricultural History Series 5. 182 p.

3. For Texas see:
HOLDEN, W. C.
    1934   The Spur Ranch. Christopher Publishing House, Boston.
SHEFFY, L. F.
    1963   The Franklyn Land and Cattle Company, a Panhandle enterprise 1882-1957. University of Texas Press, Austin.
PEARCE, W. M.
    1964   The Matador Land and Cattle Company. University of Oklahoma Press, Norman.
For the Warrego see:
HEATHCOTE, R. L.
    1965   Back of Bourke: A study of land appraisal and settlement in semi-arid Australia. Melbourne University Press, Melbourne.
BROWN, R. H.
    1948   Historical geography of the United States. Harcourt Brace and Company, New York. Chapter 20.
LEWIS, G. M.
    1966   Regional ideas and reality in the Cis-Rocky Mountain West. Institute of British Geographers, Transactions 38:135-150.
HEATHCOTE, R. L., 1965, *ibid.*, Chapter 1.

5. The minimum size of a pastoral "living area" in eastern Australia is calculated on the area sufficient to support at least 4000 sheep.
HEATHCOTE, R. L., 1965, *ibid.*, pp. 74-76.

6. The impact of blizzards of 1886-1887 on the Great Plains cattle grazing industry can be compared with the impact of the 1895-1901 drought on the cattle and sheep grazing industries of eastern Australia.
OSGOOD, E. S.
    1929   The day of the cattleman. University of Chicago Press, Chicago.
HEATHCOTE, R. L., 1965, *ibid.*

7. BRAND, D.
    1961   The early history of the range cattle industry in Northern Mexico. Agricultural History 35:132-139.
WEBB, W. P.
    1931   The Great Plains. Ginn, Boston.

8. BERTRAND, A. L., and F. L. CORTY (eds.)
    1962   Rural land tenure in the United States. Louisiana State University Press, Baton Rouge. pp. 11-12.

9. Estimates of the extent of deterioration are complicated and often based on dubious assumptions, but see:
HEATHCOTE, R. L., 1965, *ibid.*
U. S. SENATE
    1936   The Western Range. 74th Congress, 2nd Session, Document 199.
CLAWSON, M., R. B. HELD, and C. H. STODDARD
    1960   Land for the future. Johns Hopkins Press, Baltimore. Chapter 6.

10. See the evidence of the Australian pastoralists to the:
ROYAL COMMISSION to Inquire into the Condition of the Crown Tenants

    1901   Report. New South Wales Government Printer, Sydney.
In the U. S. A., the evidence provided by *The Western Range, supra,* was met by the stockmen's:
AMERICAN NATIONAL LIVESTOCK ASSOCIATION, Denver
    1938   If and when it rains: the stockman's view of the range question. 60 pp.

11. REDDING, B. B.
    1880   [Evidence before] Public Lands Commission. p. 169. Executive Document 46, 46th Congress, 2nd Session.

12. Evidence of C. S. Singleton to the *Report . . .* Royal Commission, *supra,* p. 353

13. HEATHCOTE, R. L.
    1964   Conservation or opportune use? The pastoralists' problem in semi-arid Australia. Advancement of Science 21:47-60.

14. HEATHCOTE, R. L., 1965, *ibid.*, Chapters 2 and 3.

15. COVILLE, F. V.
    1905   Report on systems of leasing large areas of grazing land. U. S. Senate Document 189, 58th Congress, 3rd Session.

16. Pastoral homesteads of 2,560 acres were allowed in Texas after 1887; their size was increased to 5,120 acres in 1906.

17. See:
HOLDEN, W. C., *ibid.*; SHEFFY, L. F., *ibid.*; PEARCE, W. M., *ibid.*

18. HEATHCOTE, R. L., 1965, *ibid.*, Chapter 3.

19. HEATHCOTE, R. L., 165, *ibid.*, Chapter 5.

20. LAWRENCE, W. D.
    1933   Henry Miller and the San Joaquin Valley. University of California, Berkeley. (M.A. Thesis in History)

21. IDRIESS, I.
    1936   The cattle king. Angus and Robertson, Sydney.

22. AUSTRALIA. NEW SOUTH WALES LEGISLATIVE ASSEMBLY
    1883   Report of inquiry into the state of the public lands. p. 133. *In* Votes and Proceedings of the N. S. W. Legislative Assembly, Vol. 3, pp. 71-248. New South Wales Government Printer, Sydney.

23. Such a generalization is probably wrong in detail, but to survive such conditions the abilities of the individuals must have been considerable.

24. INTERSTATE CONVENTION OF CATTLEMEN
    1890   Proceedings of an Interstate Convention of Cattlemen, held at Forth Worth, Texas. U. S. D. A. Bureau of Animal Husbandry, Special Bulletin. Washington, D. C. p. 48.

25. OPPENHEIMER, H. L.
    1966   Cowboy economics: rural land as an investment. Interstate Printers, Danville. p. 31.

26. HAZARD, L. L.
    1930   The American picaresque: a by-product of the frontier. *In* J. F. Willard and C. B. Goodykoontz, eds., The trans-Mississippi West. University of Colorado, Boulder.

27. HEATHCOTE, R. L.
    1967   The effects of past droughts on the national economy. pp. 27-46. *In* Report of the ANZAAS Symposium on Drought. Commonwealth of Australia, Bureau of Meteorology, Melbourne.

# LAND, WATER, AND SOCIAL INSTITUTIONS

COURTLAND L. SMITH
Department of Business and Resource Management
Carnegie-Mellon University
Pittsburgh, Pennsylvania, U. S. A.

and

HARLAND I. PADFIELD
Department of Anthropology
University of Arizona
Tucson, Arizona, U. S. A.

# ABSTRACT

The development of water resources and land areas having definite boundaries go hand in hand. The concept of "waterspace" is applied to the relationship between these two entities. Development of waterspace also involves social institutions adapted to the specific situation for developing water resources; hence, waterspace denotes the interplay of water, land, and social institutions having identical boundaries in time and space. Various types of social institutions can be used to achieve effective integration of the people within a given waterspace. The institution may be an association, board, commission, municipality or committee which integrates farmers, local communities, or even nation states.

Land-water-institution complexes are discussed cross regionally and cross nationally—Salt River Basin in Arizona, the Tennessee Valley Authority, the Metropolitan Water District of Los Angeles, the Papaloan River and Yaqui River basins in Mexico, the Gezira scheme in Sudan, the Rhône-Languedoc Region in France, the Lower Mekong River scheme, the Niger River project, and Israel.

Five major points emerge in these discussions: 1) successful development requires either one water management institution for the waterspace, or one superordinate institution with coordinating powers; 2) as development of the waterspace proceeds, a broadening in scope and interests of the institution(s) occur with it; 3) norms and cultural values vis-a-vis land and water use must be commonized or coordinated with respect to the waterspace; 4) development of waterspace sometimes carries with it mixed blessings in that it may give unequal economic advantage to one industry over others, it may even become an instrument of political power, it may over-colonize leading to privation of another sort; it may under-colonize, leading to the wholesale transplanting of populations ill-equipped for development; 5) and finally each water relevant institution has a blueprint for behavior unique to the people of that institution. For waterspace development the details of that blueprint must be taken into account.

# LAND, WATER, AND SOCIAL INSTITUTIONS

## Courtland L. Smith and Harland I. Padfield

The fact that improved utilization of water resources can lead to successful economic development is well known. Surely, economic development will not be enhanced by poorly designed dams and irrigation facilities for distributing water to the land, but why is it that not all well-designed dams and irrigation facilities promote economic development? The answer lies in examining the relation of land, water, and social institutions.

First of all, how does the concept "social institution" fit into economic development? Economic development refers both to economic growth (the increase in goods and services available to a people on a per capita basis) and to the change in social institutions that accompany this increase (1). The consideration of social institutions in the development of irrigated agriculture is not new. Elwood Mead (2) writing in 1910, a time when the social sciences in the United States were in their fetal stage, said: "The problems of irrigation are not however, of water alone, but of land and water." We would add to this statement that in part the problems of irrigated agriculture and the subsequent economic development are not of land and water alone, but of the relations of land, water, and social institutions.

A social institution can be conceptually divided into two parts. One is the "people-in-action" and the other is the "blueprint which determines their behavior" (3). The understanding of human behavior, then, requires focusing on both the people-in-action or the system of action and the blueprint for behavior or the culture of the people.

In addition to these two conceptual aspects, social institutions have two basic functions. Service (4) stated that some social institutions have an integrative function, while others have a boundary maintenance function. We would modify this statement by saying that most social institutions have both an integrative function (the integration of individuals and groups for coordinated action) and a boundary maintenance function (the maintenance or extension of specified territories). For example, a service club is a social institution which may integrate individuals from different occupations for coordinated action in community service. Some service clubs have rules requiring that the membership be from many diverse occupations; however, the activities of service clubs are geographically or territorially limited, most often to all or part of a community.

Since water availability has geographical or spatial determinants, the people concerned with water-resource management must have a territorial interest contiguous with the spatial availability of the water. We have observed that for water resource development the closeness of the unity between the territoriality of the people-in-action and the spatial availability of the water affects the success of the water development scheme. Establishing this unity often requires the integration of diverse peoples and groups. This ecological and social interplay we have called "waterspace" (4). The successful utilization of waterspace, however, requires more than just compatibility between the territoriality of the system of action and the spatial distribution of the water; successful utilization of waterspace also requires that the blueprint for behavior (the culture appurtenant to the people-in-action) also be adapted to the requirements of the specific situation for development of the water resource.

Put in a different way, our suggestion is this: when looking through a telescope or binoculars, an adjustment is necessary to bring the image to be viewed into the vision field of the glasses. This is analogous to fitting the spatial distribution of the water resource with the territoriality of the people-in-action. A second telescopic adjustment, for sharpening the image in the vision field, is analogous to the adjustments made to adapt the blueprint for behavior of the people to the requirements for effective economic development. Effective economic development of the waterspace requires both of these adjustments—the territoriality and the blueprint for behavior or culture of the people-in-action.

Our primary experience in analysis of water resource development comes from the study of a large water and power project in the arid southwestern United States (5). Here we want to couple with this experience the experience of other social scientists to indicate the wider application of the concept— waterspace or the fitting of social institutions, both in their people-in-action aspect and their cultural aspect, to the demands of water-resource development. To further elaborate the waterspace concept, examples are taken from the United States, Mexico, France, Israel, Africa, the Indian subcontinent, Southeast Asia, and Australia.

The waterspace concept does not provide a cross-cultural panacea that will make all economic development projects successful. What this concept does tell us is that each project involves a complicated set of factors regarding social institutions that must be understood and accounted for in order to achieve success. The implication of the water-space concept is that there is no cookbook, rule of thumb, or even textbook approach to the social aspects of economic development, a fact that has been recognized by those who have had experience in water-resource planning (6). The waterspace concept states that for each economic development scheme, careful and detailed study must be made of the territorial and cultural aspects of the social institutions involved.

We recognize that many economic development schemes have been successful without the conscious, detailed fitting of the social institutions of the people involved; however, fitting has taken place, if only in the common sense understanding of the situation by the planners and participants. For example, Gaitskell (7), onetime general manager of the Gezira Scheme, showed keen understanding of the importance of the exchange of British culture with the Sudanese when he said, "We should have switched our sights earlier to education, technical training and trade, for these were interests we would both need to share and to expand after independence."

Clearly there are people participating in development projects who do not articulate the concepts presented here but who succeed in their application by informally providing the institutional adjustments which are necessary. We suggest, however, that the success ratio could be improved if planners were provided with better tools to take into account the complexity of social institutions.

## WATERSPACE—A CASE STUDY

We will begin with a descriptive illustration of the waterspace concept taken from the institution we know best. This is a Southwestern water-users' association which has played an important part in the economic development of metropolitan Phoenix, Arizona, U. S. A. (5).

In the Salt River Valley of Arizona, the surface water available had stimulated the development of irrigated agriculture on over 150,000 acres of desert land by the late 1890's. The people exploiting the water resource were many individual farmers and local entrepreneurs organized into canal companies. The canal company secured the rights to the available water and maintained the works to deliver this water to the farmer. Ten essentially independent canal companies were competing for the available waterspace—that is, the water necessary to irrigate all the lands under the various canals. From 1898 to 1904 a severe drought gripped the area, and the result was too many irrigated farms and canal companies and too little water.

Leaders in the Salt River Valley recognized that the existing waterspace could be maintained and even expanded if a water-storage facility could be built to save water from high water-flow periods for use during low water-flow periods. Such an undertaking required a social institution which integrated most of the farmers in the Salt River Valley. One of the primary factors dictating such an organization was the federal government. The federal government, which would make funds available under the provisions of the newly passed National Irrigation Act of 1902, would not deal with individuals. As the farmers were to soon find out, however, the effective exploitation of the waterspace was going to require a social institution incorporating all farmers, not just most—but we are ahead of the evolutionary sequence.

In order to obtain financial aid from the federal government, most (85%) of the farmers of the Salt River Valley joined together to form a water users' association. The financial aid from the federal government enabled the water users to build a storage dam with hydroelectric generating capacity. This provided the potential for the development of additional waterspace through (a) storage of surface water from high water yield periods, and (b) the development of ground-water by using the hydroelectric energy for pumping. Third, an integrated canal system was constructed so that the water made available could be more efficiently distributed, thus reducing the loss of water. These developments increased the waterspace to the point where 250,000 acres of irrigated land could be served. The additional waterspace attracted new farmers who became members of the water users' association, and agricultural production in the Valley increased.

About the time of World War I, the waterspace again began to contract. Drought was not the cause this time; instead, waterlogging reduced the amount of land that could be effectively used for farming. In 1919 one-fourth of the farm lands under irrigation were threatened by the waterlogging problem. This brought about a reduction in farming activity, and the economy of the area was indeed in danger of drowning as a result of the rising water table, caused by wasteful water use practices.

One response to the problem was for those farmers who had not joined the water users' association to do so. The primary motivation for this was to develop the economic ability and better integration of the canal system to cope with the waterlogging problem. This response made the territoriality of the water users' association equal to the spatial distribution of the available water.

Legal rules for water allocation have served to reinforce this waterspace boundary. Also, with the territoriality of the water users' association and the spatial distribution of the water resource being syncretized by economic and legal means, other social responses to the water-availability problem could be more effectively implemented. One such social response was the development of sharper attitudes about water conservation and the importance of effective water utilization. As a result, in subsequent years less water was lost due to wasteful irrigation practices.

The method for solving the waterlogging problem was deep-well pumping. This solution created additional waterspace that could not be incorporated into the water users' association because of the limits of the canal system and the territoriality of the surface water rights. Water was therefore traded outside the boundaries of the water users' association to neighboring irrigation districts for facilities to drain waterlogged areas and for improvements to the water users' distribution facilities.

From the late 1920's to 1968, the waterspace boundary of the water users' association has not changed in any significant way. Additional pumping capacity has been installed to maintain the availability of water to lands served by the water users' association.

With the waterspace boundary fixed, the water users' association changed in a different way. It modified its structure and activities to cope with legal, economic, and social changes within the existing waterspace and to cope with changes in social institutions outside the waterspace. For example, the depression, the need for a more reliable source of electric energy, and the advantages of a tax-exempt status were a set of factors requiring modification of the water users' association. In the late 1930's the association added the activities of a municipal corporation to its complexity as a social institution. The consequences of this change were primarily economic and legal. The change enabled the refinancing of gold bonds at a lower interest rate and over longer time-periods. Under the structure of a municipality, the revenue from the sale of electric energy was tax-exempt. Later, in the 1950's, the social problems of increasing numbers of urban residents confronted the water users' association, and the institution was again modified to include a community relations and electric sales activity.

As economic development progresses, other institutions are strengthened and new institutions develop. With the urban growth in the Salt River Valley of the 1950's, the cities grew rapidly within and around the boundaries of the water users' association. The cities began to play an increasingly important role in the distribution of water to an increased number of urban residents. This placed the water users' association and the cities in competition for the existing waterspace. Within the boundaries of the existing waterspace the cities could use the water previously required for farming to meet the needs of domestic water users. For areas outside the water users' association, the cities had to create new waterspace. This has created conflict. The city officials argued that they should have the water in the existing waterspace of the water users' association for use in a waterspace which was to be spatially consistent with the territoriality of the city and the city's projected growth pattern. This conflict still hangs in the balance. To meet current needs, urban waterspace has been created by the cities by developing additional surface water which could not be economically made available to agriculture, along with the development of additional groundwater from the common groundwater basin. For the future, additional surface water from the Colorado River and treated sewage effluent could potentially change the boundaries of the existing waterspace.

The purpose of this illustration is to emphasize the interplay between land and water, and social institutions. This description has been mostly concerned with the compatibility between the spatial distribution of water and the territoriality of the social institution created to manage that waterspace. The "Comparative Observations" which follow will further illustrate this interplay. We will show that the kind of unit, a water users' association, municipality, fourth branch of government, committee, board, or commission makes no difference. The dominant requirement is that the social institution have territorial unity with the spatial availability of the water resource. Social institutions created to develop waterspace generally take on additional activities as power development, wildlife and land management, flood control, and many other activities that complement water-resource development. Conflict is likely to develop where more than one social institution is competing for waterspace. Those who control the waterspace have the advantage over others dependent on the water resource. Finally, the development of waterspace is generally initiated by an institution, such as a national government, which has jurisdiction over the area to be developed. This is primarily due to financial and legal constraints.

## COMPARATIVE OBSERVATIONS

Several comparative examples will help to better illustrate the territorial aspect of the waterspace concept. In a few cases social units have been deliberately planned to coincide with hydrologic units. Such was the case when the boundaries of the City of Los Angeles were made synonymous with the San Fernando Valley as a hydrologic unit. Later, the Metropolitan Water District was created to accomplish the same end for the entire coastal plain of Southern California (8).

The method for integrating populations may differ, but the necessity for social institutions to coordinate the use of the waterspace, if economic development is to be successful, is a general condition. In Southern California the method was a municipal corporation (the Metropolitan Water District), and in the metropolitan Phoenix area it was a water users' association. In the Tennessee Valley it was a "headless fourth branch of government" (*9*). The Tennessee Valley Authority was "the only public agency of Valley-wide jurisdiction with a general responsibility for the natural resources of the region"—for the "goodness" (*10-12*) and the "badness" (*13-15*) of TVA. The Tennessee River basin was the waterspace which dictated the area of responsibility of the Authority. The waterspace encompassed all or portions of seven states, an area of 92,000 square miles.

The Tennessee Valley Authority is of interest because of the way it developed relations with the local community. The means by which the Authority was to integrate with state and local agencies was not a part of the implementing plan. The Authority could have evolved by centralizing and building a parallel structure for all its diverse activities. But instead, the Authority's board chose to incorporate and work with existing state and local agencies in planning and implementing programs to effect economic development in the Tennessee Valley.

The Tennessee Valley Authority and the Mexican government's Papaloapan project (*9*) are illustrative of another aspect of the fitting of institutions to available waterspace. Often the institutions develop activities that are broader in scope than just the management of the water resource. In the Salt River Valley, water storage, flood control, and power development were activities of the water users' association; however, the Tennessee Valley Authority incorporated these plus agricultural and recreational development, improvement of health conditions, wildlife and forest rehabilitation, reduction of unemployment, and navigation activities.

The Tennessee Valley Authority is an integrating mechanism which coordinates the activities of many organizations having territorial interests in the same waterspace. The Authority is in a sociological sense an institutional bridge between the federal government and the local community. The Papaloapan commission was also to have engaged in a variety of activities in addition to control of the water resource—colonization, agricultural credit, school construction, health improvement, and upgrading communications. Like the Tennessee Valley Authority, the commission was a bridge between the federal government and the local social system, but lack of financial support by the federal executive prevented this bridge from bearing the traffic of programs and activities necessary for success.

For the Gezira Scheme in the Sudan, an institution composed of the colonial government, private enterprise, and the local Sudanese was created (*7, 16*). The Gezira Scheme, in addition to emphasizing the integration of different cultures for waterspace development, also illustrates the persistence of a social institution after a nearly total change in supervisory personnel. Initially, the colonial government's responsibility was to provide capital, maintain the irrigation works and canals, and secure the lands necessary for an economically feasible operation. The responsibility of private enterprise was administration and management of the scheme. The Sudanese tenants were to provide the labor inputs. In the evolution of the institution, the Sudanese national government has replaced the British government as the controller of the irrigation works and land rent system. A board of Sudanese has replaced the private company for administration and management of the scheme. The institution has continued to enhance economic growth with the same units performing the same activities; however, where British personnel once filled the positions, these positions now are filled by Sudanese.

In Southern France a "semigovernmental" institution was created to bring together both public and private interests in the Lower Rhône-Languedoc Region. Although this social institution operates like a private company, it is subject to governmental control. All actions for economic development—land acquisition, agricultural education, marketing organizations, local farmers' associations, as well as "technical, economic, psychological and financial" activities are coordinated through the national company (*17*) but:

The Company is not in sole charge of all aspects of the region's economic development. Other agencies concern themselves with related problems that supplement the work of the Company. . . . The Company cooperates fully with them . . . , and the Company is always a shareholder and member of the board of administrators of all these agencies (industrialization, tourism, and urban modernization).

This federally created institution is locally positioned and controlled. The overall agricultural plan is fitted to meet the needs of France, in terms of reducing the wine surplus from the region, yet the national company is attuned to, and a part of the specifics of, the local situation.

Economic development for some waterspace requires a social institution that integrates nation states. Development of the Lower Mekong River required creation of a social institution that would integrate Thailand, Cambodia, Laos, and South Vietnam.

The first problem was how to fashion, not a discussion group or forum for debate, but rather an instrumentality with clear powers to decide and act, and yet be so responsible to, and so integral a part of the four riparian governments that they would have no misgivings about ratifying its charter (18).

The Mekong committee which was formed in 1957 included one member from each of the participating countries. Consensus was established as the rule for decision making and the chairmanship rotated alphabetically. This example, also, shows that the successful social institution must be flexible enough to take on tasks not in the initial plan. As the responsibilities of the committee grew, an administrative unit was created to handle routine work and bring the background for major decisions before the committee. In spite of political unrest at least some progress has been made. In 1966 a dam was completed in northeastern Thailand which provides electrical energy for parts of Thailand and Laos (19).

To summarize, various types of social institutions can be used to achieve effective integration of the people within a given waterspace. The institution may be an association, board, commission, municipality or committee which integrates farmers, local communities or even nation states. An institution initially created for water-resource management often takes on other responsibilities important to the economic development of the waterspace. The generation of electric power is one very common activity associated with water resource planning for economic development; however, there are many others. The water users' association we studied had a rapidly growing power system; it was involved in water quality control, watershed rehabilitation, and recreation development. Social institutions must bring together all the resources, both natural and human, in the waterspace to fruition. As new activities emerge the institution must be able to adjust to take on new duties.

We might also look at the motives or needs behind plans for economic development. The economic development of water resources is most often initiated by a political unit responsible for an area greater than the waterspace to be developed. Most often the waterspace is being developed to relieve the economic dependence of the people of the area on the larger political unit. For example, the Mexican government has created commissions charged with the better utilization of the water resources in such river basins and drainage areas as the Yaqui (Native Zone), Valley of Mexico, Rio Fuerte, Grijalva-Usumacinta, Tepalcatepec, and Papaloapan (20). The motives for these economic development schemes through better utilization of water resources varied from political integration of the Yaqui to increased food production for a rapidly expanding population to flood control to reduce the distribution of flood relief aid.

In the Manya District of the Mysore State of South India, the state government developed irrigation to improve the economy of the area and to reduce the dependency of these peoples on the state, especially during time of drought (21). In the United States one reason for the establishment of the Tennessee Valley Authority was to alleviate rural poverty caused by the depression (10). In Australia the Murray River Area was developed with the aid of the Australian government to increase agricultural production (22). Israel worked to increase its waterspace by watering portions of the arid Negev Desert to increase agricultural production and incorporate immigrants into Israeli society (6). France initiated irrigated development in the Rhône-Languedoc Area to reduce the economic ills for the area which were due to a perennial surplus of wine (17). In Southern Rhodesia the government instituted small-scale irrigation developments in order to provide an economic base for native peoples displaced from white farms (23).

Although these examples indicate that planned change for the better utilization of water resources is initiated by a centralized political unit, this is not always the case. Both in the Salt River Valley of Arizona and the Los Angeles area of California, planned development of water resources was initiated by local action. In both areas local people worked to obtain distributions of water by developing existing water resources beyond predevelopment levels. In both cases, nevertheless, legal and financial backing from the federal government were required.

The advantage or power of people-in-action is often determined by control of the waterspace. In part the cities in the Salt River Valley were reacting to the dominance of the water users' association when they laid claim to the waterspace of the association for urban water users. In some cases it is the culture or blueprint for action that specifies who will be dominant. In other cases it is the people-in-action and their territoriality relative to the waterspace.

The Sonjo of Tanganyika (24) serve as an example of a situation in which the culture determines the dominant group. The Sonjo are not a technologically advanced group; however, irrigated agriculture is the dominant element of their economy. The established set of rules for the allocation of water favor the *wenamiji*, 17 hereditary village elders. The wenamiji have the primary right to the water for irrigation. Next after the village elders the *wakiama*, substantial families of the village without inherited rights, receive temporary rights to the water by paying the wenamiji for the water privilege. Other members of the village must deal with the wenamiji to obtain water for their crops; their priority comes after that of the wakiama. In time of shortage the privileged groups of the village are assured of water to meet their needs, while the needs of the less advantaged groups are not met. In this example the hereditary rights to the water assure the wenamiji a favored position.

In some instances the territoriality of the people in the waterspace determines the dominance of one group over another. In the Indus Valley, India was in the position to cut off water to three of West Pakistan's major canals heading in the Sutlej River. India, claiming proprietary rights, actually cut off the water supply in these canals in 1948, 1952, and 1958. Pakistan could not maintain its delicate economic balance with this situation. The partition of India and Pakistan was made on the basis of the territoriality of the existing social institutions and not the spatial distribution of the water resource. The conflict was resolved by the Indus Water Treaty of 1960. This treaty allocated the three western tributaries of the Indus to Pakistan and the three eastern tributaries to India (25, 26). Since political conflicts prevented the coordinated development of the Indus Basin, the treaty divided the existing waterspace between the two nations.

The exploitation of the potential waterspace of the Jordan river has created conflict between the Arab peoples of Syria, Jordan, and Lebanon, and Israel (27, 28). Again the boundaries of Israel were fixed by other than regard for waterspace boundaries. Agreement has not been reached, and these countries are exploiting the waters of the Jordan River on the basis of technical and economic means. Israel has built the National Water Conduit, while the Arab countries are working to develop the headwaters of the Jordan.

National governments through control of waterspace can achieve control over the peoples occupying the waterspace. The control of the Yaqui River by the Mexican government has given it the power to manipulate Yaqui institutions and secure the acceptance of government programs among the Yaqui (29). Chi-Ch'ao-ting (30), in summarizing the economic history of China from 225 B.C. to A.D. 1842, states that "the state use water-control activities as an economic weapon of political struggle and a chief means to develop and maintain a Key Economic Area as an economic base for the unified control of a group of more or less independent self-sufficient regional territories." The advantage of centralized governments who control the waterspace was the chief focus of Wittfogel who assigned the despotic advantage of hydraulic government a niche in the evolution of human social institutions (31).

Nor do the goals of economic development necessarily benefit the peoples intended by the planners. In Northwestern Mexico, Erasmus (32) has observed that the breaking up of the haciendas and creation of ejidos has not increased the distribution of wealth to the rural peasant. The ejido has not freed the peasant from labor exploitation. The bondage has merely changed in form to cope with the new system. This problem is in part due to insufficient consideration of the size of economically viable farm units and the groupings required for effective equipment utilization marketing, presentation of farm technology, and essential social services.

The motive for the institution controlling the waterspace may be economic development of the region or it may be control of the local peoples, perhaps both. Clearly those institutions, be they local or national, which control the waterspace have advantage over the other institutions which have territorial interest in the waterspace.

## CULTURAL FACTORS

To this point we have been primarily concerned with the relation between the territoriality of people-in-action and the spatial limitations of the water resource. Now, we will turn to the blueprint for behavior or the cultural aspects of social institutions to show how these must be focused to promote effective economic development.

Human social institutions are just that—human social institutions. They are often not capable of stepping back for a look at themselves. The culture or blueprint for behavior of the social institution must be understood in planning or analyzing a project. Often the blueprint for behavior of a social institution is in opposition to the water-resource development objectives. For example,

... the attitude of the Luo tribesmen presents a major obstacle ... members of the clan have cultivating and grazing rights in parcels of land within the area, as well as rights of communal grazing. Development under irrigation will involve major reallocations of land rights and property boundaries. It will further necessitate the movement of some of the inhabitants to homes elsewhere to provide holdings of economic size for those who remain and reduction in the number of livestock now maintained. The tribesmen have shown reluctance so far to accept these changes (33).

Successful merging of the planning goals with the cultural realities of an existing situation can take place if adjustments are made. For example, the French farmer is not a willing acceptor of advice. In the Lower Rhône-Languedoc program, the size of the farm agent's area was reduced in order to facilitate the establishment of a personal relationship between the agent and the farmer. In the more personal situation the farmer was more open to learning and to using new farming techniques (17).

In the interaction between the southwestern water users' association and the community the norm, "effective management of the water resource is the basis for economic growth," was one of the dominant aspects of the blueprint for behavior of this social institution. The water users' association has found that if a large segment of the community shared this norm then it was more likely to continue to retain the advantages which enable it to provide low-cost water to water users. One of these advantages was the tax-exempt status of its power resources.

In the 1950's the water users' association found that its position as the dominant water-control agency within its waterspace was challenged. The challenge was directed at the association's tax-exempt status. Twenty years before, the citizens of Arizona had granted a tax-exempt status to the power revenues of the water users' association. The reasoning for the tax-exempt status was that it was better to tax the wealth generated from water-resource development rather than the wealth required to provide water to farmers who at the time were the buttresses of the economy. In addition, initially, the power customers were the water users themselves and the mines that needed large supplies of electric energy. As more urban residents have become association electric customers, and as a much greater portion of the water operation has been financed by power revenues, the allocation of the profits to the water operation from the sale of electric energy has been increasingly questioned.

The management of the water users' association felt that the tax question was raised because of the large numbers of new people who did "not understand the water situation." The association began a program of institutional advertising and community-relations activities designed to secure acceptance for its position. This program was largely successful, for while the tax-exempt status of the association was modified, the association still enjoys a considerable tax advantage. This, then, is an example of how an organization, in this case the water users' association, works to obtain better correspondence between the cultural blueprint held by the total society and its own.

It is not unusual for social institutions to have to change their blueprint for action in order to adjust to a water resource development scheme. Epstein (21) compared the different institutional adjustments of two villages in the Mysore state of South India. One, the wet village, was incorporated within the development of a new irrigation system. The other, the dry village, was not. In order for the dry village to receive the benefits of economic development it had to adjust its traditional blueprint for behavior. In Wangala, the wet village, the traditional set of cultural patterns did not change significantly. Hereditary status and the high evaluation of farming as a way of life continued. By contrast, the culture of Dalena, the dryland village, was modified in order to become part of the regional

economy. The inhabitants developed mechanisms of achieved status, and while farming was still highly valued, village members accepted other occupational pursuits.

In Northwestern Mexico the ejido was an institution created to satisfy one of the cultural norms behind the Mexican revolution—property to the propertyless. The ejido was created to insure that the individual did not lose his land. The ownership of the land was placed with the community, rather than the individual. The ejido as an institution for the development of irrigated agriculture has several drawbacks, however. To create a viable economic unit, the lands of many ejidatarios must be combined. The collectivized type of ejido accomplishes this, but individual initiative is lacking. Institutional adjustments have been made which correct this deficiency of the ejido. First, the establishment of credit societies of fifteen to fifty persons within the ejido reduced the size of productive groupings and permitted association between ejidatarios with similar goals and work habits. Plowing and planting were still done collectively, but irrigation, cultivation, and harvesting were done by each ejidatario. The effect was that the individual effort could materially increase the harvest of an individual's ejido parcel. A second adjustment has been for ejido lands to be amalgamated by individual operators. This practice is illegal but made possible through another Mexican blueprint for behavior, mordida (31).

In some economic development the amount of waterspace created is greater than the indigenous populations of the area can feasibly utilize. In these cases the area must be colonized from outside. This is a particularly difficult problem, as existing social institutions have to be modified in both their people-in-action and blueprint-for-behavior aspects in order to successfully utilize the waterspace created.

One problem is not to overcolonize an area. The size of population which the waterspace can support must be determined. In the Murrumbridge River Basin of Southeastern Australia, planners attempted to distribute irrigated land to too many settlers. The land parcels were too small to be viable economic units. A natural correction took place with larger land parcels coming about as the successful took up the lands of the unsuccessful (22).

The Papaloapan project created waterspace which required colonization. Largely due to cultural factors, the colonization was not successful. Colonists were picked who were not familiar with farming in a tropical area. Replacements to the colonies were selected from different cultural backgrounds than the original colonists. These replacements increased the heterogeneity of the colony and the stress of intra-colony life (9).

In the Niger Project likewise, the cultural patterns of the colonists were not compatible with those required for successful economic development. The local customs of farming were not taken into account; settlers were asked to come from regions where there was already ample land for farming; farm size was not fitted to family size; the farmers felt that they were being forcibly collectivized rather than asked to cooperate voluntarily; there was local prejudice against bringing farmers from other places, especially the overpopulated areas along the northern border of Nigeria; and there were prejudices about the sale of foodstuffs to neighboring peoples (34).

Comparison of the Israeli institutions, the kibbutz and the moshav, shows how the cultural backgrounds of the colonists must be compatible with the norms and patterns of living of the settling institution. Both the kibbutz, a collective, and the moshav, a cooperative, "played a decisive part in the pioneering task of opening up barren lands and creating modern type agriculture in Israel" (35). Each was suited to a different kind of pioneer, however.

The kibbutz was important in the prestate period. It was an institution created by settlers from Europe, and it was a reaction to the European social institutions (6). The basic cultural norms which lead to creation of kibbutzim were equality, the allocation of rewards on the basis of need and the allocation of work on the basis of capacity; simplicity and modest consumption; government by democracy; and a positive evaluation of the importance of pioneering (35, 36). Initially, the kibbutzim were closely tied to the Zionist movement for statehood. Largely because of the ideological commitment of the members, the kibbutzim were able to endure the hardships of pioneering. With Israeli statehood, however, many of the elitist aspects of the kibbutz were lost. The ideals of the kibbutz were not suited to the cultural background of the wave of new immigrants, many of whom came from North Africa and Asia as well as Europe. The cultural background of the immigrants emphasized the importance of family living. Many of the immigrants came in groups of families or even villages, rather than as unassociated individuals (as in the prestate period). Because of this cultural conflict the kibbutzim found the recruiting of new members increasingly difficult. Cultural patterns of the kibbutz were modified. Some of the changes were: placing more emphasis on family functions, permitting status differentiation, and even allowing collective enterprise for profit. In spite of these adjustments of kibbutzim, the moshav institution was better suited to the cultural backgrounds of the immigrants.

The moshav placed less emphasis on ideology and elitist principles. The stress of the moshav was on "the intrinsic or stratum aspect of peasantry" (36). The work unit of the moshav was the family farm rather than a collectivity composed of egalitarian individuals, as in the kibbutz. While the kibbutz was the dominant institution in the prestate period because of its relation to the Zionist movement and its suitability for pioneering, the moshav became the dominant institution for the settlement of immigrants in the post-state period. The blueprint for behavior of the moshav was more compatible with that of the immigrants.

In these examples the culture or blueprint for behavior of a social institution was adjusted or brought into focus with the exigencies of economic development. This adjustment may be the reduction of a farm agent's area to promote a personal relation with farmers, a public relations program to foster group consensus on a critical value, the adjustment of ways of doing things to make a more viable economic unit, or settling colonists with the social institution which best incorporates their cultural background. In each case the cultural blueprint for behavior was adjusted or focused to meet the demands for successful exploitation of the waterspace by the social institution involved.

With this statement we might be criticized for being so general as to beg the real question. What are the "demands for successful exploitation of the waterspace?" To determine these each developmental scheme must be treated as a specific case study. In the French case personalization was the requirement. Within the bounds of sociological knowledge, the essential features of personalization have been identified; however, all water-resource development is not characterized by the requirement for personalization. Where there are large numbers of people, improved public relations may be required; but the attitudes of large numbers of people is not a feature of all water-resource development — nor are procedures to optimize economic efficiency the same in all economic development schemes. "Experience in Israel so far has shown that maintaining group cohesion, upsetting customs as little as possible, and by providing adequate guidance through extension personnel, it has been possible to gradually introduce the complicated methods of modern agricultural technology to people of a non-modern or traditional background" (6). Where there is more cultural uniformity the rate of change can be considerably faster. Thus, each economic development scheme must be treated as a specific case. The general case is that the cultural blueprint for behavior must be focused to promote effective economic development. The specifics on what needs to be done come from study of particular social institutions.

## CONCLUSIONS

We are suggesting that each economic development situation is a case study to which the concept of social institution should be applied. The general proposition to be used in applying the concept "social institution" is that successful economic development is enhanced where careful consideration is given to the territoriality of the existing and created social institutions. Higher degrees of success are associated with those situations where the territoriality of the social institutions responsible for management of the waterspace closely approximates the spatial distribution of the water. Often this requires the integration of many diverse peoples and groups for coordinated action to develop the waterspace. A second part of this generalization is that the blueprint for behavior of people occupying a given waterspace must complement demands of development.

In illustrating this generalization several correlates have come to light. For example, many different kinds of social institutions have been used to integrate the peoples of a given waterspace for effective utilization of that waterspace. The institution responsible for water resource development often has a broader scope of activities than just control of the water resource. Most often in economic development a unit responsible for an area greater than the waterspace, for instance a national government, initiates or at least plays an important role in the development. This may be to relieve the dependence of peoples in the area on the larger unit or it may be to gain dominance over the people, or both. Each social institution has a blueprint for behavior unique to the people of that institution. For waterspace development the details of that blueprint must be understood. When the details of the blueprint for behavior are understood, then the knowledge from the social sciences can be brought to bear on specific problems.

## REFERENCES

1. MOORE, W. E.
   1965 The impact of industry. Prentice-Hall, Englewood Cliffs. 117 pp.; pp. 5-6.
2. MEAD, E.
   1910 Irrigation institutions, a discussion of the economic and legal questions created by the growth of irrigated agriculture in the West. Macmillan Co., Ltd., London. 392 pp; p. vii.
3. PADFIELD, H., and W. E. MARTIN
   1965 Farmers, workers and machines: technological and social change in farm industries of Arizona. University of Arizona Press, Tucson. 325 pp; pp. 7-8.
4. PADFIELD, H., and C. L. SMITH
   1968 Water and culture: new decision rules for old institutions. Rocky Mountain Social Science Journal 5(2):23-32.
5. SMITH, C. L.
   1968 The Salt River Project of Arizona: its organization and integration with the community. University of Arizona, Department of Anthropology. 323 pp. (Unpublished Ph.D. dissertation)
6. WEITZ, R., and A. ROKACH
   1968 Agricultural development: planning and implementation, an Israeli case study. Frederick A. Praeger, Inc., New York. 404 pp; pp 17, 29, 144.
7. GAITSKILL, A.
   1959 Gezira, a story of development in the Sudan. Faber and Faber, London. p. 330; 372 pp.
8. OSTROM, V.
   1953 Water and politics; a study of water policies and administration in the development of Los Angeles. Haynes Foundation, Los Angeles. 297 pp.
9. POLEMAN, T. T.
   1964 The Papaloapan Project; agricultural development in the Mexican tropics. Stanford University Press. 167 pp; pp. 94, 136-138.
10. MARTIN, R. D. (ed.)
    1956 TVA: the first twenty years; a staff report. University of Alabama Press. 282 pp; p. 272.
11. SELZNICK, P.
    1949 TVA and the grass roots; a study in the sociology of formal organization. University of California Press, Berkeley. 274 pp.
12. EDISON ELECTRIC INSTITUTE
    1962 A study of the TVA power business. Edison Electric Institute, New York. 36 pp.
13. CLAPP, G. R.
    1955 The TVA, an approach to the development of a region. University of Chicago Press. 206 pp.
14. COLLINS, F. L.
    1945 Uncle Sam's billion-dollar baby, a taxpayer looks at the TVA. G. P. Putnam's Sons, New York. 174 pp.
15. DUFFUS, R. L.
    1944 The Valley and its people, a portrait of TVA. Alfred A. Knopf, New York. 167 pp.
16. VERSLUYS, J. D. N.
    1953 The Gazira scheme in the Sudan and the Russian Kolkhoz: a comparison of two experiments. Economic Development and Cultural Change 2:32-59, 216-235.
17. LAMOUR, P.
    1961 Land and water development in Southern France. pp. 227-250. In H. Jarrett, ed., Comparisons in resource management. Johns Hopkins Press, Baltimore.
18. SCHAAF, C. H., and R. H. FIFIELD
    1963 The Lower Mekong, challenge to cooperation in Southeast Asia. Van Nostrand, Princeton. 136 pp; p. 90.
19. WHITE, P. T.
    1968 The Mekong, river of terror and hope. National Geographic 134:737-787.

20. CLINE, H. F.
    1963   Mexico, revolution to evolution, 1940-1960.
           Oxford University Press, New York. 374 pp.
21. EPSTEIN, T. S.
    1962   Economic and social change in South India.
           Manchester University Press. 353 pp; p. 73.
22. LANGFORD-SMITH, T., and J. RUTHERFORD
    1966   Water and land; two case studies in irriga-
           tion. Australian National University Press,
           Canberra. 270 pp.
23. PODER, W.
    1965   The Sabi Valley irrigation projects. Uni-
           versity of Chicago, Department of Geog-
           raphy, Research Paper No. 99. 213 pp.
24. GRAY, R. F.
    1963   The Sonjo of Tanganyika, an anthropological
           study of an irrigation-based society. Pub-
           lished for the International African Institute
           by Oxford University Press, London. 181 pp.
25. JOY, C. R.
    1964   Taming Asia's Indus River, the challenge
           of desert, drought, and flood. Coward-
           McCann, Inc., New York.
26. MICHEL, A. A.
    1967   The Indus rivers, a study of the effects of
           partition. Yale University Press, New
           Haven. 595 pp.
27. STEVENS, G. G.
    1965   Jordan River partition. Stanford University,
           Hoover Institution on War, Revolution, and
           Peace. 90 pp.
28. UNITED ARAB REPUBLIC
    1965   The River Jordan and the Zionist conspiracy.
           Cairo, Information Department. 67 pp.
29. BARTELL, G. D.
    1964   Directed culture change among the Sonoran
           Yaquis. University of Arizona, Department

of Anthropology. (Unpublished Ph.D. dis-
           sertation). pp. 301-303.
30. CHI-CH'AO-TING
    1963   Key economic areas in Chinese history as
           revealed in the development of public works
           for water-control. Paragon Book Reprint
           Corporation, New York. 168 pp; p. 149.
31. WITTFOGEL, K. A.
    1957   Oriental despotism; a comparative study of
           total power. Yale University Press, New
           Haven.
    1960   Development aspects of hydraulic societies.
           pp. 43-52. In J. Steward, ed., Irrigation
           civilizations. Pan American Union, Wash-
           ington.
32. ERASMUS, C. J.
    1963   Man takes control: cultural development
           and American aid. University of Minnesota
           Press, Minneapolis. 365 pp; pp. 230-237,
           213-219.
33. THE INTERNATIONAL BANK FOR RECONSTRUCTION
    AND DEVELOPMENT
    1963   The economic development of Kenya. Johns
           Hopkins Press, Baltimore. 380 pp. p. 80.
34. BALDWIN, K. D. S.
    1957   The Niger agricultural project. Basil Black-
           well, Oxford. 221 pp; p. 9.
35. BEN-DAVID, J.
    1964   Agricultural planning and village com-
           munity in Israel. Unesco, Paris. 159 pp.
           pp. 9, 45-47.
36. EISENSTADT, S. N.
    1967   Israeli society. Basic Books, Inc., New York.
           451 pp. pp 16, 169; 451 pp.

# WATER IMPORTATION INTO ARID LANDS

A Symposium Sponsored and Organized by the
COMMITTEE ON ARID LANDS
American Association for the Advancement of Science
Dallas, Texas, U. S. A., December 30-31, 1968

papers edited by
JAY M. BAGLEY
Utah Water Research Laboratory
Utah State University
Logan, Utah, U. S. A.
and
TERAH L. SMILEY
Department of Geochronology
University of Arizona
Tucson, Arizona, U. S. A.

# INTRODUCTION

Jay M. Bagley
Utah Water Research Laboratory
Utah State University
Logan, Utah, U. S. A.

The forces of change (expanding technology, expanding population, and expanding uses of natural resources) in recent decades have required a re-examination of the approaches to development, management, and use of water. Such forces have the effect of shrinking or eliminating geographic separations, thereby increasing interdependency, interaction, and impact among and between social interests. Thus, while technology has broadened planning perspectives, project formulation and evaluation have also become more and more complex as the planning unit expands to regions. Part of this added complexity results from the sheer increase in the number of social interests to be served and the number of political and institutional entities involved. More of a factor, however, is the complex and dynamic interlinking between the physical, social, political, legal, and economic elements which enter the planning milieus. Although conceptually simple, in composite these factors are extremely difficult to quantify and relate functionally.

So while the evaluation of the social and economic consequences of water development become exponentially more difficult to predict and project, it becomes more important to predict and project accurately because of the tremendous investments involved and the "irreversible" nature of a scheme once implemented. Physical works once constructed generally translate the same "irreversibility" to the social consequences to which it is coupled. Thus, decisions about water development based on weak or flimsy linkages between the physical and social components may ultimately reveal economic and social disadvantages or sacrifice which cannot be reclaimed.

While the fundamental physical objective of water development (to bring about a new hydrologic equilibrium prerequisite to achieving some net social advantage) is independent of scale, experience with small-scale development does not necessarily provide all the necessary answers to large-scale development. Elements of the planning matrix can relate in much more subtle and devious ways and combine to create much larger impacts. If the impacts are desirable, the rate of return on such an investment can be extremely large. Alternatively, if impacts turn out to be undesirable, the losses may be of tremendous proportions.

This, then, is the basis for a symposium dealing with large-scale transfers of water between regions. The symposium was held December 1968 during the American Association for the Advancement of Science Annual Meeting, in Dallas, Texas. Such large-scale redistributions of water have potential for major and sustained impact on both exporting and importing areas. The symposium served to disseminate some of the best current thinking on this subject and provided an atmosphere for meaningful interchange of opinions and ideas developed by those who participated.

The participants represented broad interdisciplinary and regional experience from which to draw their viewpoints. They wrestled with such questions as: What is the philosophic basis for regional water transfer and how does it differ (or does it differ?) from the philosophy of water resources development within a particular river basin? What are the legal and administrative problems involved in conveying large quantities of water across state boundaries wherever doctrines, principles, and regulations of water rights vary? How are these complicated when sovereign nations are involved? What are the large-scale and long-term physical implications on local water balances and the maintenance of water quality? Can regional growth patterns be regulated by regional water transfer and should they be? What are the ecological implications and can they be predicted? Can existing institutional patterns be adapted to the considerations of large-scale water transfers? What institutional changes seem desirable? What are the intrastate, interstate, and international political realities? What are the economic impacts to regions that export and import? If a region has a surplus of water needed by neighboring regions, how can it be determined whether exported water will limit its own future economic growth? How does economic development stemming from regional water transfers compare to alternative investment plans not water-oriented? How can costs and benefits be evaluated and allocated where states and nations are involved? Are there alternatives to large-scale water imports? How can these be evaluated and compared? Will considerations of large-scale water transfer entail a painful reorientation of present philosophy on water resource development? Can this philosophy serve as an instrument for peace between neighboring countries?

These are indicative of the kind of questions implicit in the topics assigned and posed to the panel for discussion. The intent of the symposium was not to provide a forum for debate about specific pet schemes for large-scale import, although some of these were alluded to for purposes of illustrations. Rather, the intent was to encourage a frank interchange regarding the basic issues and principles universally applicable to considerations of large-scale import-export schemes.

The symposium was sponsored by the American Association for the Advancement of Science Committee on Arid Lands. The arrangers of the program were Jay M. Bagley (Utah State University) and Terah L. Smiley (University of Arizona).

# HISTORICAL BACKGROUND AND PHILOSOPHICAL BASIS OF REGIONAL WATER TRANSFER

## C. C. WARNICK
Water Resources Research Institute
University of Idaho
Moscow, Idaho, U. S. A.

## INTRODUCTION

In searching for a reasonable definition of regional water transfer, it is difficult to delimit the classification so that it is meaningful. The practice of diverting water from one hydrologic unit or basin to another offers one distinction; however, the scale of the diversion and the impact that the diversion has on the ecological, hydrologic, social, and economic system involved is certainly of concern and must be a part of the definition. An historical description of various water transfer projects will present background that hopefully clarifies the term, "regional water transfer."

## EARLY HISTORICAL WATER DEVELOPMENTS

Henri Frankfort (1) in a prehistory of ancient Near East points out that archeological evidence of the size of buildings and temples indicates a sophistication of civilization in the Mesopotamian Valley that would have required extensive irrigation in this arid land as early as 4000 B.C. Saggs (2) in his writings indicates that Lagash and Umma, two neighboring cities of the period 2500 B.C., had extensive water resource development. A canal, Shatt-el-Hai, at that time connected the Tigris and the Euphrates Rivers. There was reported conflict between these two cities over water. These irrigation diversions were extensive. Gruber (3) mentions that the Shatt-el-Hai Canal is still in use.

Delaporte (4) reported that Un-Engur, King of Ur, cut a frontier canal Nanna-gugal and made its basin equal to the waters of the sea. He also indicates that the great ruler Hammurabi, about 1950 B.C., constructed the Nar-Hammurabi to bring water to Sumer and Akkad. Gruber (3) mentions some of these same developments and gives an excellent discussion of irrigation and land use in Mesopotania. His account would certainly lead one to believe that regional water transfer was developed in early Babylonian times.

Selwyn-Brown (5), writing in the 1916 Scientific American Supplement, mentions that in the year 1000 A.D., Baghdad possessed the greatest system of irrigation canals that has ever been constructed. Canals in the vicinity of Baghdad has a length of over 3,000 miles. The largest canal was the Chosroes Canal, which was extended to a length of 290 miles by Caliph Mansur. In places the canals were 200 feet wide and 6 feet deep. About 23,600,000 acres of land was irrigated in the valley of the Tigris and Euphrates. This extensive system was destroyed to a large extent in 1258 A.D. by Hulogu and his Mongols from the north.

Mackenzie (6) notes that hieroglyphics of the Pharaohs of the Twelfth Dynasty indicate that extensive irrigation was practiced in Egypt as early as 2500 B.C. Gruber (3) writes of an early large canal, the Bohr-Housef in Upper Egypt, dating back to the time the Israelites were in Egypt. In 1913 writings of Willcocks (7) on Egyptian irrigation, it was pointed out that the population in the delta area of Egypt diminished from 12,000,000 people in 700 A.D. to 2,000,000 people in 1800 A.D., probably due to failure to meet the challenge of water development.

Fitzsimmons (8), writing in *Civil Engineering*, mentions that the Emperors Trajan (53-117 A.D.) and Hadrian (76-138 A.D.) constructed extensive aqueducts at Segovia, Tejada and Sajunto in Spain.

Pierre Paul Riquet built an impressive canal in the 1680's in Southern France. Whether these were regional water transfers may be questionable.

Clyde (9) gives an excellent summary of early water developments in the western hemisphere. In Peru there are ruins of a conduit that carried water 125 miles from the mountains to the capital of Peru. This was apparently before the beginning of the Christian era. It was certainly extensive enough to qualify as a regional water transfer.

Two early water developments in the southwestern United States are worth mentioning. Halseth (10) mentions that the Pima Indians were practicing irrigation in 1694 when Father Kino came into the Salt River Valley. Small ditches involving a dozen communities was the extent of the development. He mentions that an earlier group of Indians, referred to as "Hohokams," had a more extensive system which involved more than just flooding the bottom lands. Halseth inferred that this development probably failed due to drainage problems.

Fitzsimmons (8) reports that the acequias of San Antonio, Texas, were small canals dating to 1718. Seven such canals were built, and each appeared to serve about 500 acres. These early small water developments in themselves cannot be construed to be regional water transfers, but they do form the historical basis for the area wherein regional water transfer discussion has become so important today.

## RECENT IMPORTANT WATER TRANSFER DEVELOPMENTS

It is impossible to present information on all water developments that might qualify as present-day regional water transfers. Two important ones will be reviewed in some detail because of their importance in the evolution of today's problems. The first is concerned with the Colorado River Aqueduct, and the second with the Rocky Mountain Trans-mountain Diversions.

A publication, "California's Stake in the Colorado River," published by Colorado River Board of California (11) gives an excellent presentation of the evolutionary development that resulted in the Colorado River Aqueduct and related diversions. Irrigation development and actual diversion started with an appropriation for water by a Mr. Blythe. The original right was later succeeded by a water right issued to the Palo Verde Irrigation District. This was a small irrigation development adjacent to the Colorado River. The early water needs centered around irrigation and water for domestic use. Water development has now expanded to include six major public agencies that sponsor and own the transfer system. These evolutionary transfers now include the irrigation development of the Palo Verde Irrigation District, now irrigating about 90,000 acres; the Imperial Irrigation District started in the 1890's, now irrigating about 910,000 acres; the Coachella Valley County Water District started in 1902 and served by the All-American Canal, now irrigating 11,000 acres; the City of Los Angeles (Los Angeles Department of Water and Power), which through the foresight of William Mulholland developed the Owens River Aqueduct in 1913 and later extended it to the Mono Basin; the Metropolitan Water District of Southern California that involved thirteen cities as original members, who were incorporated to build the Colorado River Aqueduct; and the San Diego County Water Authority, formed in 1944 and later merged with Metropolitan Water District in 1946. In 1968, 95 per cent of the needs of the Metropolitan Water District were supplied from Colorado River water. Over 10,000,000 people are served by this complex of transfers. Figure 1 illustrates the physical plan for this water transfer system.

The major use of water is for irrigation. The irrigation districts of the entities have senior rights to 3,850,000 acre feet of Colorado River water. Present use is approximately 80 per cent from Colorado River water, 6 per cent from Owens-Mono basins, and 14 per cent from local sources. More recently, industrial and municipal growth has created additional high priority demands forcing the State of California to develop a State Water Plan. Construction of facilities to bring water to Southern California from Northern California is well underway. In addition, the U. S. Bureau of Reclamation has proposed the Pacific Southwest Water Plan. These water developments and diversions are regional programs that involve water transfer from Colorado River drainage and from Mono Valley and Owens Valley. The major source of water has been outside the Pacific Coastal Plain. Each effort to satisfy the demand for water has required an outreaching to more distant sources.

There is no question that this scheme of transfers has met with success. The thriving economy of Southern California could not exist if it were not for regional water transfer. Many statistics can be quoted to indicate the value of the scheme. However, such development has presented problems. The Colorado River transfers necessitated the division of the waters of the Colorado River. The division between the Upper Colorado River and the Lower Colorado River was based on a total flow over a 10-year period of 75,000,000 acre-feet to each part of the basin. However, later hydrologic data has revealed that this quantity may not be available. Thus, uncertainty has developed as to how the water would be allocated between portions of the basins and states within the Colorado River Basin.

Forty long years of controversy developed between the Lower Basin states—Arizona, California, and

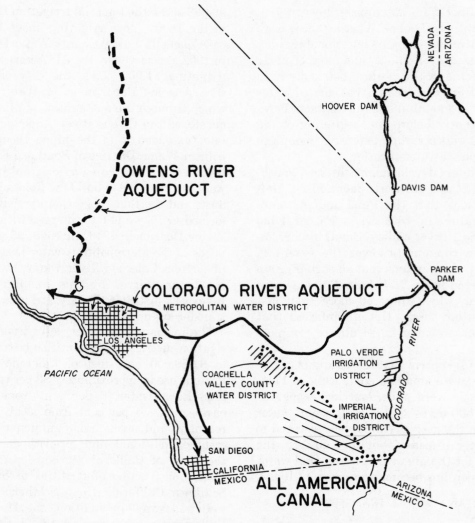

Fig. 1. Schematic map of constructed Southern California water transfer projects.

Nevada—over division of consumptive use of Colorado River water. Apportionment of Colorado River water to those states was finally decided in the Supreme Court Case (*12*) of *Arizona vs. California, et al*, 373 U. S. 546 (1963).

A second water transfer development concerns transmountain diversions through the Rocky Mountains. This is not one transfer but is again the results of a series of evolutionary developments. Prior to 1936 the City of Denver derived its entire water supply from Platte River drainage on the eastern slope of the Rocky Mountains. The need for additional water made it necessary to go out of the basin for water. A good review of this development is presented in a report by the Denver Board of Water Commissioners (*13*). This diverted water is needed for municipal and industrial water supply. A series of tunnels carrying water from the Colorado River system through the Rocky Mountains was required. The first diversion took place through the Moffat Tunnel in 1936.

As early as 1889 people were proposing diversions from the western slope to the drier eastern plains

which are more suitable for agricultural purposes. The drought years 1931-34 provided the impetus that resulted in actual construction the the Big Thompson Project in north-central Colorado. This again was a transfer of water from the Colorado River to Platte River Basin. The primary use of the water is for irrigation; however, municipal water is provided to nine communities on the eastern slope. Details on this project are described in the reports of the U. S. Bureau of Reclamation (*14*). The water is delivered to the Northern Colorado Water Conservancy District, an agency primarily organized to contract with the U. S. Bureau of Reclamation for project repayment.

Recently, under sponsorship of the U. S. Bureau of Reclamation, the Frying Pan / Arkansas Project was developed. This involves numerous conservancy districts, municipalities and power retailers. The plan provides for diverting Colorado River water from the Roaring Fork River and its sub-basins through a tunnel system into the Arkansas River on the eastern slope. Aspen Reservoir on the west slope collects and stores water that is eventually diverted through transmountain tunnels to the eastern slope. Project works in the Arkansas drainage consists of three earth-fill dams, 60 miles of power canal, and 7 hydroelectric power plants. An integrated plan for concurrent development of irrigation, power, municipal water, flood control and recreation has been developed. House Document No. 187, 83rd Congress (15) presents details.

A schematic plan of these various Rocky Mountain transmountain transfers is presented in Figure 2. These evolutionary series of water developments certainly qualify as regional water transfers. The history of the development is strangely related to the Colorado River developments in the southwest because it has been a means by which the Upper Basin State of Colorado, the state of water origin, can utilize water under the Colorado River Compact.

fornia State Water Plan now under construction is certainly a regional water transfer. This was referred to earlier in the text when the Colorado River Aqueduct was described. Detail on this can be found in publications of the California Department of Water Resources (21, 22, 23). All of these Western water developments appear to be the result of pressures from the arid Southwest to gain economic and social well being by continuously reaching out for more water.

Foreign projects that are worthy of mention that have implication of regional water transfers are the Jordan River Development Projects as reported by Sullivan (24) and American Friends of the Middle East (25) and the Indus Basin Project in West Pakistan as reported by Michel (26). The latter two have been rife with political controversy between neigh-

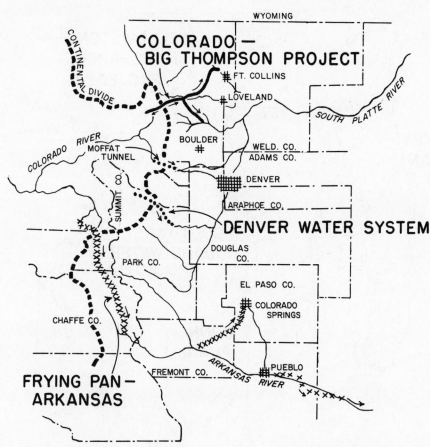

Fig. 2. Schematic map of constructed transmountain water transfer in Colorado.

A brief mention should be made of related water transfer projects that have been recently constructed or are in progress. This includes the Central Utah Project as reported by the U. S. Bureau of Reclamation (16, 17) and Central Utah Water Conservancy District (18); the San Juan / Chama and Navajo Indian projects of New Mexico as presented by the U. S. Bureau of Reclamation (19); and the Central Arizona Project recently authorized by PL90-537 and reported in U. S. Congress Hearings (20). The Cali-

boring countries and plans have been dictated to an extent by national animosity.

## RECENT CONCEPTUAL PLANS FOR REGIONAL WATER TRANSFERS

With the passage of time, the tendency has been for planning efforts to cover larger service areas and to go further geographically to obtain water supplies. In the United States this has been particularly true of the arid lands of the southwest United States.

Reading of the many programs, there appears to be one rather specific turning point. It was the decision of the U. S. Supreme Court in the case of *Arizona vs. California, et al*, 373 U. S. 546 (1963), which brought a semblance of solution to the 40-year dispute over allocation of the waters of the Lower Colorado River. This actually took 12 years of judicial proceedings. The ruling made it clear that the lower basin states, and particularly California, must look beyond the Colorado River for additional water supplies to meet the growing demands for water. A related pressure was the dramatic irrigation development in the High Plains of Texas where groundwater development has overextended the supply of water in the past decade. Time and space does not permit a careful narrative detail of the many schemes that have been proposed since 1950 and particularly since

1963. To provide a basis for comparison and emphasis as to the nature of these conceptual plans that have been advanced, Table 1 is presented giving plan name, organization or person directly concerned with the plan, year plan was first publicized, river basin concerned, countries and states concerned, and at least one literature citation. This attempts to put the material in a matrix that will be useful for those seeking information on potential regional water transfers. Figures 3, 4, 5, and 6 graphically illustrate the respective conceptual plans. A code numbering in Table 1 provides a cross reference to the plans illustrated in the figures.

It should be noted that most of these plans have only been publicized for a short period of about five years.

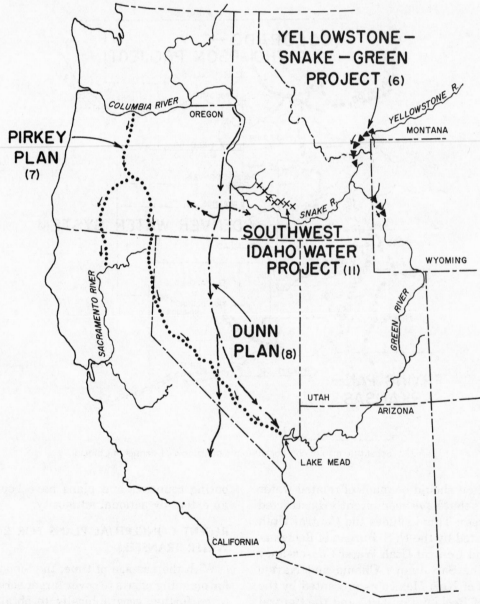

Fig. 3. Schematic map of proposed conceptual plans for regional water transfers (Pirkey Plan, Yellowstone-Snake-Green Project, Southwest Idaho Water Project, and Dunn Plan).

TABLE 1

## SUMMARY OF INFORMATION ON CONCEPTUAL PLANS
## PROPOSED FOR REGIONAL WATER TRANSFER

| Code No. | Project Name | Agency Sponsor Company Sponsor Author of Plan | Approximate Date of Proposal | River Basin(s) for Source | River Basin(s) of Use | Countries Involved | States Involved | Literature Citation No. |
|---|---|---|---|---|---|---|---|---|
| 1 | United Western | U. S. Bureau of Reclamation Rep. R. J. Welch— Calif. | 1950 | Columbia River North Pacific C Coastal Streams | Great Basin South Pacific Coastal Plain Colorado River | United States Mexico | 11 Western States | 27 |
| 2 | California Water Plan | California Department of Water Resources | 1957 | Northern California Rivers | Central Valley California South Pacific Coastal Plain | United States | California | 21, 22, 23 |
| 3 | Pacific Southwest Water Plan | U. S. Bureau of Reclamation W. I. Palmer | 1963 | Northern California Streams Colorado River | Lower Colorado River South Pacific Coastal Plain | United States Mexico | California Arizona, Nevada Utah, New Mexico | 28 |
| 4 | Snake-Colorado Project | Los Angeles Department of Water & Power S. B. Nelson | 1963 | Snake River | Colorado River South Pacific Coastal Plain | United States Mexico | Idaho, Nevada Arizona California | 29 |
| 5 | North American Power & Water Alliance (NAWAPA) | Ralph M. Parsons Company | 1964 | Alaskan & Canadian Rivers, with Columbia River | Great Lakes Basin South Pacific Coastal Plain Colorado River Texas High Plains | United States Canada Mexico | Western States Texas Lake States | 30 |
| 6 | Yellowstone-Snake-Green Project | T. M. Stetson Consulting Engineer | 1964 | Yellowstone River Snake River | Green River Colorado River | United States | Montana, Idaho Wyoming, Lower Colorado States | 31 |
| 7 | Pirkey's Plan Western Water Project | F. Z. Pirkey Consulting Engineer | 1964 | Columbia River | Colorado River Sacramento River South Pacific Coastal Plain | United States Mexico | Oregon Washington California Utah, Arizona Nevada | 32, 33 |
| 8 | Dunn Plan Modified Snake-Colorado Project | W. G. Dunn, Consulting Engineer | 1965 | Snake & Columbia Rivers | Great Basin Snake River South Pacific Coastal Plain Colorado River | United States Mexico | Idaho, Oregon Washington Utah, Arizona Nevada California | 34 |
| 9 | Sierra-Cascade Project | E. F. Miller, Consulting Engineer, Maryland | 1965 | Columbia River | Oregon Valleys Central Valley, California South Pacific Coastal Plain | United States | Oregon, Nevada California | 35, 36 |
| 10 | Undersea Aqueduct System | National Engineering Science Company F. C. Lee | 1965 | North Coast Pacific Rivers | Central Valley South Pacific Coastal Plain | United States | Oregon California | 37, 38 |
| 11 | Southwest Idaho Development Project | U. S. Bureau of Reclamation, Region 1 | 1966 | Payette River Weiser River Bruneau River | Snake River | United States | Idaho | 39 |
| 12 | Canadian Water Export | E. Kuiper | 1966 | Several Canadian Rivers | Western States (indefinite) | United States Canada | All Western States | 40 |
| 13 | Central Arizona Project | U. S. Bureau of Reclamation | 1948, 1967 | Lower Colorado River Basin | Colorado River | United States Mexico | Utah, Nevada Arizona California | 41, 42 |
| 14 | Central North American Water Project C3 NAWP | E. R. Tinney Washington State University, Professor | 1967 | Canadian Rivers | Great Lakes Entire Western States | United States Canada Mexico | Great Lake Western States | 43 |

*Table Continues*

TABLE 1 (cont)

| Code No. | Project Name | Agency Sponsor Company Sponsor Author of Plan | Approximate Date of Proposal | River Basin(s) for Source | River Basin(s) of Use | Countries Involved | States Involved | Literature Citation No. |
|---|---|---|---|---|---|---|---|---|
| 15 | Smith Plan | L. G. Smith Consulting Engineer | 1967 | Liard River McKenzie River | All river basins of 17 western states | United States Canada Mexico | 17 Western States | 44, 45 |
| 16 | Grand Canal Concept | T. W. Kierens Sudbury, Ontario | 1965 | Great Lakes and St. Lawrence River | Canadian Rivers flowing to Hudson Bay | United States Canada | Great Lake States | 46, 47 |
| 17 | Beck Plan | R. W. Beck Associates | 1967 | Missouri River | Texas High Plains | United States | South Dakota Nebraska Kansas, Colorado Oklahoma, Texas | 48 |
| 18 | West Texas and Eastern New Mexico Import Project | U. S. Bureau of Reclamation & U. S. Corps of Engineers | 1967 (1972 due) | Mississippi and Texas Rivers | High Plain of Texas and New Mexico | United States | Oklahoma, Texas New Mexico Louisiana | 49, 50 |
| 19 | Pacific-Mead Aqueduct Augmentation by Desalinization | U. S. Bureau of Reclamation | 1968 | Pacific Ocean | Colorado River | United States Mexico | California Arizona | 51 |
| 20 | Yukon-Taiya Project | Alaska Power Administration | 1968 | Yukon River | Taiya River | United States Canada | Alaska | 52 |

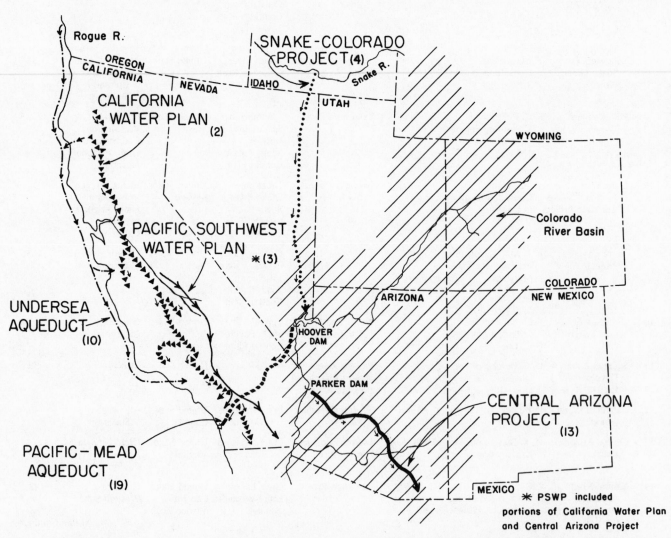

Fig. 4. Schematic map of proposed conceptual plans for regional water transfers (California water plan, Snake-Colorado project, Pacific Southwest water plan, Undersea Aquaduct, Pacific-Mead Aqueduct, Pacific-Mead Aqueduct, and Central Arizona Project).

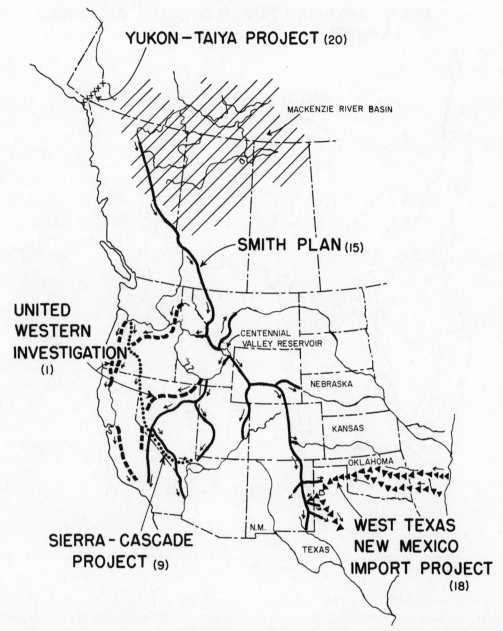

YUKON — TAIYA PROJECT (20)

MACKENZIE RIVER BASIN

SMITH PLAN (15)

UNITED
WESTERN
INVESTIGATION
(1)

CENTENNIAL
VALLEY RESERVOIR

NEBRASKA

KANSAS

OKLAHOMA

N.M.

TEXAS

SIERRA — CASCADE
PROJECT (9)

WEST TEXAS
NEW MEXICO
IMPORT PROJECT
(18)

Fig. 5. Schematic map of proposed conceptual plans for regional water transfers (United Western Investigation, Sierra-Cascade Project, West Texax-New Mexico Import Project, Smith Plan, and Yukon-Taiya Project).

## PHILOSOPHICAL BASIS FOR WATER TRANSFERS

In considering why regional water transfers have been made and potential conceptual plans proposed, it is the prerogative of the academic community to comment and speculate. Five reasons are advanced for justification: (1) regional growth and expanding of the economy; (2) social welfare; (3) wisdom of visionary leaders; (4) agency ambition; (5) professional challenge.

### Regional Growth and Expansion of Economy

An interesting comment that relates to early developments was made in lectures at Oxford University by N. F. Mackenzie (6) when he stated, "As irrigation was the foundation of Babylonia's pros-

perity, so was the want of irrigation the chief cause of her decline." Certainly the record of investment in the regional transfers of Owens River Aqueduct and the Colorado River Aqueduct, including Hoover Dam, would indicate that a regional water transfer can and does generate economic growth of a region that is in need of water. An editorial in Amarillo Sunday News-Globe by Jim Clark (52) illustrates that attitude of boosting or stabilizing economy. In this article Marvin Nichols was quoted at an organizational meeting of Water, Inc., as saying, ". . . it is practical to spend thousands to protect investments reaching billions." Another illustration of this is the words that appear on the bottom of the letterhead of the Department of Water and Power, City of Los Angeles, "Water for Life-Power for Progress."

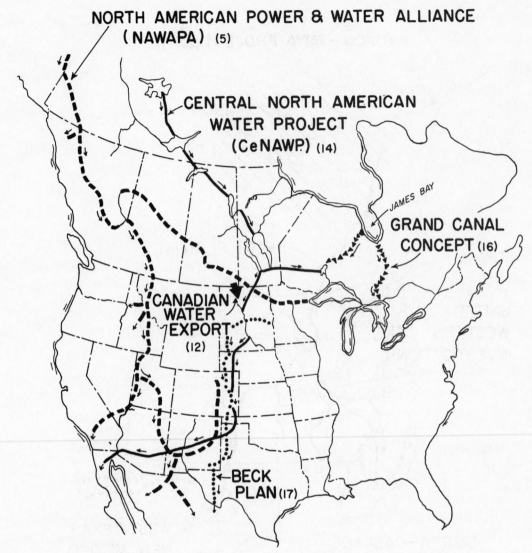

Fig. 6. Schematic map of proposed conceptual plans for regional water transfers (North American Power and Water Alliance, Central North American Water Project, Grand Canal Concept, Canadian Water Export, and Beck Plan).

Some would say this investment in regional transfer will be necessary to have an increasing gross national product, so the idea transcends regional growth and development.

*Social Welfare*

Henri Frankfort (*1*), writing on prehistory of the ancient Near East, made an interesting comment when he wrote, "In particular we know nothing about the origin of irrigation, which played so large a part in Egypt and Mesopotania, and which has been repeatedly recognized as a factor furthering social and political cohesion, since it makes each settlement dependent on its neighbors." You often hear people speak of the wisdom of making the desert blossom. This has been a goal in much of western development because it was contended to be for the best interests of the nation.

Folz (*54*) credits the writing of Irving Fox who quoted the first Commissioner of Reclamation expressing the words, "The object of the Reclamation

Act is not so much to irrigate land as it is to make homes. . . . It is to bring about a condition whereby land shall be put into hands of the small owners, whereby the man with a family can get enough land to support that family, to become a good citizen, and to have all the comforts and necessities which rightly belong to an American Citizen."

Wolman (*55*), in a George A. Miller Centennial Lecture at the University of Illinois, said, "The suggestion that the provisions of water should be geared to broad socioeconomic purposes is, of course, not new. The Homestead Act of 1964 and the Reclamation Act of 1902 were frank declarations of using land and water resources for important national social purposes. It is not surprising that similar principles have been recurrently proposed even for the extension of water lines in community water service, in order to guide and control best land use."

Religious zeal has been associated with this in the case of the Jewish people developing Israel as well as the Mormons in their pioneering in the

Great Salt Lake Basin. Lewis G. Smith, one of the proponents of the conceptual plans indicated in Table 1, appears to advocate social welfare in his presentation. Today there is much talk about the possible wisdom of population dispersion as a social goal for justification of regional water developments.

### Wisdom of Visionary Leaders

Frequently in the forward march of progress certain leaders are credited with providing the motivation and the spark for a cause or a development. Referring back to regional water transfers that have been constructed, early historians (4) appear to credit the ruler Hammurabi with much influence toward wise water development. In reviewing the southern California transfers, writers appear to credit the visionary leadership of William Mulholland of the City of Los Angeles. In the case of the Rocky Mountain trans-basin diversion, an account by the Bureau of Reclamation (56) mentions the plea of Hiram Prince in a presentation to the State Legislature as early as 1889 and also the outstanding survey of the State Engineer, K. P. Maxwell, that followed.

It takes time for historians to credit present-day proponents with vision. It is interesting to speculate whether the words of T. C. Lynch, Attorney General of California, will be credited as visionary and prophetic. In an address to the American Bar Association he stated (57): "I would like to state now, as frankly as I can, what I foresee for the future to handle that combination of a population explosion and the continuing migration to the Southwest: I foresee huge regional water projects. I envision great man-made rivers of surplus waters flowing from water-rich areas, such as the Pacific Northwest, to water deficient areas, such as the Colorado River Basin."

### Agency Ambition

Occasionally one hears the comment that agencies such as the U. S. Bureau of Reclamation and the U. S. Corps of Engineers dream up projects just to keep their program going. As an engineer who respects the work of these agencies, I only suggest this as a very minor point.

### Professional Challenge

In reviewing the various proponents of plans it is evident that consulting engineers may be undertaking the preparation of conceptual plans merely as professional challenge. There appears to be a problem that needs solving and training in the discipline of engineering planning dictates the individual give attention to the possibility of a water transfer scheme. It may be that some regional plans are prepared for the advertising value to the individual, to an organization, or to a region involved.

## OTHER PHILOSOPHICAL CONSIDERATIONS

Recent literature presents a spectrum of concern and thought on water transfers. An interesting paper was prepared by a Public Service Paper Committee of the Los Angeles Section of the American Society of Civil Engineers (58). This committee presented the comment of another group who advocates raising the price of water to reduce the demand for water. The article quotes H. H. Doe as saying, "The United States became great and comfortable because we have ever sought to fulfill the desires of people at the lowest possible cost, and not by the pursuit of a policy of planned scarcity. Water agencies must strive to provide desired service and water at the lowest possible cost. This is their charge —their reason for existence." In the American Society of Civil Engineers presentation the idea of taking people to the water is mentioned but pessimism is expressed with regard to this as a solution.

Engelbert (59), writing about Pacific Southwest Water Plan brings to focus many of the problems that revolve around water transfers and organizational issues. One of his conclusions bears quoting: "As matters now stand, western interbasin water resource developments can proceed physically and technologically in different ways. The political and economic stakes are high, and the ramifications of various plans become increasingly complex. Representatives of the public in both the legislative and executive branches of the national and state governments bear an obligation to weigh carefully all the alternatives."

An excellent discussion on water transfers is the work of Quinn (60) in which he presents information on interbasin transfers and the terms of choice. He discusses opposition due to disturbance of ecological balance, degradation of scenic beauty, and inequity of allocating water to some uses. Quinn makes a concluding statement that is worth repeating: "None of these objections, of course, ends the debate. Present transfer proposals may be lacking in consideration of ecology, efficiency, and equity; at the same time, interregional and interbasin transfers are neither good nor bad in themselves. Because the real problem behind these proposals remain, it is relevant to consider under what condition long-distance transfers may create benefits less likely to be realized in their absence."

Here one must consider total net benefits and how to identify them in the overall picture.

An interesting philosophical thought was propounded by Clyma and Young (61) in a comment on the environmental effects of irrigation in the Central Valley of Arizona. They hypothesize that the following might be the pattern of water resources development in arid lands.

1. Limited development to establish groundwater or surface water as an acceptable source.

2. Expanding irrigation and overdevelopment of the water resources.
3. Additional supplies developed by importation and/or construction of surface water storage reservoirs.
4. Additional water supplies create a rising water table, and drainage and salinity problems develop.
5. Proper drainage lowers water table and solves salinity problems.
6. Transition from irrigated agriculture to urban economy.
7. The irrigated agriculture sector of economy disappears.
8. The culture vanishes.

Certainly the future in itself does not guarantee that regional water transfers are good in themselves. Linsley (62) in testimony presented at hearings on legislation establishing a National Water Commission wrote, "Only within the last decade or two have we come face to face with the fact that an increasingly large population concentrated in great urban centers, combined with technology which finds an increasing number of uses for water, have led to demands which cannot be guaranteed from regional sources of supply available to these areas. Yet we still approach the problem of providing water with the philosophy that water is a free good which should be supplied to all who ask for it in the quantity they request without question or reservation."

The writer sees a perplexing problem developing in the request to provide more alternatives and more extensive studies. To those concerned this makes longer and longer the time between conception of the idea for a solution to water problems facing the arid-land regions and the time when the physical facilities are completed and operational. Concurrently with this is an advancing rate of technological change, which makes it dubious whether the future use and need for water from a given out-of-basin region is what it should be. However, this does not say to quit planning. Thoughtful planning by many will help provide the best solution in the long run.

## CONCLUSIONS

From the foregoing it can be concluded that a regional water transfer, although difficult to define, does entail out-of-basin diversion and has a noticeable impact on the socio-economic well being of an area.

History indicates that water transfers of a regional nature existed in the Near East 6,000 years ago. Examples of recent successful regional water transfers that show a definite favorable impact are the transfers into the South Coastal Plain of California (Colorado River Aqueduct and Owens River Aqueduct) and the transmountain diversion through the Rocky Mountains to the east slope of Colorado.

Population increase and economic growth in the southwest United States and the solving of the *Arizona v. California, et al* case relative to the allocation of the waters of the Colorado has resulted in development of many conceptual plans for rather vast regional water transfers.

As a philosophical basis or justification for regional water transfer developments and the conceptual plans that are being proposed, the following are advanced:

1. Need for regional growth and expansion of economy
2. Social welfare of a region or nation
3. Wisdom of visionary leaders
4. Agencies ambition and agency-perpetuation
5. Professional challenge

It appears that at present authorities on the subject advocate more study and are seeking more alternatives for the solution of water needs in the arid-land regions of our country. Time required for planning and accelerating technological change make it difficult to reconcile the problem of water needs at hand.

## ACKNOWLEDGMENT

The writer wishes to acknowledge the support of the Office of Water Resources Research PL 89-379 from which this project was supported in part. The following graduate students assisted in a directed studies course at the University of Idaho to provide information for the paper; J. W. Abbott, C. S. Allred, D. L. Bassett, H. T. Davis, J. S. Gladwell, and W. L. Robison.

## REFERENCES

1. Frankfort, Henri, "The Birth of Civilization in the Near East," Doubleday Anchor Books, pp. 142, (1950).
2. Saggs, H. W. F., "The Greatness That Was Babylon," A sketch of the ancient civilization of the Tigres-Euphrates Valley. Hawthorn Books, Inc., pp. 561. (1962).
3. Gruber, J. W., "Irrigation and Land Use in Ancient Mesopotania," Agricultural History Magazine, Vol. 22, pp. 69-77, May 13, 1948.
4. Delaporte, L., "Mesopotania, The Babylonian and Assyrian Civilization," Alfred R. Knopf. N. Y., 1925.
5. Selwyn-Brown, Arthur, "Ancient Mesopotania and the Irrigation System That Made it a Fertile Valley." Scientific American Supplement, (1916).
6. Mackenzie, N. F., "Notes on Irrigation Works," Constable and Company, Ltd, London, 1910.
7. Willcocks, W., and Craig, J. I., "Egyptian Irrigation," 2 vols. 3rd Edition, E & F.N. Spon Ltd., London, (1913).
8. Fitzsimmons, N., "The Acequias of San Antonio," Civil Engineering, Vol. 38, No. 10, p. 80, 1968.
9. Clyde, G. D., "Irrigation in the United States," Transactions, American Society of Civil Engineers, Vol. CT Centennial Transactions, Paper 2594, (1953).
10. Halseth, O. S., "Arizona's 1500 Years of Irrigation History," Reclamation Era, pp. 251-254, (1947).
11. Colorado River Board of California, "California's Stake in the Colorado River," Los Angeles, California, Brochure of California River Board of California, 29 p. (no date).
12. Arizona vs. California, et al, 373 U. S. 546 (1963).
13. Denver Board of Water Commissioners, "History and Description of Present and Proposed Water Sys-

tem," Report of the Denver Board of Water Commissioners, Denver Colorado, (1962).

14. U. S. BUREAU OF RECLAMATION, "Colorado-Big Thompson Project," Volume 1, Planning, Legislation, and General Description, U. S. Department of Interior, Denver, Colorado, (1957).

15. U. S. CONGRESS, HOUSE OF REPRESENTATIVES, "Frying Pan-Arkansas Project," U. S. Government Printing Office, House Document No. 187, 83rd Congress, 1st Session, (1953).

16. U. S. BUREAU OF RECLAMATION, "Central Utah Project, Initial Phase, Bonneville Unit," Definite Plan Report, U. S. Department of Interior, Bureau of Reclamation, (1964).

17. U. S. BUREAU OF RECLAMATION, "Colorado River Storage Project and Participation Projects," Report of the U. S. Department of Interior, Bureau of Reclamation, (1967).

18. CENTRAL UTAH WATER CONSERVANCY DISTRICT, "Central Utah Project News Report," Published by Central Utah Water Conservancy District, Provo, Utah, (1968).

19. U. S. BUREAU OF RECLAMATION, "San Juan-Chama Project Colorado-New Mexico," Report of the U. S. Department of Interior, Bureau of Reclamation, Region 5, Amarillo, Texas, (1955).

20. U. S. CONGRESS, "Central Arizona Project," Hearings before Subcommittee on Water and Power Resources of the Committee on Interior and Insular Affairs, U. S. Senate, Washington, D. C., (May 2-5, 1967).

21. CALIFORNIA DEPARTMENT OF WATER RESOURCES, "The California Water Plan," Bulletin No. 3, California Department of Water Resources, Sacramento, California, (1957).

22. ———, "The California State Water Project in 1966," Bulletin No. 132-66, California Department of Water Resources, Sacramento, California, (1966).

23. ———, "The California State Water Project in 1967," Bulletin No. 132-67, California Department of Water Resources, Sacramento, California, (1967).

24. SULLIVAN, J. B., "Jordan Project Becomes Focus for Old Middle East Enmities," Engineering News Record, Vol. 172, January 1964.

25. AMERICAN FRIENDS OF THE MIDDLE EAST, "The Jordan River Water Problem," American Friends of the Middle East, Washington, D. C., (1964).

26. MICHEL, A. A., "The Indus River," Yale University Press, New Haven and London, (1967).

27. U. S. BUREAU OF RECLAMATION, "United Western Investigation," Interim Report of California Section, Report of Special U. S. Bureau of Reclamation Work Group, Salt Lake City, Utah, (1950).

28. U. S. BUREAU OF RECLAMATION, "Pacific Southwest Water Plan," Departmental Task Force Report, W. I. Palmer, Chairman, Washington, D. C., 1963.

29. NELSON, S. B., "Snake-Colorado Project," Paper presented at 32nd Annual Conference, California Utilities Association, March 11, 1964, Pasadena, California, 16 p., (1964); also presented as Report of Department of Water and Power, City of Los Angeles, October, 1963.

30. RALPH M. PARSONS COMPANY, "North American Water and Power Alliance," Brochure 606-2934-19, The Ralph M. Parsons Company, Los Angeles, California, (1963).

31. STETSON, T. M., "Review of Pacific Southwest Water Plan," Consulting Engineer Report to Six Agency Committee, Los Angeles, California, p. 31-33, (1964).

32. PIRKEY, F. Z., "Water For All," Mimeographed Report, Copyrighted, 1963, Sacramento, California, (1963).

33. PIRKEY, F. Z., "Colonel F. Z. Pirkey's Plan: Western Water Project," Western Water News, Vol. 16, January, 1964, p. 2, (1964).

34. DUNN, W. G., "Modified Snake-Colorado Project," Mimeographed Report, P. O. Box T, Cupertino, California, (1965).

35. KELLY, R. P., "The Role of Water in the Development of the West," Paper of Ralph M. Parsons Company, presented, Annual Meeting of Federation of Rocky Mountain States, Sept. 11-13, 1968, Great Falls, Montana, p. 10, (1968).

36. CLARK, C. D., "Northwest-Southwest Water Diversion-Plans and Issues," Willamette Law Journal, Vol. 3, p. 215-262, (1965).

37. LEE, F. C., and STERN, D., "A Feasibility Study of Mainland Shelf Undersea Aqueduct," Report to U. S. Department of Interior, National Engineering Science Company, Pasadena, California, 161 p., (1965).

38. McCAMMON, L. B., and LEE, F. C., "Undersea Aqueduct System," Journal, American Water Works Association, Vol. 58, No. 7, (1966).

39. U. S. BUREAU OF RECLAMATION, "The Southwest Idaho Development Project," Project Planning Report, U. S. Bureau of Reclamation, Region 1, Boise, Idaho, (1966).

40. KUIPER, E., "Canadian Water Export," The Engineering Journal, Vol. 49, No. 7, p. 13-18, Montreal, Canada, July, 1966.

41. U. S. CONGRESS, "Central Arizona Project," Hearings before Sub-Committee on Water and Power Resources, of the Committee on Interior and Insular Affairs, First Session, 90th Congress, May 2-5, 1967; originally recommended to Congress by Secretary of Interior in 1948.

42. U. S. CONGRESS, "Colorado River Basin Project Act," PL 90-537, 90th Congress, September 30, 1968.

43. TINNEY, E. R., "Engineering Aspects (of NAWAPA)," Bulletin of the Atomic Scientist, September 1967, pp. 21-25, 1967.

44. SMITH, L. G., "Agricultural Water Highlights," *Irrigation Age*, p. 35-36, May 1968.

45. SMITH, L. G., "Western States Water Augmentation Concept," Revised Edition, Copyrighted, Denver, Colorado, (1968).

46. KIERANS, T. W., "The Grand Canal Concept," The Engineering Journal, Vol. 48, No. 12, p. 39-42, Montreal, Canada, Dec. 1965.

47. ALBERY, A. C. R., "Stabilization of the Great Lakes Combined With a Continental Water," Preprint No. EIC, The Engineering Institute of Canada, ASME-EIC, Fluids Engineering Conference, Denver, Colorado, April, 1966.

48. BATHEN, R. E., CUNNINGHAM, P. R., and MAYBEN, W. R., "A New Water Resource for the Great Plains," Paper presented at Annual Meeting, Mid-West Electric Consumers, Omaha, Nebraska, December 8, 1967, R. W. Bech and Associates, Denver, Colorado, (1967).

49. SCHORR, B., "Fetching A Water Plan, Texas Style," The Wall Street Journal, p. 12, Monday September 23, 1968.

50. U. S. BUREAU OF RECLAMATION, "Progress Report on West Texas and Eastern New Mexico Import Project Investigations," Report of U. S. Department of Interior, Bureau of Reclamation, Region 5, May 1968.

51. U. S. BUREAU OF RECLAMATION, "Reconnaissance Report-Augmentation of the Colorado River By Desalting Sea Water," U. S. Department of Interior, Washington, D. C. 53 pp. (1968).

52. NORWOOD, G., "Alaska Water Resources, a Strategic National Asset," Paper presented at Seminar on The Continental Use of Arctic Flowing Water," State of Washington Water Research Center, Pullman, Washington, published by Wenatchee Daily World, Wenatchee, Washington, (1963).

53. CLARK, J., "Water For The Future Across Our Wide Plains," Amarillo Sunday News Globe, Amarillo, Texas, p. 6c, July 16, 1967.

54. FOLZ, W. E., "Political and Social Barriers to Western Water Development," *Strategies for Western Regional Water Development,* Proceedings of Western Interstate Water Conference, Corvallis, Oregon, University of California Printing Department, p. 72-95, (1965).

55. WOLMAN, A., "This Environment—Friend or Foe," Special Report No. 1, University of Illinois Water Resources Center, Urbana, Illinois 15 p., (1968).

56. U. S. BUREAU OF RECLAMATION, "The Story of the Colorado-Big Thompson Project," U. S. Department of Interior, Washington, D. C., 56 p., (1962).

57. LYNCH, T. C., "Regional Water Planning: The Legal Tapestry," Proceedings, Mineral and Natural Resources Law, American Bar Association, p. 3-7, (1966).

58. CHUN, R. Y. D., and others, "Water For The West," A Public Service Paper, Los Angeles Section of American Society of Civil Engineers, Los Angeles, 12 p., October 1967.

59. ENGELBERT, E. A., "The Origin and Policy Issues of Pacific Southwest Water Plan," "New Horizons for Resources Research: Issues and Methodology," (Papers of the 1964 Western Resources Conference) University of Colorado Press, (1965).

60. QUINN, F., "Water Transfers—Must the American West Be Won Again," The Geographical Review, Volume LVIII, No. 1, p. 108-132, (1968).

61. CLYMA, W., and YOUNG, R. A., "Environmental Effects of Irrigation in Central Valley of Arizona," Journal of Irrigation and Drainage Division, American Society of Civil Engineers, (In Press).

62. U. S. CONGRESS, U. S. SENATE, "National Water Commission," Hearings before Committee on Interior and Insular Affairs, 89th Congress, 2nd Session, Washington, D. C., (1966).

# INTRASTATE, INTERSTATE, AND INTERNATIONAL LEGAL AND ADMINISTRATIVE PROBLEMS OF LARGE-SCALE WATER TRANSFER

Edward Weinberg*
Office of the Solicitor, U. S. Department of the Interior
Washington, D. C., U. S. A.

The dimensions of the legal problems of large-scale water transfer loom large or small depending upon the position and point of view taken by the observer. My point of view is that of a government lawyer for whom the primary legal problems are those centered upon statutory authority or power and the use of the law to provide methods for achieving desirable social and economic ends. My experience in government law has been primarily gained in working for twenty-five years with the Bureau of Reclamation and other agencies of the Department of the Interior in the development of water resources. What I have to say is, naturally, colored by this past experience.

In light of this experience, I would say that the degree to which the law of large-scale water importation presents problems is directly related to the difficulties of the financial, engineering and political problems involved. If such problems are capable of practical compromise, the legal problems, though real and sometimes complex in detail, are manageable.

However, I recognize that to an observer with a different background than mine and different experiences in the legal world, the legal problems of large-scale water importation may have a different

dimension. I have the feeling that what are called legal problems appear larger and of greater consequence the farther one gets away from Washington. The atmosphere of Washington is such as to lead one to believe that all things are possible if only a way can be found to achieve the necessary political compromises in the halls of Congress. This view may appear simplistic, but I believe that events over the past years have borne it out, as witnessed by the authorization of such projects as Hoover Dam and the All American Canal in 1928 made possible by the political compromise between the upper and lower Colorado River basins embodied in the Colorado River Compact of 1922 and the Central Arizona and the upper basin projects in the recently passed Colorado River Basin Project Act, which embodies a series of interrelated and delicately balanced agreements among the Colorado River states and between the states and the Federal government. Both these authorizations, of course, involve the Colorado River which many people are convinced runs deeper in legal problems than in actual water.

I do not mean to imply that the law can be ignored or that legal problems are miniscule. Rather, my point is that we operate in a legal context whose parameters are derived from those few provisions of the Constitution which make Federal legislation and international treaties the law of the land, and

*The views expressed in this paper are the personal views of the author, and do not reflect an official position of the Department of the Interior or any other Federal Agency.

which vest Congress with the authority to expend money for the public welfare, to manage and dispose of the public lands, to regulate interstate and foreign commerce and to provide for the national defense. Within these broad grants of power, the political solution enacted by Congress becomes the controlling law.

Thus, my comments on the legal problems of large-scale water transfer reflect my view of the law as primarily a method for consolidating political solutions in a manner most suitable for their future administration. The law is a tool to achieve an end and not a series of obstacles or hurdles to be surmounted.

What I propose to do today is to discuss the problems presented by large-scale water transfers both domestic and international. I will describe first some of the major legal developments in the United States concerning large-scale transfers between basins; second, I will touch on some of the legal, political, and institutional issues which must be faced in the consideration of future large-scale interbasin transfers.

The most recent, and in some respects the most significant, development in the United States concerning large-scale water transfers was the passage of the Colorado River Basin Project Act (P.L. 90-537; 82 Stat. 885) signed into law by the President on September 30, 1968.

This Act includes the long-sought-after Central Arizona Project and embodies provisions of immediate importance to the concept of large-scale water transfer. The Act is a legal document embodying as law what are fundamentally political decisions. The nod to the law given in the Act is simply that it recognizes as the law the previous statutes, treaties and judicial decisions governing the administration of the Colorado River. It adds to that previous body of law in one major respect, that is, that in relation to Central Arizona Project water supply California shall be guaranteed a priority to 4.4 million acre-feet annually of mainstream Colorado River water until the river flows have been augmented so that there is sufficient mainstream water available for release to satisfy annual consumptive use of 7.5 million acre-feet in Arizona, California and Nevada as provided in Article II (B) (1) of the decree in *Arizona* v. *California*, 376 U. S. 340, at which point the guarantee becomes academic. This was the legal method chosen to resolve the basic political dispute centering on the use of Colorado River water.

Another major dispute resolved in the Act which is of interest to us in the context of large-scale water transfer involves much broader considerations than Colorado River water. Title II of the Act authorizes full reconnaissance investigation of long-term water supply and requirements in the western United States. However, the Act also provides that for the next 10 years the Secretary of the Interior cannot undertake reconnaissance studies of any plan for the importation of water into the Colorado River Basin from any other natural river drainage basin lying outside the States of Arizona, California, Colorado, New Mexico, and those portions of Nevada, Utah, and Wyoming which are in the natural drainage basin of the Colorado River.

This provision was designed to prevent in the interim the inauguration of studies which could lead to a plan to divert water from the Columbia River Basin into the Colorado River Basin. The language, however, does more than preclude studies of transfer from the Columbia Basin alone. It prohibits, for the decade, the study of any new plan of any consequence for water importation from any other basin into the major arid areas of the United States. There is no importable water available outside of the Colorado River Basin in the States of Arizona, Colorado and New Mexico.

The provision would not prohibit the study of plans to increase diversions from northern California to southern California, but such diversions are in the planning state or are being exploited by both the United States through the Central Valley Project and the State in its State Water Plan. Moreover, there is considerable question as to whether sufficient water is available from the coastal streams in northern California surplus to California's future needs to realize the economies of scale upon which feasibility of transfer into the Colorado River Basin probably turns.

The restrictive provision of the Colorado River Basin Project Act achieves its purpose of protecting the Columbia River Basin for the time being, but in doing so it also appears to have precluded the development of water from any other natural out-of-basin source. It may appear on its face to have created an insurmountable legal obstacle to the initiation of study of any significant plan for water importation into the arid southwest for the next ten years at least. Of course, study of water augmentation by means other than interbasin transfer is not precluded. Thus, the stage is set for intensive study of the possibilities of desalting and weather modification.

I do not believe that this 10-year moratorium should be regarded as a major impediment to the study of plans for the development of additional water supplies for the arid Colorado River Basin region.

Large-scale interbasin transfers raise complex, difficult questions which have to be answered before such proposals can be either economically justified or viewed as politically feasible.

Aside from the purely financial and engineering considerations the magnitude of which are themselves enormous, proposals for large-scale transfer involve a host of other preplanning questions and affect many important interests. They call for very long-range judgments on the population and economic growth and water needs and supplies of both the area of origin and the area of import.

It is to be expected that there will be great competition for these scarce waters and that each area will have its claims for and interests in an assured supply. No realistic plan for large-scale diversion would be economically and socially justifiable if it did not assure that, in view of the large costs involved and the plans made in reliance on such supplies, there was an assured supply for a very long period of time.

On the other hand, the area of origin will understandably be apprehensive about any assurances of water for another region which in the future could lead to a lessening of supply in the export area, thus inhibiting what it believes to be its opportunities for long-term economic growth. This kind of consideration is inherent in any large-scale transfer whether it be from one country to another, from one state to another, or indeed even within the same state.

Virtually every large-scale significant interbasin transfer which has occurred in this country has been achieved only after the area of origin has been assured that over the long-term the development of export supplies would not cut into the solid muscle of the economic potential of the exporting area. Notable examples of such arrangements have been made within the State of California with respect to the Central Valley Project and the State Water Plan whereby water is transported from the watershed areas of the north to the water-short Central Valley and the Los Angeles areas; and in the State of Colorado where water is transported from the Western Slope watershed area by the Colorado-Big Thompson Project of the United States Bureau of Reclamation across the Continental Divide into the more populous and water-short areas east of the Continental Divide.

Within the context of the Colorado River Basin Project Act's 10-year moratorium on study of interbasin transfers into the Colorado, this Act too embodies the political compromise of areas-of-origin assurance and it does so by a double barrelled approach. The Act contains a legal statement to the effect that if there is an importation of water into the Colorado River Basin, the exporting area shall have a perpetual priority to the exported water as against the users in the importing area, unless the exporting States give up such right by agreement. It also provides that in any plans for works to import such water, the Secretary of the Interior shall make provision for adequate and equitable protection of the interests of the States and areas of origin, including financial assistance in meeting costs of providing local water supplies, to the end that the price of water within the States and areas of origin will not be higher than would have been the case in the absence of exportation.

These rules are to apply only to waters delivered by means of exportation works planned pursuant to Title II of the Act, but such exportation works, if they are to be effective in meeting the water requirements of the arid Southwest, must be of such size as to involve significant amounts of water. As long as these provisions are in effect, no importation plan can be considered as politically feasible unless it provides a means for compensating the exporting area for a release of the rights bestowed by this section of the Federal law. At the present time it is impossible to assess the effect which these provisions will have upon the economical feasibility of any proposed plan. It might be interesting to go back and analyze such projects as the Newlands Project (Nevada), the Central Valley Project (California), the Strawberry Project (Utah), Colorado / Big Thompson Project (Colorado), San Juan / Chama Project (New Mexico, Colorado), Frying Pan / Arkansas Project (Colorado), the Garrison Diversion Unit of the Missouri River Basin Project (North Dakota), and the All American Canal (California), all of them interbasin transfer projects to determine what effects this concept of unlimited superior rights in the exporting area would have had on their feasibility.

I do not mean to suggest that the concept of area protection necessarily will preclude consideration of any new large-scale transfers. What the concept does mean, however, is that the importing area has to recognize that it can develop its new water supplies from another area only by coming to reasonable terms with the area of export. In practical effect this will mean that the exporting area will have to receive some compensation for the assurances of supply which the import area will receive.

In terms of water development this means that the import area will have to be prepared to spend more toward projects than the cost of export facilities and will have to share in the cost of developing projects in the export areas for the utilization of their supplies. In short, an import area cannot expect to gain its assured supplies simply on the basis of the cost of export works.

In each particular case, of course, the consideration of how the cost sharing will develop will be different. Short-term versus long-term benefits in each area will have to be weighed and in many cases the ability of the more populous import region to purchase hydroelectric power in the early years may be able to provide some of the financial assistance to the export area.

An illustration of just how different application of the concept of area protection may be is found in yet another political compromise embodied in the Colorado River Basin Project Act. That compromise relates to the Colorado / Big Thompson Project which I have already mentioned.

One feature of that project is a reservoir on the western slope of the Rockies—Green Mountain Reservoir—a principal purpose of which is to provide a water supply on the western slope in replacement of the water exported to the eastern side of the Continental Divide.

Here, rather than providing for an unlimited priority for future uses in the area of origin as it does in the case of exports from the Columbia or other basins into the Colorado River Basin, the Act contains a provision which construes the authorization of the Colorado-Big Thompson Project as affording that priority only to presently existing appropriations for the storage of water in Green Mountain Reservoir.

This difference in approach to the concept of area of origin protection in the same legislation—one expansive, the other more restrictive—illustrates rather dramatically the point made earlier in these remarks that practical compromise more than antecedent law shapes the legislative destiny of proposals for large-scale water transfers.

The arrangement between the importing and exporting areas with respect to water supply and cost-sharing is only one of the many complicated matters which in the case of any large-scale transfer would require the most serious consideration.

In addition, there are extremely important conservation values which have to be carefully weighed.

The diversion of large quantities of water from the watershed of one region may have far-reaching consequences for the fish population, fish and wildlife habitat and ecological conditions along the river from which the diversion is made, including estuarine areas where fresh water flushing may be required to prevent salt water encroachment. Navigation, flood control, and water pollution aspects must also be thoroughly evaluated. Indeed, it is not unthinkable that significant geological and climatological changes may result in both the export and import basin areas. The water impoundments themselves might impinge upon scenic and wild areas. These considerations require such intensive study and may present such enormous problems that in the last analysis some other alternative to interbasin transfer would be more acceptable.

I therefore view the 10-year moratorium on studies of importations into the Colorado River Basin as a charter for the focus of attention on alternatives which, while they present their own problems, might avoid some of those of a transfer. The technology of weather modification is rapidly advancing and this interim period could, and I am sure will, be well used to intensify efforts in that

regard. It is our hope that through the refinements of such techniques the snow pack in the Colorado River watershed area can be significantly increased at reasonable cost as a means of augmenting the Colorado River system water supplies. This technique would not only involve very low cost augmentation as compared to the cost of importation but might avoid many of the political and conservation problems inherent in the basin transfers. The study of such techniques is directed at achieving results which will not cause undue harm to the areas affected or result in the diminution of natural precipitation in other areas and presents a host of legal, social and political problems of its own.

The period of the next decade will also be used for more intensive study of the possibility of recycling water for reuse and the promise of water desalting facilities which, when combined with nuclear generation of electric power, may bring the cost of water supply to acceptable levels for the areas involved without causing some of the previously mentioned problems of transfers.

I think that what I have said about the domestic problems of interbasin transfer provides a useful background to consider transfers from basins outside the United States into the United States. In discussion of this question I would like to use for illustration the Canada-to-United States import proposal of the North American Water and Power Alliance. I want to make it clear, however, that I am using this proposal for purposes of illustration only. I do not wish to be understood as necessarily endorsing the proposal or indeed as even necessarily endorsing the concept. Much study remains to be done in the United States and Canada before it can be concluded that the proposal has sufficient political vitality, engineering feasibility, and financial soundness to form the basis for judgments on the merits.

The North American Water and Power Alliance contemplates development of vast storage areas in the Canadian Rockies and movement of water from such storage areas to the arid parts of Canada, the United States, and northern Mexico. It would commence with a 500-mile long storage reservoir at an elevation of 3,000 feet in Canada in what is known as the Rocky Mountain Trench. Water from the trench would be taken by canal and pumps to the proposed Saw Tooth Reservoir in the United States. From there it would flow to all the water-short western states and to lower California and Mexico. The plan would also provide a northwest passage across Canada from the St. Lawrence Seaway to Howe Sound, British Columbia. Additional developments on the Yukon and Peace Rivers would, in addition to adding to the water supply for export, be a source of hydroelectric energy. It is estimated in the plan that the storage capacity of the NAWAPA reservoirs through flow regulation would double the firm power generating capacity of the Columbia River system. The greatest beneficiary, however, would be the arid Southwest and the Great Basin which together would receive vast quantities of water.

This description of the proposed plan is necessarily brief. A more complete description and analysis is available from the United States Government Printing Office as a Senate Committee print entitled, "Western Water Development," Committee on Public Works, United States Senate, dated October 1964.

The first major problems in the plan, assuming engineering and economical feasibility, are international. A formal relationship embodied in a treaty would have to be worked out with Canada. The plan proposes large benefits for Canada; i.e., a transcontinental shipping channel, hydroelectric power, and enhancement of the Great Lakes water supply, calling for the expenditure of at least $30 billion (1964 dollars) in Canada. In addition to the construction investment, the revenues from hydroelectric generation in Canada might be for the use of Canada. The plan does not discuss whether costs should be shared, but it is fair to assume that the United States might be called upon to make an investment for the benefit of Canada in order to insure the water supply to the United States.

As you can see, the NWAPA plan would carry with it all the considerations I have mentioned with respect to domestic interbasin transfers, i.e., water supply competition between the area of origin and the area of export, the need for special financial contribution from the area of import; scenic, ecological, water pollution, flood control, navigation and other resource management problems.

In addition, the international nature of the proposal brings with it significant additional considerations:

First, the nature of discussions between sovereign nations gives a quality to any studies and negotiations vastly different than that in domestic situations where ultimately one sovereign has authority to enact a law for the water resource.

Second, while the legal systems of the United States and Canada both have their origins in the English jurisprudence, they have, in many respects, developed differently. Federal authority to undertake water development projects in the United States is a combination of federal authority stemming from constitutional power as related to public lands, commerce, and the general welfare. Where appropriate, account must be taken of the special requirements of state riparian and appropriation doctrines.

In Canada, on the other hand, the Provincial Governments constitutionally have greater authority over the water resource developments with-

in their boundaries and any international efforts by the national government of Canada necessarily must comprehend that fact. Our experience with negotiation of the Columbia River Treaty, for example, indicated that the province of British Columbia occupied a major role in the development of the Canadian point of view.

Moreover, it must be emphasized that a plan as vast as the NAWAPA plan would have to commit the Canadian Government for a very long time to assuring water supplies for the United States in the face of a rapidly expanding Canadian population and economy.

In addition, any such arrangement would bring forth the need for a great deal of annual, monthly, and even weekly administration of the projects requiring close integration of operations of the projects in both countries. This in itself raises many interesting and as yet unexplored questions of administration. Some of the questions to be resolved are the following:

To what extent would a separate international body have to be set up to operate the projects?

To what extent would such a body result in a loss of each sovereign's independence and a diminution in state or provincial roles in the management of their own natural resources?

What would be the machinery for resolution of disagreements? What would happen in the event of an irreconcilable disagreement between two nations on important operational aspects?

How would the state, regional and local groups in the United States who used or were affected by the projects influence or participate in operations which affected them? The same question would be applicable to similar interests in Canada.

What kind of commitments would any such project entail with respect to future modifications, changes or expansions?

Many of these questions in turn raise the question of whether the domestic institutions of our country would be appropriate for development of domestic water resources affected by the plan. This calls for a discussion of the manner in which our domestic water resource projects is handled.

The present responsibility of the Government of the United States for the planning, construction, funding and operation of water resource projects is divided among a number of agencies. The principal responsibilities are as follows:

1. The United States Department of the Interior, through its Bureau of Reclamation, is responsible for Federal reclamation projects which store and distribute water for beneficial use in the 17 continental western states. The Bureau of Reclamation and the Department's Bonneville Power Administration, Southwestern Power Administration, Southeastern Power Administration and Alaska Power Administration market hydroelectric power from projects constructed by the Bureau of Reclamation and the United States Army Corps of Engineers.

2. The United States Army Corps of Engineers builds dams and conducts dredging operations whose underlying rationale involve flood control and navigation, although other purposes may, as is the case with the Bureau of Reclamation projects, also be served.
3. The Department of Agriculture conducts programs of watershed management for soil conservation.
4. The Federal Power Commission licenses non-Federal entities, including private utilities, to construct and operate hydroelectric facilities on navigable rivers and public lands.
5. Other agencies; for example, Interior's Fish and Wildlife Service, the Federal Water Pollution Control Administration and the National Park Service, also have a voice in project formulation and review.
6. The Water Resources Council, composed of the heads of various Federal departments and agencies with water resource responsibilities, attempts to develop common principles, standards and procedures as guidelines for the various Federal agencies.
7. The newly created National Water Commission is a temporary body with the duty of preparing a national water assessment.
8. The Bureau of the Budget reviews all the proposals for Federal authorization of and appropriations for water resource projects.

The construction and operating agencies must take into account in the development and operation of their water resource programs not only other values which might be involved such as protection and enhancement of fish and wildlife, but aesthetics, wilderness preservation, recreation and pollution control.

Of course, any Federal water resource project of magnitude must be reviewed and authorized by the Congress and may be constructed only upon Congressional appropriation.

It is to be expected that there will be competition and sometimes conflict between the various Federal agencies since each, understandably, has special confidence in its own abilities, special enthusiasm for its own missions and special interest in the development of its own organization.

The fact that this structure may and sometimes does generate competition and conflict has both positive and negative effects. It is often proposed that this potential is sufficiently bad to justify the establishment of a single unified national agency for the purpose of planning, constructing and operating all Federal water resource projects and for the licensing of non-Federal projects. On the other hand, there are those who argue that this potential for competition is insufficient in itself to justify a wholesale departure from our past practice for these reasons:

First, many of the conflicts are substantial conflicts of interest which will have to be dealt with and reconciled whether or not a single agency is the administrator as, for example, a conflict between priority of water for irrigation versus recreation uses.

Second, the scope of water resource activities is so vast that any super-agency created will soon have to divide itself into special areas of competence, and any conflicts and competition will necessarily reappear among constituent divisions of the umbrella agency.

Third, the adversary process has great utility in challenging each side to present its best case and this process may be dulled by covering too many interests within one agency.

However that debate is ultimately resolved, there is little doubt that the current difficult machinery must be improved, even if its basic framework is retained, if it is to be equal to the administrative challenge posed by large-scale water transfers. Working agreements between the various executive departments for high-level resolution of conflicts between their respective bureaus will have to be more intensively pursued. Such an arrangement was worked out between the Secretary of the Interior and the Secretary of the Army a year or two ago for reconciling differences between Interior and the Corps of Engineers with respect to dredging which might impair conservation values. Arrangements might also be worked out for closer coordination of Federal Power Commission licensing activities with proposals of the executive departments for Federal construction of hydroelectric projects.

I should also mention that the Federal process for water resource development must be increasingly cognizant of state and local government agencies which have an interest and sometimes participate directly in Federal project development.

Here also there have been suggestions that even state and local water resource development be shifted to a national agency.

I am not convinced that such a proposal is justified.

So long as the United States has the legal authority to control development on the navigable waters of the United States and those on the lands of the United States (subject to compensation for the established property rights) there is no reason why the states should not be free to exercise their water resource responsibilities within their sphere of constitutional powers.

There is adequate legal machinery to protect the rights accruing to the beneficiaries of both Federal and state projects and only rarely will the case arise where there will be unanimity on the question of the best use of a scarce water resource. This leads us to a regimen of sometimes competing, sometimes independent resource activities which can hardly be labeled tidy.

On the other hand, as the population increases and the needs for water multiply and diversify in the face of a relatively fixed supply, it is too much to expect that any meaningful government machinery for handling this problem can be a completely tidy one. The diversity of interests reflected in a diversity of water resource agencies may in the end turn out to be a pretty good guarantee of development approaching the optimum. A statement by the late Judge Learned Hand on the value of competition in the newspaper business may be applicable to competition in the water resource field. Speaking of the interest in a dissemination of news from many different sources, Judge Hand said*

. . . it presupposes that right conclusions are more likely to be gathered out of a multitude of tongues, than through any kind of authoritative selection. To many this is, and always will be, folly; but we have staked upon it our all.

As I have said, there is little doubt that if we are to retain a multiplicity of water resource agencies, there must be continuing improvements in coordina-

*United States v. Associated Press, 52 F. Supp. 362, 372 (SDNY) 1942.

tion. The question is whether, even with such improvements, that system poses substantial obstacles to any international arrangements for interbasin transfer. In view of the indefinite nature of such international arrangements, I can offer no definite views on whether the structure would have to be drastically altered for implementation of such arrangements. My hunch however, is that it will not be workable to completely supersede the domestic agencies and their special competence in the planning of any international water transfer. Let me emphasize again, however, that a higher degree of coordination than has previously been required will have to be worked out so that each piece of the plan will fit together with each other piece.

And particularly in the field of administration and operation we will be challenged to devise regimes which combine efficiency with reasonable responsiveness to local as well as to national and international needs and issues.

# PHYSICAL IMPLICATIONS OF LARGE-SCALE WATER TRANSFERS

Percy Harold McGauhey
University of California
Sanitary Engineering Research Laboratory
Richmond Field Station
Richmond, California, U. S. A.

It is characteristic of most every effort man has made to shape his environment more nearly to his heart's desire that he has viewed his objective more as an isolated task rather than as one facet of a complex system existing in a dynamic equilibrium. Consequently he is continually surprised to find that the evil he set out to overcome was in reality the device that held a whole set ot other evils in balance; and that by his well-intentioned action he has triggered a scramble for a new equilibrium in nature. This new balance may be either more or less desirable than the old in relation to the original objective of the action. The important fact is that by failing to think things through, whether as a result of shortsightedness or the obscurity of nature's interrelationships, man often finds that sheer probability determines whether his purposeful scientific and engineering endeavors lead ultimately to benefit or to disaster. Here his science may tell him how to accomplish a task without evaluating the wisdom of doing it at all. It is in this context that one must approach the subject of the physical implications of large-scale water transfers.

Concerning the equilibria which might be disturbed by large-scale water transfers, it is difficult to separate the physical from the social, economic,

philosophical, legal, political, institutional, and ecological factors. On the other hand, to deal simultaneously with all such interrelated factors taxes the resolving power of the human mind. In what follows, the author directs attention primarily to the physical implications of water transfers, leaving to others on the program a discussion of most of the dislocations such implications might generate.

There is no question that the subject of the physical implications of large-scale water transfers is in itself an intriguing one, or that there is an urgent need to pursue it. However, even this simplification of the overall subject of water redistribution does not bring it within the range of our ability to identify all the equilibria involved. Unlike Samson, who knew that if he pulled down a column the roof would fall, the author can only tug at a few guideposts and speculate on what may be their significance.

Several guideposts seem worthy of examination for evidence of the implications of large scale interregional transfers of water such as have been suggested or proposed in recent years. They include:

1. Natural large scale transfer (river) systems.
2. Man-made versus natural systems.
3. Inherent land-water relationships of arid lands.

4. Past experience of man, both in water transfers and in living with limited water resources.

## NATURAL TRANSFER SYSTEM

Although everyone, if only in journeys beyond the arid lands, has seen a river, not all of us have reflected on its role as a natural transfer system. In reality, there are several aspects of rivers, and the way man has exploited them, that either by similarity or by contrast lead to an understanding of man-made systems for transfer of water.

If for purposes of this discussion that a river is defined as a surface stream in which water flows the year round, as contrasted with the sometimes dusty defile of arid lands, a number of pertinent factors may be noted. First, the river exists because of a favorable combination of geographical, hydrological, meteorological, climatological, topographical, and geological factors. Characteristically, it collects surface runoff in times of storm and accomplishes a large-scale transfer of this water to the ocean. In quality, such flood waters are characterized by suspended silt and soil particles washed from the earth's surface or created by grinding and eroding of stones. They are low in dissolved minerals and carry organic debris both in a degraded and an undegraded state. In dry weather the stream collects water from the ground at many points of outcrop of ground water and grows in size as these inputs are assembled into a single main stream. The origin of this dry weather flow is precipitation just as is the flood, but during residence in the ground the water has picked up minerals. Hence the river water in dry weather is clear and mineralized to a higher degree than in flood season.

A river performs many functions for nature along its route which need not be catalogued here. Recharge of ground water in areas where no springs outcrop is one such function. This recharged water, however, may return to the stream in some lower reach by way of springs.

In utilizing and managing a river, man has followed certain concepts and practices which have relevance to the subject at hand. These, too, however obvious, are worthy of summary for subsequent reference. Historically, the United States has largely subscribed to the ancient riparian concepts of British Common Law which permitted the holder of riparian rights to utilize water as long as he did not diminish it in quantity or quality. In the unsophisticated system to which this concept was particularly applicable, the riparian owner drew his modest water supply from a well or spring and discharged his bodily wastes to the soil. The quality changes wrought by the wastes of his farm animals reaching the stream was of no consequence; and his industry, commonly grist or other mills directly

driven by water wheels, involved only brief diversion of water before it was returned to the water course.

Control of natural streams first took the form of impoundment of flood waters by storage. Thus the government in effect said to the riparian owner, (a) we will protect your person and your installations from the wild fluctuations of nature, and (b) we may also reduce the size of the stream to which you have rights, but your activities will not thereby be constrained.

As civilization became more complex and sophisticated, the concept of no diminution of quality and quantity became patently unworkable for several reasons which are germane to the subject of this discussion. First, water was diverted for public water supply of cities. The result was a modest 25 to 30 percent reduction in quantity returned, but a major degradation of quality. Next, water withdrawn for some industries came back only warmed in temperature and modestly depleted in quantity. But from other industries 50 percent or more disappeared. The remainder came back with its original mineral content multiplied in concentration, and often with a wide spectrum of quality degrading chemicals added as well. Finally, water withdrawn for agriculture was reduced in quantity by evapotranspiration to the extent of some 70 percent. The remainder came back via surface or ground-water flow heavily mineralized—often by a concentration factor of from 5 to 10 (1).

Although this more varied and intense use of water introduced a deviation from the ancient riparian concept, its major effects were to force society to impose quality restrictions on urban and industrial return waters, and to delegate to government the right to decide the degree to which *quantity* might be diminished for irrigation or other purposes. Of great importance to the physical implications of large scale water transfers, however, is the fact that *no control of the quality of agricultural returnwaters is considered feasible.*

The final significant aspect of the natural transfer system is that throughout its length the river continues as a resource for sequential use, including use by riparian owners which may be continuous from its origin to its mouth.

## MAN-MADE TRANSPORT SYSTEMS

Bearing in mind the foregoing characteristics of rivers and of their use by man, attention is now directed to man-made systems; and so to the principal subject under discussion. Perhaps the first factor worthy of consideration is the nature of the water to be transferred. Since a large amount is involved it seems evident that impoundment rather than simple diversion at the point of origin will be necessary. The first physical implication is a loss in quality through buildup of total dissolved solids in impounded water due to evaporative losses.

The next important aspect of the man-made transport system is that it is a river in reverse. The man-made river begins with a maximum quantity and flows "uphill," as it were. Along the way it loses both volume and quality by evaporation and evapotranspiration. It loses volume by seepage to the ground water, although unlike in the natural river this loss goes off to strata which do not outcrop to feed again the source stream. The stream gets smaller as tributary users purchase water. And finally, since the major purpose of large scale transport is irrigated agriculture, the water is spread upon the soil at myriad points and disappears into it, thus completing in fact the concept of the river in reverse.

The contrasts between natural and man-made transport systems are, however, profound and not simply a case of reversal of nature. Some of them have physical implications; some have social and institutional implications which can hardly be ignored in this discussion. For one thing there are no riparian owners along the man-made river. Ownership of all land along the river must first be acquired before the river is built. The river, or transport system, thus belongs to some agency of government rather than to individual freeholders. It is therefore somewhat analogous to a railway right of way. Moreover, the water being transported is likewise the property of the agency. No one along this new artificial watercourse may appropriate water from the system any more than he may take goods from a passing freight train.

Government ownership of large areas of right of way has policy-related physical implications, albeit in the realm of increased govermental control. For example, as proposed by the California State Department of Water Resources in relation to its water transfer plans, the right of way offers vast opportunity for the development of a planted strip with a large number of picnic, camping, and related recreational areas. In California, where the population of 20 million is expected to double in two or three decades, and land use planners consider it imperative to spread out population and to create new cities in more appropriate geographical locations, the physical works for large-scale water transport might work to a betterment of man's physical environment.

This is to say that the public control of large areas of land stretched out along the route of water transfer systems might serve the recreational needs of many more people than a natural river of equal length confined within a shell of riparian owners. Evidence to support such a possibility may be found in statistics on the nature of recreation. It has been noted (2) that seven of the ten most common recreational activities, representing some 60 percent of the total, take place in landscaped areas which were planned and developed by man and are man maintained. The importance of this facet of man-made rivers is suggested by estimates that the demand for recreation in California, a semiarid region, is growing at three times the national average rate and in the year 2000 may be some nine times the 1966 demand.

## LAND-WATER RELATIONSHIPS OF ARID LANDS

Further contrasts between natural and man-made rivers and their implications might best be considered in relation to the realities of aridity. Just as the natural river, as noted in a previous section, is a phenomenon of many favorable factors, so is aridity a result of geographical, topographical, climatological, and other factors of nature. Aridity is not simply the result of a lack of rainfall; it is a lack of all the factors which lead to rainfall. In the strictest sense this means that the natural river is the result of a multiplicity of causes and that creation of a man-made river can only bring to the arid land a symptom of some other climate without changing the climate itself.

On a microscale this idea is supported by past experience, but it may be worthwhile to consider at what point "large scale" transfers of water might become large enough to have a climatological effect. The answer is beyond the scope of this paper. However, as to the possibility of changing an arid area to a humid one some discouraging factors are evident. For example, in the so-called "humid" areas of the United States, only about 33 percent of the annual precipitation comes in from the ocean by planetary circulation. The remainder is recycled between the earth and sky by evaporation and evapotranspiration. In most of this area the annual rainfall is not too much greater than the amount of water applied in irrigated agriculture. However, it covers the entire land surface, whereas irrigation is selective. It occurs as rain with low dissolved solids, hence the one-third that reaches the ground water is less saline than the underflow from agricultural lands; the soil itself has less soluble solids; and the evaporation from free water surfaces is one-half or less of that in the semiarid sectors of the United States.

In arid lands evapotranspiration is not generally repreciptated locally. It does, however, lower the temperature and increase the humidity. In the Coachella Valley of California in summer, it is not unusual for the air temperature over an irrigated section to be 10 degrees lower than that over the desert. A rise in relative humidity from 3 or 4 percent to 10-15 percent or more, however, does not give cause for optimism concerning climatic changes on a macroscale. It seems likely that where local topographic conditions are favorable the micro-climate might well respond to large-scale imports of water spread upon the land.

Here, however, many factors relating to aridity come into play. For example, the west side of the San Joaquin Valley in California is expected (1) to remain hot and dry even after water from the north is delivered there for agricultural use. At the same time, because of the direction of air mass movement, climate to the east is expected to be adversely affected by the increased transpiration on the downwind side of the valley. However, when one attempts to evaluate such physical implications of large scale water transfers into arid areas, the available evidence appears inconclusive. It has been said of places like Phoenix, Arizona, and of Fresno and Imperial, California, that climatic changes, generally for the worse, have occurred since irrigation was begun, largely as a result of irrigation. The basis of such statements is not clear. It was too hot for people to live in such places prior to the coming of water; or perhaps it should be said that those who lived there prior to irrigation were not particularly concerned to amass climatological data. In the case of Phoenix, the climatic complaints of today seem to date back to Indian days.

The soil-water relationships of arid lands is a matter of concern. The author has suggested elsewhere (3) that although our irrigated agriculture does not seem to be in danger of failure by salting up the land, it may in the long run fail from a build-up of salts in the water resource. The extent of this danger, of course, depends upon the source of water imported on a large scale. Domestic use of water adds some 300 milligrams per liter to the dissolved solids content; industrial use may more than double the original salt concentration; irrigation return waters are observed to contain from 5 to 10 times the salinity of the applied irrigation water. The degree of recycling from groundwater to surface water, and the content of waste water return flows, affect the original supply. Evaporation during impoundment and open channel transport further concentrate salts. Application to the land to grow crops and to leach salts from the root zone further increases salinity of water.

A striking example exists in the case of the Colorado River. A generation ago its content of dissolved solids was some 200-300 mg/l. In 1968 it is in the 700-800 mg/l range at Hoover Dam. With agricultural return waters added, it reaches Mexico at some 1100 mg/l—already marginal for further use. Here is evidence to support the fear that in the long run the physical management of water, which is a necessary part of large-scale transfers, will seriously impair its usefulness. It should be pointed out, however, that the Colorado involves both a natural river situation and an arid area phenomenon. As a natural river, upstream riparian users have had the opportunity to utilize Colorado River water in agriculture and so to increase its salinity by return flows from irrigation. Thereafter the main storage of the river occurs in an arid area where evaporative losses are high. Its use likewise occurs in the arid region and, more important, it becomes the drain as well as the water source for the arid land it supports.

Implications of large-scale transfers of water are less foreboding than the Colorado situation since the only source of large volumes of water are the vast areas to the north where the climatological and hydrological phenomena produce water which is high in quality as regards dissolved impurities. Thus "large scale," as far as the arid lands of the United States are concerned, means "long distance" as well. The question then is whether long-distance transport, and the storage needed along the way, will degrade the quality of water to the extent experienced in a natural river developed by man for multiple uses. Logic tells us that such will not be the case for the reason that much of the water used locally along the route of transfer will appear as return flows in the local water courses rather than in the transfer system. A strict separation would, of course, not be likely because it would seem necessary from an engineering and economic viewpoint to make use of some natural river channels in the transport scheme. From these sectors, return flows might not be excluded. Nevertheless, the overall effect should be a less serious and more readily controllable degradation of the quality of transferred water. This same set of circumstances—a high-quality water delivered to an arid region with minimum impairment of quality—should in general preclude most of the problems of chemical incompatibility of imported and local waters or soils. However, the reverse could be the case if the water should be applied to some types of soils previously irrigated with a high sodium water.

A final factor for consideration in relation to large-scale imports of water into arid lands is the ultimate fate of that water, or the ability of the land to get rid of water by means other than evaporation.

Applied to the soil in amounts sufficient to provide the leaching needed by agriculture, water acquires dissolved solids. If collected by underdrains it may, as in the case of the Salton Sea, be isolated from usable land. Otherwise it may go down to become saline ground water. Picked up by pumps and re-used in irrigation, it undergoes further decrease in quality. Thus in many localities the long-term result of large-scale transfers could be a buildup of saline ground waters to such levels as to reach the root zone and destroy agriculture.

## PAST EXPERIENCE

Experience in water transfers in the western United States has largely concerned the capture and redistribution of water, either because of a desire for particularly high-quality water or for the support of agriculture or large urban populations. Typical of the high-quality motive is the East Bay Municipal Utilities District and the City of San Francisco, both in the San Francisco Bay area. In each case the public chose at one point in time to go to the Sierra Nevada to capture water and to transport it entirely across the Central Valley. Closed conduits constitute the transport system in which the owner agency transports its property past a number of other communities, each of which make their own arrangements elsewhere for water. Thus is illustrated the concept cited in a previous section—the transport system as a device to move property as against the enroute usage of water in a natural system.

A similar situation is involved in the Los Angeles water supply from Owens Valley and in the Colorado River supply of the Metropolitan Water District of Southern California. Because these examples are all limited objective agencies, the physical implication of water transfer is simply that it removes the aspect of sequential reuse of water as it flows from high to low elevations. This same possibility is implicit in large-scale transfers, except that, as previously noted, water in transport might be used to provide physical facilities for recreation along the route of transport.

Capture and transfer of water primarily for irrigation is a common practice in the western United States. The Central Valley Project of the Bureau of Reclamation in California is a typical example involving movement of water over considerable distances. The California Water Plan, the Central Arizona Project, and the Texas Water Plan are equally ambitious undertakings already in various stages of development.

As shown in Table 1, experience such as herein cited is relatively modest in comparison with such proposals as that of the North American Water and Power Alliance (NAWPA) which envision water transfers from Alaska and Canada to Arizona and, perhaps, even to Mexico.

From experience in transferring water in the West, particularly for irrigation, several specific facts have been well documented. For example, the problems of salt buildup in the overall water resource have been demonstrated; mosquito control problems of a serious public health import have been created; and some of the problems of management of reservoirs under population pressure for recreation have been brought into sharp focus.

In exploring the physical implications of truly large-scale water transfers it might be revealing to consider the physical effects of not making such transfers. Wolman (4) in 1960 estimated that by 1980 water demands will exceed the supply in the entire Upper Missouri, the Great Basin, the South Pacific, the Colorado, and the Upper Rio Grande and Pecos basins. Peterson (5) notes that the 500-million-

dollar annual agricultural industry of Southern Arizona has been built by mining groundwater. Of a daily pumpage of $5 \times 10^9$ gallons, 98 percent goes to agriculture. Even the Central Arizona Project's 1.2 million acre-feet of water from the Colorado River will not meet one-half of the annual ground water overdraft. Such overdraft is characteristic of numerous areas of the west from the Dakotas to Texas and from Kansas to California. In Central California, for example, the groundwater level has declined to the extent of 150 to 300 feet over a distance of more than 75 miles.

Surely if water is not conveyed long distances in large quantities the physical consequences will include disappearance of agriculture and more intensive urbanization probably in the present population centers of arid areas. Because of the appeal of climate, it is not likely that migration would depopulate present cities, although wage earners might give way to retired people. Most important, however, is the likelihood that without large transfers of water the opportunity to dissaggregate a burgeoning population by building new cities will be lost in arid areas. Conversely, there are precedents for the very rapid development of cities in rather unlikely arid places when water becomes available. Thus it might result that a socially and culturally desirable population shift would occur with a minimum of government planning and interference as a by-product of interregional or interstate water transfers.

Not all men are agreed that population shifts are desirable. Critics of such large scale interregional

TABLE 1

## SOME WATER TRANSFER SYSTEMS IN CALIFORNIA

| Project | Approx. Distance of Transfer (miles) | Approx. Capacity (cubic feet per second) |
|---|---|---|
| Los Angeles Aqueduct (Owens Valley) | 240 | 440 |
| Metropolitan Water District of Southern California | 242 | 1600 |
| San Francisco, Hetch Hetchy | 150 | |
| East Bay Municipal Utilities District | 94 | 285 |
| Central Valley Project (Partial Listing Only) Delta Mendota Canal | 117 | 4600 |
| Friant-Kevin Canal | 153 | 4500 |
| Feather River Project (Oroville Dam) | 450 | 450 (Various) |

water transfer schemes as the North American Water and Power Alliance view the resulting support of people in arid areas as one of the adverse physical effects of water redistribution. Included are:

1. Inundation of large areas of land.
2. Loss of physical bases for economic opportunity.
3. Distribution of biological environments by such factors as change in stream size, temperature, and chemical characteristics.
4. Loss of reaches of open stream to reservoirs.
5. Encouraging growth and concentration of population which may threaten man's survival.
6. Decline in environmental quality through air, water, and land pollution intensified by the human activity made possible by water transfers.

While to a considerable extent the implications of large-scale interregional or interstate transfers of water must remain speculative, particularly as regards local situations and microclimates, there is little doubt that they will bring people to arid areas. Perhaps it should be said that the desire of people to enjoy the climate of the Southwest, together with the agricultural productivity which water makes possible, is in reality the justification for water transfers. Certainly the nature of the transfer systems involved is such that it might serve as an instrument of public policy with which government, for good or evil, may come to guide land use planning and economic growth; to control the pattern of urbanization; and to meet some of man's recreational needs along the route of transfer.

### REFERENCES

1. ELDRIDGE, E. F., "Return Irrigation Water – Characteristics and Effects," U. S. Public Health Service, Region IX, May 1, 1960.
2. SMITH, S. C., and M. F. BREWER, "California's Man-made Rivers," University of California, Division of Agricultural Science, California Agricultural Experiment Station, Extension Service, June 1961.
3. McGAUHEY, P. H., "Engineering Management of Water Quality," McGraw-Hill Book Company (Publishers), 1968.
4. WOLMAN, A., "Water Resources," National Academy of Sciences, Washington, D. C., Rational Research Council Publication 1000-B, 1962.
5. PETERSON, D. F., "Man and His Water Resource," Thirty-second Faculty Honor Lecture, The Faculty Association, Utah State University, February 1966.

# SOCIAL AND ECOLOGICAL IMPLICATIONS OF WATER IMPORTATION INTO ARID LANDS*

Gerald W. Thomas and Thadis W. Box
Dean of Agriculture (Thomas) and Director at Large (Box)
International Center for Arid and Semi-Arid Land Studies
Texas Technological College, Lubbock, Texas

As the world's human population grows, the demand for water steadily increases. Not only is water in short supply, but it is poorly distributed over the earth's surface. Few areas are blessed with just the right amount of water. Throughout the world today projects are either being implemented or in the planning stage which necessitate large-scale movement of water from areas of surplus to areas where water is in short supply. We do not argue that this large scale movement of water is not inevitable or unnecessary, but we do believe that, before further action is taken more careful investigations should be made of the ecological and social implications of water transport.

Twenty seven years ago the late Aldo Leopold wrote at a conference on hydro-biology (1):

Mechanized man, having rebuilt the landscape, is now rebuilding the waters. The sober citizen who would never submit his watch or his motor to amateur tamperings freely submits his lakes to drainings, fillings, dredgings, pollutions, stabilizations, mosquito control, algae control, swimmer's itch control, and the planting of any fish able to swim. So also with rivers. We constrict them with levees and dams, and then flush them with dredgings, channelizations, and floods and silt of bad farming.

The willingness of the public to accept and pay for these contradictory tamperings with the natural order arises, I think, from at least three fallacies in thought. First, each of these tamperings is regarded as a separate project because it is carried out by a separate bureau or profession, and as expertly executed because its proponents are trained, each in his own narrow field. The public does not know that bureaus and professions may cancel one another, and that expertness may cancel understanding. Second, any constructed mechanism is assumed to be superior to a natural one. Steel and concrete have wrought much good, therefore anything built with them must be good. Third, we perceive organic behavior only in those organisms which we have built. We know that engines and governments are organisms; that tampering with a part may affect the whole. We do not yet know that this is true of soils and water.

Thus, men too wise to tolerate hasty tinkering with our political constitution accept without a qualm the most radical amendment to our biotic constitution.

Leopold's words should be heeded today. We cannot risk tampering with large-scale water transport

*ICASALS Contribution Number 63.

until we make an assessment of the ecological and sociological changes that are likely to take place— both in the areas of origin of the water and in the areas of eventual distribution.

## THE PRESSURE FOR WATER IMPORTATION

Large-scale water movement to the arid zones appears to be inevitable. There are several reasons why the pressure will increase for these projects:

First, large-scale water transfer projects are technologically possible. Engineering feats of continental magnitude cannot only be envisioned but executed. Bigger and better machines are available. Nuclear energy can be harnessed for power on a competitive basis (2). Mountains can literally be moved. The engineer and the scientists have risen to the task. It is interesting to note that, of the so-called "seven wonders of the engineering world" as described by the National Association of Professional Engineers, two of these are water movement plans— the California State Water Plan and the Snowy Mountain scheme of Australia.

Secondly, many water transfer projects are economically feasible—particularly if adequate consideration is given to the "multiplier" effects on the economy and the added value of water based recreation. Refinements of our economic analyses techniques clearly show that the burden for payment cannot—nor should not—rest exclusively on the first, or primary, user of water. For example, studies in the Texas High Plains show that irrigation water (separated from dry-land crop production) in 1965 generated $440 million to the local area, of which $340 million were off-farm values to the supply sector, to the processors of farm crops, and to the tertiary sectors of the local economy (3). This type of analysis was also used to help establish a "Zone of Benefits" tax for financing the California Water Plan. A Nebraska study showed that for every one dollar net increase in crop production due to irrigation, a total of $6.68 in new business activity was generated throughout Nebraska (4). Likewise, recent studies have shown tremendous benefits from water-associated activities such as recreation. In the 1964 proposed half-billion dollar Potomac projects, $140 million was anticipated from recreation (5). In another study, Grubb (6) estimated that the primary recreation benefits accruing to the 54 proposed reservoirs in the preliminary Texas Water Plan would exceed $3.8 billion over a 100-year period.

Thirdly, water movement into the arid zones may be necessary for society to survive and grow. *Unless significant breakthroughs are made in water efficiency for crop and livestock production, water, rather than land, will be the first limiting factor in world food production.* Under our present farming practices we are "spending" about 3,000 kilograms

of water to produce a kilo of wheat. As we introduce animal protein or other essentials for a balanced diet, our water expenditures are vastly increased. Texas Tech studies indicate that, on some brush-infested West Texas range areas, over 200,000 kilograms of water are used (or dissipated) in the production of 1 kilo of beef (7).

And lastly, but not least in importance, political pressures will be brought to bear to move water to the arid and semiarid lands. A public that has the technology available to move water and the economic power to organize vast schemes simply cannot stand to see water "wasted" in the high rainfall areas when more "efficient use" could be made in another area. Regardless of the biological problems involved, great political pressures will be exerted to save areas that are now in their height of economic development but will decline unless outside sources of water are tapped. Such an example is the High Plains of Texas where dependence rests on a "depletable" underground water resource. Additional demands will be put upon countries to open new areas for food production through schemes such as the Aswan High Dam, the irrigation projects of the Nile, or the Australian Snowy River Project. If and when peace comes to the world, gigantic work projects may become desirable to keep the economy going. Therefore, we can expect to see an increasing number of projects involving large-scale water movement.

## SOME EXAMPLES OF LARGE-SCALE WATER MOVEMENT

Previous papers have listed some of the major projects involving movement of water into arid and semiarid areas. While these illustrate the nature and scope of our technological planning to date it would be in order to again review some of the features which have ecological implications. The California State Water Project is the largest water movement scheme that has progressed through the planning and major construction phase. This project will deliver more than 4.2 million acre-feet of water per year from water surplus areas of Northern California to water-deficient areas throughout the state—from the Feather River area to San Diego County. Major features of this massive project include: a group of reservoirs located in the upper tributaries of the Feather River; Oroville Dam, with a storage capacity of 3.5 million acre-feet; a peripheral canal to carry the project water around the Sacramento-San Joaquin Delta; the North Bay Aqueduct; the South Bay Aqueduct; the California Aqueduct, moving water over the Tehachapi Mountains, 600 meters in a single lift; the Coastal Branch Aqueduct; the West Branch Aqueduct, extending into the northwest corner of Los Angeles County; and the San Joaquin Master Drain to remove agricultural waste waters from the San Joaquin Valley. Subsequent features include a proposal to tap the waters of the Upper Eel River for conveyance to central and southern California (8).

The proposed Texas Water Plan is an even more imaginative scheme in many respects than the California Water Project. The proposal was released in mid-December, 1968 (9). Several major features are pertinent to our discussion of ecological effects. The plan calls for 68 new dams and reservoirs and modification of several existing reservoirs (Fig. 1). Sixteen of these reservoirs would have a storage capacity in excess of 1 million acre-feet. The largest would be about 5.4 million acre-feet. Four individual reservoirs would have a capacity in excess of the Oroville Dam in California. The total storage (in-

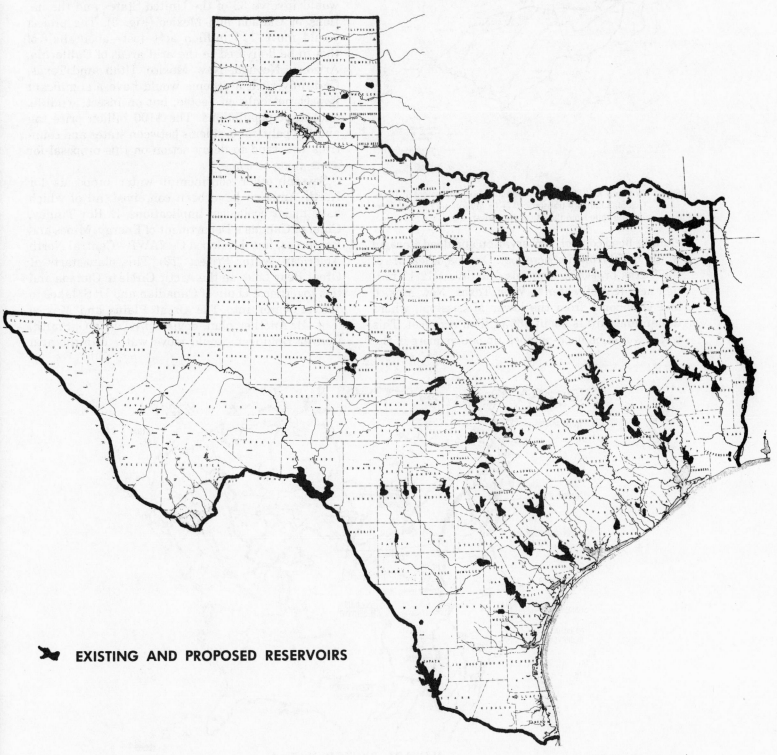

**EXISTING AND PROPOSED RESERVOIRS**

Fig. 1. Plan for surface-water development in Texas as proposed by the Texas Water Development Board in December 1968. The 68 proposed dams and reservoirs will have a profound influence on plant and animal populations—particularly on the patterns of migratory waterfowl movement.

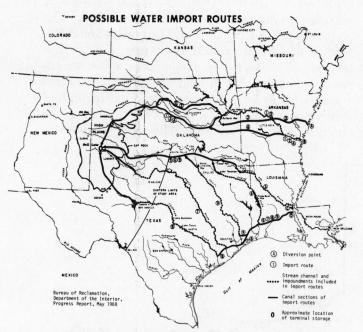

Fig. 2. Proposed route for water importation and movement in Texas. This plan calls for transfer of 12–13 million acre-feet of water into the state from the Mississippi River.

cluding flood control, conservation and dead storage) would be 52.5 million acre-feet. One major feature of the Texas project is a proposal to divert 12–13 million acre-feet of water to West Texas, New Mexico and Oklahoma from the lower Mississippi, where the annual flow is about 365 million acre-feet. This water would be back-pumped up one or more of the

major river systems. Figure 2 shows the route which has been proposed for this water movement. This route covers a distance in excess of 600 miles with a total lift of about 1,000 meters (3,500 feet) (10).

Probably one of the largest continental schemes envisioned to date is the NAWAPA (North American Water and Power Alliance) concept developed by Ralph M. Parsons Company (11). This proposal would involve 33 of the United States and the nations of Canada and Mexico (Fig. 3). The project would move 110 million acre feet—about half of this to be delivered to the arid areas of California, Arizona, Nevada, New Mexico, Utah, and Texas. Certainly such a scheme would have a significant impact not only on people, but on insect, wildlife, and plant populations. The $100 billion price tag and the political problems between states and countries will tend to delay action on this proposal for some years to come.

Several other continental water proposals for North America have been conceived, all of which have major ecological implications. E. Roy Tinney, Chief of Canada's Department of Energy, Mines and Resources, has developed CeNAWP—Central North American Water Project (12). This plan starts at Great Bear Lake on the Arctic Circle in Canada and ties together a chain of Canadian and U. S. lakes to deliver water into The Great Plains and Mexico (Fig. 4). Another plan, devised by Lewis G. Smith, Denver Engineer, would move water into the arid

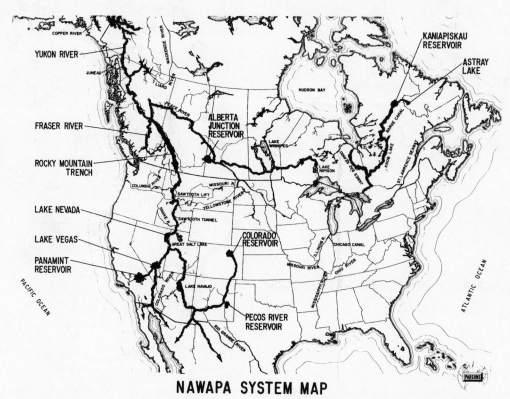

**NAWAPA SYSTEM MAP**

Fig. 3. The Nawapa concept for the development of North American water resources. From Ralph M. Parsons Company, New York.

Fig. 4. The Cenawp Plan for the Development of North American water resources, as proposed by Mr. E. Roy Tinney of Canada

and semiarid Great Plains and Southwest through the Rocky Mountain Trench and a series of rivers, lakes and canals (13). Many of these projects emphasize that the good irrigable land—the great "bread basket" of North America—lies in the Southern Great Plains and in the Southwest.

Australia's Snowy Mountain scheme is a regional approach to divert waters from the eastern flowing Snowy and Eucumbene Rivers, through a series of dams and tunnels as much as 1,000 meters below the surface and into the arid western flowing Murray and Murrumbidgee Rivers (14). The scheme will consist of 128 kilometers of tunnel through rocks and mountains, 9 major dams, many smaller dams, 11 power stations, and 320 kilometers of aqueducts. Upon completion it will divert 1.8 million acre-feet of water annually into the arid region of Australia. Approximately 1 million acre-feet per year will be diverted into the Murrumbidgee and 0.8 m.a.f. into the Murray (15). The cost of the Snowy Mountain Project is justified primarily through the development of hydro power. The diversion of waters into the western flowing rivers for irrigation is only a by-product. This is only one of several projects being planned for Australia.

In South America one of the most interesting concepts—and possibly the one with the greatest potential ecologic and social implications—is the Great Lakes Plan for the Amazon (Fig. 5). Using low dams, certain sections of the Amazon would be closed off creating seven large lakes. The combined area covered by the lakes would be about twice as large as that of the Great Lakes in North America (16). Other plans are envisioned for South America which

would divert water over the Andes to the vast desert areas of the west.

One large-scale water plan that has received world-wide recognition is that of the Aswan High Dam and the accompanying development of the Nile. The Nile basin covers approximately 3 million square kilometers and includes parts of Kenya, Tanganyika, the Congo, part of Ethiopia, most of the Sudan and the cultivated part of Egypt (19). The Aswan Dam will have a capacity of 164 billion cubic meters and will divert water for agriculture and industry in the arid areas of several countries.

The scope of water projects completed to date in the United States—mostly in the West—is illustrated by the following statement from Bureau of Reclamation Assistant Commissioner Gilbert Stamm (17).

In 65 years Reclamation has completed or rehabilitated 196 reservoirs and has 14 under construction with a total storage capacity of over 134 million acre-feet of water to serve people of the west and the Nation. Total storage and conveyance features, including those constructed by others, in operation on Federal Reclamation projects, include 276 storage reservoirs, 302 diversion dams, 13,180 miles of canals, 420 miles of pipeline, 160 miles of tunnels, 30,840 miles of laterals, 103 major pumping plants, and 13,230 miles of project drains. The aggregate value of all crops produced since 1906 now totals $24.9 billion, over 5 times the total Federal investment in completed plant-in-service for all functions of the Reclamation program.

We have reviewed only a few examples of water plans under construction or in the planning stage.

Fig. 5. The Great Lakes Plan for the Amazon. Adapted from Kovaly, Industrial Research, September, 1967.

Additional status reports are contained in the vast array of "Country Situation" papers presented at the International Conference on Water for Peace held in Washington, D. C., in May, 1967. Parenthetically, we might add that, unfortunately, in this, the largest worldwide conference on water, practically no attention was given to ecology as related to water planning.

## SOCIAL AND ECOLOGICAL IMPLICATIONS

Any large-scale water movement project will have profound ecological implications in the area of origin and the region of final distribution. Some changes can also be anticipated along the transport routes.

One of the immediate effects of major water projects is the flooding of the impoundment area. Generally some of the most fertile lands in the area of origin are covered with flood waters. Breeding grounds of wildlife may be destroyed, fertile farm lands covered, or even entire villages moved. During the construction of the Snowy Mountain Scheme in Australia several villages were picked up and moved to new locations (14). Nearly every new impoundment will displace people and change land-use patterns. It is easy to visualize the adjustments in people that would follow continental schemes such as NAWAPA or the Great Lakes Plan of the Amazon.

All aquatic flora and fauna are affected by water impoundment. Some changes may be desirable— others very undesirable. If a river is dammed to form deep lakes, lower water temperatures will usually result, and problems of pollution may be intensified. Ingram and Mackenthum (18), aquatic biologists, state:

Aquatic organisms respond to the aquatic environment by producing a crop that is best suited to the particular environment in which they exist. Organisms respond to changes that take place within their environment with shifts in species dominance in the aquatic community and with sometimes dramatic changes in the population numbers of a single species or a group of species with similar habitat requirements. . . . a fish, an oyster, a duck and a caddisfly each deserve to be considered in water resources conservation and in quality improvement and preservation along with water usages for humans, wildlife, domestic animals, industries, municipalities and agriculture.

Fish and wildlife supposedly became a partner in water-resources development after the passage of a bill on August 12, 1958, amending the Fish and Wildlife Coordination Act, which states that one of the purposes of the act is "to provide that fish and wildlife conservation shall receive equal consideration . . . with other features of water-resource development programs." But, "Strangely enough, the saying of this phrase by Acts of Congress does not automatically make it so" (19).

Many times changes take place far down river from the areas of impoundment or diversion. Estuaries at the coast line normally suffer due to the restriction of fresh water flowing into them or from pollution due to changes in water quality. Fortunately, in the formulation of the Texas Water Plan a serious attempt was made to determine the fresh water inflow requirements of the estuaries. These studies are summarized in a publication entitled "Return Flows—Impact on Texas Bay Systems" (21). The authors made a search for existing ecological information, developed an approximate physical exchange model and a biological degradation model with projections of in-flow requirements to the year 2020. There are over 1.3 million acres of estuaries and bays behind the barrier islands along the Texas

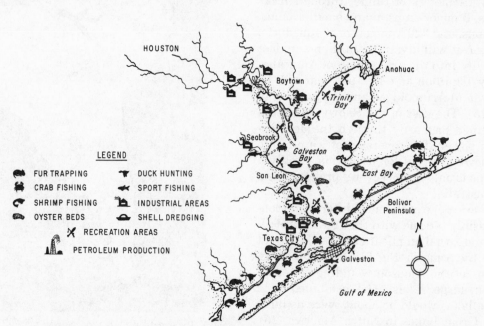

Fig. 6. Wildlife, recreational and commercial activities in the Galveston Bay estuaries. From a special report prepared for the Texas Water Development Board by Currington et al., January, 1966.

Coast. Studies of the ecology—the vegetation, flora and fauna—of the bays and estuaries should be an important part of planning and decision making for all water movement schemes (Fig. 6).

In a recent report submitted to President Johnson by his Environmental Pollution Panel, estuaries are described as follows (19):

This zone of interplay between the margins of the sea and the land is the environment for a remarkable assemblage of terrestrial and aquatic life. Large populations of birds, including such game species as ducks, geese, swans, rails, and snipe, concentrate in the water-logged lowlands—wetlands—associated with estuaries, bays, sounds and keys. Waterfowl come there chiefly during the winter to feed on the lush vegetation or on the brackish water invertebrate animals that abound in the zone. Many of our most valued commercial and game species such as prawns, menhaden, bluefish, weakfish, croaker, mullet, and channel bass, spend their juvenile stages in the protected inside waters of the estuarine zone. Oyster, soft clams, blue crabs, and diamond back terrapins are all residents of estuaries. Fishes that divide their lives between fresh water and salt such as salmon, striped bass, shad, river herring, and eels, pause for a sojourn between coastal waters and their upstream or oceanic spawning grounds.

The sociological pattern in the area of origin normally changes drastically during the construction years relating to the scheme. Most of us are familiar with the new communities that were formed in California during and after the construction of the Oroville Dam and other features of the State-wide project. Another example of how such major construction activities can change the sociology of an area is represented in the Snowy Mountain Project in Australia. A number of new towns were formed. Cooma, the town nearest to the major construction activity tripled from approximately 8,000 to more than 25,000 in a few years. Not only was there an influx of people to the construction area, but new immigrants were brought into the country to furnish the manpower necessary for the vast engineering project. One third of the city of Cooma are new Australians or people who have immigrated to the country since the construction project began (14). The Australian government conducted a special recruitment of migrant tradesmen for the Snowy Mountain Scheme. These people will continue to have an impact on the social structure of Australia long after the Snowy Mountains project is completed.

The area through which the water passes from its area of origin to its final point of distribution will not be affected as drastically as either the source or the area of final use. However, certain changes are bound to occur. These will be most noticeable when it's necessary to flood large natural troughs or depressions, or to add water to existing river channels. Here any influx of new waters or movement of water upstream as opposed to natural flow will cause changes in the limnological patterns of the rivers and profoundly alter the aquatic flora and fauna.

New channels or impoundments for irrigation or other purposes may cross the natural migration paths of wild animals. In some cases direct losses are a result. In other cases such blockage of migration routes may force big game animals into already overused range or into other areas not meeting the habitat requirements. Certainly, new systems of game management must be developed. Klemmedson (22) recently reported on this problem and stated further that, "Unfortunately, the most critical big game winter range is located in valleys and canyon bottoms where water development projects are constructed. The program for Idaho serves as an example of the impact that these projects can have on winter range. In southern Idaho alone an estimated 50 square miles of winter range have been flooded by 10 dams built since 1940."

Marked ecological changes can be expected in the area where water is distributed following large scale movement. The first and most apparent change will be a variation in the aquatic populations. The water distribution scheme will increase the number of water sources in the arid lands. This distributional pattern, consisting of lakes, canals and drainages, will be designed primarily for engineering efficiency rather than biological importance.

The seasonal availability of water may be altered. For instance, irrigation canals in northern Victoria contain water in the summer and none in the winter. This is exactly the reverse of the natural streams where seasonal flows are dependent upon precipitation patterns (23).

The distribution pattern and physical characteristics of lakes or reservoirs in the area of distribution are changed. In most arid areas, natural lakes are shallow and ephemeral. On the other hand, artificially constructed ones may be permanent and deep. These impoundments may have thermal stratification that was unknown prior to water transport. The chemical composition of the waters of the region may be altered. All of these factors may result in a complete shift in ecological habitats in the area where water is distributed. These changes in environment result in a new succession of biological organisms, both plant and animal, in the waters and shorelines of the area.

Examples of the extent to which marine life is changed can be drawn from the biological changes in the Murray and Murrumbidgee Rivers in Australia following completions of dams in the Snowy Mountain scheme. The extent, duration and nature of flooding on the two rivers changed. Lower water temperatures were encountered for considerable distances downstream from the large dams. The dams themselves acted as barriers to fish movements. The dissolved solids turbidity and total alkalinity of the rivers waters were reduced and the pH of the waters lowered.

The native fish of inland Australian rivers breed to a stimulus of rising flood waters (24). The most suitable conditions for favorable spawning is a season of extensive flooding with turbid waters covering the low lying areas adjacent to the river. Construction of dams has caused a more uniform and stable environment that was not suitable for many of the native species. The golden perch is now extinct in many of its original areas. However, the newly created habitats, although not particularly desirable for certain native species, form new niches in which introduced species flourish (25). The English perch and the tench have become common game fish in New South Wales. Brown and rainbow trout have replaced the native cod and golden perch in the colder water areas caused by the dams. Ingram and Mackenthum (1966) report similar changes in aquatic populations in the United States as water quality varies from the natural conditions (18).

Changes in terrestrial populations also occur in association with water transport schemes. Water distribution systems create dams, lakes, irrigation channels, and water catchments that provide water fowl habitat in an otherwise dry area. These new sources of water may change the migratory patterns of waterfowl. Frith (26) found that the migratory movement of wild ducks in Australia was associated with water sources, and new permanent water sources could possibly alter migration patterns.

In North America, Vaught and Kirsch (27) reported that winter waterfowl populations, particularly the Canada Goose (Branta canadinsis) could be altered as habitat was made available by water impoundment and food plantings along their migratory routes. They report substantial increases, with inland water impoundments, in the percent of geese wintering north of their former wintering grounds. In fact, they predict that the entire Eastern Prairie flock may eventually winter in the "new" area—much to the concern of goose hunters farther south. Waterfowl biologists refer to this phenomenon as "short stopping" the population.

Problems, such as disease, may arise in the new locations because of the extraordinary concentrations of birds in the new impoundments. A quick look at the proposed Texas Water Plan showing the many planned reservoirs could cause much concern over migratory waterfowl movement. The Gulf Coast serves as the southern boundary for certain migratory fowl for the North American Continent as well as the northern boundary for other migratory fowl from the South American Continent (28).

Two studies in Utah also relate to this subject. The Ogden Bay Refuge on the east shore of the Great Salt Lake was the first waterfowl project founded by Pittman-Robertson Funds for the restoration of wildlife. Impounding the influx of fresh water at the mouth of the Weber River improved the habitat and reduced botulism losses in waterfowl from 100,000 birds in some years to about 2,000 birds per year (29). Fur harvest from muskrats also increased from about 2,000 pelts per year to 12,000 per year. Vegetation changes along the banks and in the water also were significant with the reduction in salinity and water stabilization. Also, Bolen (30) concluded that management aimed at impoundment would reduce soil salinity and increase carrying capacity of the spring fed salt marshes in Western Utah.

Some species of water birds breed on the storage lakes behind dams, but these deep lakes usually are not particularly good waterfowl habitat and do not generally support large populations. They do, however, form temporary refuges in time of drought in areas such as arid Australia (31). In Australia's center all birds tend to shift their range, in time of drought, to permanent waters. If temperatures are extreme and long in duration, large numbers of birds may die. Keast (32) reported that 60,000 birds dropped dead at one small stock water reservoir. Large storage dams or irrigation canals lined with trees might prevent such widespread bird loss. Even irrigation channels can serve as focal points for wood ducks (Chenonetta jubata) and frequently support small numbers of black ducks (Anas superciliosla) and gray teal (Anas giberfrons) (26). All of these studies show the profound effects of water impoundments on waterfowl movement and productivity. Some of these effects are desirable—at least in the short run—but some also have very profound undesirable implications in the long run.

Many other types of terrestrial fauna will be affected in the distribution area when water is moved to the arid and semiarid lands. Some species will be affected directly, but most biota will be influenced more indirectly through vegetation change or change in the land-use patterns. With the coming of irrigation to "make the desert bloom" the native vegetation will be destroyed and cultivated crops—mostly annuals—will be planted. The habitat of many smaller birds and animals frequently is destroyed—at least in intensively farmed areas. Some species may not survive the new environment, while other species may benefit from the change. Figure 7 shows the rapid growth of irrigation in Texas. Over 2.8 million hectares of land have gone under irrigation in the State since World War II, eliminating vast areas of native vegetation.

With new irrigation there is a tendency to develop large acreages of single crops—or monoculture. As this occurs vast changes in the insect populations can be anticipated. Bosch and Telford (33) state

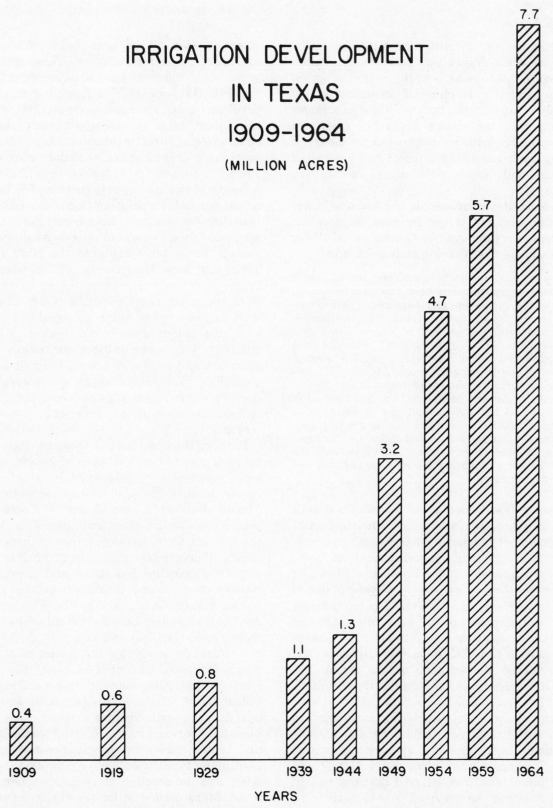

# IRRIGATION DEVELOPMENT
# IN TEXAS
# 1909-1964

## (MILLION ACRES)

YEARS

Fig. 7. Chart illustrating the rapid growth of irrigation in Texas. The state is now studying the possibility of water importation to sustain or supplement this development.

that, "In many cases replacement of natural biota or diversified agriculture with large monocultures has caused general faunal impoverishment, yet certain species of phytophagous arthropods have become extremely abundant." Many similar references are cited by these and other investigators and others on the subject. Furthermore, as insect problems develop on the new cropland, chemical control techniques are initiated, new mutations may emerge, and a complete adjustment in the ecosystem results. In some instances it may become expedient to treat vast acreages of land, both cultivated and uncultivated, with chemicals. The largest such project was probably the cotton boll weevil (*Anthonomus grandis*) control program in Texas involving the chemical treatment by air of 80,000 hectares in the Rolling Plains to attempt to prevent the establishment of the boll weevil on the adjoining High Plains. Reports on the effects of this program on the target insect and associated "desirable" and "undesirable" insect species are available (*34*):

A more direct and noticeable effect of new water distribution systems is the increase in mosquito populations. An Inter-Agency Committee on Water Resources made the following statement (*35*):

In many areas, particularly in the West, the expansion in irrigation agriculture has been accompanied by an increase in the production of mosquitoes. These blood-sucking insects have a serious impact on the health, comfort, and economic welfare of people. They also hinder agricultural, and recreational activities and greatly reduce the overall benefits of irrigation developments. The irrigation-mosquito problem is being intensified by the continuing expansion of irrigation agriculture, the increase in human population, the acceleration of urbanization and industrialization, the development of insecticide resistance in mosquitoes, the growing concern over insecticide residues in food and water, the rapid expansion in outdoor recreational activities, and the public's increasing demands for a more comfortable and healthful environment.

Changes in insect populations accompanied with changes in plant populations and other biological phenomena must be a reflection of changes in microclimate. Yet, physical measurements of such changes are difficult. There is much evidence of "desert encroachment" on several continents due to changes in the vegetation as a result of overgrazing or other mismanagement of range areas. On the other hand, to measure the "retreat of the desert" is much harder. In a discussion of the Russian view of the influence of man upon nature, Burke (*36*) stated, with reference to a gigantic tree planting scheme, "Implicit in this plan was the idea that man can control climate. Not only was it expected that the planting of trees and shrubs, by altering the micro-climate in the protected areas, would protect Russian farmlands against the several physical hazards, but it was also believed that the macro-climate . . . would be changed significantly. . . ." We cannot substantiate this Russian view about the "macroclimate"—possibly because we do not have adequate techniques for measurement. But, we do believe that "microclimate" is changed as large volumes of water are distributed into the arid zones and new ecosystems become observable. Indeed, the biotic communities themselves are a measure of microenvironmental change.

Furthermore, we know that tremendous amounts of water will be evaporated or transpired as new

water is used in the arid zones for storage, for agriculture and for industry. These evaporation losses could influence the microclimate of the area. Lof and Hardison (*37*) estimated that, with full development of streamflow regulation projects in the United States, storage would reach 4,439 billion cubic meters. Direct evaporation from those reservoirs in the arid zones can exceed 3 meters per year. The California Water Plan alone calls for 6.9 million acre-feet of storage capacity covering 58,708 surface acres. Annual evaporation losses are likely to exceed 400,000 acre feet. Reeves and Perry (*38*) have estimated that, in order to provide adequate surface storage for water moving to the High Plains of Texas and New Mexico, evaporation losses at the terminal delivery points would be much greater than those for the California Water Plan. Even with a conservative value attached to this water, such evaporation losses would cost over $10 million annually. This is one of the major reasons that our scientists and engineers are studying underground storage in West Texas as an alternative. Such a decision on terminal storage would certainly pose different considerations from the standpoint of ecology.

Finally, since movement of water into the arid zones is justified on the basis of benefit to people, social systems obviously must be affected. There seems to be evidence, if we can use Arizona, New Mexico, California, and Texas as examples, that people like the arid zones for living. Water is always the limiting factor in development of these environments. Water made available to agriculture will require supporting industries and complete community development including schools, churches, banks, market facilities, etc. Water storage at the distribution centers can, and usually does, generate recreational activities and new political forces.

A planning study by the government of New South Wales (Department of Local Government, Sydney, 1960) indicated that the development of the Coleambally irrigation district with some 1,500 new farmers would require considerable additional industrial services at the existing towns of Griffith and Leaton (5,000 pop. range), the development of Darlington Point from a village of 500 to a town of 5,000 and the creation of an entirely new town of 5,000. Many other similar reports are available on demographic patterns and social changes from major water projects.

Examples given here come from five continents and illustrate, to some extent, the types of ecological and social changes that can be expected from large scale water movement. It should be obvious that there are rather dramatic implications from these continental or regional projects. And yet, we cannot point to a single example of good advance planning involving the entire scientific community. We believe that such multidisciplinary studies are essential to decision making as we look toward proper utilization of our renewable resources. In a time when quality of the environment is receiving so much emphasis we should demand that ecologists and sociologists be appointed to those high level planning commissions whose decisions help determine the kind of world in which we live.

## SUMMARY

Movement of vast amounts of water from areas of "surplus" to areas of deficiency appears to be inevitable. Gigantic regional or continental schemes are technologically possible and in many instances can be justified from an economic, sociological or political standpoint.

Data are presented in this paper which give some indication of the effects of water impoundment, diversion and distribution on aquatic populations, on terrestrial plant and animal communities, on man and his environment. There *is* evidence of significant change in certain ecosystems as water projects are developed. There *is* evidence of loss of habitat for certain biota. There *is* also evidence of changes in the local environment which call for new approaches to wildlife or plant management.

We do not argue against such large scale water transport *per se*. Rather, we present an urgent plea for a better understanding of the ecological implications of these schemes. We urge that sound ecological studies be incorporated in the initial planning for large scale water movement. Indeed, it is highly possible that ecological understanding could be the *major* factor in decision making—even overriding political, social or economic consideration.

## REFERENCES

1. ALDO LEOPOLD, Lakes in relation to terrestial life patterns. A Symposium on Hydro-Biology. pp. 17-22. (University of Wisconsin Press, 1941).
2. EDWARD TELLER, Application of Nuclear Energy to Water Resources Problems. Proceedings of the Sixth West Texas Water Conference. (Texas Technological College, Lubbock, Texas, February 2, 1968).
3. HERBERT W. GRUBB, Importance of Irrigation Water to the Economy of the Texas High Plains. Texas Water Development Board Report 11. (1966).
4. NEWSLETTER, Water, Inc., Value of Irrigation. (Lubbock, Texas, November, 1968).
5. ALLEN V. KNEESE and STEPHEN C. SMITH, Introduction: New Directions in Water Resources Research. In Water Research Published for Resources for the Future, Inc. (Johns Hopkins Press, 1965).
6. HERBERT W. GRUBB and JAMES T. GOODWIN, Economic Evaluation of Water-Oriented Recreation in the Preliminary Texas Water Plan. Texas Water Development Board Report 84, (1968).
7. GERALD W. THOMAS, Toward Improvement of Water Use on Arid and Semi-Arid Range Lands. Contribution No. 44. International Center for Arid and Semi-Arid Land Studies. (Texas Technological College, Lubbock, Texas, 1968).
8. DEPARTMENT OF WATER RESOURCES, STATE OF CALIFORNIA. California's State Water Project (1968).
9. TEXAS WATER DEVELOPMENT BOARD. Water for Texas—A Plan for the Future (Preliminary), (1966).
10. JOHN J. VANDERTULIP, Water Importation—Current Outlook. Proceedings of the Sixth West Texas Water Conference, (Texas Technological College, February 2, 1968).
11. RALPH M. PARSONS COMPANY, NAWAPA—North American Water and Power Alliance. Brochure 606-2934-19, (1965).
12. GLENN LORANG, New "River" for U. S. Farmers. Farm Journal, (August, 1968).
13. STAFF REPORT. Agricultural Water Highlights. Irrigation Age, (May, 1968).
14. ROBERS COGGIN, The Snowy Mountain Scheme, (Longmans Greene and Co., Croydon, Victoria, Australia). 40 pp., (1963).
15. W. H. HUDSON, Opening address, proceedings of Bankers Residential Conference on the use of the Snowy Mountain waters for irrigation of New South Wales, Canberra, pp. v-x, (1961).
16. KENNETH A. KOVALY, Great Lakes Plan for Amazon. Industrial Research, (September, 1967).
17. GILBERT G. STAMM, Water Development and Society, Proceedings of the Workshop for Sociological Aspects of Water Resources Research. (Utah State University, Logan, Utah. April 18-19, 1968).
18. WILLIAM M. INGRAM and KENNETH M. MACKENTHUM, The Pollution Environment, Proceedings of the Second Annual American Water Resources Conference. (University of Chicago, November 20-22, 1966).
19. STANLEY A. CAIN, Coordination of Fish and Wildlife Values with Water Resources Development Goals. Proceedings of the Second Annual American Water Resources Conference. (University of Chicago, November 20-22, 1966).
20. Y. M. SIMARKA, Multi-Purpose Development of the Nile Basin. International Conference on Water for Peace. Washington, D. C. May 23-31, 1967.
21. P. E. CURRINGTON, D. M. WELLS, F. D. MASCH, B. J. COPELAND and D. E. GLOYNA, Return Flows—Impact on Texas Bay Systems. Technical Report to the Texas Water Development Board. (Bryant—Currington, Inc., Austin, Texas).
22. JAMES O. KLEMMEDSON, Big Game Winter Range—A Diminishing Resource. Transactions of the 32 North American Wildlife and Natural Resources Conference. Washington, D. C. March 13-15, 1967.
23. W. D. WILLIAMS, The changing limnology seen in Victoria *in* Australian inland waters and their fauna. (Australian National University Press, Canberra, 1967) pp. 240-251.
24. J. S. LAKE, Principal fishes of the Murray-Darling River system *in* Australian waters and their fauna. (Australian National University Press, Canberra, 1967) pp. 192-213.
25. A. H. WEATHERBY and J. S. LAKE, Introduced fish species *in* Australian inland waters and their fauna. (Australian National University Press, Canberra, 1967) pp. 217-239.
26. H. J. FRITH, The ecology of wild ducks in Inland New South Wales. I Waterfowl habitats, CSIRO Wildlife research 4:47-107, (1959).
27. R. W. VAUGHT and L. M. KIRSCH, Canada Geese of the Eastern Prairie Population, With Special Reference to the Swan Lake Flock. Missouri Department of Conservation Tech. Bull. No. 3, (1966).

28. ERIC G. BOLEN, B. McDANIEL and C. COTTAM, Natural History of the Blackbellied Tree Duck (*Dendrocygna autumnales*) in Southern Texas. Southwestern Naturalist Vol. 9. (2), pp. 78-88 (1964).

29. N. F. NELSON, Factors in the Development and Restoration of Waterfowl Habitat at Ogden Bay Refuge. (Utah State Dept. of Fish and Game Publication No. 6, 1954).

30. ERIC G. BOLEN, Plant Ecology of Spring-Fed Salt Marshes in Western Utah. Ecological Monographs 34:143-166, (1964).

31. H. J. FRITH, The ecology of wild ducks in inland New South Wales, II. Movements, CSIRO wildlife research 4:108-130, (1959).

32. ALLEN KEAST, Australian birds: Their biogeography and adaptation to an arid continent. Monographiae Biologicae 8:89-113, (1959).

33. R. VAN DEN BOSCH and A. D. TELFORD. Environmental Modification and Biological Control *in* Biological Control of Insect Pests and Weeds. (Reinhold Publishing Company, New York, 1965).

34. E. W. HUDDLESTON, C. R. WARD, and J. L. PULLEY, Spring Population Trends on Non-target Insects Following the Fourth Year of the High Plains Boll Weevil Control Program. Entomology Report No. 68-2. (Texas Technological College, July, 1968).

35. COMMITTEE ON VECTOR CONTROL, Interagency Committee on Water Resources, "Mosquito Prevention on Irrigated Lands," U. S. Department of Agriculture. Agriculture Handbook No. 319, (February, 1967).

36. ALBERT E. BURKE, Influence of Man Upon Nature— The Russian View: A Case Study *in* Man's Role in Changing the Face of the Earth. (University of Chicago Press, 1962).

37. GEORGE O. G. LOF and C. H. HARDISON, Storage Requirements for Water in the United States. Water Resources Research, Vol. 2, No. 3, Third Quarter. (Published by American Geophysical Union, 1966).

38. C. C. REEVES, JR. and N. T. PERRY, Geology of Water Importation, The Cross Section Vol. 14, No. 3, Lubbock, Texas. (August, 1967).

# ECONOMICS OF LARGE-SCALE TRANSFERS

Charles W. Howe
Water Resources Program
Resources for the Future, Inc.
Washington, D. C., U. S. A.

Interbasin transfers are no novelty in the United States. The Los Angeles aqueduct was completed in 1913 to carry about 150,000 acre-feet* of water per year from Owens Valley on the eastern slopes of the Sierra Nevada to Los Angeles. In 1928, the

*An acre-foot is the volume of water which would cover one acre of land to a depth of one foot, equaling approximately 326,000 gallons.

Metropolitan Water District was organized in Southern California to build a 240 mile aqueduct from the Colorado River to the Los Angeles metropolitan area. Today, this aqueduct carries approximately 1.2 million acre-feet. The latest major transfer within California is the State Water Project currently under construction and intended eventually to transfer 4.2 million acre-feet from the Feather

River south to the Central Valley and Southern California.

New York City developed the Croton River in steps from 1842 to 1904, the Catskill reservoirs from 1915 to 1924, and finally the Delaware Basin system from 1936 to the present. The entire system can deliver about 2 million acre-feet per year.

The Colorado / Big Thompson project which collects water from the upper Colorado River delivers about 230,000 acre-feet per year to the eastern slope of the Rockies. Denver is developing its system so as to have an eventual capacity for diverting 250,000 acre-feet per year from the western slopes of the Rockies.

In spite of experience with these transfers and others not named above, the plans for interbasin transfers currently being made take us outside relevant experience in several ways. The first major difference is *size*. Current proposals range upward from a Columbia-Colorado transfer of 2.4 million acre-feet per year (costing perhaps $1.4 billion) to the huge North American Water and Power Alliance which might involve transporting 110 million acre-feet from the far northwestern portion of the continent to various parts of Canada, the United States, and Mexico at costs of the order of magnitude of $100 billion.*

Greatly increased physical and financial size implies new challenges and difficulties. What institutional framework can handle the planning and management of such schemes? How would they be financed? How could the huge increments of water supply be phased in with the gradual growth of demand? Can the scaling-up of engineering works really be done with a guarantee of high reliability and efficiency in the final system? These questions cannot at present be answered.

Another factor differentiating prospective from past transfers is the higher potential for conflict between exporting and importing regions. The Northwestern states can foresee full use of their supplies for various purposes by the turn of the century. Proposed Mississippi diversions imply possible difficulties of salt water intrusion into the Delta. Missouri River diversions would directly diminish irrigation and navigation. For these reasons, exporting regions will have to receive guarantees of future replacement water which imply future obligations which are very difficult to evaluate.

Turning to the immediate economic issues, there are two major questions about the economics of large-scale transfers which should be clearly answered before such projects are adopted as the central components of regional water plans:

1. Is there an economic demand for the additional water in the potential importing regions?
2. What are the costs of large-scale interbasin transfers and their alternatives?

The first question above concerning the existence of a demand for water must not be interpreted as "Do people want more water?" (which of course they do, as they do of nearly everything) but in the form of the following two questions:

1. Are total social benefits from the transfer sufficient to offset the total social costs of the transfer?
2. Will water users be able and willing to pay a price for water which will cover project costs?

The first of these questions inquires about the economic feasibility of the project. The second asks whether or not the direct beneficiaries will be able to cover the capial and operating charges or whether the project will require continuing subsidy from other sources.

In answering these questions, it seems reasonable to center on the demands of the agricultural sector since that sector uses about 90 per cent of all water consumed in the arid and semi-arid areas of the Southwest. Physically, agriculture would have to absorb a large part of the proposed increases in supply, at least for a long period of time, if the water were to be put to use at all. There is also evidence that the willingness to pay for additional water is much lower in irrigated agriculture than in municipal and industrial uses. This evidence is found in the prices now being paid for water by the agricultural sector. If water is freely available at a given price so that farms (or any user) may adjust to their most profitable rate of water application, the result should be an application such that the last acre-foot applied yields a profit (gross of water costs) just equal to the water price. Under such circumstances, the price of water provides an upper bound on the value of additional water in terms of its direct use on the farm and an upper limit on what a farmer would pay for more water. It happens that in some irrigation districts where farmers pay extremely low prices for water, the amount available is also constrained, so that price might be much less than the marginal value of water. However, in areas which rely heavily on ground water, there are generally no constraints on water use, and the "common-pool" rationale leads pumpers to apply water until the marginal net profit per acre-foot equals the cost of pumping—the analog of price in this case. Thus, pumping costs may be used as a measure of willingness to pay in some of the major agricultural areas and may be contrasted with prices paid by commerical and industrial users in

---

*It should be emphasized that these proposals have been only sketched out, not engineered, and that the costs involved are exceedingly speculative.

urban areas—another situation in which users are free to use whatever amounts they want.

Pumping costs are known for some major areas. Taylor (1964) quotes pumping costs averaging $5 per acre-foot (with a variable component of $1.96) in the Madera (California) Irrigation District in 1961-62. Young and Martin (1967) state that variable pumping costs in Central Arizona range from $7 to $12 per acre-foot. In the Texas High Plains, Grubb (January, 1966) shows combined pumping and delivery costs of $8.74 per acre-foot. Thus, in these *major potential importing* areas, marginal values of water have been pushed quite low.

In contrast, one can look at the rates paid by commercial and industrial users in urban areas which run from $35 to $100 per acre-foot. Table 1 presents several examples. It should be emphasized that the disparity between urban and agricultural water prices does not in itself imply a present misallocation of water. The prices indicate the maximum amounts which would be paid for more water and, therefore, the maximum cost which could be justified in terms of direct benefits in bringing more water to the user.

TABLE 1

### URBAN COMMERCIAL AND INDUSTRIAL WATER RATES

| City | Applicable to Use Over . . . (cu. ft./mo.) | Rate per Acre Foot ($) | |
|---|---|---|---|
| | | In City | Outside City |
| (California) | | | |
| Los Angeles | 33,000 | 61 | 92 |
| Santa Barbara | 60,000 | 70 | 140 |
| Santa Ana | 25,000 | 61 | — |
| Milpitas | 3,300 | 74 | — |
| Ontario | 10,000 | 52 | — |
| Provo, Utah | 50,000 | 35 | 52 |
| Denver, Colo. | 13,500 | 61 | 96 |
| Ft. Worth, Tex. | 50,000 | 100 | — |
| San Antonio, Tex. | 25,000 | 35 | 48 |

Source: *Modern Water Rates*, Published by American City Magazine, 757 Third Avenue, New York, N. Y. 10017, 1966.

Thus, for physical and economic reasons, it appears that agriculture is the crucial sector whose benefits, direct and indirect, must justify the costs of large-scale inter-basin transfers or other large-scale alternatives. We therefore turn to a survey of the evidence regarding the value of water to the agricultural sector in some of the arid and semiarid areas of the country.

## EMPIRICAL STUDIES OF DIRECT BENEFITS

Brown and McGuire (1967) have studied the problem of optimizing the distribution of surface water among the constituent districts of the Kern County Water Agency. As part of the analysis, the marginal values of water in the several districts were derived. Hartman and Anderson (1962) studied the value of supplemental irrigation water in Northeastern Colorado through the analysis of farm sales data, while Anderson (1961) has studied the rental prices for seasonal water on an interim basis in the South Platte Basin. In Arizona, Young and Martin (1967) studied the value of water by constructing farm budget studies for typical farms, and Goss and Young (July, 1967) have compiled data on the pricing policies of the major water distributing agencies in the State. Stults studied the potential impact on agriculture in Pinal County, Arizona, of the falling water table and, as a result, was able to compute the total loss of income per acre-foot of water as acreage was dropped from production. This figure indicates the value of replacement water to agriculture. Grubb (1966) has studied the value, direct and indirect, of irrigation water to the High Plains of Texas. That region, perhaps along with the Imperial Valley of California, is unique in that it has very little non-agriculturally related activity and little prospect of attracting significant amounts of alternative activities. In such areas, the slack created by a decline in agriculture (through the exhaustion of ground water) cannot be quickly absorbed in other expanding industries. There may be, in that case, substantial secondary effects.

In Utah, Johnson (1966) has studied the productivity of irrigation water in the Milford area, and Gardner and Fullerton (1968) have analyzed the operation of the water rental market in the Delta area of the State. It should be noted that rental markets provide an excellent chance for determining water values, especially when transfers can take place between as well as within irrigation companies, for the prices established represent the outcome of a highly competitive and well informed process of bidding.

The results of all of these studies are summarized in Table 2.

TABLE 2

### SUMMARY OF DIRECT BENEFITS PER ACRE-FOOT AT THE MARGIN OF APPLICATION: WESTERN U. S.

| | | |
|---|---|---|
| California: | KCWA (after State Water Plan) | $ 19 |
| Colorado: | N. Poudre Irrigation Co. (rental) | 3 |
| | S. Platte Basin | 3-8 |
| Arizona: | Central Arizona, short-run | 21 |
| | long-run | 13 |
| | Major distributing agencies, Central Arizona | 0-10 |
| | Pinal County | 9 |
| Texas: | High Plains, average, now | 27 |
| | 1990 | 36 |
| | S. High Plains, hard land soils | 10 |
| | mixed soils | 45 |
| Utah: | Milford area | 0-15 |
| | Delta | 12 |

In addition to the benefits accruing to the first user of the water, there can be, under the proper physical circumstances, reuse of part of the water through return flows with consequent benefits.* Hartman and Seastone (1965) have shown that a sequence of $n$ water users each experiencing direct benefits of (db) per acre-foot of water applied and having a percentage return flow to the irrigation stream of $r$ can ultimately generate direct benefits of

$$DB = db \left\{ \frac{1-r^n}{1-r} \right\}$$

For the particular case of $r = 0.5$ and $n = 5$, the "multiplier" above takes the value 1.97. The derivation of the return flow multiplier assumes, however, that return flows occur during the irrigation season. This would hold, in fact, only where continuous cropping takes place. In many areas irrigating from surface supplies, the intensive irrigation season is short and return flows are delayed past the end of the season. Bittinger (1964) indicates that in a narrow, highly permeable aquifer along a stream, about 50 per cent of the non-consumptive application will return to the stream within 1 month of *reaching the water table*, 80 per cent within 2 months, and practically all within 4 months. Under less ideal circumstances, return times can well increase to a year or more. In some areas where groundwater withdrawals exceed recharge with consequent falling water tables, excess irrigation resulting in deep percolation may never reach the water table and may never be recoverable.

For these reasons, return flow multipliers are not likely to exceed 1.5 under the best of circumstances. The values of Table 2 could be increased by factors up to 50 per cent in appropriate areas.

## CONCLUSIONS ON DIRECT BENEFITS

The evidence presented to date points to a range of direct benefits at the margin of application from essentially zero to 45 dollars per acre-foot. In some areas, the values may be somewhat increased by the value of return flows, but this would not apply to Central Arizona and probably not to the Texas High Plains, since ground water tables are receding rapidly in these areas.

Direct benefits appear to be fairly low compared to prospective costs of new water supplies. This has two implications: (1) that the economic feasibility of new supplies will depend upon the existence of indirect or secondary benefits; (2) that regardless of economic feasibility, it will probably not be possible to impose user charges sufficiently high to

cover the costs of providing the water. The latter implies a financing program of public subsidy and/or imposing part of the cost on secondary beneficiaries through special tax assessments.

## THE ISSUE OF SECONDARY BENEFITS

The expansion of water using activities (or the contraction of those activities if water is being withdrawn or exhausted) will have effects on two related groups of activities: those which supply inputs to the water user, and those which further process the water user's output. These have been described as impacts "induced by" and those "stemming from" the direct water-caused expansion (or contraction). The "Green Book"† has stated in clear but seductively simple sounding terms the conditions under which legitimate secondary benefits may be generated:

Secondary benefits as defined above are the increase in net incomes or other beneficial effects as a result of the project activities stemming from or induced by the project. . . . Secondary benefits are not attributable to the project from a national public viewpoint unless it can be shown that there is an increase in net incomes in such activities as a result of the project compared with conditions to be expected *in the absence of the project.*‡

"Beneficial effects" other than net income increases presumably refer to quantifiable benefits which fail to be valued directly by any market process, e.g. water quality improvements, esthetic enhancement of the environment, etc. The operational difficulty with the definition is to be able to predict with meaningful accuracy the "with and without" development paths of the market related activities.

There appear to be four cases in which secondary benefits may be significant:

1. When there exist economies of larger scale production in the related industries and when such economies do not exist in any activities forced to contract.
2. When the expansion of the water-using activities serves to relieve a bottleneck critical to the expansion of other activities which would otherwise not be relieved.
3. When the expansion of related industries draws largely upon unemployed labor and industrial capacity.
4. When, in considering areas in which water using activities will have to contract unless replacement water is provided, capital goods and labor will be slow in moving to alternative occupations or have to move into distinctly less remunerative occupations.

In the first case, the cost saving on the prior (without project) level of output in project related activities plus any net benefits (roughly profits) on the increment of output would be counted as secondary

---

*Lyle T. Alexander has pointed out that there is usually some degradation of water quality in the return flows. In extreme cases, e.g., Yuma Mesa, the return flow may be unusable or even injurious.

†*Proposed Practices for Economic Analysis of River Basin Projects,* Report to the Inter-Agency Committee on Water Resources, prepared by the Subcommittee on Evaluation Standards, Revised May, 1958, p. 9.

‡Italics provided.

benefits, *provided* that the same increase in output would not have occurred in the absence of the water project. In the second case, the increase in the profits of related industries plus any measurable "consumers' surplus"* should be counted as secondary benefits, again *provided* that the bottleneck would not have been broken in other ways. In the third situation, any increase in related profits *plus* the incomes of labor or returns to capacity which otherwise would have *remained* unemployed should be counted. Finally, in the fourth case, the loss of incomes to those factors which are immobile and unable to move into other occupations should be counted *over the period of the immobility only*. This period will be quite short for certain kinds of capital goods, the remaining economic life for other capital goods, and at least the period needed for retraining, moving, and adjusting to new occupations for labor.

It must be admitted that much of the information required to determine the existence and magnitudes of secondary benefits is not readily available and that little research has been done in this difficult area. There exists a fairly strong practical argument for forgetting secondary benefits altogether, namely that the secondary *disbenefits* caused by the reductions in expenditures following from the financing of a project (e.g. through taxation of funds otherwise spent) are not counted in project costs and can be expected to be about the same as the forthcoming secondary benefits, unless there is evidence to the contrary. As a practical guide to widely dispersed public expenditure programs this argument seems quite valid, but for situations in which project benefits are highly concentrated while project costs are widely dispersed, it would seem worthwhile to look carefully at secondary benefits.

To get some grasp on possible magnitudes of secondary benefits, we now look at several case studies of *total benefits* attributable to water, i.e. direct plus secondary benefits.

## EMPIRICAL STUDIES OF TOTAL BENEFITS

Hartman and Seastone (1966) have analyzed the total income losses which might result from the transfer of water from agriculture to municipal and industrial losses in Northeastern Colorado. This transfer of water out of agriculture can be thought of as the obverse of providing new irrigation water to replace the water transferred. For the several-county area studied, the gross value of products produced per acre-foot of water applied was $27. To allow for the return flow in the area, this figure was doubled to $54.† The first cost of phasing out this

agricultural production would be the profits and land rents lost directly in agriculture (obversely, these would constitute the direct long-term benefits from preventing this water withdrawal), an amount of $12.90 per acre-foot.‡ If all of the industries in the area which provide inputs to agriculture experience reductions in output as implied by the input-output model of the area, and if these reductions are not offset by an expansion of other demands created by the new users of the water, then an additional loss of $1.60 per acre-foot in profits and rents in market-related industries would occur. Thus, if it is assumed that agricultural and related capital is completely immobile and that the land has no alternative use, but that labor generally can move into other occupations, the benefits from agricultural replacement water in the short and long run would be $14.50 per acre-foot.

The direct wage and salary income in agriculture related to the gross output of $54 was estimated to be $10.25, with indirect wage and salary payments totalling $6.95. If agricultural labor was totally immobile, $10.25 per acre-foot would have to be added to the benefits figure, with an additional $6.95 if the immobility extended to all input-related sectors. Clearly, such benefits can be counted *only over the period of labor immobility* – a period which needs to be estimated for each particular area. Thus, depending on the degree of labor and capital immobility, the total benefits from agricultural water as determined by Hartman and Seastone for Northeastern Colorado range from $12.90 to $31.70 per acre-foot. Depending upon the assumptions which are appropriate, the secondary benefits component ranges from zero to $18.80 per acre-foot.

Grubb (1966) has studied the incomes generated directly and indirectly by irrigation agriculture in the High Plains of Texas, an area dependent upon a falling ground water table. He found direct benefits per acre-foot of $18 at present and $27 after 1980 (due to assumed increases in efficiency). Because the High Plains area is highly specialized in agriculture and agricultural processing, there is substantial additional value-added (i.e. incomes generated) in the processing and marketing of agricultural output. There are also substantial incomes generated in the provision of consumer goods which would not be sold in the absence of irrigated agriculture. Grubb has found the total benefits per acre-foot to range from $81 at present to $119 after 1980. Again the secondary benefits would be attributable to replacement water only over the period of labor and capital immobility, but this might be a long period for so highly specialized an area.

Wollman (1962) has studied the incomes gen-

---

*Consumers' surplus refers to values to buyers in excess of what they have to pay in the market. It can be measured only if the demand curve for the commodity is known.

†The prior discussion would indicate that this return flow multiplier of 2 is likely to be an overstatement.

‡A multi-county input-output study was the basis of the analysis.

erated by water directly and by the "first round" of purchases in agriculture, recreation, and industry in the San Juan and Rio Grande Basins of New Mexico. He has found that the values per acre-foot range from $28 to $51 in agriculture, from $200 to $300 in recreational uses, and from $1300 to $3700 in municipal and industrial uses. It is clear that water shortages are *not* a bottleneck to the expansion of municipal and industrial uses, for groundwater remains plentiful and sales of agricultural water rights to municipalities and industry take place regularly. Thus, the value-added figures in industry are largely irrelevant to our present interests, since plenty of water is available for these high valued uses. It is agriculture which is constrained by water availability. While the methods used by Wollman may fail to capture some secondary benefits from agriculture, it is clear that total benefits per acre-foot in agriculture will not greatly exceed the $28 to $51 range stated.

The situation in Central Arizona has been analyzed by Young and Martin (1967) in terms of personal income generated per acre-foot of water intake in agriculture and other sectors. Personal income includes wages, salaries, rents, interest, and profits generated within each sector. Thus, if one uses personal income as a measure of benefits, one is assuming either complete immobility of all the related resources out of agriculture (in the case of potential contraction of agriculture) or a complete lack of alternative opportunities for available inputs (in the case of prospective expansion of agriculture). Both are rather extreme assumptions. Under these conditions, an estimate of the value of replacement water can be determined by adding full pumping costs per acre-foot to the personal income figures. Using an average pumping depth of 300 feet at 3.5 cents per foot ($10.50 per acre-foot), one arrives at $27.50 per acre-foot for food and feed grains, $29.50 for forage crops, and $103.50 for high value intensive crops. Since the low value crops are the ones which would be phased out eventually if replacement water isn't provided, the lower values are the relevant ones for comparing benefits to the costs of importation of water.

Howe and Easter (1968) have considered the potential magnitudes of losses stemming from a possible reduction in water availability for irrigation in Southern California. Basing the phase-out of agricultural acreage on the profitability of various crops at the farm level, the total impacts on outputs and incomes generated throughout Southern California were estimated for several sets of assumptions concerning the importance of "forward linkages" ("induced by" effects). Adjustments were also made for anticipated major price changes.

Under the most *extreme* assumptions of no capital or labor mobility and no substitution possibilities

of external inputs for those produced in Southern California, a total benefit figure of $317 per acre-foot was estimated. Under the more realistic assumption that substitutes would be available (making the forward linkages less important) the range of total benefits became $103 to $143 per acre-foot. This assumes a high degree of immobility of resources out of agriculture — probably a good assumption for the Imperial and Coachella Valleys which are completely specialized in agricultural pursuits.

Table 3 summarizes the ranges of total benefits per acre-foot found in these studies, but the reader must keep in mind the various sets of underlying assumptions mentioned above and elaborated in the original sources.

TABLE 3

### TOTAL BENEFITS PER ACRE-FOOT IN AGRICULTURE*

|  | Total Benefits |
| --- | --- |
| Northeastern Colorado (Hartman-Seastone) | $15- 32 |
| Imperial Valley (Howe-Easter) | 103-317 |
| Texas High Plains (Grubb) | 81-119 |
| New Mexico (Wollman) | 28- 51 |
| Arizona (Young and Martin) | 28-104 |

*The figures are based on widely differing conditions which must be understood for valid interpretation. The reader should consult the text and also the original sources before using these figures.

### SUMMARY OF EMPIRICAL STUDIES OF TOTAL BENEFITS

From Table 3 it can be seen that total benefits can range substantially above direct benefits. Especially in developed areas which are highly specialized in agriculture and related processing and which have scant prospects for generating other types of activity to replace agriculture, the degree of immobility of labor and capital can be high, the period of immobility potentially long, and the resultant benefits from replacement water high. Thus figures in excess of $100 per acre-foot for the Texas High Plains and the Imperial Valley of California could obtain for "fairly long" periods. In areas like Central Arizona, New Mexico, or Northeastern Colorado where alternative activities are rapidly growing, the lower values from Table 3 are probably more realistic.

Before figures such as those in Table 3 can be compared with costs of importing water or other sources of supply, the extent of potential forward linkages from agriculture to other sectors must be carefully studied for new agricultural areas, and the detailed structure of potential capital and labor immobility must be determined for established areas in which the demand for water is a replacement demand. To illustrate the importance of the period of immobility in the latter case, let us consider a hypothetical situation using the Texas High

Plains figures. It is assumed for illustrative purposes that complete capital and labor immobility in agriculture and related industries obtains from the time water is withdrawn or exhausted until year $T$, at which time all those inputs except those which would remain engaged directly and indirectly in dry-farming would be able to move into new activities. In other words, benefits from replacement water will be taken to be $119 per acre-foot from year zero to year $T$ and then $27 per acre-foot from year $T$ until year 50 (the assumed life of a water transfer project). The question is, "For different values of $T$, what would the average annual value* per acre-foot of water be over the 50 year life of a water import project?" The figures in Table 4 were computed using a 5 per cent rate of interest. Table 4 makes it extremely clear that the period of immobility is a crucial factor which must be studied carefully for each potential rescue operation.

### TABLE 4

### AVERAGE ANNUAL TOTAL BENEFITS PER ACRE-FOOT AS A FUNCTION OF THE PERIOD OF CAPITAL AND LABOR IMMOBILITY†
(50 Year Project Life, 5 Per Cent Interest)

| Period of Immobility | | | |
|---|---|---|---|
| $T = 2$ | $T = 5$ | $T = 10$ | $T = 20$ |
| $36 | $49 | $66 | $90 |

†For the period of immobility, the computations used the present value of an annuity for $T$ years, the annual annuity being the higher value of $119. For the period beyond, the value of an annuity for (50-$T$) years was used with the lower value of $27. The latter value was then discounted $T$ years to time zero and added to the former. This present value total was then transformed to an equivalent annual value for 50 years.

## WHAT ARE THE COSTS OF LARGE-SCALE INTERBASIN TRANSFERS AND THEIR POTENTIAL ALTERNATIVES?

The first objective of this paper has been accomplished: to set forth the benefits likely to be forthcoming over the short and long terms from water importations. The second objective is to compare the costs of interbasin transfers of water to the potential benefits and to the costs of alternative sources of water. Since Mr. Young will be considering the costs of some of the important alternatives in another paper, this section will be restricted to estimates of the cost of several proposed large-scale transfers.

The California State Water Project is planned to deliver 4.23 million acre-feet per year from Northern California to the San Joaquin Valley and Southern Coastal area. The total costs of this large diversion may vary from approximately $18 per acre-foot for deliveries in the Northern San Joaquin

Valley to about $65 for deliveries to the Metropolitan Water District.†

The Pacific Southwest Water Plan (which is now outdated) proposed transfers from Northern California of 1.2 million acre-feet to Southern California and 1.2 to Lake Havasu on the Colorado River. The average total cost implied by the figures in the report came to about $44 per acre-foot, but substitution of a 5 per cent discount rate and a 50 year life increase this figure to $90.

Nelson's proposal (1963) for diverting 2.4 million acre-feet (maf) from the Snake River into the Colorado at Lake Havasu contained an estimated cost of $32 per acre-foot. The use of a 5 per cent interest rate and a 50 year life increases this to $46.

Dunn (1965) has proposed a modified Snake-Colorado Project to divert initially 5 maf per year from the Columbia into Lake Mead. The estimated delivered cost at Lake Mead was $38, but a 5 per cent interest rate and 50 year life increase this to $54.

The Sierra-Cascade Project proposed by E. Frank Miller (ASCE, 1967) would divert initially 7.5 maf per year from below Bonneville on the Columbia to Lake Mead, later being extended to 30 maf. According to the designer of this project, the average costs for the first increment could be as low as $9 per acre-foot. Average costs would rise to $20 when the project had been increased to 30 maf. Cost details are not available on this project to permit a closer analysis nor to adjust to uniform interest rate and lifetime.

Pirkey (1963) has proposed a Western Water Project diverting 15 maf from the lower Columbia at the Dalles to Lake Mead and along the lines of Owens Valley to Southern California. The present author's best estimate of costs per acre-foot is $58, based on 5 per cent interest and a 50 year life.

Three *caveats* must be noted about these preliminary cost estimates: (1) they are not based on detailed engineering studies; (2) all of the Columbia-Colorado plans include a large amount of power recovery which substantially reduces the ultimate cost (such advantages would not accrue to plans in which the ultimate destination is the highest point en route, e.g. the Texas High Plains); (3) except for plans which will provide water to replace existing *surface* sources, there will be substantial additional costs in delivering the water to the customer from the end point of the transfer facility.

It thus appears that preliminary estimates of transfer costs from the Columbia to the Colorado would, after adjustment for contemporary discount

*Computed on an equivalent present worth basis. See footnote to Table 4.

†For cost figures on components of the system, see the Department of Water Resources Bulletin No. 132-65, State of California. Also Gardner M. Brown, Jr. (1964), Appendix Table 3A, p. 204 and Table 4.4, p. 68.

rates and an economic lifetime of 50 years, run in the $50-60 range at present cost levels. Diversions of similar length from the Mississippi would not benefit from possibilities of power recovery and would probably cost as least half again as much: $75 to $90 per acre-foot.* The other much larger proposals such as NAWAPA are far too speculative to warrant hazarding any guesses about costs.

## CONCLUSION

The estimates of costs and benefits assembled above suggest that large-scale transfers are at best marginal for all areas at present, but that their use in providing water for already established, highly specialized agricultural areas which will be exhausting their water supplies is an area of potential application which warrants more careful study.† If some projects are found, upon further study, to exhibit economic feasibility (benefits exceeding costs per acre-foot), it must be remembered that there is an optimal *time* for construction which in general occurs after a project is *first* found to be feasible. Premature construction can be needlessly costly.

The estimates of direct benefits indicate that the direct users of water will be able to pay only a portion of the total cost of imported water, a small portion in most cases. Thus, taxes on the communities to which the secondary benefits accrue or on the general public would have to cover a large part of the bill.

Any regional plan for water development must naturally consider a wide range of alternatives to large-scale transfers. Howe and Easter (1968) indicate that a surprising range of possibilities exists, some with demonstrably attractive costs. Some very relevant alternatives lie quite outside the traditional water resources area in terms of program content. One such alternative is suggested by the potentially large difference between the physical life of transfer projects (50 years or more) and the period of immobility of the capital and labor they are supposed to rescue (perhaps 10 to 20 years). This alternative would be to utilize public resources to facilitate capital and particularly labor mobility through retraining programs and partial payment of moving expenses. The consideration of such alternatives will clearly require a new institutional

and agency orientation. Here lie some of our greatest challenges.

## BIBLIOGRAPHY

AMERICAN SOCIETY OF CIVIL ENGINEERS, Los Angeles Section, *Water for the West, Summary of Regional Water Plans*, October, 1967.

ANDERSON, RAYMOND L., "The Irrigation Water Rental Market: A Case Study," *Agricultural Economics Research*, U. S. Department of Agriculture, Vol. XIII, No. 2, April, 1961.

BITTINGER, M. W., "The Problem of Integrating Ground-Water and Surface Water Use," *Ground Water*, Vol. 2, No. 3, July, 1964.

BROWN, GARDNER, M., JR., "Distribution of Benefits and Costs: A Case Study of the San Joaquin Valley-Southern California Aqueduct System," Ph.D. Thesis, Berkeley, June, 1964.

—— and C. B. McGuire, "A Socially Optimum Pricing Policy for a Public Water Agency," *Water Resources Research*, Vol. 3, No. 1, 1967.

DUNN, WILLIAM G., "Statement on Modified Snake-Colorado Project (Dunn Plan)," a statement prepared for presentation before the Subcommittee on Irrigation and Reclamation, Committee on Interior and Insular Affairs, U. S. House of Representatives, mimeo., dated October, 1965.

GRUBB, HERBERT W., *Importance of Irrigation Water to the Economy of the Texas High Plains*, Texas Water Development Board Report 11, January, 1966.

HARTMAN, L. M., and R. L. ANDERSON, "Estimating the Value of Irrigation Water from Farm Sales Data in Northeastern Colorado," *Journal of Farm Economics*, Vol. XLIV, No. 1, February 1962.

—— and D. A. SEASTONE, "Efficiency Criteria for Market Transfers of Water," *Water Resources Research*, Vol. 1, No. 2, 1965.

————, "Regional Economic Interdependencies and Water Use," in (eds.) Allen V. Kneese and Stephen C. Smith, *Water Research*, Johns Hopkins Press, 1966.

HOWE, CHARLES W., and K. WILLIAM EASTER, *Interbasin Transfers of Water: Economic Issues and Impacts* (in manuscript, Resources for the Future, Inc., November 1968).

JOHNSON, RICHARD L., *An Investigation of Methods for Estimating Marginal Values of Irrigation Water*, Master's Thesis, Utah State University, Logan, 1966.

NELSON, SAMUEL B., *Snake-Colorado Project: A Plan to Transport Surplus Columbia River Basin Water to the Arid Pacific Southwest*, Department of Water and Power, City of Los Angeles, October, 1963.

PIRKEY, F. Z., "Water for All," mimeo. with enclosed diagrams and maps, private distribution, 1963.

STULTS, HAROLD M., "Predicting Farmer Response to a Falling Water Table: An Arizona Case Study," in *Water Resources and Economic Development of the West*, Report No. 15, Committee on the Economics of Water Resources Development of the Western Agricultural Economics Research Council, Las Vegas, December 1966.

TAYLOR, GARY C., *Economic Planning of Water Supply with Particular Reference to Water Conveyance*, Ph.D. thesis, University of California, Berkeley, 1964.

WOLLMAN, NATHANIEL et al., *The Value of Water in Alternative Uses with Special Application to Water Use*

---

*A major unsolved problem in some proposed transfer systems is provision of sufficient storage to permit the costly pumps and aqueducts to operate year 'round while servicing seasonal irrigation demands. Evaporation losses after delivery at the project terminal point are not taken into account in the above costs.

†The critical importance of secondary benefits in the total must be kept in mind. In the long term, most of the secondary benefits will wash out. The period constituting the "long term" is thus crucial, as has been indicated by Table 4.

*in the San Juan and Rio Grande Basins of New Mexico,* University of New Mexico Press, Albuquerque, 1962.

YOUNG, ROBERT A., and WILLIAM E. MARTIN, "The Economics of Arizona's Water Problem," *Arizona Review,* Vol. 16, No. 3, March, 1967.

GARDNER, B. DELWORTH and HERBERT H. FULLERTON, "Transfer Restrictions and Misallocation of Irriga-

tion Water," *American Journal of Agricultural Economics,* Vol. 50, No. 3, August, 1968, pp. 556-571.

GOSS, JAMES W., and ROBERT A. YOUNG, "Organization and Pricing Policy of Major Water-Distributing Organizations in Central Arizona," File Report 67-5, Department of Agricultural Economics, University of Arizona, July, 1967 (dittoed).

# IMPORT ALTERNATIVES*

## Gale Young

Oak Ridge National Laboratory
Oak Ridge, Tennessee, U. S. A.

I at first thought of calling this talk "Alice in Waterland" or "Through the Drinking Glass." Fortunately, in view of the austere nature of the audience that I see before me, this frivolous idea was quickly abandoned. It did, however, leave me with an opening quotation as follows:

"The rule is," said the White Queen, "jam tomorrow and jam yesterday—but never jam *today.*"

"It *must* come sometimes to 'jam today'," Alice objected.

"No, it can't," said the Queen. "It's jam every *other* day: today isn't any other day, you know." (*1*)

Once upon a time there was plenty of good water and sometime this may be so again, what with all the fine plans that folks have—it is just that we are a mite short today, like in Arizona and California and Texas and Lake Erie and New York and halfway around the world.

### THE RAINS CAME

And God said, Let the waters under the heaven be gathered together unto one place, and let the dry land appear; and it was so. And God called the dry land Earth; and the gathering together of the waters called he Seas; and God saw that it was good. . . . And the evening and the morning were the third day. (*2*)

The world, it is said, was once very hot—too hot for water or life. Then slowly it cooled off, and its surface began to freeze to form a crust of land; and the water which was dissolved in the molten magma came out as the rock crystalized, to make the oceans. This was perhaps some four billion years ago. And the rains came and the waters moved upon the face of the earth.

And then after a while—say another billion years or so—there was life in the waters. That life has since changed a lot but still stays close to water. "A population map of the world would follow seashore and river course so closely in its darker shading as to give an impression that 'homo sapiens' was an amphibious animal" (*3*).

*Research sponsored by the U. S. Atomic Energy Commission under contract with the Union Carbide Corporation.

Beginning with the blue-green algae some three billion years ago, all life has sprung from the waters and is still very dependent on them. Man is no exception: the Euphrates watered the Garden of Eden, and the canals of Chaldea are older than history. The vision of "green pastures . . . beside the still waters" is almost a racial memory.

The waters which move upon the face of the earth are older even than life and have been at work even longer. Mountains are lifted up by shiftings of the earth's crust and are then worn down again and carried away by the ceaseless waters. The history of the land has been largely written by water. As proclaimed long ago by the prophet Isaiah, and heard again in our time against the majesty of Handel's musical background in the Messiah, "Every valley shall be exalted, and every mountain and hill shall be made low." The Ganges will someday conquer Everest.

Those rains which began so long ago—and were probably of unimaginable torrential violence for ages—have never ceased to fall. Their rate in our time is shown in Table 1 (*4*). You see that on the whole there is on the land an excess of rainfall over evaporation, so that there is a continual runoff of fresh water into the seas. Half of this runoff takes place in 68 major river systems; one-fifth in the Amazon alone.

TABLE 1

**WORLD WATER CYCLE**
$10^9$ **gallons per day**

| | |
|---|---:|
| Rainfall on land | 72,000 |
| Rainfall on sea | 268,000 |
| Total rainfall | 340,000 |
| Evaporation from land | 48,000 |
| Evaporation from sea | 292,000 |
| Total evaporation | 340,000 |
| Runoff from land to sea | 24,000 |
| U. S. rainfall | 4,300 |
| U. S. runoff | 1,100 |

The world is, in effect, an enormous desalination plant, continually recycling the water it possesses. "All the rivers run into the sea; yet the sea is not full; unto the place from which the rivers come, thither they return again" (5).

Table 2 lists some of the famous rivers of the world and their average flow rates (6-9).

The Jordan is a small river. Its flow is less than that into New York City. There is a spring in Florida (Silver Springs) that gushes over half as much. On the subject of the Jordan, the normally staid Encyclopaedia Britannica waxes almost lyrical: "On the whole it is an unpleasant foul stream running between poisonous banks . . . the Hebrew poets did not sing its praises." This is the river that wars have been fought over.

The Clinch is the local stream that flows past Oak Ridge, to disappear shortly into the Tennessee River. It is a neat enough little bundle of branch water, but not particularly impressive as a river per se. Folks take a second look at it, however, when you tell them it is over three times the size of the famous Jordan which they have been hearing and reading about all their lives. And they become downright fascinated when you add that if they had it in the right place and could sell it to cities for ten cents per thousand gallons, it would be worth over 100 million dollars per year. Actually, the Clinch carries nearly enough water for our three largest cities.

The Hudson is currently described as "an open running sewer" flowing beneath the windows of water-short New York City.

The Colorado has long been a bone of contention

TABLE 2

**AVERAGE RIVER FLOWS**
**10⁹ gallons per day**

| | |
|---|---|
| Jordan | 0.9 |
| Clinch | 2.9 |
| Delaware | 12 |
| Hudson | 14 |
| Tigris | 14 |
| Colorado | 15 |
| Euphrates | 18 |
| Rhine | 50 |
| Nile | 65 |
| Hwang Ho (Yellow) | 75 |
| Yukon | 120 |
| Indus | 130 |
| Niger | 139 |
| Danube | 141 |
| Columbia | 170 |
| Mekong | 252 |
| Ob | 290 |
| Mississippi | 400 |
| Yenisei | 400 |
| Ganges | 426 |
| Brahmaputra | 452 |
| Yangtze | 497 |
| Congo | 900 |
| Amazon | 4800 |

among the states of the dry Southwest and Mexico. As I understand it, one problem is that various compacts and treaties have legally awarded more water to various people than there is in the river to start with (10).

If you were to look at a population map of Africa you would see a sparsely settled continent—roughly half desert and half jungle. There is, however, because of a sort of accident, a little spot of high population density in the upper right-hand corner. The accident is the River Nile which brings water from thousands of miles to the south to irrigate that little patch of desert which men call Egypt. This little spot feeds a tenth of all the people in Africa.

The total amount of water in the seas is such that it would take about 3,500 years to evaporate it, at the rate shown in Table 1. If we could go back this far in history, to a time such that the oceans have since evaporated just once, we would see the conquering Aryans pushing eastward with their war chariots to come at last to the bank of the mightiest river they had ever seen. Being somewhat economical in language (their name for war merely expressed a desire for "more cows"), or perhaps a bit speechless before this stream, they did not give it a name but called it merely by their word for river.

Oh, mighty Indus, "river" is thy name—and the subcontinent below that men call India, "land of the river." And then in a linguistic chain reaching across the centuries and around the earth we have the Indian Ocean, the fabulous spice islands of the Indies, Columbus looking for these but finding our continent instead and calling the natives Indians, and so on right down to the last 500-mile race at Indianapolis. From India to Indiana has been 3,000 years and 10,000 miles, but the river without a name has written its name around the earth. "And so the currents of that far-off river of Hindustan, coursing through the obscure channels of nomenclature, have girdled the world" (11).

The Amazon is incomparably the mightiest river in the world—its flow exceeds all the rainfall on the entire United States. So powerful is its current that it turns the sea fresh for some 200 miles out. Ships sailing there, out of drinking water and out of sight of land, have been saved by this miracle of finding fresh water in the ocean. Though three orders of magnitude smaller, the Jordan, because of its religious significance and the critical water supply situation in the near East, is said to be of greater significance to more people than the Amazon.

## MOVING THE WATERS

The water from rainfall is not evenly distributed. Figure 1 shows the resulting distribution of arid areas (12). Please note, for later reference, that northern India is not shown as an arid area. Dry climates are the most extensively developed over the land surface of the earth of any of the great climatic groups. Dry, barren, and practically lifeless deserts and semideserts cover about a third of all the land. We will return later to these "vast rainless belts of land that lay across the continental masses, from Gobi to Sahara and along the backbone of America, with their perfect air, their daily baths of blazing sunshine, their nights of cool serenity and glowing stars, and . . . yet only desolations of fear and death to the common imagination" (3).

Since the natural water supply is not evenly distributed, and even if it were, people would still want it some other way, there are always projects for moving water somewhere else. Figure 2 shows one of the great classical examples. In the heyday of the Roman empire there flowed into the Imperial City a half-billion gallons of water per day via many aqueducts, of which the largest brought 90 million gallons per day from a distance of 150 miles (13). The per capita consumption was high (450 gallons per day), presumably due to the popularity of Roman baths, but is exceeded in our day by the city of Beverly Hills with its swimming pools.

In Russia the Ob and Yenisei Rivers daily put several hundred billion gallons of fresh water uselessly into the Arctic Ocean, and a plan exists for intercepting and using some of this water (4). This Russian plan would create a lake nearly the size of Italy, and would irrigate over $100 \times 10^6$ acres of croplands and pastures.

A similar situation exists in Alaska, and a corresponding plan also exists (14, 15). This Alaska plan would also create a lake (in two countries) and would irrigate $50 \times 10^6$ crop acres.

Now going ahead in California is a project for bringing water south from the north end of the state. Beginning to receive serious study is a larger plan for conveying water from the huge Columbia River to the several drier and highly populated Southwest states. The Alaska plan mentioned earlier is the largest in this ascending sequence of schemes.

The economies of scale in conveyance by pipe line are noteworthy. The unit cost of transporting water (horizontally) varies inversely as the 0.4 power of the total flow rate capacity (16).

It is not yet known what kind of plan may develop in response to the recent drought situation in the northeast. In New York, water has only recently become a household topic; in Los Angeles it has long been a religion. In July of 1965 New York City was ordered to reduce its consumption of Delaware River water by $150 \times 10^6$ gallons per day, and to start releasing nearly $200 \times 10^6$ gallons per day of already impounded water from its dams in the western Catskills in order to help stave off salt contamination of the Philadelphia water supply. It is reported that the New York City Water Commissioner "made angry remarks" at the meeting.

Meanwhile the Hudson flows serenely by, emptying alongside into the sea enough water for a dozen New York Cities. But the water is by then opaque and poisonous from uncontrolled upstream dumping of raw sewage and industrial wastes by a small fraction of this number of people. A typical idyllic upriver scene shows sewage pouring into the Hudson in the shadow of the stately Tappan Zee Bridge. It is things like this which have led to a current description of Uncle Sam as a man worried about food and water, standing knee deep in sewage, and shooting rockets at the moon.

Another classical mode of water conveyance is the pumping of underground water up to the surface. Enshrined in story and fable as the farmyard pump, this process has in some places become a highly developed art. Thus Phoenix is lowering its water table several feet per year, and the foundations of Mexico City—one of the largest cities of the world—are sinking as water is withdrawn from beneath them (17).

## DRY LANDS AROUND THE WORLD

A third of the world's land is dry and virtually unoccupied, while half of the world's people are jammed—impoverished and undernourished—into a tenth of the land area. It may be noted that the fastest growing region in our country is the Southwest desert into which Americans move happily, taking their water and power requirements with them. This suggests that the warm dry areas of the world may be more appealing for human occupancy than the rain forests or the frozen tundras as earth's population soars, if the water and power needs can be met. It may also be that nuclear energy—not tied by any umbilical cord to coal or petroleum deposits—will play a significant role in opening up these arid areas for human living space. Since much arid land lies relatively near the sea, and the aggregate length of the coastal deserts would stretch nearly around the world, desalination is a fresh water source of broad potential applicability.

I know of no better desert overlook than the magnificent little poem by Shelley (18).

I met a traveler from an antique land
Who said: Two vast and trunkless legs of stone
Stand in the desert. Near them on the sand,
Half sunk, a shatter'd visage lies, whose frown
And wrinkled lip and sneer of cold command
Tell that its sculptor well those passions read
Which yet survive, stamp'd on these lifeless things,
The hand that mock'd them and the heart that fed;
And on the pedestal these words appear:
'My name is Ozymandias, king of kings:
Look on my works, ye Mighty, and despair!'

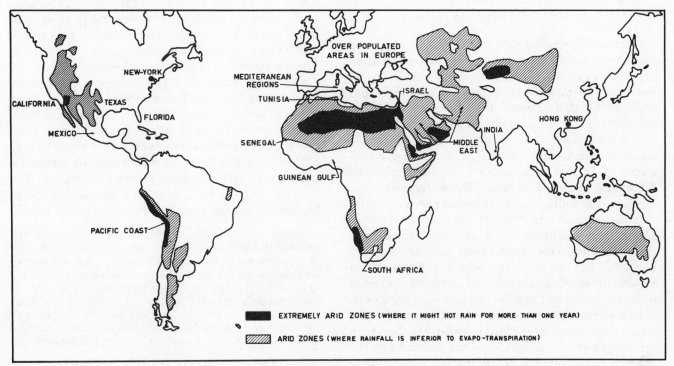

Fig. 1. Arid zones of the world.

Fig. 2. Roman aqueducts. Painting by Zeno Diemer in the German Museum, Munich.

Nothing beside remains. Round the decay
Of that colossal wreck, boundless and bare,
The lone and level sands stretch far away.

In speaking of areas such as the Northern Sahara, "deserted" would perhaps be a better word than "desert." For "some 15 civilizations and cultures were able to produce their food in the sands that drift about their ruined monuments," including the Romans "whose engineering genius made North Africa the richest granary in their vast empire. . . . Archaeological study [shows] that the real monuments of ancient greatness are not the tombs and triumphal arches, but rather the waterworks, the terraces, the wells, and the irrigation systems" (19).

"One could say that of all the aspects of civilization that the Romans left us, their agricultural theory and practice are the most valuable and important," but meanwhile "tens of thousands of square miles of cultivated land, painfully wrested from the desert by the patience, skill, and hard work of men almost 2,000 years ago when the actual population need for space and food were far less than they are today have been abandoned and lost through the characteristic process of destroying one culture without replacing it with another. . . . The climate

in Roman times was no different from what it is today, so that the fertility of the land had nothing to do with meteorology, but was the prize of hard work" (20). It was the Third Augusta Legion — 12,000 men for two and a half centuries — that built Rome's agricultural empire in the Sahara.

Perhaps nowhere in all the vast sector of the world once washed by the Roman tide is there a better place in which to meditate upon the fall of empire than the Sahara ghost city of Leptis Magna. Here dwelt many thousands of people with every luxury of their age. The city still stands as if its occupants had just stepped outside, to return in a moment. The theatre (Fig. 3) holding 40,000 people "seems to be awaiting the next production of Euripides."

As the visitor walks down the Via Triomphale and through the arch of Septimius, he has the feeling that he is in a city not of a dead civilization . . . the warehouses line the quays, the customhouse is open for business, the great markets are strategically placed . . . the sky so madonna-blue, the air so clear . . . the vast buildings sparkle with life, and the long streets with their latticed fountains and marble benches lead the visitor enchanted to the sea through a series of triumphal arches. (20)

Yes, the city still waits — as it has waited for so long — in vain. For its people are really gone, and the

Fig. 3. Theatre at Leptis Magna (*The Great Sahara* by James Wellard. Reproduced by permission of E. P. Dutton & Co.).

myriads of farm acres behind it—both to the dust returned. The channel that once linked its harbor to the sea is covered over by the sifting sand, and the ships for Europe laden with grain and olive oil will never sail again.

TABLE 3

**AGRICULTURAL USE OF LAND**
$10^9$ acres

| | | |
|---|---|---:|
| Grain | | 1.6 |
| Other major crops | | 0.7 |
| Minor crop use | | 1.1 |
| Permanent pastures and meadows | | 6.4 |
| Irrigated | | 0.3 |
| Arid | | 12.1 |
|   Within 500 miles of sea | | 8.0 |
|   Within 300 miles of sea | | 5.2 |
|     Africa | 1.8 | |
|     Asia | 1.5 | |
|     Australia | 0.9 | |
|     South America | 0.6 | |
|     North America | 0.4 | |

We saw earlier (Fig. 1) that there is much arid land in the world. This is repeated in more quantitative manner in Table 3 (21), along with the present world agricultural use of land (22). Note that the arid area substantially exceeds the total area now used for crops. Most human food comes directly or indirectly from grain, and there is over three times as much arid land within 300 miles of the sea as the whole world now employs in grain production.

The desert is man's future land bank. Fortunately, it is a large one, offering eight million square miles of space for human occupation. It is also fortunate that it is a wondrously rich bank, which may turn green when man someday taps distilled seawater for irrigation. Bridging the gap from sea to desert will be greatly facilitated by the geographical nearness of most of the world's deserts to the oceans. When this occurs it will surely be one of the greatest transformations made by man in his persistent and successful role in changing the face of the planet. (23)

## GANGETIC INTERLUDE

It was noted earlier that northern India did not show up on the map as an arid region. Figure 4 is a picture taken a year ago in the Ganges Valley near New Delhi. Because of the seasonal nature of the monsoon rain, the area is flooded once a year and is a desert the rest of the time. Thus a hundred million acres among the best soil on earth, with a year-round growing season, shielded from cold air masses from the north by the towering Himalayan mountains, is one of the great famine areas of the world.

The buckling in the earth's crust that lifted up the mountains also created a giant trough or foredeep in front of them to the south. So recent is this event that the Himalayas have risen appreciably since man has been on earth, and they are still rising slightly even today. This trough has been filled with river-borne soil from Central Asia, to form one of the greatest stretches of flat alluvium in the world.

Potentially the Ganges Valley is one of the finest agricultural regions on earth, like a giant Imperial Valley; yet the people are starving, caught in the Malthusian famine stranglehold. There is good ground water underfoot, and the real irony of the present suffering is that food is short because of the lack of power to lift this water to the parched surface. Bullocks turning Persian wheels or men running up and down well poles (Fig. 5) cannot suffice for a large-scale expansion in food production. They do not much more than feed themselves, and they represent an expensive energy input obtained from consumption of an expensive fuel, namely food. The energy needed to fix the nitrogen required by the crops is several times larger than that for pumping water, and even farther beyond the reach of muscle power.

The need for energy input from nonagricultural sources has been vividly pointed out by Perry Stout (24) and Dean Peterson (25). In well-managed experimental areas with improved crop strains, fully irrigated and fertilized, it has been found possible to go as high as four crops per year (24) and a food output more than ten times the normal indigenous level (24, 26). A 1000-Mwe (megawatts electric) central station with its power output devoted to running tube wells and producing ammonia via electrolysis of water could lift the entire population of the Ganges Valley (which equals the population of the United States) from a semistarvation condition to a healthful diet (27, 53).

If the water outflow (Table 2) at the mouth of the Ganges (fed partly from the monsoon rain and partly from melting snow in the mountains) could be spread evenly over the hundred million arable acres in its valley, it would amount to nearly 5 feet of water per year.

## A PIPE TO THE SEA

Cities in the United States typically use water at the rate of about 150 gallons/day per person. This includes use in homes and industry. It does not include rivers used to carry garbage and wastes from the city. As illustrated by the Hudson River, it appears that in some cases more water is in some sense used up or destroyed by throwing waste into it than by all other uses combined. I should like to call to your attention the words of a well-known writer, Wallace Stegner, who referred in a recent article to man being like a colony of bacteria planted in a dish of nutrient agar. When it reaches the edge of the dish it must either die of starvation or strangle on its own wastes. We are getting ever nearer the

Fig. 4. Caravan resting.

edge of the dish. We are coming closer and closer to strangling on our own wastes.

Pollution of streams and lakes has indeed become a major problem (28) but it may be that some brakes are about to be applied to it as a result of the new Federal Pollution Control Administration. Conceivably, for example, some cities and industries may have to compact their wastes and haul them away by other means for disposal. Which reminds me that I have been saving a delightful statistic which maybe I can get in here—namely, a projection by the U. S. Public Health Service shows the United States garbage production rate by 1980 to be sufficient (according to an authoritative Eastern newspaper) to cover the Los Angeles metropolitan district to a depth of three feet annually.

I don't have a material flow balance sheet for a big city, but there are lots of trucks and trains carrying things in and some going away empty that might instead carry away treated solid waste. Carry it to where? Maybe to land fill or gorges or empty mines or to refuse heaps or spreads somewhere else "far enough away." Which leads to questions such as "how far" and "far away from whom" and "what is the cost?" Or maybe we would eventually carry the wastes to the sea for disposal, which sort of brings us full circle because that is mainly what we do now.

There is a great transportation system that carries away the rainfall runoff. From practically any nonarid place in the land there is a nearby stream that winds eventually to the sea. We locate our cities either at the seashore or along the interior rivers. Thus one way or another the cities all have a most important thing in common—a pipe to the sea. Into this pipe we throw our garbage and sewage

Fig. 5. Human-powered irrigation well (Photographed by Stephen J. Mech).

and industrial wastes of all kinds, to be carried away. Our biggest use of water is for transport of wastes, and our rivers are mostly sewers—the lowest cost outward bound pipelines available. This is a fine system, but trouble comes because some people would also like to use these river-sewers for fresh water sources or to swim in and other odd purposes, and they object to the garbage and eels and what not. Thus the Hudson is everybody's sewer, while New York City reaches out farther and farther into the countryside to construct rain catchment basins for fresh water supply. As with the aqueducts of Rome, the life of NYC depends upon the water pipes. Chicago draws its water supply from Lake Michigan (which is getting more and more polluted from industrial plants south of Chicago) and flushes its wastes down the Illinois Waterway to the Mississippi. Los Angeles brings water overland from the somewhat salty Colorado River, and discharges its wastes directly into the Pacific.

If it is only a matter of water supply, it is cheaper for cities sharing a large lake or river to treat their intake rather than their discharge (29). Thus the additional cost for advanced waste discharge treatment would in such cases have to be charged off against the esthetic and other improvements in the waterways themselves.

## WASTE WATER RECYCLE

When the river sewers become too contaminated to serve as a fresh water supply, or when an area's water need outgrows the fresh supply available, then somebody has to do some cleaning up of something or else haul in fresh water from farther away. The latter is the principal theme of this symposium, while I guess I am of the nature of a minority report. A conventional water flow sheet for a city is shown in Figure 6. The capital letters refer to flow rates and the small letters at the right to salt concentrations. In this simple case $W = N$ and $k = n$. Full primary and secondary treatment for all the waste water is rather better than most cities actually

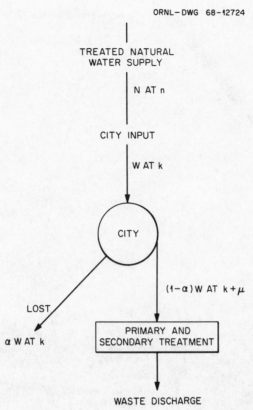

Fig. 6. Diagram of conventional city water system.

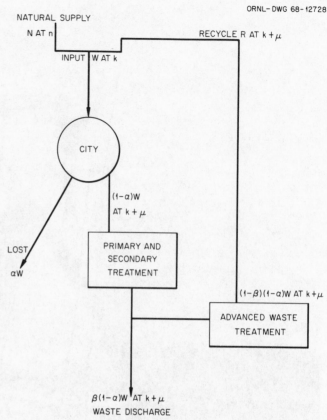

Fig. 7. Diagram of city water system with recycle.

do today, but is in the direction in which things are moving.

Some important hydrological aspects of a city are seen here. First, a fraction $\alpha$ of the water taken in gets lost somewhere somehow, such as in leaks and evaporation from lawns or gardens. Second, the salt content of the emerging water is increased by an amount $\mu$. Typical values are $\alpha = \frac{1}{3}$ and $\mu = 250$ parts per million (30). If the input water contains some salt ($k > 0$), that in the lost fraction $\alpha$ gets deposited somewhere around or under the city; i.e., the salt may be building up under the lawns, in the groundwater, or somewhere else. Just where it goes is a good question, and should not be forgotten about, but we will not pursue it further today.

If the city needs to increase its water supply rate $W$, we may consider adding tertiary or advanced waste treatment (30, 31) and shifting to a recycle system, as indicated in Figure 7. Limits on the amount of recycle water are

$$\frac{R}{W} \leq 1 - \alpha, \tag{1}$$

set by the water balance; and

$$\frac{R}{W} \leq \frac{k_0 - n}{k_0 - n + \mu} \tag{2}$$

set by the salt balance. In (2) $k_0$ is taken to be the recommended maximum value of $k$ for municipal

water supply; Public Health Service standards set this at 500 ppm.

With $\alpha = \frac{1}{3}$, equation (1) would permit $\frac{R}{W} = \frac{2}{3}$ or $W/N = 3$; i.e., recycle would enable a city to multiply its natural water supply threefold. If the natural water supply is free of salt, as in NYC, equation (2) permits $\frac{R}{W} = \frac{500}{500 + 250} = \frac{2}{3}$ which is the same as the value allowed by (1). But for Los Angeles with $n = 350$, (2) permits only $\frac{R}{W} = \frac{3}{8}$ or $\frac{W}{N} = 1.6$.

The cost of the advanced waste treatment in Figure 7 has been estimated (30) at 8¢ per 1,000 gallon in very large treatment plants. As the city grows and increases its recycle flow, it may become necessary in some cases to also enlarge the primary and secondary treatment facilities. This would add another 4¢ per 1,000 gallon to the cost of additional recycle water beyond that point.

## DESALINATION

The Bible speaks of sweetening the waters of Marah, which were too bitter to drink (32). I do not know if this is a reference to desalination or not. If so, the process used (casting a tree into the waters) does not seem to be one that is accessible today.

The process most in use at present is based on a technology indigenous to Tennessee, as shown in

Fig. 8. Mountaineer evaporator plant.

Figure 8. This is how Oak Ridge came to be in the act.

This process — evaporation — is also the one used by nature in the earth's water cycle described earlier. Aristotle taught his students "Vapor produced from seawater, when condensed, was no longer salty" and, as early as 1791, Thomas Jefferson issued to the United States merchant seamen instructions on how to build and operate a seawater distillation device (33).

Ships and other small units around the world have been employing evaporators for seawater desalination for a long time. Large evaporators have been under study for several years, with the emphasis on multistage flash plants using plain smooth tubes, for which the most experience and development work is available at present. It will be quite some while before large-scale city recycle comes into use, and it is expected that evaporator technology will have advanced appreciably in the meantime. A recent advanced design (34, 51) embodies a combined multistage flash and vertical tube plant, using doubly-fluted or doubly-ribbed tubes for heat transfer enhancement (35, 36).

For low salt concentrations, membrane processes are of interest. Electrodialysis units are available commercially, while hyperfiltration has reached the pilot plant stage. Particularly interesting — at least scientifically — are hyperfiltration membranes formed in place on porous surfaces by deposition of materials from a solution (37). Very high product outputs per unit area can be obtained — so high that turbulence promotion is desirable to combat concentration polarization at the surface. Many examples of significantly salt-rejecting membranes formed on various porous bodies have been observed, with product flow rates up to several hundred gallons per day per square foot. It is beginning to look as if high flux salt rejecting barriers may have existed all around us, rendered invisible by concentration polarization.

Just as in the New York City example considered previously, salt-free water supplied to a city from a seawater evaporator can be tripled by recycle. Thus, if the desalted water supply costs $25¢/10^3$ gallon, and the recycling costs $12¢/10^3$ gallon, the average cost of the water used by the city is $16.3¢/10^3$ gallon.

It is also possible to incorporate desalting into the city cycle itself, as illustrated in Figure 9, with a combination of evaporator and membrane processes (38). This, in effect, allows threefold multiplication of water supplied to the city at $n = 500$ ppm salt content. The estimated cost of this advanced waste treatment with partial salt removal is $16¢/10^3$ gallon.*

*The system pictured in Figure 9 reduces the salt content from 800 to 470 ppm. For convenience in our numerical examples we imagine the performance slightly downgraded, namely from 750 to 500 ppm, without changing the cost.

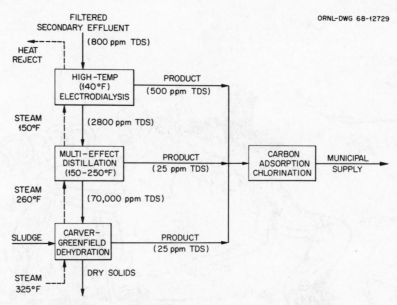

Fig. 9. Diagram illustrating recycle concept with salt removal.

Now, if you were Los Angeles and needed to triple your water supply with no increase in the natural source, would you desalt seawater or waste water? For convenience let us (incorrectly) assume that the city now has complete primary and secondary treatment capacity for all its waste water, as in Figure 6. For definiteness let us put down specific numbers, as in Figure 10.

To triple the water input via "internal" desalting

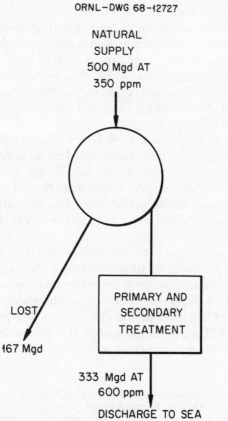

Fig. 10. Diagram illustrating water supply in Los Angeles in 1968.

in the city waste cycle, we pass to the configuration of Figure 11. The added costs are:

| Mgd | Process | ¢/10³ gal. | Cost |
|---|---|---|---|
| 667 | Primary plus secondary | 4 | 2668 |
| 300 | Advanced waste treatment | 8 | 2400 |
| 700 | AWT plus desalting | 16 | 11200 |
| | | | 16268 |

This amounts to 16.3¢/10³ gal. for the 1,000 Mgd incremental water supply.

To accomplish the same end with "external" desalting of seawater we go to Figure 12. The added costs are now:

| Mgd | Process | ¢/10³ gal. | Cost |
|---|---|---|---|
| 667 | Primary plus secondary | 4 | 2668 |
| 767 | Advanced waste treatment | 8 | 6136 |
| 233 | Seawater distillation | 25 | 5825 |
| | | | 14629 |

This comes to 14.6¢/10³ gal. for the 1,000 Mgd incremental water supply. While this is slightly lower than the cost obtained with internal desalting, the difference is probably not decisive in view of the rough nature of the cost estimates.

It should be remarked that, while Figure 7 for $\beta = 0$ and Figure 11 show no waste discharge streams, there are in all cases residues from the treatment plants to be disposed of via, for example, a pipe to the sea with at least enough flow to move them as slurries.

## FOOD FACTORIES

Farming in dry regions contrasts markedly with cities. Most of the water is lost to the atmosphere via evapotranspiration, and there is little potential recycle. Thus one meets the full cost of new water head on.

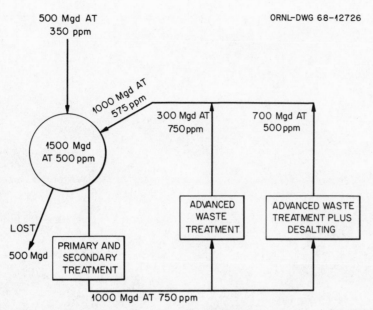

Fig. 11. Diagram illustrating enlargement of water-transfer facilities in Los Angeles, with recycle desalination.

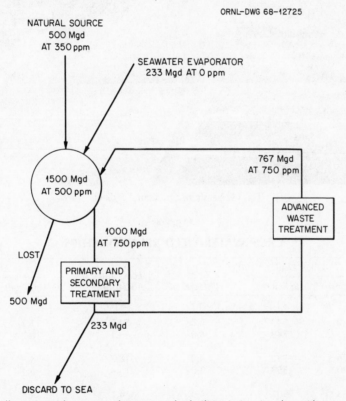

Fig. 12. Diagram illustrating enlargement of water-transfer facilities in Los Angeles, with seawater desalination.

One of the dreams for massive use of desalination is to open up the vast arid areas for food production and living space as the earth's population soars. Figure 13 shows an artist's conception of an agricultural project stretching along the coast of a desert (39). It is important to note that agricultural research and improved crop varieties to reduce the amount of water needed will be just as important as research on reducing the cost of the fresh water itself. About where we are now is indicated in Table 4 (40). The yields are those being obtained by the top 20 per cent of cultivators in major producing areas. Note that grains and potatoes have reached yield levels where they can supply a man's calorie needs with an average water requirement similar to the per capita use in our cities.

There are quite a few things hiding behind the word "average" in the preceding sentence. Figure 14 shows water-use rates for a number of crops as a function of time in the southeast Mediterranean area. It is seen that there is a general seasonal variation, peaking in the hot summer months. Thus,

Fig. 13. Seashore agricultural project.

TABLE 4

## CROP WATER-YIELD RELATIONSHIPS

| Crop Type | Crop | Water Use (in.) | Water Use (gal/acre) | Yield (lb/acre) | Food Value (Cal/lb) | Efficiency of Water Use (lb/1000 gal) | Efficiency of Water Use (Cal/gal) | Efficiency of Water Use (gal/2500 Cal) |
|---|---|---|---|---|---|---|---|---|
| | | | $\times 10^3$ | $\times 10^3$ | $\times 10^2$ | | | |
| Grain | Wheat | 20 | 543 | 6.0 | 14.8 | 11.1 | 16.4 | 152 |
| | Sorghum | 27.6 | 749 | 8.0 | 15.1 | 10.7 | 16.2 | 154 |
| Legume | Peanuts | 34.5 | 937 | 4.0 | 18.7 | 4.3 | 8.0 | 313 |
| | Dry beans | 20.6 | 559 | 3.0 | 15.4 | 5.4 | 8.3 | 302 |
| Oil | Safflower | 33.4 | 907 | 4.0 | 14.2 | 4.4 | 6.2 | 404 |
| | Soybeans | 33.4 | 907 | 3.6 | 18.3 | 4.0 | 7.3 | 343 |
| Vegetables | Potatoes | 16 | 434 | 48 | 2.79 | 111 | 31 | 81 |
| | Tomatoes | 19 | 516 | 60 | 0.95 | 116 | 11.0 | 227 |
| Citrus fruit | Oranges | 53.1 | 1442 | 44 | 1.31 | 30.5 | 4.0 | 628 |
| Fiber | Cotton | 34.5 | 937 | {1.75 (lint) / 2.8 (seed)} | | 4.9 (total) | | |

to match farm requirements to a steady water input, one has to resort to such measures as (1) storing water in underground aquifers at certain times of the year and pumping it back up again at others and (2) making seasonal adjustments in the size of the irrigated area. The problem is further complicated by the fact that the individual crops have seasonal variations of their own; note, for example, how the water use of wheat falls off before that of sorghum starts to rise, leaving a gap between. With all these

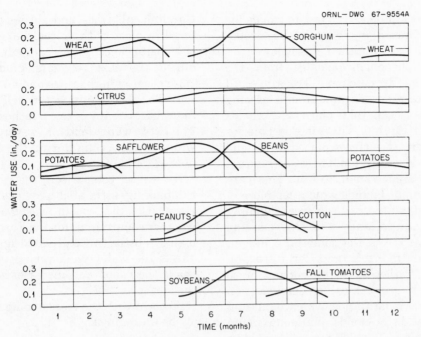

Fig. 14. Graph showing evapotranspiration, El Arish.

climatic and crop seasonal variations, plus differing food or monetary values for the various crops, it is seen that the problem of working out an optimal pattern of crop rotations, acreage variation, and water storage is a complex one.

It should be pointed out that such a farm with a complete irrigation and well point system, together with an expensive water source such as a desalination plant or a long pipeline, is a highly capital intensive undertaking and hence has to be worked hard the year around. The extra cost for the expensive water may be at least partially offset by the opportunity to conduct intensive scientifically-managed farming in an area where year-round production is possible under such highly controlled conditions that the enterprise might be called a "food factory."

## THE RAINMAKERS

One appealing aspect of desalination is that it makes fresh water that is legally new and unencumbered by prior rights, precedents, doctrine, down stream effects, interference with natural courses of events, property boundaries, political boundaries, and so on. Surface water encounters these head-on, with the legal base varying from place to place and time to time (28) so as to minimize unemployment among lawyers.

We stipulate that rainmaking does in fact work,* and estimates of the achievable increase in rainfall are available for various areas. I have not seen a corresponding estimate for the increase in lawyers— this is presumably a function also of the density and

disposition of the population downstream in the air current, not to mention that fraction of the indigenous folks who maybe don't want any more rain on them. Beyond the fact that several states have passed laws asserting their sovereignty over clouds, I know nothing of the legal situation confronting the much heralded "rivers in the sky." I suppose that when the rain reaches the surface it passes out of the jurisdiction of the space-age aerial attorneys and into the clutches of the old-fashioned ground or underground types.

Perhaps the simplest thing to say about rainmaking is that it has passed beyond the rain dance and early experimental phases and is now being carried out on a routine countrywide basis as a normal part of the national day-to-day activity. The average increase in precipitation, accomplished by cloud seeding from airplanes, has been 19 per cent over a 5-year period with, however, substantial variations from year to year. The country I am talking about is Israel (42, 43).

In the United States it is hoped to achieve by the mid-1970's a 15 per cent increase in winter precipitation over the mountain area of the Upper Colorado River basin, and to obtain thereby an average runoff increase of 5/3 billion gallons per day (44, 45).

Rainmaking is typically accomplished by seeding clouds, which are already near to conditions for natural rainfall, with additional condensation nuclei such as silver iodide smoke.* You are not apt to make rain in a flat desert where it never rains now, though some work has been done on creating "thermal mountains" by black-topping desert coasts.

---

*A recent article reviews some of the controversies and uncertainties which have attended the subject (41).

*Positive results also are reported using particles of common salt, namely an increase in rainfall near Delhi of about 20% or 5″ per year (46).

Seeding has also been proposed to create clouds for shade (rather than for rain) in order to reduce evapotranspirative water loss from the fields below.

Quite separate phenomenologically from the "rainmakers" are the "fog catchers." Trees on a hill top can strain quite a lot of water from a passing cloud cover, and this can take place with clouds which are far from raining. The same is true for mesh or other extended surfaces placed in a cloud touching the ground or a hill top. These catchers get fog particles which are too small to rain. There are places in the Hawaiian Islands which get more of their irrigation water from hill top trees catching fog than from actual rainfall (47). One exciting speculation concerns the use of something like an electrical precipitron to discharge a large fraction of a cloud's water, rather than the percent or so obtained in rain or by mechanical fog catchers.

Still different is dew formation which occurs on clear nights without fog or clouds. While maximum deposits of 0.45 mm/night (which rate, if it could be sustained, would amount to 6.5" in a year) have been observed, it is said that "on the available evidence it is unlikely that dew is of great physiological importance, either in humid climates or in the desert" (48). Sometimes plants can also absorb water vapor directly from clear humid air without the formation of visible water.

## THE BASIN BUILDERS

The runoff from catchment basins, like that from the land as a whole (Table 1), may be only a third or so of the rain that falls. Thus if one could pave over the entire basin surface and construct big underground tanks to hold the water, one could triple the water yield. In fact, part of this is just what the Romans did in the Sahara; namely, they built underground cisterns to which occasional rain was carefully conducted (19).

Now we are in fact paving over substantial areas of the country, but their location is not generally chosen with an eye to promoting useful runoff. The principal result in many cities is merely to overload the sewage system and back it up into the streets. To complete the circle, the building of large underground cisterns for hold-up storage of this rain-augmented garbage has been suggested.

While we are not about to pave our catchment basins—at least not until the population explosion gets a bit further along—their surface can be changed in other ways. The typical basin is forested, and trees use up lots of water. Presto, let us cut down all the trees and substitute instead some low ground cover that does not evaporate as much water but is adequate to hold the soil against erosion.

Like rainmaking, this happy idea also works. In one of the most favorable of several experiments conducted in the Coweeta watershed area in North

Carolina, all trees and shrubs were cut and scattered in place with no removal and with minimum soil disturbance. The regrowth was cut annually during the next 15 years, gradually producing a close cover of herbaceous and low shrubbery growth (49). The increase in water yield was 17 inches the first year, declining to 11 inches after a few years. The extra water tends, however, to show up in the winter, rather than during the summer when it is needed (52).

By way of comparison, the runoff increase from rainmaking estimated previously for part of the Upper Colorado basin was 2.5 inches per year. Another comparison (for the Eastern United States) is that substituting cultivated agriculture for forests affects the water yield very little (50), although it peaks the runoff and increases sedimentation from the surface.

So down with the trees and up with the runoff; let joy be unconfined.

## REFERENCES

1. CARROLL, LEWIS, *Through the Looking Glass* (1872).
2. GENESIS.
3. WELLS, H. G., *The World Set Free,* Dutton (1914).
4. REVELLE, ROGER, "Water," *Scientific American,* Sept. (1963).
5. ECCLESIASTES.
6. U. S. GEOLOGICAL SURVEY, "Large Rivers of the World," data card prepared for the American Geological Institute (1961). The values in Table 2, starting with the Rhine, are from this reference, except for the last entry which has been revised on the basis of the USGS Amazon River expedition (1963), as per private communication from Eric Meyer of USGS to F. L. Parker of ORNL. Flows are given at the mouths of the rivers.
7. U. S. GEOLOGICAL SURVEY, *Large Rivers of the United States,* Geological Survey Circular 44 (1949). Delaware and Hudson values are flows at the mouths; the Colorado value is an estimate for the virgin flow at the Mexican border.
8. TENNESSEE VALLEY AUTHORITY, *Clinch-Powell Valley Summary of Resources,* Knoxville, July (1963). Flow cited is that at Scarboro station in Oak Ridge.
9. GARBELL, MAURICE A., "The Jordan River Plan," *Scientific American,* March 1965.
10. MEHREN, LAWRENCE, "Big Red," talk presented at a meeting of the National Reclamation Association, Kansas City, November 10, 1965.
11. LANE, F. C., *Earth's Grandest Rivers,* Doubleday (1949).
12. INTERNATIONAL ATOMIC ENERGY AGENCY, *Desalination of Water Using Conventional and Nuclear Energy,* Technical Reports Series No. 24, Vienna (1964).
13. HANSSON, KARL-ERICK, "Economics of Conveyance of Water," United Nations paper (1965).
14. *NAWAPA—North American Water and Power Alliance,* Brochure 606-2934-18, Ralph M. Parsons Co.
15. *Bulletin of Atomic Scientists,* Sept. (1967) and Sept. (1968).
16. BURWELL, C. C., "Effect of Low Cost Power on Water Transmission by Pipeline," WATER for PEACE conference paper, May (1967).
17. STROTHER, ROBERT S., "A City Sinking in a Sea of Mud," *Reader's Digest,* August 1966.

18. SHELLEY, PERCY BYSSHE (1792-1822), "Ozymandias."
19. BRITTAIN, ROBERT, *Let There be Bread,* Simon and Schuster (1952).
20. WELLARD, JAMES, *The Great Sahara,* Dutton (1965).
21. MOYERS, JOHN, and HELEN KUHNS, "World Arid-Area Summary," Internal ORNL Memo, Oct. (1963). Results obtained by use of planimeter on maps given in "Reviews of Research on Arid Zone Hydrology," UNESCO (1953).
22. BROWN, LESTER R. *Man, Land, and Food,* U. S. Dept. of Agriculture, Foreign Agricultural Economic Report No. 11 (1963).
23. LOWE, JR., CHARLES H., in the Introduction to *The Desert,* LIFE Nature Library (1961).
24. STOUT, PERRY, *Potential Agricultural Production from Nuclear-Powered Agro-Industrial Complexes Designed for the Upper Indo-Gangetic Plain,* ORNL-4292, (November 1968). An earlier draft was given informal distribution in August (1967). Also, "Power: The Key to Food Sufficiency in India?" *Bul. Atomic Scientists,* Nov. (1969).
25. U. S. PRESIDENT'S SCIENCE ADVISORY COMMITTEE, *The World Food Problem,* Vol. 2, p. 448 (1967).
26. DAVIS, R. B., "Farming Success in India," Letter to the Editor, *Science,* 159, 153 (1968).
27. YOUNG, GALE, "Apollo over the Ganges," *IEEE Transactions on Nuclear Science,* Vol. NS-16, No. 1, pp. 9-18, February 1969.
28. CARR, DONALD E., *Death of the Sweet Waters,* Norton and Co. (1966).
29. *Resources,* No. 29, Sept. 1968 (Resources for the Future, Inc.) "A Drink of Water."
30. SCHAFFER, JR., W. F., I. SPIEWAK, R. E. BLANCO, and W. J. BOEGLY, JR., *Survey of Potential Use of Nuclear Energy in Waste Water Treatment,* ORNL-TM-2160 (June 1968).
31. WEINBERGER, L. W., and D. G. STEPHAN, "Technology of Advanced Waste-Treatment," Presented at International Conference on WATER for PEACE, May 23-31, 1967, Washington, D. C. Paper No. 695.
32. EXODUS.
33. STOUGHTON, R. W., "Desalination and Water Recovery," American Chemical Society Lecture Tours Nos. 82 and 83, April (1966).
34. HAMMOND, R. P., I. SPIEWAK, and E. C. HISE, "Design of a 250-Mgd Vertical Tube Evaporator for Seawater Desalting," presented at the Western Water and Power Symposium, Los Angeles, April 8-9, 1968.
35. THOMAS, DAVID G., "Enhancement of Film Condensation Heat Transfer Rates on Vertical Tubes by Vertical Wires," *Ind. Eng. Chem. Fundamentals,* Vol. 6, p. 97, Feb. (1967). Mentioned also in New Scientist, *16,* March (1967).
36. THOMAS, DAVID G., "Enhancement of Film Condensation Rate on Vertical Tubes by Longitudinal Fins," *AIChE J.,* Vol. 14, No. 4, pp. 644-649, July (1968).
37. KRAUS, K. A., A. J. SHOR, and J. S. JOHNSON, JR., "Hyperfiltration with Dynamically-formed Membranes," *Desalination, 2,* (1967) pp. 243-266.
38. SPIEWAK, IRVING, "Application of Low-cost Energy to Processing of Sewage for Reuse," presented at Conference on Abundant Nuclear Energy, Gatlinburg, Tennessee, August 26-29, 1968.
39. Cover Picture, *Nuclear News,* May (1966).
40. *Nuclear Energy Centers: Industrial and Agro-Industrial Complexes,* Summary Report, ORNL-4291 (July 1968); main report ORNL-4290 (Nov. 1968).
41. MACDONALD, GORDON J. F., "Science and Politics of Rainmaking," *Bulletin of the Atomic Scientists,* Oct. (1968).
42. NEUMANN, J., K. R. GABRIEL, and A. GAGIN, "Cloud Seeding and Cloud Physics in Israel: Results and Problems," presented at International Conference on WATER for PEACE, May 23-31, 1967, Washington, D. C.
43. GABRIEL, K. R., "Recent Results of the Israeli Artificial Rainfall Stimulation Experiment," *J. of Applied Meteorology,* Vol. 6, pp. 437-8, April (1967).
44. DOMINY, FLOYD E., "The Last Third of the Century," a Luncheon Address before the National Reclamation Association in Honolulu, Hawaii, November 14, 1967. (U. S. Dept. of the Interior News Release).
45. HURLEY, P. A., *Augmenting Upper Colorado River Basin Water Supply by Weather Modification,* Office of Atmospheric Water Resources, Bureau of Reclamation, U. S. Dept. of the Interior, Denver, Colorado, Oct. 18, 1967.
46. BISWAS, KAPOOR, KANUGA, and MURTY, "Cloud Seeding Experiment using Common Salt," *J. of Applied Meteorology,* Vol. 6, p. 914 (1967).
47. EKERN, PAUL C., University of Hawaii, private communication.
48. MONTIETH, J. L., "Dew: Facts and Fallacies" in *The Water Relations of Plants,* Wiley (1963).
49. HEWLETT, JOHN D., and ALDEN R. HIBBERT, "Increases in Water Yield After Several Types of Forest Cutting," *Quart. Bul. Internatl. Assoc. Sci. Hydrol.,* Louvain, Belgium, Sept. 1961, pp. 5-17.
50. CURLIN, J. W., and D. J. NELSON, *Walker Branch Watershed Project: Objectives, Facilities, and Ecological Characteristics,* ORNL-TM-2271 (Sept. 1968).
51. SIEDER, E. N., and I. SPIEWAK, "Development of Technology Applicable to 50-Mgd and Larger Seawater Distillation Plants," presented at the International Atomic Energy Agency Symposium on Nuclear Desalination, Madrid, Spain, November 18-22, 1968.
52. HEWLETT, JOHN D., "Will Water Demand Dominate Forest Management in the East?" Meeting of Society of American Foresters, Seattle, Washington (1966); Proceedings, Division of Watershed Management, p. 154.
53. YOUNG, GALE, "Dry Land and a Hungry World," *Transactions of the New York Academy of Sciences,* Ser. II, Vol. 31, pp. 145-187. February 1969.

# PHILOSOPHICAL CONSIDERATIONS OF WATER-RESOURCE IMPORTATION

Dean F. Peterson*
Office of Water for Peace
U. S. Department of State
Washington, D. C., U. S. A.

Nothing is good in itself, for goodness is a matter of subjective judgment, and therefore what everybody thinks good will in fact be good—for them.

To form an opinion, the problem before us is to calculate correctly probabilities of good and evil results; this is something which the wiser among us find difficult and the less wise can do standing on their heads.

*Bertrand De Jouvenal (1)*

United States policies on water-resources development reflect in large measure our constitutional concepts of individual political equality and opportunity, responsibility for the general welfare, and the sovereign powers of the individual states. But the conflicting concerns for regional development, correction of maldistribution of wealth and opportunity, and various aspects of conservation make rational choice of alternatives difficult. A more complete philosophy based on ethical considerations could assist in a wise resolution of differences in connection with large-scale water transfers in arid lands; whether or not such refinements can be found and agreed upon remains to be seen.

One path man took in his transition from food-gatherer to member of a highly organized society was through the desert (2). At the *margin* of desert life he ekes out a precarious livelihood, a nomad— quite fully adapted to the basic ecology of the variable desert environment and in equilibrium with it. On the other hand, the quick and profitable returns made possible by irrigation spawned the great civilizations of the Middle East six millenia ago— highly organized and stratified, authoritarian societies first designed to manage the water supply. These were not, as today, relatively minor parts of the civilized world; for, before the commercial and industrial revolution, the *hydraulic* civilizations— essentially *hydroagricultural*—contained the majority of living human beings (3). Available mechanical power—human and animal—constrained the physical works of these societies. Today there is no such constraint. No longer is the desert society hydroagricultural—it is *hydroindustrial* as well. Man's ability to introduce water into the desert is now virtually unlimited, if he chooses to do so.

But there are conflicts in this choice. Those living in the watersheds of origin set values of their own on the waters to be diverted. There is a growing concern over the disruption of ecological systems resulting from large-scale changes in the flow of rivers. On the other hand, large numbers of people see their futures and those of their children dependent upon obtaining large new sources of water.

How are choices to be made? The several dimensions of the problem have been explored by the members of this symposium. Clearly, more is involved than the conventional economics, the physical feasibility, or the problem of finding a path through the tortuous maze of custom, law, and administrative procedure. Somehow we, as a society, are uneasy about making the decisions. We are not confident that we have assessed all of the consequences—nor do we know how to value in relative scale the advantages and disadvantages.

Large-scale water transfers are only one class of water-resource development activity for which we need to seek an improved ethical basis. Five per cent of all of the legislation introduced into the Congress deals with water resources. Large-scale transfers, however, may involve the largest financial commitments and may be the least reversible.

When I agreed to try the assignment given me on this program, I was captivated by the thought that I might say something really new and useful, or might bring a message that would be inspiring, or perhaps even arrive at some breakthrough of significant new wisdom. Clearly this is not to be. I find the ground well plowed by better philosophers, economists, and political scientists (4). My only hope is that I may somehow be able to relate my surface-scratching to the task at hand in a complex field in which I have no special competence.

Two relevant and related topics, currently very much under discussion, and which definitely have ethical dimensions, are the concepts of *minimizing man's interference with nature in contrast to his desire to conquer nature* and a mounting concern over *depreciation of the environment*. These two topics will be discussed in the sections that follow. This will be followed by a discussion of ethical concepts related to our historical water policies, and an attempt will be made to relate these considerations to the concept of large-scale water transfers.

## MAN-NATURE THEME versus CONQUEST OF NATURE

Resource-development philosophy lies somewhere between the monolithic extremes which view, on one hand, that any disruption of nature is bad, to

*On leave from Utah State University, Logan, Utah, U. S. A.

the ideal that man's mission is to conquer and dominate the earth, on the other. The watchword at the beginning of the nineteenth century was *develop the West*. The concern was that the land westward to the Pacific be occupied rapidly to provide family-sized farms worked by sturdy, independent, democratic-minded farmers (5). But nature is viewed as the aesthetically desirable—as a unity, providing a fundamental social value—these ideals were advanced by Thoreau, Emerson, Rousseau. Undisturbed nature is the ideal form of scenery as viewed by Western European artists, and man's communion with nature, the favorite subject of poetry.

Man's use of the land for his own material and economic advancement has been quite overwhelming in America. The rich resource of timber, soil fertility, and minerals, much of it stored as the result of unharvested biological processes through the millenia—and quite irreplaceable in the short term, provided the economic margin to activate the social goals implied in the concept of political equality for every man set by the emerging nation. Having now completed much of the harvest, perhaps because of the harvest, man's population now appears to be seriously out of control. Given extrapolations of future population, we are impelled to make plans now to meet the needs for twice as many as our already teeming 200 million by the year 2020.

There is a personal commitment to growth by each of us. Each feels in his bones—and it's no doubt true—that, as his community grows, so grow his vested interests; lands will become more valuable. My business will have more customers; my university or my bureau will become larger and my own advancement, faster. As the population grows, so grows the economy.

Economic growth also must provide the key to improved social equity. If the economically underprivileged are to become less so, new wealth needs to be created; redistribution of existing wealth is not a generally desirable answer. Economic growth can continue to remove man ever farther from the necessity to spend all of his efforts against starvation and disease, provide the resources for his intellectual advancement and his aesthetic enjoyment— to enhance his human destiny.

But the suddeness of the change in the environment is shattering to many. We who have lived half a century or more need only recall nature as we saw it in our own childhoods and recall how different it is now. Can any sensitive man do this without feeling insecure, or somehow sad? Lately the environmental change has been ". . . enormous and swift . . .— slightly more than a century separates the observations of Chateaubriand that the forests through which he walked and the clearings he saw had been untouched since the creation, from the Governor's conference on the conservation of natural resources

called by President Roosevelt in the White House, May 13-15, 1908" (6).

But perhaps our feelings are only subjective, and our children, having never known nature as it was, will suffer no loss.

What of the longer-range future? Species *man* has been around for a million years or so. But in less than half a millenium he has multiplied his population by ten times and his economy by a great deal more. The additional energy which drives the new system comes from the combustion of fossil fuels. The future will bring new sources of energy—particularly nuclear. But non-solar sources are likely finite—and known reserves will last only a few hundred years. What then of species *man?* Technology may well provide new energy sources or, as suggested by Boulding (7), invent an *anti-entropic* economy such that high-level energy and information input are not dependent on fossil fuels. As population grows, the difficulty transcends the problem of energy; more important to a human being is the quality of life.

But these considerations may be irrelevant to us today; natural species, including more lately a few tens of hominids besides man, have waxed and waned in the earth for more than a billion years, and those faceless future generations—once again limited to husbanding photosynthesis, their populations limited by the ravages of disease—will be far removed from the lives of those who may live in the desert cities of the immediate tomorrow. Nevertheless, sensitive men are disturbed.

Man is not unique in the fact that his exploitation of his ecological niche is at the expense of other species. Man likely has been by far the most destructive of all species. In commenting on the utilitarian anthropocentrism of human economics, Glacken (6) says "It was seen long ago that, even if man were to all appearances at the apex of creation, it does not follow that all things were created in order to satisfy his wants; the earth with its plant and animal life may well have its own rationale, transcending its highest living form." Glacken notes that Aristotle made this statement: "The deer has its own τελοσ—it was not made for the lion!"

Schweitzer felt the problem of ethics too narrowly constructed; that it should include man-to-nature relations—not solely man-to-man ones (6). But Boulding asks if it is better to save the lives of a few thousand people or preserve a few species, considering that nature itself is not conservative and will destroy species whether man is here or not. The "moral impact of the ecological point of view needs to be based on a dynamic ethic which will accept the role of man as an evolutionary agent." But "nature is neutral—in this light human attitudes are aberrant and selective; they show a respect for life which is irrelevant in nature as a whole. Respect for life is subsumed in the dignity of human life." (7)

Such is the gist of the man-nature argument. Whether one is among those who feel uneasy in the face of the argument or not, it will have to be faced: Echo Park, Marble Canyon, and Storm King make this clear.

## SPACE SHIP EARTH

All biological activity produces waste materials. Man's technology, however, has increased his capability to do this far, far beyond that of any other species. The substance of waste is not created; the matter or the energy which is at one moment waste existed in some other form before it became waste. There seems to be some innate malice in the apparent law that man's use degrades material occurring in nature in forms that are aesthetic, or at least neutral, into forms that are ugly and even actively harmful. Roger Revelle has said that the earth is a space ship which must carry all of man's accumulating waste with it forever. The danger is that our space ship could become non-habitable.

The only way we have of reducing the impact of ultimate waste is to dilute it—mix it with our air, water, or earth. We may make more efficient use of our resources by finding ways to utilize what might have been wasted; but to restore waste back to the shape and form it had before it served us would be more costly than foregoing the use in the first place. But much can be done to reduce both the harmful elements in waste and its amount. Even so, the input of waste materials into the atmosphere clearly poses serious threats to human health, and the uses of biocides have resulted in massive kills of aquatic life. No one can assess the less dramatic but subtle damages of reduced health of both man and other species.

The ethical responsibility of waste management is clear. The first question is where the responsibility lies. The problem is that those who produce the wastes seldom suffer the disbenefits caused, nor do these damages accrue specifically to individuals, but to the public at large (8). The cost of the disbenefits can be deferred—they appear as profits in the interim. The second question is at what level we are willing to trade off the quality of our environment for other forms of material wealth.

But what is the relevance of waste management to large-scale diversion? In the past, diversions were made primarily to grow crops—to be consumptively used up. Future large diversions may provide largely municipal and industrial (M. and I.) water—this water is used primarily to wash away or dilute man's wastes. To the extent that the waste loads on our water supplies can be reduced, or waste-bearing waters cleaned up and recirculated, the need for large-scale diversions can be reduced.

## POLICY AS A REFLECTION OF PHILOSOPHY

We would be in error to assume that the United States has no philosophical basis for its water-resource development. This may be something less than formal and inadequate for the future, but a great deal of thought, study, and public debate have gone into the development of public water and related land policy in this country for a century and a half. These efforts have been greatly accelerated since World War II. Policy is not, of course, philosophy; but a reflection of philosophy—a decided-upon course of action after weighing alternatives. Very likely few areas of public action have been more earnestly subjected to De Jouvenal's criterion—"to calculate correctly probabilities of good and evil" than have public water-resource projects.

But the needs of our country are rapidly changing. An essentially rural nation has been quite suddenly transformed into an urban one—at a level of industrial productivity far beyond our wildest dreams of a few decades ago. This has not only placed new strains on our water resources, it has led to identification of new values and changed the priority of the old ones.

Our basic constitutional concepts—individual political equality, responsibility for the general welfare, and the sovereign powers of states—provide the underpinnings for our water-resources policy. The riparian doctrine of water rights (9) inherited from our Anglo-Saxon origins was the basic water law—but in the West this was modified in more or less degree by introduction of the Arabic-Spanish concept of appropriation for beneficial use, using constitutional, statutory and even judicial action. Relevant also is the assumption by the federal government of responsibility for maintenance of navigation and for flood control largely at federal expense. Federal financing may be used for the development of hydroelectricity where this is part of a federal multi-purpose project.

Reclamation policy was forged in heated controversy during the latter part of the past century and well into this one. American farmers, experienced in humid region agriculture, were unable to comprehend the consequences of sub-humid climate; the dream of bountiful nature, perpetuated by occasional success in good years, turned into bitter reality as crops failed because of inadequate moisture. Land speculators perpetuated the dream well into the 1930's. Even today the dream continues, though tracts for retirement rather than for agriculture are its currency.

Few have understood nearly as well, even today, a proper use of the water and land resources of the sub-humid West, as John Wesley Powell, whose centennial is being celebrated in 1970. This remarkable man, primarily a geologist, but bureaucrat, administrator, explorer, ecologist, and humanist also, gained his insight from a rigorous decade of personal study and observation on the ground of the land and water resources of the West and their settlement (10). He realized that successful colonization would require man to come to some workable terms with nature. He understood that agriculture would depend on irrigation, that irrigation development would require public effort, and that the limited water supply would need to be shared equitably. He correctly visualized a proper use of grazing lands and the needs to conserve water in reservoirs. Largely because of pressure to exploit the lands—timber, cattle, mining, it was not until 1902 that the Reclamation Act was passed; it was another three decades before concepts advocated by Powell were implemented on the public grazing lands.

While favoring public support of water-resource development, Powell could not have foreseen the technological possibility for large-scale diversions into the desert which have already taken place, let alone those under discussion. We cannot know what his opinion would be where large metropolitan areas are the beneficiaries. We do know that he was greatly concerned about the welfare of the individual settler—"Here individual farmers, being poor men, cannot undertake the task" (11). In those days, the social goals, and a way to relate them to nature considering the current technology, were clearly seen by Powell. Today the social goals now to be achieved are seen less clearly—nor can one see a way to relate them so elegantly to nature as Powell did in the context of a century ago.

Western water-use and land-use policy is not the total story of American resources policy, although most of the concepts appear to have been reflected there. Caulfield (12) has incisively identified three main thrusts—(a) *the development thrust,* the strong desire for collective and individual economic development; (b) *the progressive thrust,* concern with correcting maldistribution of wealth or income; and (c) *the conservation thrust,* reaction of the educated elite to resource exploitation generally. He points out, however, that Pinchot's concept of conservation, the "greatest good for the largest number for the longest time," may well be at variance with the ideal that "any disruption of nature is bad," attributed to some ecologists. Principles guiding American water-resource policy have been set forth succinctly in Senate Document 97 of the 87th Congress, 2nd Session (13). Water resources are to be developed for the benefit of people. All uses and needs must be weighed and evaluated. National economic efficiency—expressed as the ratio of benefit to cost—is to be used as the economic criterion. Multiple use is strongly advocated, and conflicts and alternatives must be decided in the political market place.

Besides considering the harnessing of nature for man's use, national economic efficiency, and im-provement of environmental quality, a National Academy of Science Report (14) adds *a strong concern for regional economic growth* and *political equity* as national aims of large-scale, water-resource development projects. *Income redistribution* also is such an aim.

*Political equity* refers to an assumed entitlement growing out of earlier negotiations or decisions; or concepts of fairness. Certainly there is an obligation to honor solemn compacts which allocate water between states or nations. Likewise, fairness dictates that actions of the politically powerful should not unreasonably foreclose opportunities for the politically weak. *Income redistribution* is action designed to give larger shares to persons of different income groups, specific areas, or those exposed to economic or natural hazard.

In recent years a trend toward federal domination of water-resource development—divisive in itself because of the single-purpose orientation of the various agencies—has been reversed. The roles of the states and the communities are receiving greater recognition, particularly under the Water Resources Planning Act, designed to develop comprehensive, regional planning in which states play major roles. The necessary role which states must play in water-resources planning cannot be overemphasized. In the balance of sovereignty, the state must be the advocate of the communities within its boundaries and the guarantor of political equity among them. The state bears the final responsibility for assuring that the separate roles of the concerned Federal agencies are integrated in the final design.

Water-resource policy must remain dynamic, matching the dynamism of our society. Fisher (5) mentions the need for casting our water-resources policy in a new mold in which "these kinds of water-resource policies should find a place: multiple-purpose management, frequently on a regional basis; sustained and increased production and use of land and water-resources; minimum economic waste of the stock resources; wide diffusion of benefits from resource-development projects with equitable sharing of costs; adequate consideration of the qualitative, non-monetary aspects of resources such as clean air and unspoiled scenery; enhancement of both national defense and world peace; and sustained contribution to economic growth and stability in this country and elsewhere."

This then is an outline of our policy. No one can quarrel with the basic guideposts. Clearly these have been initiated in the interest of a social "good." Clearly there are conflicts between some of them, also. These may have to be resolved project by project in the political market place. More hopefully, our guidelines might be greatly refined to achieve improved efficiency, both in the decision-making process, and in the results.

## REGIONAL DIFFERENCES IN BENEFITS FROM WATER IMPORTATION

There is danger in assuming that needs and likely results of water-importation proposals can be generalized. These will be different region by region and each should stand on its own merits. Benefits which might eventually be realized in the Upper Colorado River are quite different from those accruing to Southern California. Benefits of rescuing an economy, still largely agricultural, based on ground water overdraft in the Texas high plains, will be different from those in Southern Arizona, now a relatively small part of an economy which has many opportunities to expand in industry and commerce. One of the great difficulties in economic analysis is the assessment of intangibles and social goals.

Israel provides an interesting example. Here, a talented and relatively well-educated group of people, refugees without financial resources but strongly motivated toward specific social goals, developed a water resource, largely ground water, to increase irrigated area from 300,000 dunums* less than 20 years ago to 1,650,000 dunums in 1966-67 (15); the crop value† increased from 44 million Israeli pounds (1966 prices) to 1,423 million Israeli pounds; and annual consumption of water increased from 257 million cubic meters (MCM) to 1,418 MCM. In order to offset long-term ground-water overdrafts, 250 MCM of water is now imported annually from the Jordan River Basin by pumping from Lake Kinneret (Tiberias) into a national water carrier extending through the heart of the country into the Negev.

But the national carrier does more than transport Jordan River water. It dispenses water in some areas and collects it in others, not only to provide water when and where needed, but to equalize the burden of salinity as well. While it diverts 250 MCM, it exercises management control over 1000 MCM, or about 80 per cent of the annual supply. Through the carrier also flows much of the subsidy which supports the country's social commitment to a strong, dispersed population on the land.

The benefits accruing from investments in water-resources development in Israel can hardly be determined. Who can assess the value of Israel's existence or decide what share to attribute to water? Land and water are allocated, and the price of water fixed by the state. The benefits of water development may be quite different today than they were during the period of colonization. Then the opportunity costs of labor were low—almost nil. This is not so now. Today, there are many opportunities for investment in industry and commerce which

* A dunum is approximately one-fourth of an acre.
† Includes all farming—irrigation not separated.

may well be more profitable, and the diversion of water to municipal and industrial use at the margin from irrigated agriculture may not be excessively costly economically or in terms of social goals now largely achieved. We cannot question the wisdom of water-resource investments during the past development stage, nor can we judge the total result in monetary terms alone. Clearly, these investments made available the one resource which was by far the most effective under the circumstances. This may or may not be less true of the future, or even the present; the vital force played by water for irrigation is not forgotten.

The National Academy of Science study (17) raises the question that investment in water development may no longer be the best alternative for public investment in Southern Arizona, or that rescuing agriculture in the Phoenix area, the best alternative even if water-resource-development investments are specified. Here the need to extend the life of part of the agriculture in a regional economy which has many other outlets may be quite different from the needs that might be met by future importations into the Upper Colorado Basin. In the latter case, the greatest need may be associated with the development of the vast mineral and fuel reserves of that area. The scarce resource could be municipal and industrial water and dilution water necessary to preserve environmental quality. The Academy report states that a 2-million-barrel-a-day, oil-shale industry would consume between 110,000 and 200,000 acre-feet per year directly and another 55,000 in supplemental industrial and municipal use. A recently proposed 5,000 megawatt, mine-mouth, electrical plant on the Kaiparowitz Plateau would consume 100,000 acre-feet per year in its cooling towers. Not only will the benefits of water importation vary regionally but also timewise in any region depending on the stage and nature of development.

## MAJOR AREAS OF CONFLICT TO IMPORTATION

Four major areas of conflict particularly relevant to the importation question are discussed briefly in the following paragraphs. In these areas, judgments will need to be made based on discussions which transcend both natural science and technology and conventional economic analysis.

### Area of Origin versus Area of Importation

It seems reasonable and fair that those who live nearest to valuable natural resources should have some priority of claim to the benefits from their use. Resources should not be exploited adversely to local interests by absentees. On the other hand, it would seem wrong for the use of surplus resources, of little local value now or in the foreseeable future, to be unreasonably foreclosed in the face of great need elsewhere.

These basic positions are in conflict; there can be no general solution. Their resolution may well depend on the structure of the political market place. At the moment, for example, the interests of the Northwest are well defended at the Congressional level. On the other hand, had the boundaries of California been drawn differently or the population distributed differently, the bond issue for the California Water Plan might never have been passed.

In many cases we may not now be able to "*calculate correctly probabilities of good and evil results*" between the mutually exclusive alternatives of development and non development in order to make a wise decision. With each side equally convinced of the rightness and wisdom of its position, the result may be an impasse, or a decision depending too largely on political power.

### Development versus Conservation

The conflict between the interests of development and conservation is similar, but the battle lines are drawn differently. These are no longer regional, but idealistic. Areas of origin might receive some form of monetary compensation; but money is not a practical means of compensation for a conservation loss.

### Alternatives to Importation

Where water importation is advocated in order to maintain regional growth, one alternative would be to promote growth instead in the areas where water supplies are plentiful. Another alternative would be to reallocate water supplies used in agriculture to municipal and industrial uses. Either of these alternatives may cost less than importation of new water, but the relative benefits are difficult to assess. Powell foresaw a pattern in which society and individuals would cooperate in the development of family-sized farms in a water-scarce area. Is this ideal still valid in those areas where there are now diverse opportunities in commerce and industry and farming may be largely corporate on a large scale? One argument in reply is the desirability of maintaining and developing irrigation agriculture, which now produces 25 per cent of the nation's crop value on 10 per cent of the cropped land and is the principal commercial source of many important crops (16).

Desalting may provide an alternative which could be effective in coastal areas. For rivers arising in the interior, water rights could be transferred upstream benefitting upstream users also. I am inclined to believe, however, that estimated future water costs now appearing in the literature may be very optimistic.

What of areas developed by exploitation of non-renewable supplies of ground water? Little is known of *rescue economics*.

Now and in the foreseeable future there will be pressing alternatives for the investment of limited public capital. Large-scale importations will face competition from such things as social and economic rehabilitation in the United States and economic development overseas and pollution control, transportation and education. Many of these programs may command strong support on moral grounds.

### How shall the cost be assessed?

Most of the financing of water-resource development projects is by public funds from the federal government, communities, or states. Actual costs are borne by the federal government where benefits are to the public generally and cannot logically be assessed against individuals. Interest costs of federal irrigation projects are not repaid under reclamation legislation, and revenues from hydroelectric power may be used to offset the costs of multiple-purpose projects. Large-scale importations will raise new questions of financing and cost allocation and estimation of benefits solely because of their size if not for other reasons. Present doctrines of cost-sharing probably need to be reviewed. In doing so, however, we should make sure that commitments involving political equities already made or implied are not voided.

### REAL OBJECTIVES VERSUS PROCEDURES IN DECISION-MAKING

Procedural rules and guidelines are necessary to assure an orderly process of decision-making. We should be careful, however, that the procedure does not become the game—which ought to be to identify the best course to be taken considering limited availability of resources. The benefit-cost ratio is a valid concept, but it does not tell the whole story, for example nothing is said about the distribution of benefits. Economic evaluation designed to compare national or regional alternatives for the allocation of scarce resources should be based on opportunity costs of labor and capital and opportunity value of the product. The concept of marginal costs and values ought to be applied to each project increment. Likewise concepts and methods leading to more valid estimates of benefits need to be developed. Conceptual economic models which properly identify and assess indirect and intangible benefits and illuminate their distribution apparently are very poorly developed. Financial analysis with the purpose of allocating costs among beneficiaries is an entirely different matter. Social goals involving the redistribution of income or costs to be borne by the general public plus political bargaining are among the considerations on such allo-

404

cations. I do not argue that economic analysis should provide the sole basis for decision – simply that this is a good place to start the process.

We should distinguish between ecology, which is a scientific discipline, and the *ethic* of environmental conservation. The *goal* of environmental conservation as I understand it is to insure that all impacts of proposed actions on the biosphere are considered. The *function* of ecology is to provide the scientific basis by which the results of alterations of the biosphere may be predicted.

The process of decision making is faced with both internal and external conflicts whose resolution may tend to obscure objective planning. Internally, there is the dilemma of achieving excellence in planning, which must be done by highly skilled specialists; and, at the same time, to communicate with and take into account the wishes of the beneficiaries – the *participatory democrats* – who do not desire to have plans in which they have not participated imposed on them from above. Externally, there is the need for administrative and public support and choice among alternatives (*17*). Decision-making is a complicated process. Complex plans must be either rejected or approved at a high public level, and this is one level of decision making. On the other hand, the decision process begins even when the planner starts to collect the data. The final plan is an accumulation of countless decisions, large and small. But final plans are the only ones legislators and executives see. They cannot make good decisions unless good plans are laid before them. Thus, the degree to which plans are objective or "good" may well depend upon how well the planner understands the social goals involved more than upon how well he understands the established procedure.

Planners cannot draw plans for a preconceived *utopia*. Neither can the objectives be fully separated from the rules of the game. Our strength lies in the broad opportunity for innovation by planners, not only in achieving objectives, but in refining and sharpening opportunities at every step in the planning process within broad goals and procedures which assure some balance of equity. We will need to make the best possible use of the common resources at our disposal. What we need are improved concepts of what those best uses are.

## EXERCISE OF SOVEREIGNTY

What is different about a *large-scale* importation into the desert? The first diversion of a stream, no matter how small, begins an ecological disruption. If there is more than one riparian interest on the stream – that is the day on which social conflict starts. The distinction then must be a matter of *scale*. But there is a scale in the sovereignty also. The rub comes when more than one sovereignty is involved particularly if the importing sovereignty is non-riparian to the proposed supply. In this event, if plans are to be consummated, one sovereignty must prevail, or else all sovereignties must receive significant benefits from the action.

Historically, there is much in our water-development philosophy – particularly in the more arid West – that would argue for water importations, especially the doctrine of appropriation of unused waters for beneficial use. Regardless of the extent of the benefits of such a diversion to a non-riparian state, this doctrine would not likely prevail against the sovereignty of riparian states; or against that of another nation. It is relatively easy for a downstream riparian user to appropriate water against an upstream riparian state on the same river or, likewise, to appropriate water from one watershed for use in another within the same state on the basis of the best common good – or because the majority of the voters want it that way. It is quite a different matter for a non-riparian state to do so by a transbasin diversion; this does not yet appear to have been done.

It may be politically possible for a powerful region of water shortage to prevail even against the sovereignty of a riparian state; but, in our system, negotiations would always proceed under strong federal legislative auspices. These will always strive mightily to reach a reasonable consensus before action is taken. It is hard to conceive of Congress authorizing a diversion to a non-riparian state against the strong opposition of the riparian ones, but this could happen.

## CAN A PHILOSOPHICAL BASIS BE FOUND?

Generalized solutions of conflicts based on ethical considerations do not appear possible. Nevertheless every effort should be made to construct and debate ethical arguments in both generalities and for specific cases. By doing so our decisions will be far better; we might avoid some very bad or even disastrous ones.

In the absence of a system of values which may moderate it, man's social evolution proceeds as an extension of biological evolution. The fittest survive. In a political ecology, the measure of fitness is political power. Political power can result in economic advantage which can cumulatively add to political power. But man's intellectual powers and his foresight may help him avoid some of the traps. Nonhuman species cannot see into the future, and opportunities are blindly exercised without consideration of that future. Hopefully, man may continue to be at least as different as his welfare as a human being and his ultimate survival demands. In the words of Aldo Leopold (*18*):

All ethics so far evolved rest upon a single premise: that the individual is a member of a community of independent parts. His instincts prompt him to compete for his place in the community, but his *ethics* prompt him also to co-operate, perhaps in order that there may be a place to compete for.

## REFERENCES AND NOTES

1. DE JOUVENAL, BERTRAND
   1957  Sovereignty—an inquiry into the political good (Translated by J. F. Huntington.) The University of Chicago Press. Chicago, Ill. 1957. 320 pp.
2. The other has been identified as through feudal communities relying on rain-fed agriculture.
3. WITTFOGEL, KARL A.
   1955  The hydraulic civilizations. *In* Man's role in changing the face of the earth, ed. William L. Thomas, Jr. The University of Chicago Press. Chicago, Ill. pp. 152-164.
4. ROELOFS, ROBERT
   1968  Values, ethics and policy in relation to resource development and conservation: a selected bibliography. Center for Water Resources Research. Desert Research Institute. University of Nevada. Reno, Nevada. 18 p.
5. FISHER, JOSEPH L.
   1966  Natural resources and economic development: the web of events, policies and policy objectives. *In* Future environments of North America: transformation of a continent, ed. F. Fraser Darling and John P. Milton. The Natural History Press. Garden City, N. Y. pp. 261-277.
6. GLACKEN, CLARENCE J.
   1966  Reflections on the man-nature theme as a subject for study. *In* Future environments of North America: transformation of a continent, ed. F. Fraser Darling and John P. Milton. The Natural History Press. Garden City, N. Y. pp. 342-355.
7. BOULDING, KENNETH E.
   1966  Economics and ecology. *In* Future environments of North America: transformation of a continent, ed. F. Fraser Darling and John P. Milton. The Natural History Press. Garden City, N. Y. pp. 225-235.
8. For a discussion of the economic implications of environmental pollution, see Orris J. Herfendahl and Allen V. Kneese. Quality of the environment (Re-

sources for the Future. Washington, D. C. 1965, distributed by the Johns Hopkins Press, Baltimore, Md.)
9. Under this doctrine, each adjacent-to-the-river (riparian) landowner is entitled to have the river flow by undiminished in quantity and unpolluted in quality.
10. UDALL, STEWART L.
    1963  The quiet crisis. Holt, Rinehart, and Winston, Inc. N. Y. 244 pp.
11. POWELL, JOHN WESLEY
    1962  Report on the arid region of the United States, ed. Wallace Stegner. Belknap Press of the Harvard University Press. Cambridge, Mass.
12. CAULFIELD, HENRY P., JR.
    1965  Welfare, economics, and resources development. *In* Readings in resource management and conservation, ed. Ian Burton and Robert W. Kates. The University of Chicago Press. Chicago, Ill. pp. 571-575.
13. UNITED STATES SENATE
    1962  Policies, standards, and procedures in the formulation, evaluation and review of plans for use and development of water and related land resources. 87th Congress, 2nd Session, Document No. 97, U. S. Government Printing Office, Washington, D. C.
14. NATIONAL ACADEMY OF SCIENCE
    1968  Water and choice in the Colorado Basin, a report by the Committee on Water of the National Research Council. Publication 1689.
15. CENTRAL BUREAU OF STATISTICS, GOVERNMENT OF ISRAEL
    1967  Statistical abstract of Israel 1967. The Government Press. Jerusalem. pp. 318, 358.
16. FEDERAL WATER POLLUTION CONTROL ADMINISTRATION
    1968  Water quality criteria. Report of the National Technical Advisory Committee to the Secretary of the Interior. Washington, D. C. pp. 144.
17. NATIONAL ACADEMY OF SCIENCES—NATIONAL RESEARCH COUNCIL
    1966  Alternatives in water management. A report of the Committee on Water. Publ 1408. Washington, D. C.
18. As quoted by John C. Weaver *in* Conservation: more ethics than economics. *In* Readings in resource management and conservation, ed. Ian Burton and Robert W. Kates. The University of Chicago Press.

# PANEL DISCUSSION

## INTRODUCTORY REMARKS

Henry P. Caulfield, Jr.
United States Government
Water Resources Council
Washington, D. C., U. S. A.

Although members of the panel will no doubt approach the topic of major water transfers within the North American continent from different points of view, I doubt that I will have much to moderate — at least in the sense of keeping peace. Thus, because I am unaccustomed to an inactive role and because I believe that it would be helpful to lay before you certain important background information that will not otherwise be included in the symposium, I would like to make a brief presentation to you with the use of slides before I introduce each of the panel members present for their introductory prepared remarks. This background information relates to the First National Assessment of the Nation's Water Resources by the Water Resources Council and to the Federal-State comprehensive framework planning program now underway and projected for the United States under the aegis of the Water Resources Council.* In the time available I can only give sufficient information, I trust, to invite you to explore this area of information more fully on your own.

The Water Resources Council was created by Title I of the Water Resources Planning Act of 1965 (PL 89-80). Its members are the Secretaries of Agriculture; Army; Health, Education and Welfare; Interior; Transportation; and the Chairman of the Federal Power Commission. The Secretaries of Commerce and of Housing and Urban Development are Associate Members, and the Director of the Bureau of the Budget and the Attorney General are Observers. President Johnson designated the Secretary of the Interior, Stewart L. Udall, to serve as Chairman. I serve as Executive Director at the pleasure of the Council.

Most of the top officials of the Federal departments and agencies having authority and responsibilities concerning the Nation's water and related

land resources are Members, Associate Members, or Observers. Thus the Congress, in passing the Water Resources Planning Act, decided that it was most appropriate for the Water Resources Council to be responsible, among other responsibilities, for maintaining a continuing study and for preparing periodically an assessment of the adequacy of supplies of water necessary to meet the water requirements in each water resources region in the United States and of the national interest therein.

Specifically, the requirement for national assessment by the Council is that the Council shall (1) maintain a continuing study of water requirements, adequacy of water supplies, problems in each river basin, and relation to larger regions of the nation; (2) assess the national interest in regional problems; and (3) report its findings to the President and Congress. The First National Assessment was transmitted by the President to the President of the Senate and the Speaker of the House of Representatives on November 12, 1968.

Eight objectives for projected conditions in each of twenty major basins or regions are covered in the First National Assessment on the basis of readily available data throughout the nation and available analytical techniques and resources. The objectives are listed as follows:

1. Estimate requirements for water and related land.
2. Identify water supply problems.
3. Identify water quality problems.
4. Identify flood problems.
5. Identify drought problems.
6. Identify institutional and organization problems.
7. Inventory plans and investment schedules.
8. Analyze use of water and related land resources.

The need for improvement in support of national assessments was recognized at the outset of preparation of the First National Assessment by the Council, and a three-phase plan of improvement was adopted in January 1967. Phase I was the initial assessment in 1967, involving emerging problems, estimates of withdrawals, and estimates of con-

---

*The Nation's Water Resources, the First National Assessment of the Water Resources Council, Washington, D. C., 1968. (Library of Congress Catalog Card Number 68-62779) The full report and a summary report are on sale through the Superintendent of Documents, U. S. Government Printing Office, Washington, D. C. 20402 at $4.25 and 65 cents, respectively. For further information on the Council's comprehensive framework planning progress, see pp 5-9-7 to 5-9-11 of the Assessment.

sumptive uses. Phase II includes a more fundamental analytical framework—that is, more detailed measures of adequacy and substantial data inputs. Phase III is a continuing refinement of phase II, involving relation of water supply to economic factors, differences in productivity, and institutional constraints.

The need for research in support of phase II and phase III was discussed with the Committee on Water Resources Research of the Federal Council on Science and Technology and with a group of Federal resource research administrators. Agreement was reached that research leading to the development of a central analytical system proceed immediately and that the Office of Water Resources Research (OWRR) in the Department of the Interior was the appropriate agency to support such research. A seminar on research needs for national assessments was jointly sponsored and was attended by over 60 interested Federal and State agencies, universities, research organizations, and individuals. On the basis of an evaluation of proposals for such research submitted to OWRR, a contract funded under Title II of the Water Resources Research Act was consumated with A. D. Little, Inc., Cambridge, Massachusetts in mid-1968.

Figure 1 illustrates the relation between the increase in population and gross national product from the year 1910, extended to the year 2020. Fertility rates in recent years suggest that in future years a more appropriate annual rate of population increase will be less than 1.6 percent. Such a change downward in future national assessments would also affect the projection of gross national product through the year 2020. Both long-term projections, which are unique as guidelines for public programs, reflect the fact that the Nation has no policy or plan at the present time to slow down or stabilize either population or economic growth at any future time. As brought out previously in the symposium, the need for such a policy or plan in response to concern with human ecology possibly should be the subject of widespread national discussion and of official concern. A high public official, Secretary of the Interior Stewart L. Udall, argues the need for a Population Policy Commission to suggest an optimum population and the methods and norms necessary to achieve such optimum.* Until such a policy is formulated and implemented, responsible public planning within a functional area, such as water and related land resources, must be based upon

what appears to be future reality under established policies, mores, and conditions.

Consistent with the above projections of population and gross national product, but climbing even faster, is the projection of energy production (Fig. 2). All present commercial forms of energy production, including atomic energy production, require large water withdrawals for cooling purposes. Even with substantial allowance for increasing use of cooling towers and other forms of recirculation, increased water withdrawals for steam-electric production are projected to be, as indicated above, very great by 2020. Fortunately, use of sea water,

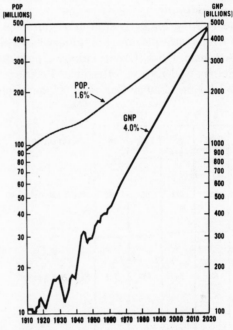

Fig. 1. Graph showing relation between population and gross national product, on the basis of the 1954 dollar value, in the United States from 1910 projected to the year 2020.

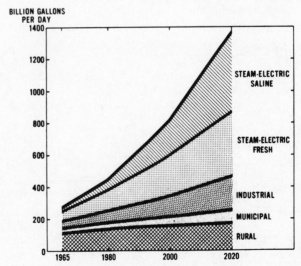

Fig. 2. Graph showing national water requirements for various forms of energy production from 1965 projected to the year 2020.

*Stewart L. Udall, *1976—Agenda for Tomorrow*, Harcourt, Brace, and World, New York, 1968: p. 124. See also, especially, Chapter 3, *Population: Less is More;* and Chapter 7, *Population, Parenthood and the Quality of Life*. See also *Population Crisis*, Hearings before the Subcommittee on Foreign Aid Expenditures, Committee on Government Operations; United States Senate, intermittently 1965 through 1968, under the Chairmanship of former Senator Earnest Gweing of Alaska.

combined with economies in long-distance transmission of power, will reduce withdrawals of fresh water which would otherwise be required.

This chart (Fig. 3) of projected national land use change does little more than confirm what might be expected from common knowledge. The complexities of land use change are many. The Council's National Assessment endeavors to explore these as fully as possible in its first effort.

This map in Figure 4 depicts the twenty major river basins and regions adopted by the Water Resources Council for purposes of national assessment and for Federal-State comprehensive framework planning. A further breakdown into some 160 or more sub-basins is presently being considered for official adoption.

Of the twenty major basins or regions, comprehensive framework planning is underway in eleven: North Atlantic, Ohio, Great Lakes, Upper Mississippi, Souris-Red-Rainy, Missouri, Pacific Northwest, California, Lower Colorado, Upper Colorado

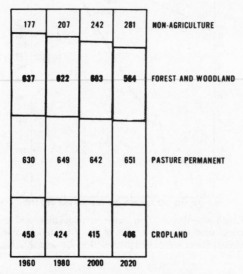

Fig. 3. Chart showing trend in land use, in coterminous million acres, in the United States.

Fig. 4. Map showing the twenty major river basins and regions of the United States.

and Great Basin. A Pacific Southwest Analytical Report is to be prepared covering the last four of these areas. The Ohio study is nearing completion. Several others will be completed within the next few years.

The major topics covered in the development of comprehensive framework plans for the use of water and related land resources are: alternatives and strategy for future development, projections of economic activity, demands for water and related resources, projections of water-resource availability, projections of land-resource availability, current and projected resource problems, alternative solutions, and a program of detailed planning and action.*

Preparing and keeping up to date such comprehensive framework plans (Type 1), together with more-detailed Federal-State plans for sub-basins or regions (Type 2), specific Federal and Federal-assisted project plans (Type 3), and non-Federal plans (Type 4) is the presently structured function of the Nation's official "water community"—Federal, State, and local.

Title II of the Water Resources Planning Act authorizes the establishment of Federal-State river basin commissions for the conduct of comprehensive planning and the coordination of all Federal, State, local, and private water and related land resource planning in a basin or region. Federal-State commissions established so far under Title II cover New England, the Great Lakes basin, the Souris-Red-Rainy basins, and the Pacific Northwest.

Title III of the Water Resources Planning Act authorizes $5 million annually for ten years beginning in Fiscal Year 1967 to States for comprehensive water and related land resource planning. The major purpose of these grants, as emphasized by its provision with regard to training of State employees, is to encourage increased State competence in solution of water and related land resource problems.

Figure 5 provides information on population projections according to major basins or regions. Figures 6, 7, 8 and 9 indicate the general form in which regional water resources are related in the First National Assessment to projected withdrawal use—total and consumptive. The annual natural runoff as measured by its mean value is used as general index of availability if the water resource is fully regulated by provision of necessary inter-year storage capacity and if unaffected by consumptive use. Annual water supplies available 90 and 95 years out of 100 for natural water conditions are also indicated. The 95 percentile flow has been used in the

*For further information, see *Guidelines for Framework Studies,* Water Resources Council, 1967. A copy may be obtained free of charge by writing the Executive Director, Water Resources Council, 1025 Vermont Avenue, N. W., Washington, D. C. 20005.

assessment as an indicator of the low flow. Water supplies, of course, also include such things as annual recharge of aquifers, basin imports, and desalting facilities. Water uses include offstream uses as measured by withdrawals and instream uses. Instream uses have been discussed only qualitatively in the First National Assessment.

The Arkansas-White-Red Basins include arid as well as humid areas. Thus, the chart in Figure 7 — as all similar charts, in varying degrees — marks substantial differences within major regions.

On the basis of the chart for the Texas-Gulf region (Fig. 8) one can readily see in a general way why the State of Texas is so very much concerned with development and acceptance of a State Water Plan and is interested in the possibilities of large-scale water transfer from the Mississippi River.

The Rio Grande region, as indicated in Figure 9, has a most severe projected water quantity problem.

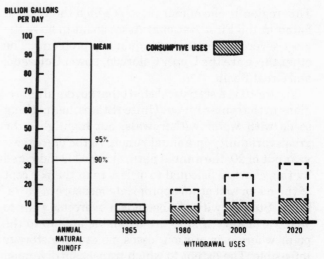

Fig. 7. Graph indicating quantity of water supply and use in the Texas-Gulf Region.

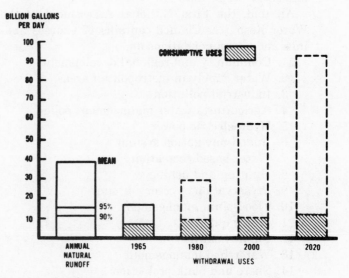

Fig. 8. Graph indicating quantity of water supply and use in the Rio Grande Region.

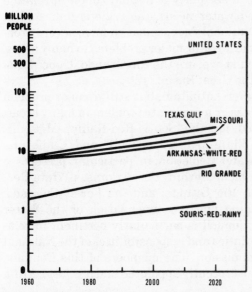

Fig. 5. Chart indicating population projections relative to six major basins or regions in the United States.

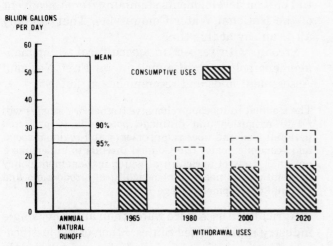

Fig. 6. Graph indicating quantity of water supply and use in the Missouri Region.

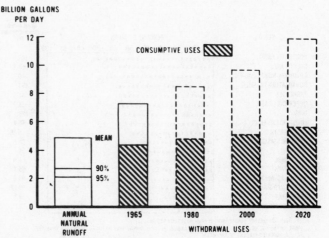

Fig. 9. Graph indicating quantity of water supply and use in the Arkansas-White-Red Region.

This region is one of four regions which the Council found in the First National Assessment to have the most severe projected water quantity problems. The other three are the Upper Colorado, Lower Colorado, and Great Basin.

Figure 10 is a statistical chart that covers all 17 regions in the conterminous United States, has nothing to do with water withdrawals, but highlights the great variability in annual runoff. In the one dryest year out of 20, the annual natural runoff is indicated by this chart to be equal to or less than the per cent of the mean and as an appropriate measure of variability. Storage within a basin can overcome year to year variability if sites are available and up to the point where evaporation losses make more storage infeasible. The extent to which transbasin diversion will ever meet within-basin problems of variability can be answered only on the basis of future analyses, not just in engineering terms, but in economic and social terms also.

All told, the First National Assessment of the Water Resources Council contains 17 express findings and recommendations on:

1. Community and regional development
2. Water supply in metropolitan areas
3. Industrial pollution
4. Agricultural water management policy
5. Hydroelectric power
6. Inland navigation system
7. Water-based recreation
8. Hunting and fishing
9. Preservation of reservoir sites
10. Flood-plain management
11. Water quality
12. Public health
13. Watershed management
14. Shore and bank protection
15. Land drainage
16. Wilderness and scenic rivers
17. Water-short regions

All of these findings and recommendations cannot

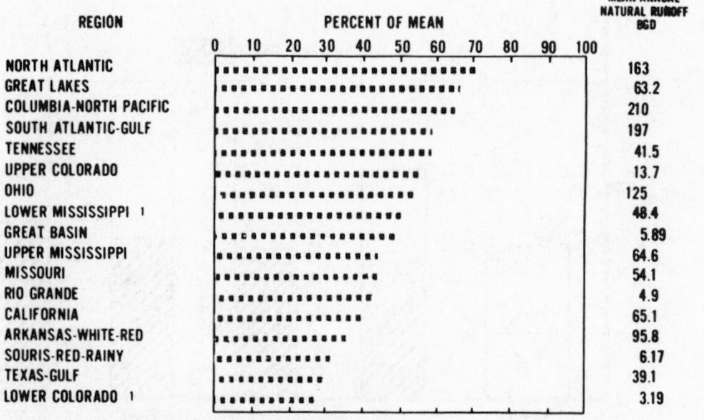

be discussed here, any more than I could summarize the Council's water quantity analyses for all regions.

Two of the Council's findings and recommendations are most pertinent to the issue of large-scale interstate inter-basin transfers of water. *First,* with respect to water-short regions, the Council in its First National Assessment finds and recommends as follows:

In view of the present and projected inadequacy over the next 50 years of the water supplies available to meet requirements in several regions, the Council finds that alternative means of alleviating such water shortages and the national interest therein should be identified and considered. Any study of such alternative means should include consideration of transfer of water between basins, interstate and international terms and conditions of such possible transfers, protection of areas of origin, economic alternatives available to the areas of need, and the national interest in both areas of origin and need. (Page 1-32)

In terms of adequacy of annual runoff in relation to projected water needs, the regions indicated in the First National Assessment to have severe problems in some areas or major problems in many areas, as indicated above, are the Rio Grande, Upper Colorado, and the Great Basin.

A less severe situation, but still a major problem in some areas or a moderate problem in many areas, is indicated for the Souris-Red-Rainy, Missouri, Arkansas-White-Red, Texas-Gulf, and California.

Groundwater depletion in particular presents a severe problem within the Arkansas-White-Red, Texas-Gulf, Rio Grande, and the Lower Colorado.

This finding and recommendation of the Water Resources Council is particularly pertinent in relation to the duties and responsibilities of the National Water Commission. The members of this Commission were just recently appointed and it is just getting under way.

What action follows from this Council view in its First National Assessment will depend in substantial part on developments stemming from the advent of the National Water Commission. That is really all I can say at this time.

*Second,* with respect to agricultural water management policy, the Council in its First National Assessment finds and recommends as follows:

The Council in its comprehensive framework studies will relate irrigation and drainage potentials to projected national food and fiber requirements and regional social and economic goals. The Council intends to undertake a national study of agricultural water management policy in relation to needs for agricultural production and regional development. (Page 1-13)

Agriculture uses more water than any other single industry in the United States. Much of the justification from a national point of view of large-scale transfers of water will turn on this issue of national

| REGION | PERCENT OF MEAN | MEAN ANNUAL NATURAL RUNOFF BGD |
|---|---|---|
| | 0  10  20  30  40  50  60  70  80  90  100 | |
| NORTH ATLANTIC | | 163 |
| GREAT LAKES | | 63.2 |
| COLUMBIA-NORTH PACIFIC | | 210 |
| SOUTH ATLANTIC-GULF | | 197 |
| TENNESSEE | | 41.5 |
| UPPER COLORADO | | 13.7 |
| OHIO | | 125 |
| LOWER MISSISSIPPI ¹ | | 48.4 |
| GREAT BASIN | | 5.89 |
| UPPER MISSISSIPPI | | 64.6 |
| MISSOURI | | 54.1 |
| RIO GRANDE | | 4.9 |
| CALIFORNIA | | 65.1 |
| ARKANSAS-WHITE-RED | | 95.8 |
| SOURIS-RED-RAINY | | 6.17 |
| TEXAS-GULF | | 39.1 |
| LOWER COLORADO ¹ | | 3.19 |

¹ DOES NOT INCLUDE RUNOFF DERIVED FROM UPSTREAM REGIONS
NOTE IN THE ONE DRYEST YEAR OUT OF 20 THE ANNUAL NATURAL RUNOFF WILL BE EQUAL TO OR LESS THAN THE INDICATED PERCENT OF THE MEAN

Fig. 10. Chart showing the variability in annual runoff in seventeen regions of the United States.

interest in preserving and extending the present regional pattern of agricultural production. The most difficult analytical task faced by the Council in preparing its First National Assessment was in relating projected national food and fiber requirements to irrigation and drainage trends and potentials. Present data, analytic techniques, and bases for needed assumption, were found to leave much to be desired; one basic assumption that needs to be made, for example, is that the future rate of application and saturation point of presently known new technology (e.g., fertilizers, pesticides, fungicides, etc.) and likewise the now unknown technological developments of the future will be of critical importance in relating future agricultural demand to agricultural water and land needs. The Council's dissatisfaction in this area leads directly to its view that a special national study was needed in fully adequate depth.

Exactly how and when the Water Resources Council will be able to obtain adequate financing and embark upon its proposed national study of agricultural water management policy has not been determined. This intention of the Council will have to be related, of course, to work in this same area that the National Water Commission will want to undertake. The Congress has enjoined the Commission and the Council to cooperate closely during their period of coexistence. At this point, it is too early to be able to say how this matter will be worked out between the Council and the Commission.

The Council's First National Assessment and its comprehensive planning program is clearly laying a foundation for analyses of large scale transfer needs, if any. It is also providing an analytical base for discussion about water problems at any appropriate time with our neighbors on the North American continent.

# CANADIAN WATERS AND ARID LANDS

E. Roy Tinney and Frank J. Quinn
Ottawa, Ontario, Canada
Policy and Planning Branch
Department of Energy, Mines and Resources
Government of Canada

In a strict sense Canada has no arid lands, as you are probably all aware. But we do have considerable territory, roughly 100,000 square miles, in our Southern Prairies and in Interior British Columbia, which is semi-arid. Elsewhere on the continent, we recognize that there are truly arid lands, both in western United States and in northern Mexico. Unfortunately, no representative from the Mexican Republic is on this panel. We hope that doesn't mean that this session is based on the assumption that importation to arid lands is simply a matter of transferring Canadian waters into the United States Southwest. Controversy on this international issue has continued unabated for several years. Myths abound and fiction confuses fact. Presumably, we are on this panel to review Canadian thought on this matter.

Almost twenty years before Powell's magnificent report on the arid lands of the United States, Captain John Palliser identified the bulk of Canadian dry lands in his Western expedition from 1857 to 1860. For the growing water requirements of these lands, studies are already well underway. The Canadian federal government and the provincial governments of Alberta, Saskatchewan and Manitoba are jointly engaged in a $5 million survey of the water supplies of the Saskatchewan-Nelson Basin, a sur-

vey which includes provision for importing from Northern rivers into this basin. Meanwhile, the Alberta government is working on its own program, called PRIME, with a view to redistributing flows, in turn, from the North Saskatchewan, Athabaska, and Peace rivers toward the southern parts of the province; and the British Columbia government has investigated the merits of a proposal to transfer spring runoff from the Shuswap River in the Fraser Basin into the Okanagan Valley of the Columbia. It should be noted that Canada has reserved the right, in the Columbia River Treaty, to divert water out of that basin for consumptive use. We shall return to Canadian studies in more detail later.

Long-distance water diversions have been altering continental geography for a long time. Interestingly enough, of all the man-made diversions which presently carry water outside the basin of origin, none cross provincial, state, or national boundaries. It appears that political lines, drawn as arbitrarily as they often were in North America, have contained such movements more effectively than have the mountains, deserts, and other barriers erected by nature. We are not suggesting that political entities should build protective walls around their water resources and defy outside "interference" indefinitely. But any breakthrough must ultimately be made in legislation before Parliament and Congress, not on the drawing board.

The escalation of continental water thinking, it would seem, began with a succession of critical events dating from 1963: the United States Supreme Court's decision in Arizona v. California; the Secretary of the Interior's Pacific Southwest Water Plan; a rash of suggestions for more distant sources of supply, with NAWAPA in the lead; a series of Colorado River Basin Project bills in Congress determined to include importation studies; and, simultaneously, a serious drought diagonally across the country in the Northeast. By this time it appeared that every consulting engineer and even some academics with the time to spare were making paper projections. It became fashionable to speak of coordinating governments at all levels for the purpose of conveying millions of acre-feet of water over distances of thousands of miles and pumping lifts of thousands of feet for costs in the billions of dollars.

Canada seems to figure prominently in at least eight of these large-scale, long-distance water diversions schemes. To the extent that they have collectively focused public attention on water problems and opportunities and have provoked debate on our goals and constraints, we must applaud these contributions. But none of them tell us all we need to know. These are engineering schemes, privately sponsored; whether they speak to human needs and public goals, whether they provide the most efficient system for economic growth, allow for its equitable distribution interregionally and internationally, and protect environmental qualities for other kinds of human fulfillment, remains unknown. And to the extent that some of these continental thinkers promote their schemes as panaceas, insensitive to other public needs and political sovereignties, they clearly hurt their own cause.

If water importation into arid lands is as obviously beneficial as many promoters would have us believe, we wonder why better-watered regions within the United States are not more eager to share their supplies. Northern Californian counties of origin struggled in vain to preserve their waters from the South; the states of the Pacific Northwest have emerged from an interregional conflict in Congress with more success, at least for another ten years; some Missouri Basin representatives are now organizing similar resistance against a raid of their supplies; and even the state of Alaska seems unwilling to declare all out support for exportation.

Apparently, not all water authorities in the United States have been convinced of the need for international diversion. One of the more thoughtful reviews we have seen of United States water-management problems, completed in 1966 by the Committee on Water of the National Academy of Sciences, begins:

This report is not prompted by a national shortage of water, for there is no nationwide shortage, and no imminent danger of one. . . .

Indeed, the impression one gets north of the border is that the United States government is more concerned with increasing efficiency in water use and in alleviating pollution levels. When asked whether he had any designs on Canada's waters, Secretary Udall made this decisive statement:

We've suddenly begun to realize in the United States that if we do the right job in pollution control, we are going to increase our water resources enormously . . . we are not looking hungrily at Canada's water resources, we are looking at our own.

To be quite clear on this matter, the United States government has made no offer to the government of Canada, formally or informally, either to buy water or to initiate a joint discussion of the question.

These are only our observations and impressions, of course, of what is happening in your country. We certainly do not propose to tell the United States what it should do in this situation; that is your business, not ours. And in return, we ask that no one misrepresent our position or tell us what is in our national interest; that is the business of the Canadian people and their governments.

What response has Canada made vis-à-vis these privately-sponsored diversion proposals? Let us summarize the statements made by the responsible ministers and officials in the federal and provincial governments of Canada over the past few years. We think you will agree these statements have been clear, consistent and reasonable. From federal ministers' remarks, both inside the House of Commons (September 2, 1964; June 28, 1966; April 3, 1967; and October 10, 1968) and outside the House, several points continually reappear:

1) Canadian waters are not a continental resource: they are as Canadian as any other resource found with the national boundaries.
2) There is no identifiable market as yet for Canadian water in the United States.
3) Canada would be unwilling to negotiate any sale of water at present even if there were a market, because Canadian water supplies have not yet been adequately inventoried and Canadian water requirements into the future have not been assessed. Canada must satisfy its own requirements first.
4) An accelerated effort is underway in Canada to this end, but it will take at least several years to complete.
5) Federal and provincial governments in Canada must both agree before international negotiations can begin.
6) Canadian waters will never be sold under conditions which would jeopardize their permanent ownership and their repatriation if and when needed in Canada.

These, then, are the major points that continue

to be repeated by the government of Canada. The provincial governments have been in accord with the federal position. Earlier this month in Victoria the Co-Chairman of the British Columbia Hydro and Power Authority, Dr. Hugh Keenleyside, re-iterated our common opposition to consider export at present and rejected the claim sometimes made that Canada has a "moral responsibility" to supply water-short areas of the United States.

It is essential to understand both federal and provincial attitudes toward water export in Canada, because, under the B.N.A. Act, jurisdiction over resources is divided. The provinces own their waters; it is therefore quite improbable that any export could take place without the active support of the province(s) from whence the water would come. Conversely, no province could arrange an export without the approval of the federal government, in view of the latter's jurisdiction over boundary waters, and over the regulation of interprovincial and external trade. What this comes down to, then, is a veto power at either level of government on water export.

If we are not prepared in Canada to consider export at this time, however, neither are we sticking our heads in the sand. The Water Sector of the federal Department of Energy, Mines and Resources, in cooperation with provincial agencies, is accelerating and expanding its inventory of water supplies and existing uses, and is considering means of assessing future requirements for all purposes in the major basins and regions of Canada.

Figure 1 will give you some idea of our progress in measuring the resource. Hydrometric coverage naturally began in the more populated southern regions and is now being extended into the far north. Several years from now we should have reliable average flow data for the whole country. In the meantime, an order of their magnitude can be appreciated from some of the major rivers illustrated in Figure 2 and Table 1.

Information of existing water uses is similarly incomplete on a national basis. Some provinces maintain a partial inventory through their licensing procedures; detailed use information has been collected in a few major basin studies; and the federal Water Sector publishes annually a record of developed and potential hydropower sites. Obviously, a systematic inventory of all existing water uses will be required before we can make an adequate assess-

TABLE 1

### AVERAGE FLOW OF TEN MAJOR CANADIAN RIVERS
#### (millions of acre-feet per year)

| | |
|---|---|
| Fraser River at Hope, B. C. | 70 |
| Columbia river at International Boundary | 72 |
| Yukon River at Dawson, Y. T. | 55 |
| Mackenzie River at Fort Simpson, N. W. T. | 185 |
| Nelson River at Cross Lake, Man. | 53 |
| Churchill River at Granville Falls, Man. | 20 |
| Ottawa River at Grenville, Que. | 49 |
| St. Lawrence at Cornwall, Ont. | 174 |
| Hamilton River at Muskrat Falls, Lab. | 40 |
| St. John River at Pokiok, N. B. | 19 |

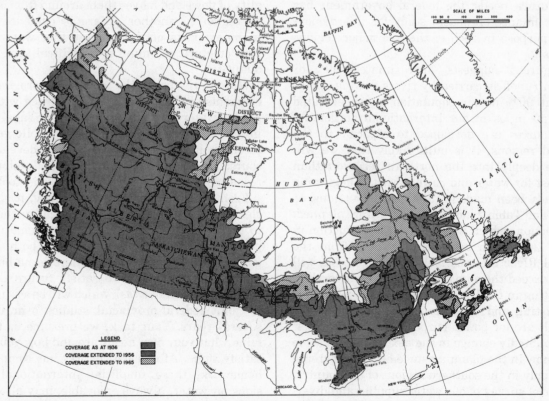

Fig. 1. Hydrometric coverage in Canada.

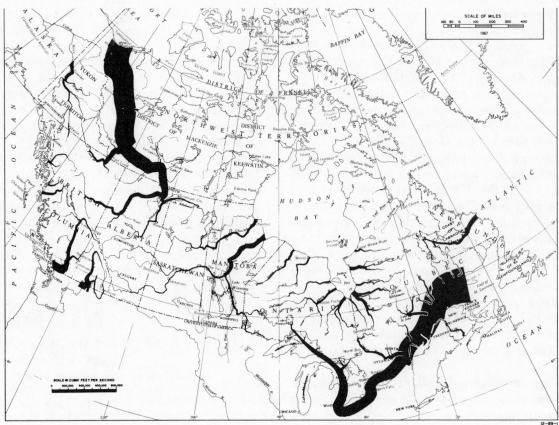

Fig. 2. Runoff from major Canadian rivers.

ment of Canada's future requirements. Federal-provincial discussions have been initiated toward developing efficient tools for a comprehensive evaluation of future needs; the federal government has offered to finance one basin study for this demonstration purpose in each of the five major regions of Canada.

As Figure 2 suggests, approximately 60% of Canadian runoff is carried by rivers flowing northward, but 90% of our population is concentrated within 150 miles of our international border. As local resources are developed to capacity in the populated regions, it is quite likely that we will dip more deeply into the northern reservoir than in the past for water and power needs. A beginning has already been made in interbasin diversions, as Figure 3 and Table 2 indicate, and more are planned. One diversion, which will be under construction in 1969, of 24-million acre-feet annually, from the Churchill River into the Nelson in Manitoba, will by itself exceed the sum of all interbasin diversions now existing in the United States. Of the few diversions existing and under construction in Canada, almost all are for power purposes, a situation that will undoubtedly change in the next decade as some projects now in the planning stage begin to increase water supply in the southern regions of the country.

This brief survey of our water supplies and diversion studies is not intended to suggest that Canada's

problems and opportunities in water are simply matters of quantity, and thereby amenable to an engineering approach which evens out the seasonality of flows and moves them around from one basin or region to another. There is evidence already in many parts of the country that water quality problems are becoming more critical than those of quantity.

More to the point, the federal and provincial governments of Canada are not concerned with either kind of water development for its own sake, but only insofar as it can satisfy broader public goals. We see no point in a philosophy which develops a resource merely because it is there, or develops a new resource merely because the old one looks used.

From all that we have said today, it should be clear that the question of exporting Canadian water to the United States arid lands is premature. It is premature from the viewpoint of both national governments. Quite simply, there is no buyer and no seller. Our task is to satisfy future Canadian requirements first, a task which will be accomplished by joint federal-provincial studies over the next several years. Your task, we presume, is to determine what your own needs are and how you want to satisfy them. After each country has completed its homework, these unofficial international discussions on water, always enjoyable, may also become more definitive and official.

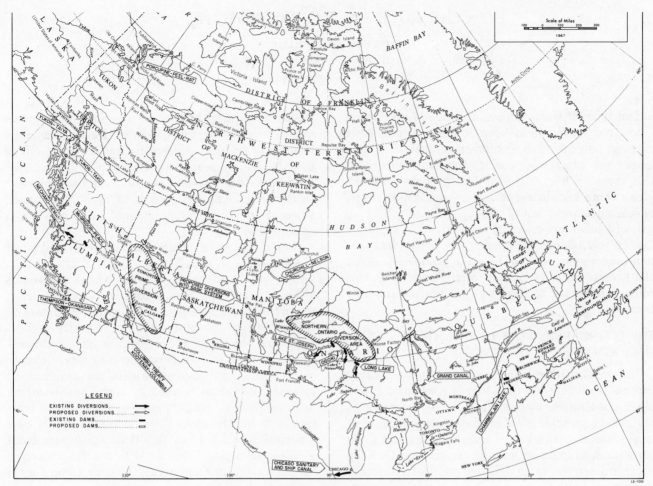

Fig. 3. Interbasin water diversions in Canada.

TABLE 2

**INTERBASIN WATER DIVISIONS: CANADA**

| River Diversion | Location | Average Annual Amount mill. acre-feet (or sq.mi.dr.area) | | Purpose (formerly) now | Year Established |
|---|---|---|---|---|---|
| *A. Affecting Canada* | | | | | |
| 1. Allagash to Penobscot | Maine | ? | (270) | (log driving) power | 1841 |
| 2. L. Michigan to Illinois R. | Illinois | 2.31 | | (navigation) supply, dilution | (1848) 1910 |
| 3. St. Mary R. to Milk R. | Montana | .18 | | irrigation | 1917 |
| *B. Existing in Canada* | | | | | |
| 1. Nechako R. to Kemano R. | B. C. | 2.17 | | power | 1952 |
| 2. L. St. Joseph to Winnipeg R. | Ontario | 2.02 | | power | 1957 |
| 3. Long Lake to L. Superior | Ontario | 1.01 | | power | 1939 |
| 4. Ogoki R. to L. Superior | Ontario | 2.82 | | power | 1943 |
| 5. Megiscane R. to St. Maurice R. | Quebec | ? | (263) | power | 1953 |
| 6. Indian R. to Humber R. | Newfoundland | .15 | | power | |
| 7. Grey R. to Salmon R. | Newfoundland | .69 | | power | 1967 |
| *C. Under Construction in Canada* | | | | | |
| 1. Churchill R. to Nelson R. | Manitoba | 24.55 | | power | 1969— |
| 2. Naskaupi & Canairiktok to Hamilton R. | Labrador | ? | (4384) | power | 1968— |
| 3. Victoria & White Bear to Grey R. | Newfoundland | 1.87 | | power | 1968— |
| *D. Under Consideration in Canada* | | | | | |
| 1. Porcupine to Peel to Rat R. | Y. T., N. W. T. | | | power | |
| 2. Yukon to Taiya or Taku | B. C., Alaska | | | power | |
| 3. McGregor R. to Peace R. | B. C. | | | flood control & power | |
| 4. Shuswap R. to Okanagan R. | B. C. | | | supply, dilution, etc. | |
| 5. Alberta's PRIME Study | Alberta | | | supply, etc. | |
| 6. Fed-Prov. Sask.-Nelson Study | Prairie Prov. | | | supply, etc. | |
| 7. Northern Ontario Study | Ontario | | | supply, navigation, etc. | |

# THE ROLE OF THE ECONOMIST

Bruce R. Beattie
Department of Agricultural Economics, Oregon State University
Corvallis, Oregon, U. S. A.

The topic of this symposium, Water Importation Into Arid Lands, is of keen interest to me. As a starting point for today's discussion, I would like to briefly relay to you some of my thoughts concerning this complex issue. In so doing, perhaps you will glean some insight regarding my particular frame of reference and point of view.

I will focus my opening remarks on two points: (a) the role of the economist in the social decision-making process, and (b) how the economist, given the present state of his art, might best fulfill this role with regard to the issue of interbasin water transfer.

## THE ECONOMIST'S ROLE IN THE SOCIAL DECISION-MAKING PROCESS

The economist's role in the social decision-making process is no different than that of a scientist of any other academic discipline. That is, the economist is obligated to provide positive analysis of the economic implications of alternative social actions and to be careful not to imply or leave the impression that the optimum arrived at, given the technical criteria of economics, has any more or less normative significance regarding the total problem than do optima arrived at, given the technical criteria of other disciplines.* The economist must guard against any desire or tendency to identify (or to be misinterpreted as identifying) a particular technical optimum with a social optimum. He may act as an expert consultant to decision-makers, but he can not replace decision-makers. It is my belief that the economist can best serve the decision-making process by limiting his role to explaining "what is" or "what will be" the consequence of alternative social actions (as judged according to specific discipline criteria), rather than attempting to prescribe what social action "ought to be."

## ECONOMIC CONSIDERATIONS RELEVANT TO THE ISSUE OF INTERBASIN WATER TRANSFER

Assuming that the role of the economist is as described above, i.e., an information provider, or merely an input in the decision process, we are now prepared to ask the second question: How can the economist, given the present state of his art, best fulfill this role? In this regard I will comment on two fundamental economic issues: (a) the efficiency consequences, and (b) the distributive consequences of interbasin water transfers. The efficiency consequences relate to the question: Do the economic benefits of transfer exceed the economic costs? The distributive consequences relate to the question: Who bears the costs and who are the beneficiaries? In the first, a national point of view is assumed, whereas in the latter a regional point of view is appropriate.

Economists have traditionally concentrated much of their effort on the efficiency implications of alternative social actions. In fact, some have accused us of being preoccupied with such matters. However, for the most part we have been rather negligent in the provision of information to decision-makers regarding the distributive implications of alternative actions. If we desire to increase the relevance of information provided to the social decision-making process, then data related to both considerations are required. In reality, most of those individuals involved in the social decision-making process (political process) have less than a national point of view. In a paper delivered to the first Western Interstate Water Conference, held in 1964, Castle commented:

The national point of view is often implicitly adopted for water resource problems. Special interest groups, either on a use or geographic basis, are ignored as such. It is appropriate that someone adopt this point of view and follow it through to its logical conclusion. However, most of the people participating in the decision-making process are concerned about a point of view which is somewhat less comprehensive. They are concerned with a state, a particular use, or an agency of government. Economic analysis based on the national point of view may specify the consequences in terms of national wealth or riches of different regional development plans, but the best plan from a national viewpoint may be a poor predictor of the way development will occur. An unusually large number of economists have been surprised and indignant to find that many of our water resources programs do not always advance the national interest. Rather than being surprised at the existence of a gap, they should have been surprised if one did not exist. I do not wish to be misunderstood—such analyses are important and should be encouraged, both before projects are built and in measuring past performance. But we should not be surprised if the "optimum" or "best" plan, as defined in this way, fails to be the most popular and is not adopted (1, p. 2).

Consequently, in our research at Oregon State we have elected to consider not only the efficiency implications of possible water transfer schemes, that is, a comparison of benefits and costs from a national point of view, but we are also considering the dis-

*This position is consistent with that of Fox (3) and Smith and Castle (4).

tributive implications of such schemes, that is, the income redistributive effects from a regional point of view.

Basically, what we have done is to conceptualize a theoretical model such that the efficiency implications of a specific transfer scheme may be deduced, and that will simultaneously permit the identification of benefits and costs by regions. From this model the following necessary conditions for efficiency in interregional water transfer were derived:

(a) The transfer scheme in question must be less costly than alternative sources of supply in the area of destination.

(b) The marginal value productivity of water in the area of destination must exceed the marginal value productivity less the marginal factor cost of water in the area of origin, plus the marginal cost of transferring water from the area of origin to the area of destination.

Furthermore, the relevant marginal value productivities for economic analysis should relate to the least valuable use of water in both areas.

For those not familiar with the technical language of economics, what is being said, basically, is this: If a transfer of water from one region to another is to be in the national interest from an economic efficiency point of view (that is, if the dollar benefits are to exceed the dollar costs), then the value of water in its least valuable use in the area of destination (probably agriculture), less the costs of transfer, must exceed the value of water in its least valuable use in the area of origin (again, probably agriculture) less the costs of using water in the area of origin. That is, net gains must exceed net losses. Furthermore, if we desire to enhance the national interest as much as possible via interregional water transfers, then the transfer scheme employed should be the least costly. Of course, this model is oversimplified, but it provides an excellent framework upon which one can expand, once confronted with a specific transfer proposal.*

The key economic variables, then, are the value productivity of water in the area of destination, the value productivity of water in the area of origin, and the costs of transfer. In our empirical work at Oregon State (at least that related to my thesis), we are

concentrating on methodology for obtaining reliable estimates of the marginal value productivity of water in irrigated agriculture.

As indicated earlier, the efficiency consideration is one, and only one, piece of technical information that the economist might provide concerning the issue of interregional water transfer. Given the way that our decision-making process is structured, certainly the income redistributive implications are of paramount importance. What is proposed in our research is but a very modest beginning toward providing relevant information concerning the income redistributive impacts of interregional water transfers. We are proposing merely that income redistributive impacts be described by region, and thereby made explicit. By incorporating information derived from the basic efficiency model into a regional input-output framework, the total economic impact on a particular region could be estimated. From a regional point of view, indirect effects may be substantial, and must certainly be considered. As Castle has noted, ". . . indirect regional benefits are a potent political factor in the current scene" (2, p. 5). In addition to recounting the direct and indirect effects, an evaluation of effects due to changes in product and factor prices (pecuniary externalities) and determination of proportionate share of the Federal tax burden will be required to provide a complete description of regional income redistributive impacts of water transfers.

If economists can accomplish these two tasks— describe the efficiency and distributive consequences—then we will have taken a step forward in fulfilling our role as consultants to decision-makers concerned with whether or not to import water into arid lands.

### References

1. CASTLE, EMERY N., "Water Resources of the West: Economic and Social Aspects", Paper presented to the Western Interstate Water Conference at Las Vegas, Nevada, September 1964.

2. CASTLE, EMERY N., "The Economics of Massive Interbasin Water Transfers", Paper presented to a Seminar, Washington State University, Pullman, Washington, April 1968.

3. FOX, IRVING K., "Policy Problems in the Field of Water Resources", *Water Research*, The Johns Hopkins Press, Baltimore, 1966.

4. SMITH, STEPHEN C., and EMERY N. CASTLE, "Introduction", *Economics and Public Policy in Water Resource Development*, Iowa State University Press, Ames, 1964.

*The model may be readily expanded to include value in transit as well as indirect benefits, when appropriate.

# LARGE-SCALE TRANSFERS OF WATER AND ALTERNATIVES

Sol Resnick
University of Arizona
Water Resources Research Center
Tucson, Arizona, U. S. A.

In my opinion, large-scale transfers of water between regions as of the present time do not appear to be economically favorable for agricultural uses.

And fortunately, again in my opinion, water-related research at the University of Arizona at present, for the most part, pertains to the development of alternatives to large-scale water imports. Besides weather modification and desalination research programs underway through the Institute of Atmospheric Physics, studies are being conducted concerned with evaporation suppression, artificial ground water recharge, water harvesting, reuse of domestic and industrial waste waters, and increase of irrigation efficiencies. Further, the above and broader water resources management studies are being conducted using an interdisciplinary approach engaging both the social and physical scientists.

## ALTERNATIVES TO WATER IMPORTATION

Several of these studies are described in more detail below.

### Evaporation Suppression

Evaporation from reservoirs and stock tanks results in large losses of water in the Southwest. The loss from the Colorado River just from Lake Mead and Lake Powell, behind Hoover and Glen Canyon Dams respectively, is estimated to average about 1,500,000 acre-feet yearly.

Our research initially involved testing various long-chain alcohols in small ponds for their efficiency in evaporation control. A wind-activated dispensing system for application of diluted pre-emulsified alcohol was developed. One of the conclusions reached as a result of the investigation, supported essentially with U. S. Bureau of Reclamation funds, was that evaporation reduction using long-chain fatty alcohols on less than one-acre ponds is not economically feasible, as costs averaged about $1,000 per acre-foot of water saved. (It should be noted that costs decrease rapidly with increase of the size of surface area of the reservoir.)

Our recent efforts involving the use of floating styrofoam rafts are much more encouraging, as costs are averaging about $65 per acre-foot of water saved, which probably puts it in the same range as costs for large-scale water transfers.

Of course one can demonstrate almost any economic result he desires by his approach to the problem. We believe our approach is realistic; at least more so, we hope, than that of my neighbor who drives to Nogales, Mexico, from Tucson to drink beer because it is five cents cheaper per bottle. Since this involves a round trip of about 130 miles, the economics of the situation set me wondering; but when I queried him about it, he said, "Oh that is easy, I just drink until I show a profit."

### Artificial Ground Water Recharge

One of the most critical problems in the Southwest today is the diminishing ground water supply. The average yearly overdraft on the ground water reserves in Arizona alone is approximately three million acre-feet.

By means of artificial ground water recharge, flood waters, largely lost by nonbeneficial evapotranspiration, and industrial and domestic waste waters, usually treated to some extent, can be stored in the underground reservoir. There are two general methods of artificial recharge: (1) Water spreading on permeable surfaces, such as natural stream channels, basins, or trenches; and (2) direct injection to underground storage through recharge wells, shafts, or lined pits.

The principal objectives of our project, operating essentially with funds granted by the Office of Saline Water of the U. S. Department of the Interior, are to evaluate the feasibility of conservation and inland disposal of industrial waste effluent by dilution with higher quality water and to examine the role of artificial recharge methods in a conservation-disposal program. Cooling tower blowdown effluent from the Tucson Gas & Electric Co.'s Grant Road plant is being recharged at present by means of a pit and injection well located at the Water Resources Research Center Field Laboratory. Water movement is monitored during recharge by the neutron logging technique; changes in chemical quality of the percolating water are determined by sampling from observation wells; and the hydrogeological and geochemical characteristics of the alluvial fill being recharged are being examined in detail. In a 120-day test, about 43 acre-feet of the effluent were recharged by means of the pit with no attempt made to maintain the initial infiltration rate of about 70 feet per day.

When coupled with flood control, wherein the facilities can be mutually beneficial, the economics of artificial ground water recharge looks very favorable.

## Water Harvesting

One of our Eastern winter visitors to Tucson asked an old cowhand if it ever rained in Arizona. He thought awhile and then said, "Remember the time in the Bible when it rained for forty days and forty nights? Well, we got one-half inch that time." Actually, of course, Arizona receives quite substantial amounts of precipitation, namely, about 80 million acre-feet annually on an average. Our problem is that most of it is lost by nonbeneficial evapotranspiration, and very little runoff (about three per cent from the desert areas around Tucson) occurs.

Hence one aspect of our research program is devoted to a continuing effort toward increasing the runoff from desert watersheds. Two approaches are being investigated. One involves rather limited catchments to provide stock or municipal supplies and includes a storage scheme. The other is for a much broader treatment of the desert watershed with rock salt.

In the limited approach, plastic covered with pea gravel is used as a catchment. A one-half acre area was graded and treated. Coupled with the catchment is a 100,000 gallon storage reservoir built by lining an excavation with 6 mil polyethylene and covering it with 30 mil butyl rubber. This system should provide a firm water supply of 300 gallons per day in a 12-inch rainfall area at a cost of less than $3,000 for the complete system. This obviously can not be of importance in producing irrigation water but could be of great assistance in more limited municipal and ranch situations.

The broader approach for producing irrigation water is also under study. This is based on evidence produced by Dr. W. D. Kemper of Colorado State University that the characteristic reduction of infiltration due to sodium salts might be helpful in increasing runoff. In our experiment, common rock salt, sodium chloride, which is spread on the surface, is used to increase the exchangeable sodium concentration in the soil. Initial experiments in pans demonstrated the feasibility of the idea and the experiment was moved on to one-acre and sixteen-acre plots in the Atterbury Experimental Watershed at Tucson. The cost of treatment is estimated to range from $6 to $12 per acre, depending on the amount of salt required.

From the results obtained to date—namely, about a four-fold increase of runoff from the treated plots from precipitation in winter, the critical runoff period—and if the treatment has a reasonable life, it appears that additional water could possibly be produced at prices attractive for irrigation. It is recognized that a number of unknowns exist, such as increased erosion, effect on plant life, and duration of the treatment. However, the initial promise appeared so good that a major effort in this technique is justified.

## Reuse of Waste Waters

Sewage effluent exceeds 50 percent of the total water consumption of municipalities in Arizona. Hence a sizeable amount of water exists (by 1975, estimates for the Phoenix and Tucson areas are 423 and 111 acre feet per day, respectively) that with proper treatment can be made available for even domestic use at prices less than that for water brought in by means of large-scale transfers. Further attention should be given to the possible use of industrial waste water effluents for agricultural and recreational purposes.

One investigation seeks to evaluate the feasibility of grass filtration for treating oxidation pond effluent prior to artificial recharge.

The City of Tucson, in cooperation with The University of Arizona, is now engaged in a research project wherein effluent from the Tucson Sewage Treatment Plant receives further treatment by controlled percolation through gravel and soil. The resulting effluent then will be examined in detail to determine if it is satisfactory for recreational purposes. This project is patterned somewhat after the successful Santee, California, recreation project.

The University of Arizona and the Tucson Airport Authority have a cooperative project wherein 84 acres at the Tucson International Airport will be developed as an environmental research and recreation area. It is proposed to develop this facility using only waste water—flood, industrial, and perhaps sanitary—which is otherwise lost. This research includes operation of greenhouses for growing food, as discussed in the following paragraph.

An exciting research investigation, which may classify as reuse of wastewater, being conducted mainly by the Institute of Atmospheric Physics with funds from the Rockefeller Foundation, is the following: In an area like Puerto Peñasco, Mexico, where a diesel generator is used to produce electricity for the community, the waste heat energy (about two-thirds of the total) from the generator can be used to heat sea water. The heated sea water then produces vapor in the airstream of an evaporator, and the vapor is then condensed by the cooling power of the incoming sea water. Thus, a fresh water supply is provided for the community. Further, the heated poor quality water from the condenser is used in low-cost plastic greenhouses, providing heat and a high humidity, so crops may be grown with practically no fresh water required. In fact, condensation in the greenhouse may provide the necessary fresh water. Further the carbon dioxide required by the plants growing in the greenhouse is obtained from the above-mentioned diesel generator. Such research may show that electricity, fresh water and food may all be produced economically in a package unit, which would be a blessing for underdeveloped areas like Puerto Peñasco.

*Increasing Irrigation Efficiencies*

By far the largest single use of water in Arizona is for irrigation. Irrigation efficiencies, while as high as anywhere in the United States, are estimated at about 50 percent. With irrigation water use estimated to be at least seven million acre-feet per year, a 17 percent increase in irrigation efficiency would save about 1.2 million acre-feet, or about the amount of water that the Central Arizona Project will produce.

To increase irrigation efficiencies, improvements such as the following are required: Use better land preparation; prepare better irrigation systems, including lined ditches, etc.; determine how and when to irrigate and how much water to apply; measure water being applied; develop varieties of plants which are more efficient water users; and attempt to get maximum yields per unit of water rather than per unit of land.

Several important research investigations are underway; for example, results of experiments through the Department of Agricultural Engineering on the Yuma Mesa, Arizona, show a need for only 4.8 acre feet of water per acre for sprinkler plots as compared with 9.5 for flooded areas, almost *a 50 percent saving* in water. Further, the Department of Agronomy has received a $120,000 grant from the U. S. Department of Agriculture to study water conservation through development of crop varieties which use water more efficiently.

However more extensive research and improvements concerning the technical, social, legal, institutional and political aspects of the entire sphere of irrigation are urgently needed.

## LARGE-SCALE WATER TRANSFER STUDIES

While large-scale water transfer does not appear to be economically favorable at the present time and alternatives should be continuously investigated, studies regarding water importation into arid lands should be undertaken and reviewed at suitable intervals to determine if and when a favorable change in the economic, social, legal, philosophic, institutional, or political status has taken place.

This change to at least a favorable economic status with regard to water importation into arid lands will occur in time, in my opinion, for the following reasons.

1. The tremendous rise in world population—

projected estimation is six billion people by the year 2000—will necessitate in time the production of food in warm arid regions like the Southwest with its favorable soils and year-round growing season.

2. The Southwest is at least one of the fastest growing areas in the United States, and hence provision will have to be made to produce the required domestic and industrial water supply.

3. Crop yields are constantly increasing with almost the same consumptive use of water; the value per unit of crop is also constantly increasing; yet unit pumping costs, which would be a large factor in the expense of large-scale water transfers, are not increasing and may in fact actually decrease with further development of nuclear energy.

4. Then also the economists should with time learn more about the processes of economic growth of an area with import of water or decay of an area as it runs out of water. Perhaps they can develop a measure for these factors which would then permit a more realistic economic analyses for large-scale water transfer. Quoting Dr. M. M. Kelso, Department of Agricultural Economics, University of Arizona: "The process of understanding of economic growth or decay is not mature at present."

Other important reasons, in my opinion, for perhaps immediately starting studies concerning water importation into arid lands are the following.

1. It takes many years between the conception and the completion of a water resources development plan; for example, it required about 40 years for the California State Water Plan to reach fruition.

2. At times it may be necessary to push for a project before it is economically feasible in order to prevent institutional rigidity from locking out the water forever. In this same vein, the possible effects of political factors have been amply demonstrated: (a) The cutting off of the United Western Investigations 18 years ago—these preliminary studies, incidentally, showed that large-scale transfers of water from the Northwest to water-short areas was economically feasible; and (b) the present 10-year moratorium on undertaking reconnaissance studies of any plan for the importation of water into the Colorado River Basin from any other natural river drainage basin.

Very encouraging, in the sense that the entire water resources picture for the United States is being looked at, are the studies underway by the Water Resources Council and the Western States Water Council, the studies authorized under the Colorado River Basin Project Act, and the studies proposed by the National Water Commission. The Western States Water Council, of course, provides the Western States with the opportunity to work out at least some of their water problems on their own.

If the above and, one wishes, other studies show

that excess water does exist in some regions and that some large-scale water transfers are feasible from economic, social, legal, philosophic, institutional and political viewpoints, it should be hoped that agreements to transfer water could be worked out to the satisfaction of everyone.

Incidentally, my other neighbor drinks bourbon at a cost of two million dollars per acre-foot (and this is at Nogales, Mexico, prices); hence, one can only wonder as to what he would pay for water if he got thirsty enough.